AA008774

# Processes in Isotopes and Molecules 2009

Journal of Physics: Conference Series Volume 182

## Cluj-Napoca, Romania
## 24-26 September 2009

ISBN: 978-1-61738-302-1

**Printed from e-media with permission by:**

Curran Associates, Inc.
57 Morehouse Lane
Red Hook, NY 12571
www.proceedings.com

**Some format issues inherent in the e-media version may also appear in this print version.**

Copyright© (2009) by the Institute of Physics
All rights reserved.

Printed by Curran Associates, Inc. (2010)

For permission requests, please contact the Institute of Physics
at the address below.

Institute of Physics
Dirac House, Temple Back
Bristol BS1 6BE UK

Phone: 44 1 17 929 7481
Fax: 44 1 17 920 0979

techtracking@iop.org

# TABLE OF CONTENTS

## MOLECULAR AND BIOMOLECULAR SYSTEMS


MOLECULAR MODELING OF THE WEAKLY BOUNDED DIMERS OF SOME PHENOTHIAZINE DERIVATIVES ........ 1
*Attila Bende, Ioan Turcu*

$^1$H NMR INVESTIGATION OF SELF-ASSOCIATION OF VANILLIN IN AQUEOUS SOLUTION ........ 5
*Mircea Bogdan, Calin G Floare, Adrian Pîrnau*

INCLUSION COMPOUND OF VITAMIN B6 IN β-CD. PHYSICO-CHEMICAL AND STRUCTURAL INVESTIGATIONS ........ 10
*Gheorghe Borodi, Irina Kacso, Sorin I Farca?, Ioan Bratu*

SPECTROSCOPIC INVESTIGATION OF THE INTERACTION BETWEEN β-CYCLODEXTRIN AND ASCORBIC ACID ........ 15
*Ioan Bratu, Marieta Muresan-Pop, Irina Kacso, Sorin I F?rca?*

SHORT RANGE CORRELATION OF THE ERYTHROCYTE MEMBRANE FLUCTUATIONS ........ 19
*Luiza Buimag?-Iarinca, Vasile V Morariu*

DETERMINATION OF COBALT IN HUMAN BIOLOGICAL LIQUIDS FROM ELECTROTHERMAL ATOMIC ABSORPTION SPECTROMETRY ........ 23
*Felicia Dragan, Lucian Hîncu, Ioan Bratu*

STRUCTURAL INVESTIGATION OF LISINOPRIL BY POWDER X-RAY DIFFRACTION AND SOLID-STATE NMR ........ 29
*Xenia Filip, Carmen Tripon, Gheorghe Borodi, Lumini?a Oprean, Claudiu Filip*

ANALYSIS OF THE SECONDARY STRUCTURE OF A PROTEIN'S N-TERMINAL ........ 33
*C G Floare, M Bogdan, O Horovitz, A Mocanu, M Tomoaia-Cotisel*

INCLUSION COMPOUND OF VITAMIN B13 IN β-CYCLODEXTRIN. STRUCTURAL INVESTIGATIONS ........ 38
*Irina Kacso, Gheorghe Borodi, Sorin I F?rca?, Ioan Bratu*

NEW IONIC LIQUIDS AS INNOVATIVE REACTION MEDIA ........ 42
*J Liebscher, J Shah, Z Yacob, S Sadiq, S Hanelt, H Blumenthal*

ASSESSMENT OF MOLECULAR INTERACTION IN A CYCLURON-CYCLODEXTRIN INCLUSION COMPLEX ........ 45
*Mihaela Mic, Dyanne Cruickshank, Ioan Turcu*

NUCLEAR MAGNETIC RESONANCE STUDIES OF LIQUIDS MORPHOLOGY INSIDE PARTIALLY SATURATED POROUS MEDIA ........ 49
*Ruben Nechifor, Codruta Badea, Ioan Ardelean*

NMR SPECTROSCOPIC CHARACTERIZATION OF β-CYCLODEXTRIN INCLUSION COMPLEX WITH VANILLIN ........ 53
*Adrian Pîrnau, Mircea Bogdan, Calin G Floare*

PREPARATION OF 1D NANOSTRUCTURES USING BIOMOLECULES ........ 58
*Stela Pruneanu, Liliana Olenic, Lucian Barbu Tudoran, Irina Kacso, Said A Farha Al-Said, Reda Hassanien, Andrew Houlton, Benjamin R Horrocks*

ATOMIC FORCE MICROSCOPY STUDY OF NANOCRYSTALLINE CERIA THIN FILMS ........ 63
*T Ristoiu, T Petrisor Jr, M S Gabor, M Nasui, B Mos, L Ciontea, T Petrisor*

THE EXTRACTION AND CHROMATOGRAPHIC DETERMINATION OF THE ESSENTIALS OILS FROM OCIMUM BASILICUM L. BY DIFFERENT TECHNIQUES ........ 67
*Maria Loredana Soran, Simona Codruta Cobzac, Codruta Varodi, Ildiko Lung, Emanoil Surducan, Vasile Surducan*

DIELECTRIC PROPERTIES OF ZEA MAYS KERNELS – STUDIES FOR MICROWAVE POWER PROCESSING APPLICATIONS ........ 72
*Emanoil Surducan, Camelia Neamtu, Vasile Surducan*

THE EFFECT OF COLOURED LIGHT ON IPOMOEA PURPUREA GROWTH ........ 76
*Vasile Surducan, Ildiko Lung, Emanoil Surducan*

INFRARED AND UV-VIS SPECTROSCOPIC STUDY OF 3,7,10-SUBSTITUTED-PHENOTHIAZINE DERIVATIVES ADSORBED ON GOLD NANOPARTICLES ........ 80
*N Tosa, L Olenic, I Bratu, R Turdeanu, I Turcu*

## MODERN TECHNIQUES AND TECHNOLOGIES


**THE SKIN'S ELECTRICAL ASYMMETRY** .......... 85
*N M Bîrlea, S I Bîrlea, V To?a*

**LIQUID AS TEMPLATE FOR NEXT GENERATION MICRO DEVICES** .......... 89
*Jérôme Charmet, Henri Haquette, Edith Laux, Ganna Gorodyska, Marcus Textor, Guido Spinola Durante, Erwin Portuondo-Campa, Helmut Knapp, Roland Bitterli, Wilfried Noell, Herbert Keppner*

**ION MOBILITY ANALYZER – QUADRUPOLE MASS SPECTROMETER SYSTEM DESIGN** .......... 94
*C Cuna, M Leuca, N Lupsa, V Mirel, V Bocos-Bintintan, Stela Cuna, V Cosma, Florina Tusa*

**ON THE SENSITIVITY OF FPPE – TWRC METHOD IN THERMAL EFFUSIVITY INVESTIGATIONS OF SOLIDS** .......... 99
*Dorin Dadarlat, Mihaela Streza, Mircea N Pop, Valer Tosa*

**MASS SPECTROMETRY OF TOLUENE CLUSTERS USING A LASER PRODUCED PLASMA FOR TUNABLE SINGLE PHOTON IONIZATION** .......... 103
*Tonia M Di Palma, Antonio Borghese*

**INCREASING THE ACCURACY OF STRUCTURAL INVESTIGATIONS BY MAS SPIN-ECHO SOLID-STATE NMR EXPERIMENTS** .......... 107
*Xenia Filip, Carmen Tripon, Steven P Brown, Claudiu Filip*

**UNEXPECTED SYNCHRONIZATION** .......... 111
*Szabolcs Horvát, E Ágnes Horváth, Gabriell Máté, Erna Káptalan, Zoltán Néda*

**CHARACTERIZATION OF SOME EXTRACTS FOR THERAPEUTIC USE BY GC/MS** .......... 115
*Andreea Maria Iordache, Monica Culea, Onuc Cozar*

**TRACING QUANTUM TRAJECTORIES OF ELECTRONS IN INTERACTION WITH ARBITRARY-SHAPE LASER PULSES** .......... 120
*Katalin Kovács, Valer To?a*

**GENERAL ASPECTS OF SOLID ON LIQUID GROWTH MECHANISMS** .......... 124
*E Laux, J Charmet, H Haquette, O Banakh, L Jeandupeux, B Graf, H Keppner*

**SYNTHESIS AND SPECTRAL CHARACTERISATION OF A NEW METAL COMPLEX WITH A BIDENTATE LIGAND** .......... 128
*Violeta Niculescu, Radu Tamaian, Nadia P?un, Adrian Pîrn?u, Laszlo Szabo*

**A STUDY ABOUT ILLUMINATION AND COLORS VISION** .......... 133
*Florin Toadere*

**RAMAN AND EPR STUDIES OF CALCIUM-PHOSPHATE GLASSES DOPED WITH MANGANESE IONS** .......... 137
*Dana Toloman, A R Biris, Adriana Popa, Oana Raita, L M Giurgiu, I Ardelean*

**NEW SOLID-STATE METHODS FOR INVESTIGATING MOLECULAR STRUCTURE: THE CHCC APPROACH** .......... 141
*Carmen Tripon, Xenia Filip, Vasile V Morariu, Claudiu Filip*

**SIMPLIFIED PORTABLE 4 MHZ RF PLASMA DEMONSTRATION UNIT** .......... 145
*Cristian D Tudoran*

**QUASI-BALLISTIC LIGHT SCATTERING – ANALYTICAL MODELS VERSUS MONTE CARLO SIMULATIONS** .......... 149
*Ioan Turcu, Mikhail Kirillin*

**METHOD VALIDATION FOR DETERMINATION OF HEAVY METALS IN WINE AND SLIGHTLY ALCOHOLIC BEVERAGES BY ICP-MS** .......... 154
*Cezara Voica, Adriana Dehelean, A Pamula*


## ENVIRONMENTAL MOLECULAR PROCESSES


**OPTICAL SOUNDINGS OF CARS' EXHAUST AND URBAN ATMOSPHERE WITH LASER-PLASMA LIGHT AND OPEN-PATH SPECTROPHOTOMETRY** .......... 159
*Antonio Borghese, Luciano Speranza*

**STABLE OXYGEN AND HYDROGEN ISOTOPES MEASUREMENT BY CF-IRMS WITH APPLICATIONS IN HYDROLOGY STUDIES** .......... 163
*Diana Costinel, Voicu V Grecu, Raluca Vremera, Stela Cuna*

**PHOTODEGRADATION OF THE INDOOR ORGANIC POLLUTANTS BY UV IRRADIATION USING $TIO_2$ CATALYSTS** .......... 167
*Ioana C Glajar, Z Moldovan*

**THE IMPROVEMENT OF REMOVAL EFFECTS ON ORGANIC POLLUTANTS IN WASTEWATER TREATMENT PLANTS (WWTP)** .......... 170
*O Marincas, P Petrov, T Ternes, V Avram, Z Moldovan*

STRUCTURAL ELUCIDATION OF ORGANIC CONTAMINANTS BY CHEMICAL IONISATION MASS SPECTROMETRY ........ 174
*Zaharie Moldovan*
NUMERICAL MODEL TO SOLVE IMPURITIES' MIGRATION IN WATER PIPES ........ 178
*Valer To?a, Katalin Kovács*
IDENTIFICATION AND MEASUREMENT OF PESTICIDE CONTAMINANTS IN FOOD PRODUCTS BY ELECTRON IMPACT GC/MS ........ 182
*Florina Tusa, Zaharie Moldovan, Mircea Vlassa*
NON-DESTRUCTIVE ANALYSIS IN A STUDY OF THE RELIGIOUS ART OBJECTS ........ 186
*Nicoleta Vornicu, Maria Geba, Cristina Bibire*

## HYDROGEN AND RENEWABLE SOURCES OF ENERGY

HYDROGEN STORAGE IN PRNIIN AND RELATED ALLOYS ........ 190
*O Ardelean, G Blanita, G Filoti, P Palade*
A PHOTOELECTROCHEMICAL CELL FOR THE STUDY OF THE PHOTOSENSITIVE MATERIALS USED IN SOLAR-HYDROGEN ENERGY ........ 194
*P Ardelean, E Indrea, T D Silipas, C Ardelean, G H Mihailescu, Ramona-Crina Suciu, Simina Virginia Dreve, Z Moldovan, Marcela Corina Rosu*
THE EFFECT OF SOLUTION/FREE VOLUME RATIO ON THE MOF-5 CHARACTERISTICS ........ 198
*G Blanita, D Lupu, M Lazar, A R Biris, V Pascalau, O Ardelean, I Coldea, I Misan, G Popeneciu, M Vlassa*
GOLD NANOPARTICLES BIND TO PORPHYRINS. A POTENTIAL DYE-SENSITIZED SOLAR CELL ........ 201
*Sorina Garabagiu, Gheorghe Mihailescu*
PREPARATION AND CHARACTERIZATION OF NICKEL BASED MULTICOMPONENT CATALYSTS ........ 205
*Mihaela Lazar, Maria Mihet, Monica Dan, Valer Almasan, Petru Marginean*
HYDROGEN DESORPTION FROM $NAALH_4$ CATALYZED BY BALL-MILLING WITH CARBON NANOFIBERS ........ 209
*D Lupu, G Blanita, I Misan, O Ardelean, I Coldea, G Popeneciu, A R Biris*
MOBILITY OF HYDROGEN SPECIES ON NI SUPPORTED CATALYSTS ........ 213
*Maria Mihe?, Mihaela Lazar, Valer Alm??an*
TEMPERATURE INFLUENCE ON THE GLOBAL AND LOCAL STRUCTURE OF THE CHROMIA SUPPORTED NICKEL CATALYSTS ........ 217
*Stelian Pintea, Vasile Rednic, Petru M?rginean, Xie Yaning, Nicolae Aldea*
INVESTIGATION ON A THREE-STAGE HYDROGEN THERMAL COMPRESSOR BASED ON METAL HYDRIDES ........ 222
*G Popeneciu, V Almasan, I Coldea, D Lupu, I Misan, O Ardelean*
METAL HYDRIDES REACTORS WITH IMPROVED DYNAMIC CHARACTERISTICS FOR A FAST CYCLING HYDROGEN COMPRESSOR ........ 227
*G Popeneciu, I Coldea, D Lupu, I Misan, O Ardelean*
$TIO_2$ – BASED SYSTEMS FOR PHOTOELECTROCHEMICAL GENERATION OF SOLAR HYDROGEN ........ 232
*Teofil D Silipas, Emil Indrea, Simina Dreve, Ramona-Crina Suciu, Marcela Corina Rosu, Virginia Danciu, Veronica Cosoveanu, Violeta Popescu*

## NANOSTRUCTURED MATERIALS AND NANOCOMPOSITES

LOCAL STRUCTURE INFORMATION BY EXAFS ANALYSIS USING TWO ALGORITHMS FOR FOURIER TRANSFORM CALCULATION ........ 236
*N Aldea, S Pintea, V Rednic, F Matei, Hu Tiandou, Xie Yaning*
SYNTHESIS OF CARBON NANOTUBES FROM ACETYLENE ON THE FECOMGO CATALYTIC SYSTEM OBTAINED BY BALL MILLING ........ 241
*A R Biris, A S Biris, E Dervishi, Z Li, F Watanabe, S Simon, D Lupu, I Misan, M Lucaci*
AB INITIO STUDY OF DNA NUCLEOTIDES SANDWICHED BETWEEN AU(111) ELECTRODES ........ 245
*Diana Bogdan, Radu Bratfalean, Radu Isai, Cristian Morari*
SYNTHESIS, CHARACTERIZATION AND DRUG DELIVERY APPLICATION OF THE TEMPERATURE RESPONSIVE PNIPA HYDROGEL ........ 249
*I Craciunescu, A Nan, R Turcu, I Kacso, I Bratu, C Leostean, L Vekas*

**FTIR SPECTROSCOPIC STUDY OF SOME LEAD GERMANATE GLASSES** .......... 253
*E Culea, L Pop, M Bosca, T Rusu, P Pascuta, S Rada*

**STRUCTURAL AND PHYSICAL CHARACTERISTICS OF $XGD_2O_3(100-X)[BI_2O_3\ B_2O_3]$ GLASSES** .......... 257
*Eugen Culea, Lidia Pop, Maria Bosca, Viorel Dan, Petru Pascuta, Simona Rada*

**MAGNETIC STUDIES OF $TEO_2 - FE_2O_3$ GLASS SYSTEMS OBTAINED BY THE SOL – GEL METHOD** .......... 262
*Adriana Dehelean, Eugen Culea*

**MAGNETIC BEHAVIOUR OF EUROPIUM IONS IN SOME TELLURITE GLASSES OBTAINED BY THE SOL – GEL METHOD** .......... 265
*Adriana Dehelean, Eugen Culea*

**CHITOSAN-BASED DELIVERY SYSTEMS FOR DICLOFENAC DELIVERY: PREPARATION AND CHARACTERIZATION** .......... 269
*Simina Dreve, Irina Kacso, Ioan Bratu, Emil Indrea*

**STRUCTURAL CHARACTERISATION OF BINARY $SIO_2/TIO_2$ NANOPARTICLE AEROGELS BY X-RAY SCATTERING** .......... 273
*Emil Indrea, Anca Peter, Danut T Silipas, Simina Dreve, Ramona-Crina Suciu, Marcela Corina Rosu, Virginia Danciu, Veronica Cosoveanu*

**SYNTHESIS OF NEW PYRROLE-CONTAINING BIOMOLECULES AS BUILDING BLOCKS FOR FUNCTIONALIZED POLYPYRROLES IN NANOBIOTECHNOLOGY** .......... 277
*S Karsten, A Nan, R Turcu, J Liebscher*

**SURFACE INITIATED RING-OPENING POLYMERIZATION OF LACTONES ON IRON OXIDE NANOPARTICLES** .......... 281
*Alexandrina Nan, Rodica Turcu, Izabell Craciunescu, Cristian Leostean, Ioan Bratu, Jürgen Liebscher*

**SYNTHESIS AND CHARACTERIZATION OF LSMO NANOPARTICLES COVERED WITH AU HAVING A CORE-SHELL STRUCTURE** .......... 285
*O Pana, R Turcu, M L Soran, S Macavei, C Leostean*

**PREPARATION AND STRUCTURAL CHARACTERIZATION OF SOME $FE_2O_3$-$B_2O_3$-ZNO GLASSES AND GLASS CERAMICS** .......... 288
*Petru Pascuta, Gheorghe Borodi, Maria Bosca, Lidia Pop, Simona Rada, Eugen Culea*

**FERROMAGNETIC RESONANCE MEASUREMENTS ON CO NANOWIRES** .......... 292
*A Popa, M Jivanescu, D Toloman, O Raita, L M Giurgiu*

**THE LOCAL STRUCTURE OF GADOLINIUM-BORATE-TELLURATE VITROCERAMICS INVESTIGATED BY FTIR AND EPR SPECTROSCOPY** .......... 296
*M Rada, E Culea, S Rada, P Pascuta, M Culea, V Dan, T Rusu, V Maties, I Bratu*

**IMMOBILIZATION OF GADOLINIUM IN BORATE-TELLURATE GLASSES** .......... 300
*S Rada, E Culea, M Rada, V Maties, M Bosca, L Pop, R Fechete, R Chelcea, D Moldovan*

**EPR INVESTIGATIONS OF $SN_{1-X}FE_XO_2$ NANOPOWDERS** .......... 304
*O Raita, D Toloman, A Popa, M Stan, Al Darabont, L M Giurgiu*

**X-RAY PHOTOELECTRON SPECTROSCOPY AND MAGNETISM OF ALDYNI, $ALDYNI_4$ AND $ALDY_3NI_8$ COMPOUNDS** .......... 307
*V Rednic, M Coldea, L Rednic, L G Pascut, N Aldea, S Pintea, M Neumann*

**A SPECTROSCOPIC STUDY OF DYES DECOMPOSITION BY IRRADIATED NANOCRYSTALLINE $TIO_2$** .......... 312
*M C Rosu, R C Suciu, I Kasco, S V Dreve, E Indrea, T D Silipas*

**DISPERSION OF CARBON NANOTUBES BY SINGLE-STRANDED DNA WRAPPING FOR ADVANCED BIOMEDICAL APPLICATIONS** .......... 316
*S Simon, A R Biris, D M Lupu, I Misan, S Clichici, T Mocan, A S Biris*

**$TIO_2$ THIN FILMS PREPARED BY SOL – GEL METHOD** .......... 320
*R C Suciu, E Indrea, T D Silipas, S Dreve, M C Rosu, V Popescu, G Popescu, H I Nascu*

**SMART COMPOSITES BASED ON MAGNETIC NANOPARTICLES AND RESPONSIVE POLYMERS** .......... 324
*R Turcu, A Nan, I Craciunescu, O Pana, C Leostean, S Macavei*

**CERAMICS WITH DECORATIVE ASPECT** .......... 328
*Cezara Voica*

**Author Index**

# PREFACE

These Proceedings present some of the Invited Lectures and Contributed Papers of the International Conference 'Processes in Isotopes and Molecules' (PIM), held in Cluj-Napoca, Romania, 24–26 September 2009. The PIM conference, started in 1999 as a local event, is now an international conference organized every two years by the National Institute for R&D of Isotopic and Molecular Technologies in Cluj-Napoca, the capital city of Transylvania, Romania. The meetings are attended by researchers in the field of atomic and molecular physics as well as those developing new materials and technologies. The scientific subjects are at the cross-roads of three fundamental research areas: physics, chemistry, and biology.

The papers here are grouped according to the five conference topics:
T1 – Molecular and biomolecular systems
T2 – Modern techniques and technologies
T3 – Environmental molecular processes
T4 – Hydrogen and renewable sources of energy
T5 – Nanostructured materials and nanocomposites

We gratefully acknowledge the contribution of our colleagues from the Scientific Committee and Program Committee who contributed their time, energy and expertise to the refereeing process. Finally, we would like to thank people from IOP Publishing for their friendly advice and prompt help during the editing process, as well as for their efforts making the *Journal of Physics: Conference Series* available to the scientific community.

**Diana Bogdan and Valer Tosa**
National Institute for R&D Isotopic and Molecular Technologies, Cluj-Napoca

All papers published in this volume of *Journal of Physics: Conference Series* have been peer reviewed through processes administered by the proceedings Editors. Reviews were conducted by expert referees to the professional and scientific standards expected of a proceedings journal published by IOP Publishing.

# Molecular modeling of the weakly bounded dimers of some phenothiazine derivatives

**Attila Bende and Ioan Turcu**

National Institute for Research and Development of Isotopic and Molecular Technologies, 65-103 Donath, 400293 Cluj-Napoca, Romania

E-mail: attila.bende@itim-cj.ro

**Abstract**. The intermolecular $\pi$-$\pi$ stacking interaction between some phenothiazine derivatives has been investigated using the Hartree-Fock (HF) and the local second order Møller-Plesset perturbation theory (LMP2) methods. Two different conformations (parallel and antiparallel stacking) of ethyl-phenothiazine and the parallel stacking of thio-methyl-phenothiazine were investigated. In all three cases the HF results show nearly unbounded dimers, while at LMP2 level they become stable structures, mainly due to the presence of large dispersion forces. The intermolecular optimal distances were also calculated by both methods the values obtained by HF being significantly larger as compared with those obtained by LMP2. Considering the LMP2 geometry, the HF contribution in the intermolecular interaction energy gives positive values acting as a repulsion forces and slacking the dimer interaction. This effect is compensated by the large contribution of the dispersion component which finally keeps together the binary stacking structures. These results show the major role of the dispersion effects in stabilizing stacking structures of the aromatic rings and the feasibility of self-assembling in molecular two-dimensional supra-structures.

## 1. Introduction

In the last two decades, it is becoming more and more unambiguous that non-covalent interactions are crucial for determining the structures and properties of the supramolecular and biomolecular systems. Despite the fact that parallel-displaced $\pi$-$\pi$ stacking interactions have been recognized to be an important force in stabilizing the double helical structure of DNA and the tertiary structure of proteins, very few features are known about their roles in self-assembled monolayers of other large $\pi$ molecules such as aromatic thiols [1]. In the class of aromatic thiols we are interested in several molecular species from the phenothiazine family which are potentials candidates for establishing $\pi$-$\pi$ stacking interactions. The final target is to bring some insight for an optimal design of molecular systems able to self-assemble on gold atomically flat surfaces by sulfur bonds.

The phenothiazine (PT) derivatives have been selected due to their wide applicability. Due to their properties, PT derivatives are interesting candidates for the obtaining new materials (e.g. via mercapto or thioacetate derivatives) by deposition as self-assembled monolayers (SAMs) on Au(111) surface. Some recent publications revealed special properties for the aggregates obtained by the deposition of phenothiazine units on gold nanoparticules via mercapto derivatives [2].

The main goal of the paper is to use the local expansion of second order Møller-Plesset perturbation theory (LMP2) method for checking the stability and for optimizing the geometry of several dimers containing PT derivatives. The paper is organized as follows: the first two chapters

Journal of Physics: Conference Series **182** (2009) 012001 doi:10.1088/1742-6596/182/1/012001

present the molecular systems and the computational methods. The results are presented and discussed in the chapter 3 while the conclusions and the references follow in the ending part of the paper.

## 2. Molecular systems and computational methods

The optimized geometry structures of two (parallel and antiparallel) stacking configurations of the ethyl-phenothiazine (EPT) and the parallel stacking configuration of the thio-methyl-phenothiazine (TMPT) were carried out using the Hartree-Fock (HF) and the local expansion of second order Møller-Plesset perturbation theory (LMP2) methods [3-6]. The investigation techniques were implemented in the MOLPRO program package suite [7] considering the cc-pVDZ basis set. Local correlation methods are, by construction, free from the basis set superposition errors (BSSE) and in this way the LMP2 method is a suitable method for the calculation of intermolecular interactions. Two of the most important advantages of the method are that potentially expensive counterpoise (CP) corrections are avoided and the dispersion effects, which are crucial for stacking interaction of the aromatic rings, are also included. Further reductions in the computational cost are provided by combining LMP2 with the density fitting (DF) approximation of electron repulsion integrals in a method denoted DF-LMP2 [8,9].

## 3. Results and discussion

The main results of our work are presented in a condensed way in table 1 and are shown graphically in the figure 1. Our results consist mainly in the intermolecular interaction energies as well as in some relevant inter-atomic distances. The quantitative estimations has bee done for $\Delta E^{LMP2}$ – the total LMP2 energy, $\Delta E^{HF}$ – the HF component and $\Delta E^{disp}$ – the dispersion component, values obtained for LMP2 geometry and $\Delta E^{HF}$ – the HF energy value obtained for HF geometry. Concerning the geometrical aspects of our molecular systems the most important have been considered to be the distances between S atoms $d^{S\text{-}S}$, N atoms $d^{N\text{-}N}$ or S–N atoms $d^{S\text{-}N}$, each of them obtained for both LMP2 and HF geometries. In case of EPT both the parallel and antiparallel dimers were investigated. As our final purpose is to use the affinity between the sulfur and gold as a tool for self-assembling TMPT molecules on atomically flat Au(111) surfaces, in the TMPT case only the parallel dimer was investigated.

Figure 1 shows the parallel and antiparallel EPT dimer geometry considering two different viewing directions, as well as the TMPT parallel conformation. As concerning the color code the gray is assigned to carbon, the yellow to sulfur, and the dark and light blue to nitrogen and to hydrogen respectively.

All investigated geometry show bonded dimer structures at LMP2 level, the strongest bonding being obtained in case of TMPT dimmer. A little bit smaller is the bonding energy of parallel EPT dimer, while the weakest bonding is found for the antiparallel EPT dimer. In order to find out the contribution of electron correlation to the intermolecular interaction energy, the HF values were also calculated at the given LMP2 geometry. In a very surprising manner the obtained results show positive values which means that the HF counterpart of LMP2 energy has a repulsive character (more than 6.0 kcal/mol in all three cases) and only the correlation correction stabilize the dimer geometry. This fact presents a very important handicap because about 50% of the correlation correction is required to compensate the HF repulsion: -8.667 kcal/mol *versus* +7.416 for parallel EPT dimer, -7.600 kcal/mol *versus* +8.849 for antiparallel EPT dimer and -10.250 kcal/mol *versus* +6.042 for parallel TMPT dimer. Additionally, it is well-known that in case of aromatic rings the most part of the electron correlation contribution of the intermolecular interaction energy is given by the dispersion effects [30]. They seem to have a crucial role also in our case, because the major part of the correlation effects is given by these dispersion contributions: -13.906 kcal./mol for parallel EPT dimer, -14.198 kcal/mol for antiparallel EPT dimer and -14.173 kcal/mol for TMPT dimer, respectively. As regarding the basis set dependency, it was confirmed that the double-zeta split valence basis set (cc-pVDZ) gives a slightly large intermolecular energy [10,11]. Taking into account the relative large number of atoms contained in the investigated molecular systems our feeling is that we are near the upper limit of applicability for the MP2 method implemented in Molpro. Moreover, using basis sets completed with diffuse functions the amount of interaction energy will be further a little bit reduced [10]. Getting so strange behavior of HF component we have performed the geometry optimization at HF level in order to check whether the HF geometry show a bonded structure or not. We have found negligible negative intermolecular interaction energies for all three molecular systems: -0.168 kcal/mol for parallel EPT dimer, -0.367 kcal/mol for antiparallel EPT dimer and -1.217 kcal/mol for parallel TMPT dimer.

In the last case the HF energy has a slightly larger negative value, attributed to the H–S attractive interaction of the thiol groups.

**Table1.** The total intermolecular interaction energies, (the HF and the dispersion components) and several relevant atomic distances (see figures) for EPT parallel and antiparallel dimers of and for the TMPT parallel dimer respectively.

| Parameter | Value | Method | Parameter | Value | Method | Parameter | Value | Method |
|---|---|---|---|---|---|---|---|---|
| EPT-parallel | | | EPT-antiparallel | | | TMPT-parallel | | |
| $\Delta E^{LMP2}$ | -8.667 | DF-LMP2 | $\Delta E^{LMP2}$ | -7.600 | DF-LMP2 | $\Delta E^{LMP2}$ | -10.250 | DF-LMP2 |
| $\Delta E^{HF}$ | +7.416 | DF-LMP2 | $\Delta E^{HF}$ | +8.849 | DF-LMP2 | $\Delta E^{HF}$ | +6.042 | DF-LMP2 |
| $\Delta E^{HF}$ | -0.168 | DF-HF | $\Delta E^{HF}$ | -0.367 | DF-HF | $\Delta E^{HF}$ | -1.217 | DF-HF |
| $\Delta E^{disp}$ | -13.906 | DF-LMP2 | $\Delta E^{disp}$ | -14.198 | DF-LMP2 | $\Delta E^{disp}$ | -14.173 | DF-LMP2 |
| $d^{S\text{-}S}$ | 3.891 | DF-LMP2 | $d^{S1\text{-}N2}$ | 3.704 | DF-LMP2 | $d^{S\text{-}S}$ | 3.867 | DF-LMP2 |
| $d^{S\text{-}S}$ | 5.129 | DF-HF | $d^{S1\text{-}N2}$ | 5.007 | DF-HF | $d^{S\text{-}S}$ | 5.367 | DF-HF |
| $d^{N\text{-}N}$ | 4.054 | DF-LMP2 | $d^{N1\text{-}S2}$ | 3.925 | DF-LMP2 | $d^{N\text{-}N}$ | 4.114 | DF-LMP2 |
| $d^{N\text{-}N}$ | 5.426 | DF-HF | $d^{N1\text{-}S2}$ | 5.523 | DF-HF | $d^{N\text{-}N}$ | 5.436 | DF-HF |
| | | | | | | $d^{S'\text{-}S'}$ | 4.594 | DF-LMP2 |
| | | | | | | $d^{S'\text{-}S'}$ | 5.412 | DF-HF |

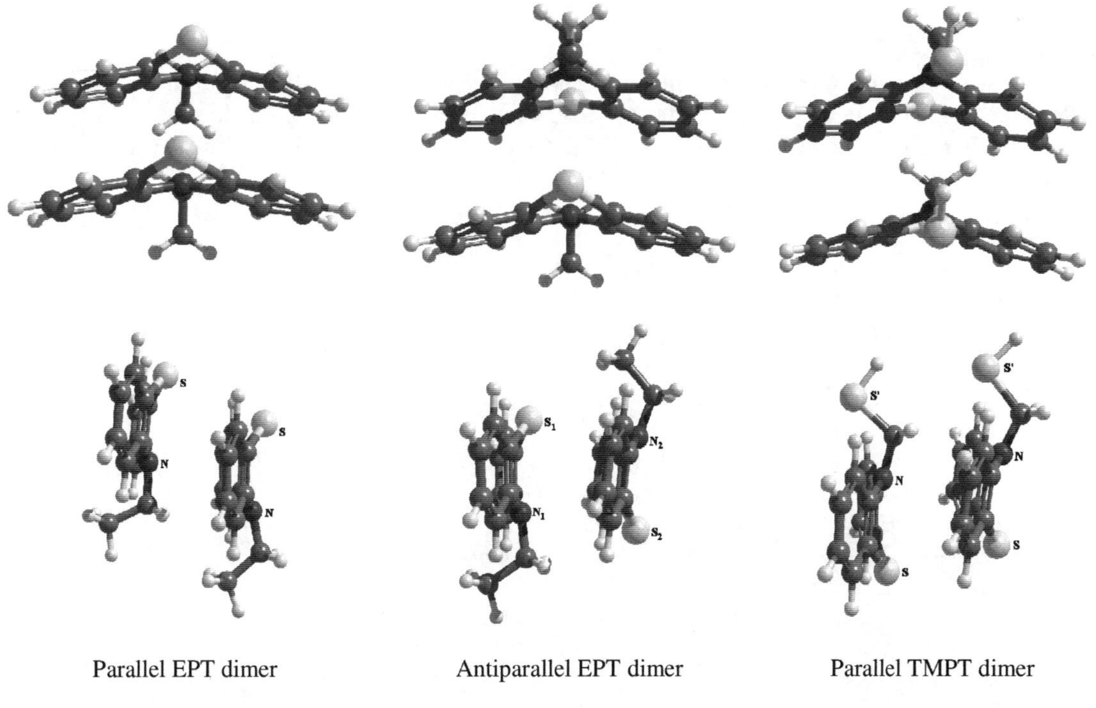


Parallel EPT dimer Antiparallel EPT dimer Parallel TMPT dimer

**Figure 1.** The geometry structures of parallel and antiparallel EPT and parallel TMPT.

The phenothiazine molecule has a three-ring structure in which two benzene rings are joined by a sulfur and nitrogen atom at nonadjacent positions, due to the "umbrella" form of the N orbitals. The angles between the planes of the benzene rings vary between 140°-150°, for different parallel or antiparallel dimers. This special non-planar configuration of the – let it call "V" shape – allows a more stable dimer structure as compared to the planar case. The monomers degrees of freedom for the relative displacement are reduced to the line defined by the S and N atoms, since in the perpendicular directions the spatial shift is much more difficult.

Analyzing the results described previously one observes that the calculated values of inter-atomic distances S–S and N–N are correlated with the energy values. Namely, if in LMP2 geometry they are about 3.8 Å and 4.0 Å, respectively; in case of HF geometry they become longer with about 1.2 Å and 1.4 Å, respectively (see figure 1). This fact is due to the strong attractive character of the dispersive interaction given by the electron correlation. The superposition of benzene rings for different types of monomers presents also some particularities. While, for EPT and TMPT parallel structures, one of the benzene rings is markedly shifted relative to the corresponding benzene ring from the other monomer, for antiparallel conformation they remain almost unshifted (see figure 1) but presenting a small planar angle ($\approx 7°$) between the rings.

## 4. Conclusions

The intermolecular $\pi$-$\pi$ stacking interaction between some phenothiazine derivatives has been investigated using the Hartree-Fock and the local second order Møller-Plesset perturbation theory methods combined with the density fitting approximation of electron repulsion integrals and considering the cc-pVDZ basis set. Two hypothetical parallel and antiparallel stacking conformations of ethyl-phenothiazine and the parallel stacking conformation of thio-methyl-phenothiazine were investigated. In all cases the stacking structures joint together in a characteristic "V" form. The results show that the dispersion effects have a crucial role in stabilizing stacking structures of the aromatic rings and consequently our prediction is that the process of self-assembling in molecular supra-structures (such as monolayers for example) having as the basic unit some phenothiazine derivatives might occur with a high probability.

## Acknowledgement

This research was supported by the Romanian Education and Research Ministry in the framework of NIRDIMT Core Program (project PN 09-440101, contract no 44N/2009). All the calculations were performed in the Data Center of NIRDIMT Cluj-Napoca. We wish to thank to Professor Ion Grosu for helpful discussions.

## References

[1] Dou R F *et al* 2006 *Langmuir* **22** 3049
[2] Komine Y, Ueda Y, Goto T and Fujihara H 2006 *Chem. Commun.* **3** 302
[3] Pulay P 1983 *Chem. Phys. Lett.* **100** 151
[4] Hetzer G, Pulay P and Werner H-J 1998 *Chem. Phys. Lett.* **290** 143
[5] Schütz M, Hetzer G and Werner H-J 1999 *J. Chem. Phys.* **111** 5691
[6] Hetzer G, Schütz M, Stoll H and Werner H-J 2000 *J. Chem. Phys.* **113** 9443
[7] MOLPRO, version 2008.1, a package of *ab initio* programs, Werner H-J, Knowles P J, Lindh R, Manby F R, Schütz M, and others, see http://www.molpro.net
[8] Vahtras O, Amlöf J and Feyereisen M W 1993 *Chem. Phys. Lett.* **213** 514
[9] Werner H-J, Manby F R and Knowles P 2003 *J. Chem. Phys.* **118** 8149
[10] Šponer J, Riley K E and Hobza P 2008 *Phys. Chem. Chem. Phys.* **10** 2595
[11] Bende A and Almásy L 2008 *Chem. Phys.* **354** 202.

# $^1$H NMR investigation of self-association of vanillin in aqueous solution

**Mircea Bogdan, Călin G Floare and Adrian Pîrnău**

National Institute for Research and Development of Isotopic and Molecular Technologies, 65-103 Donath, 400293 Cluj-Napoca, Romania

E-mail: mircea.bogdan@itim-cj.ro

**Abstract.** A self-association of vanillin have been studied by $^1$H NMR spectroscopy using the analysis of proton chemical shifts changes in aqueous solution as a function of concentration. The experimental results have been analysed using indefinite non-cooperative and cooperative models of molecular self-association, enabling the determination of equilibrium constants, parameters of cooperativity and the limiting values of vanillin proton chemical shifts in the complex. It was found that the dimer formation creates energetically favourable conditions for subsequent molecular association.

## 1. Introduction

The process of molecular association induces local and global changes to the molecules involved, which alter their physical and chemical behaviour in solution. The π-π stacked aromatic interactions have bee found to govern diverse supramolecular organizations and aggregates in both the solution and the solid state [1]. In comparison to more conventional interactions, such as hydrophobic interactions and hydrogen bonds, the π-π stacked interaction is not so clear and no commonly used model has been built to interpret the experimental observations. Therefore, the study on the variety of systems involved in the π-π stacked interaction is important to fully understand this type of noncovalent interaction.

Vanillin, on 4-hydroxy-3-methoxybenzaldehide is the primary component of the extract of the vanilla bean. Natural "vanilla extract" is a mixture of several hundred different compounds in addition to vanillin. Artificial vanilla is a solution of pure vanillin, usually of synthetic origin. Because of the scarcity and expense of natural vanilla extract, there has long been interest in the synthetic preparation of its predominant component. Today, most artificial vanillin is produced from the petrochemical raw material guaiacol [2]. The largest use of vanillin is as flavouring, usually in sweet foods. The ice cream and chocolate industries together comprises 75% of the market of vanillin as flavouring. Vanillin displays antioxidant and antimicrobial properties and may be used as a food preservative [3]. It is also used in the fragrance industry, in perfumes, and the mask unpleasant odours or tastes of drugs. Vanillin has been encapsulated in oxidized starches [4], and in β-cyclodextrin [5] in order to enhance the protection towards oxidation.

In this paper, the concentration dependences of chemical shifts of the non-exchange able protons of vanillin have been investigated, for a quantitative determination of the equilibrium constants of self-association, as well as the limiting values of proton chemical shifts of the molecular complexes.

## 2. Experimental

### *1.1. Materials*
Vanillin was purchased from Alfa Aesar GmbH&CoKG, Germany and used without any further purification. Deuterium oxide (99.7% D) was obtained from Heavy Water Plant Romag-Prod, Romania.

### *1.2. Apparatus*
The NMR experiments were performed on a Bruker Avance III 500 MHz spectrometer. The $^1$H NMR spectra were recorded in $D_2O$ solution at 295 K and all chemical shifts were measured relative to TMS. Typical conditions were as follows: 32 K data point, sweep width 3500 Hz, giving a digital resolution of 0.1 Hz/point. The $90^0$ pulse width was 10 μsec and the spectra were collected by co-addition of maximum 128 scans.

## 3. Results and discussion
The $^1$H NMR spectrum of vanillin in aqueous solution is shown in figure 1. Assignment of vanillin protons was obtained from 1D and 2D (COSY) experiments and is presented in table 1.

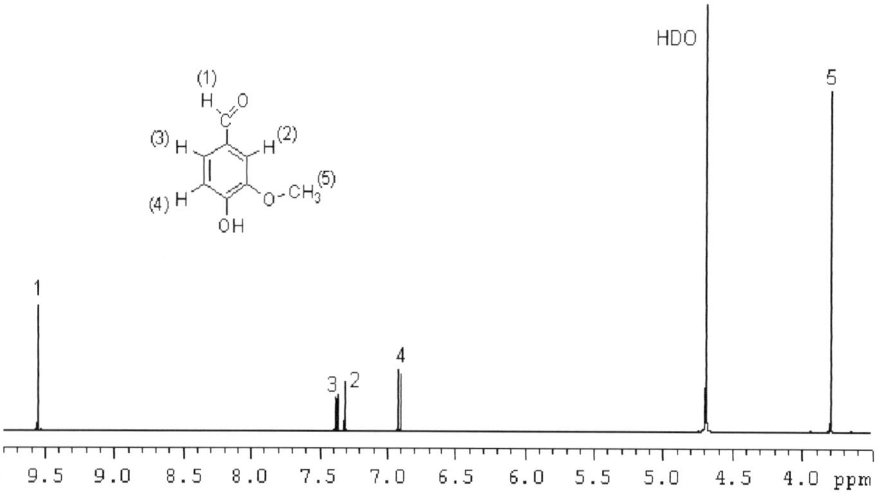


**Figure 1.** $^1$H NMR spectrum of 2 mM vanillin in $D_2O$.

**Table 1.** $^1$H NMR chemical shift values and coupling constants, J, of 2 mM vanillin in $D_2O$.

| Proton | δ (ppm) | J (Hz) |
|---|---|---|
| H-1 | 9.6400 | |
| H-2 | 7.4644 | $J_{2\text{-}3} = 2.0$ |
| H-3 | 7.4741 | $J_{3\text{-}4} = 8.2$ |
| H-4 | 6.9890 | |
| $-CH_3$ | 3.8641 | |

In order to obtain information about the solution behaviour of vanillin, the dependence of proton chemical shift on concentration was analysed. The dependence of proton chemical shift on concentration of vanillin is reported in figure 2. A general decrease in proton chemical shift values was observed with increasing concentration, suggesting that vanillin is involved in a self-association process through stacking interactions [6]. This interaction causes a shielding term modulation due to change in intensity of the anisotropic magnetic field arising from electron motion in the aromatic rings. The experimental results have been analysed using indefinite non-cooperative and cooperative models of molecular self-association [7], enabling the determination of equilibrium constants, parameter of proton chemical shifts in the complex.

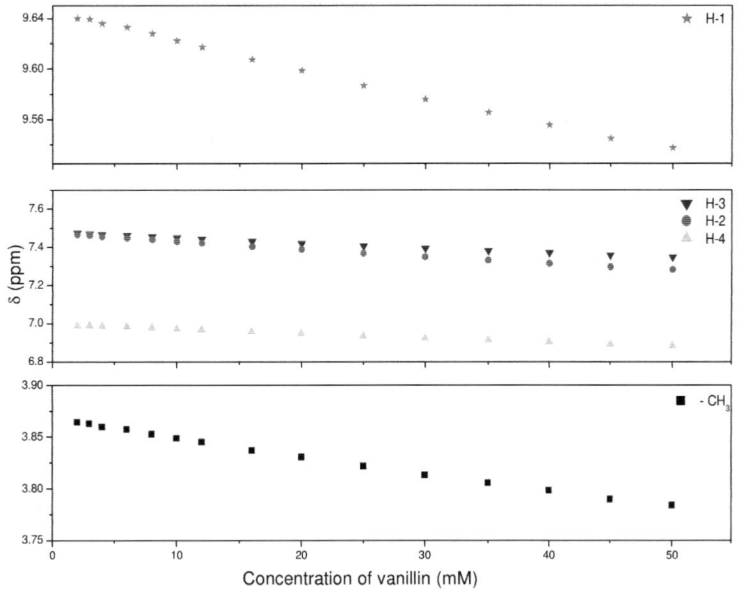


**Figure 2.** Experimental concentration dependences of chemical shifts of vanillin non-exchangeable protons.

*1.3. Analysis by indefinite non-cooperative (isodesmic) model*

The isodesmic model of solute self-association is based on the assumption that solute molecules associate to from stacks (dimers, trimers etc.) where the equilibrium constant K for each step is the same.

Thus

$$A_j + A \xleftrightarrow{K} A_{j+1} \tag{1}$$

It can be shown that:

$$[A_0] = \frac{[A]}{(1 - K[A])^2} \tag{2}$$

where $[A_0]$ is the total concentration of solute A and [A] is the concentration of the monomer in solution. Further it is assumed that the proton chemical shifts changes observed for a solute molecule binding at the end of a stack are half those observed for a molecule associating in the middle of a stack [8]:

$$\delta_j = \frac{(j-1)\delta_i + \delta_m}{j} \tag{3}$$

where $\delta_j$ is the proton chemical shift of the j-mer, $\delta_m$ is the proton chemical shift of the monomer, *i. e.* at infinite dilution and $\delta_j$ is the maximal proton chemical shift of solute *A* present in a stack.

Owing to rapid exchange in self-association reactions, the observed chemical shift of all species present in solution is given by:

$$\delta_{obs} = \frac{[A]\delta_m + 2[A_2]\delta_d + \ldots + i[A_i]\delta_{A_i} + ..}{[A_0]} = \frac{[A]}{[A_0]}\left\{\frac{\delta_m}{1-K[A]} + \frac{\delta_i K[A]}{(1-K[A])^2}\right\} \tag{4}$$

Substituting equation (2) into equation (4) gives:

$$\delta_{obs} = \left(\frac{\sqrt{1+4K[A_0]}-1}{2K[A_0]}\right)\delta_m + \left(1 + \frac{1-\sqrt{1+4K[A_0]}}{2K[A_0]}\right)\delta_i \tag{5}$$

We note that a dimerization model assuming only the shift values $\delta_m$ and $\delta_{dimer}$, with an association constant $K_{dimer}$, gives an identical equation the only differences being that K is replaced by 2 $K_{dimer}$ and ($\delta_i$ - $\delta_m$) is replaced by ($\delta_{dimer}$ - $\delta_m$) which is equivalent to ($\delta_i$ - $\delta_m$)/2, since a molecule in a dimer is effectively at the end of a stack of two.

Therefore, NMR chemical shift data cannot be used to distinguish dimerization from stacking, if the isodesmic model is applicable. The parameters $\delta_m$, $\delta_i$, and K in equation (5) were calculated from a least-squares fitting of the data to the isodesmic model. Table 2 presents the obtained results. In order to estimate the probability of formation a vanillin complexes of higher order than dimers, the experimental results have also been analysed using the indefinite cooperative model of molecular self-association.

*1.4. Analysis by indefinite cooperative (semi-isodesmic) model*

This model allow for the possibility that the equilibrium constant for the first step in a self-association dimer formation differs from the remaining constants. In this model the equilibrium constants of reactions (1) are assumed to be equal for all $j \geq 2$ ($K = K_3 = K_4 = \cdots = K_i$) and $K_{dimer} = \sigma K$. If the dimer forms more easily than subsequent additions, $\sigma > 1$, while if the dimer forms with more difficulty, $\sigma < 1$. The case $\sigma = 1$ corresponds to the isodesmic model considered above.

Taking into account that the mass conservation law for reactions (1) is given by:

$$[A_0] = \sum_{j=1}^{\infty} j[A_j] \tag{6}$$

and:

$$[A_2] = \sigma K[A]^2 \, ; \, [A_j] = \sigma K^{j-1}[A]^j \, ; \; j \geq 2 \tag{7}$$

the summation in equation (6) leads to:

$$[A_0] = (1-\sigma)[A] + \frac{\sigma[A]}{(1-K[A])^2} \tag{8}$$

The observed proton chemical shift, $\delta_{obs}$ is given by:

$$\delta_{obs} = \frac{1}{[A_0]}\sum_{j=1}^{\infty} j[A_j]\delta_j \tag{9}$$

Substituting equations (3) and (7) in equation (9) we obtain:

$$\delta_{obs} = \frac{[A]}{[A_0]}\left\{\left(1 + \frac{\sigma K[A]}{1-K[A]}\right)\delta_m + \frac{\sigma K[A]}{(1-K[A])^2}\delta_i\right\} \tag{10}$$

Equation (8) is cubic in *[A]*, and was solved analytically. Equation (10) contains there unknown parameters $\delta_i$, $\sigma$ and K to by determined from experimental concentration dependences of vanillin proton chemical shifts. We developed a computer programme based on an iteration procedure, in order to fit the experimental values of $\delta_{obs}$ to the approximate equation. Each iteration sets up a quadratic programme to determine the direction of search and the loss function:

$$E = \sum_{i=1}^{N} \left( \delta_{i,obs}^{\exp} - \delta_{i,obs}^{calc} \right)^2 \quad (11)$$

until the search converges. N is the number of experimental points. The values of E obtained in the present calculations are in all cases centred arround $6 \cdot 10^{-5}$ indicating a very good fit to the experimental data. The mean values of the calculated parameters of equation (10) are presented in table 2.

**Table 2.** Calculated self-association parameters of vanillin.

| | Isodesmic model | | | Semi-isodesmic model | | |
|---|---|---|---|---|---|---|
| Proton | $\delta_m$ (ppm) | $\delta_i$ (ppm) | $\bar{K}$ $(M^{-1})$ | $\delta_i$ (ppm) | $\bar{K}$ $(M^{-1})$ | $\bar{\sigma}$ |
| H-1 | 9.6466 | 8.0686 | | 9.1753 | | |
| H-2 | 7.4756 | 5.6040 | | 6.6400 | | |
| H-3 | 7.4801 | 6.2404 | 2.15±0.46 | 6.8000 | 10.62±1.73 | 0.50±0.06 |
| H-4 | 6.9960 | 5.3486 | | 6.5620 | | |
| $-CH_3$ | 3.8682 | 2.9745 | | 3.4870 | | |
| | Dimerization<br>$\delta_{dimer} = \delta_i$<br>$K_{dimer} = 1.07 \pm 0.23 M^{-1}$ | | | $K_{dimer} = 5.31 \pm 0.86\ M^{-1}$ | | |

## 2. Conclusion

In the present study, experimentally measured chemical shifts of vanillin up to a concentration of 50 mM, provide clear evidence that vanillin undergoes self-association in solution as the concentration increases.

Two modes which account for the self-associative properties of vanillin were tested. The model which gave the most satisfactory fit to the experimental data was one in which self-association occurs according to an indefinite cooperative scheme. The value of $\sigma \cong 0.5$ for vanillin self-associative proves that this complexation process is cooperative. Hence one may assume that formation of j-mers ($j > 2$) complexes of vanillin in aqueous solution is favourable from an energetic point of view and the cooperative model of molecular association is realized in the conditions of the experiment.

## References

[1] Hunter C A, Lawson K R, Perkins J and Urch C J 2001 *J. Chem. Soc. Perkin Trans.* **2** 651
[2] Esposito L J, Formanek K, Kientz G, Mauger F, Maureaux V, Robert G and Truchet F 1997 Vanillin *Kirk-Othmer Encyclopedia of Chemical Technology* vol 24 4th edition (New York: Wiley–Interscience) pp 812-25
[3] Davidson P M and Naidu A S 2000 *Phyto-phenols Natural food antimicrobial systems* ed A S Naidu (Boca Raton: CRC Press) pp 265-94
[4] Chattopadhyaya S, Singhal R S and Kulkarni 1998 *Carbohydrate Polymers* **37** 143
[5] Karathanos V T, Mourtzinos I, Yannakopoulou K and Andrikopoulos N K 2007 *Food Chemistry* **101** 652
[6] Sindkhedkar M D, Mulla H R and Cammers-Goodwin A J 2000 *J. Am. Chem Soc.* **122** 9271
[7] Martin R B 1996 *Chem. Rev.* **96** 3043
[8] Baxter N J, Williamson M P, Lilley T H and Haslam E 1996 *J. Chem. Soc. Faraday Trans* **92** 231.

Processes in Isotopes and Molecules IOP Publishing
Journal of Physics: Conference Series **182** (2009) 012003 doi:10.1088/1742-6596/182/1/012003

# Inclusion compound of vitamin B6 in β-CD. Physico-chemical and structural investigations

**Gheorghe Borodi, Irina Kacso, Sorin I Fărcaş and Ioan Bratu**

National Institute for Research and Development of Isotopic and Molecular Technologies, 65-103 Donath, 400293 Cluj-Napoca, Romania

E-mail: ioan.bratu@itim-cj.ro

**Abstract**. Structural and physico-chemical characterization of supramolecular assembly of vitamin B6 with *β*-cyclodextrin (*β*-CD) prepared by different methods (kneading, co-precipitation and freeze-drying) has been performed by using several spectroscopic techniques (FTIR, $^1$H NMR, UV-Vis), powder X-ray diffraction and DSC in order to evidence the inclusion compound formation. An analysis of the chemical shifts observed in the $^1$H-NMR spectra and of the vibrational frequency shifts led to the tentative conclusion that the vitamin B6 probably enters the cyclodextrin torus when forming the *β*-CD–vitamin B6 inclusion complex.

## 1. Introduction

A bioactive compound act orally, parentherally or locally depending on the administration form (solid or liquid); in function of the auxiliary substances, its biodisponibility is influenced [1]. In order to solve this problem, the most employed method is the complexation with cyclodextrins (CD) [2-4]. Cyclodextrins are cyclic oligosaccharides having a hydrophobic external surface and an inner hydrophilic surface. There are different CD types depending on the glucopyranose unit number [α-six; β-seven, γ-eight, etc.] (see figure 1a). Cyclodextrins and their derivatives are used to increase the solubility and the solving rate of the bioactive substances.

(a) (b)

**Figure 1.** (a) β-Cyclodextrin molecule. (b) Pyridoxine molecule.

They are capable to form more soluble guest-host systems through non-covalent bonds, Van der Waals interactions, hydrophobic effect, solvent molecules reorganization and hydrogen bonds [5].

Vitamin B6 (pyridoxine) (figure 1b) is a water-soluble vitamin and is part of the vitamin B complex group. Pyridoxal phosphate (PLP) is the active form and is a cofactor in many reactions of amino acid metabolism, including transamination, deamination, and decarboxylation. PLP is necessary also for the enzymatic reaction governing the release of glucose from glycogen. The primary role of vitamin B6 is to act as a coenzyme to many other enzymes in the body that are involved predominantly in metabolism. Vitamin B6 is involved in the following metabolic processes: amino acid, glucose and lipid metabolism, neurotransmitter synthesis, histamine synthesis, hemoglobin synthesis and function and gene expression.

The aim of this paper was to investigate the molecular encapsulation of the vitamin B6 in β-CD through different spectroscopic methods (FTIR, NMR, UV-Vis), DSC and powder X-ray diffraction.

## 2. Experimental

The preparation of the inclusion compounds was done by kneading (*kn*), co-precipitation (*co*) and freeze-drying (*fd*) methods. Also, physical mixtures (*pm*) were prepared in order to be compared with the inclusion compounds.

To evidence the formation of the inclusion compounds, FTIR, DSC and X-ray powder diffraction (XRD) were employed. FTIR measurements were performed with JASCO 6100 spectrometer in the 4000 to 400 $cm^{-1}$ spectral region with a resolution of 4 $cm^{-1}$ using KBr pellet technique. DSC thermograms were obtained with DSC 60 Shimadzu calorimeter and X-ray diffraction patterns were obtained with a Bruker D8 Advance diffractometer in the $2\theta = 2\text{-}50°$ angular domain using Cu $K\alpha_1$ radiation.

The inclusion complexes of *β*-CD (named host, H) with the active substance as guest (G) were prepared in aqueous solutions starting from milimolar "mother solutions" of G and H in $D_2O$ for $^1H$ NMR and in $H_2O$ for UV-Vis, respectively. $^1H$ NMR spectra were collected with Bruker Avance 500 spectrometer after 15 min. of thermal equilibration at 25°C. The spectrometer was operated at 500.1325 MHz with the following parameters: 32000 data points, pulses of 90°, 2 seconds delay between scans (16 scans) and a digital resolution of 0.588 Hz/point. The chemical shifts (in ppm) are relative to the HOD signal located at 4.6897 ppm. UV-Vis spectra were obtained with JASCO V-550 spectrometer.

## 3. Results and discussion

In order to evidence the vitamin B6–*β*-CD inclusion compound formation, several methods were employed: FTIR, DSC and XRD. As concerned FTIR spectroscopy, the spectra of pure vitamin B6, of *β*-CD, their 1:1 physical mixture and of the products obtained by co-precipitation, kneading and freeze-drying procedures were compared.

### *3.1. FTIR*

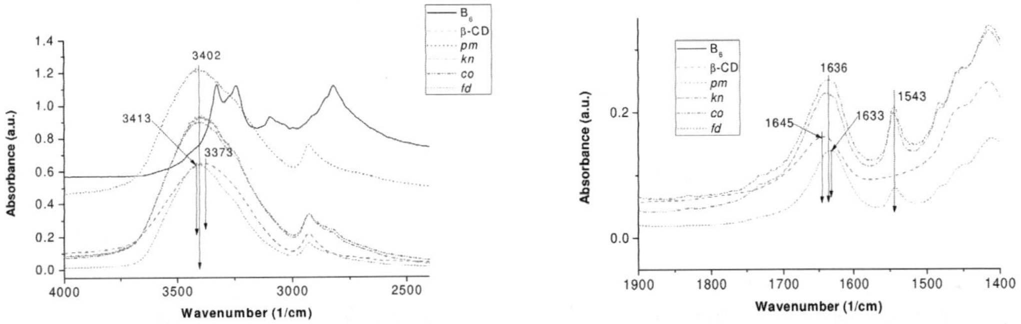


**Figure 2.** FTIR spectra of: vitamin B6, *β*-CD, their 1:1 *pm* and the products obtained by *kn*, *co*, and *fd* procedures, 4000-2500 $cm^{-1}$ (left) and 1900-1400 $cm^{-1}$ (right) spectral ranges.

As concerned the O–H stretching vibration, its maximum, located at ~3373 cm$^{-1}$ in β-CD FTIR spectrum (see figure 2), is shifted to 3402 cm$^{-1}$ for *kn* and *co* and to 3413 cm$^{-1}$ for *fd* products spectra, respectively. The frequency shift of the O–H bending vibration located at 1645 cm$^{-1}$ in pure *β*-CD, is shifted to 1636 cm$^{-1}$ for *kn* and *co* products and to 1633 cm$^{-1}$ for *fd* product. This frequency shift can be related to the appearance of hydrogen bonds during the complexation process, especially for the *fd* product. This complexation mechanism must be taken into account for the *kn* and *co* products, also.

### *3.2. X-ray powder diffraction*

X-ray powder diffraction patterns of vitamin B6, β-CD and of the products obtained by *co*, *kn* and *fd* procedures were compared (figure 3a). From the powder diffraction patterns one can see that inclusion compounds are formed in different degree. Both crystalline and amorphous phases are present in the obtained compounds. The difference between these patterns is due to different ratios between crystalline and amorphous phases. The highest degree of formation is obtained for kneading procedure.

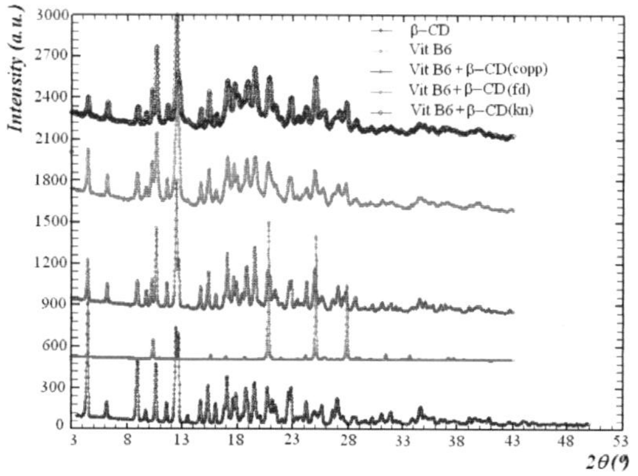


(a)

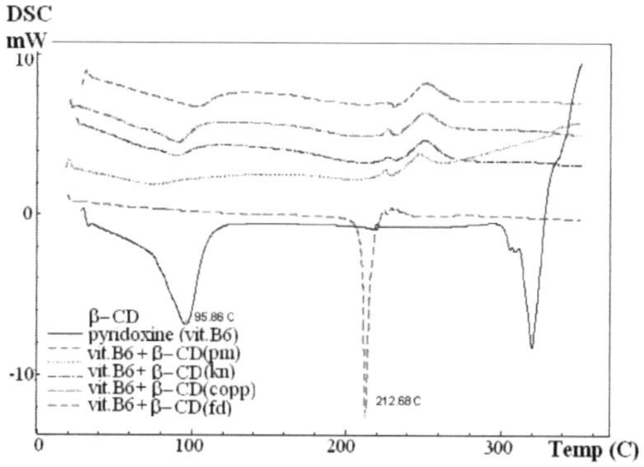


(b)

**Figure 3.** (a) X-ray patterns of: vitamin B6, *β*-CD, and the products obtained by *kn*, *co* and *fd* procedures; (b) DSC thermograms of the vitamin B6, *β*-CD and of the obtained inclusion compounds.

### *3.3. DSC*

DSC thermograms of vitamin B6, *β*-CD and of the products obtained by *kn*, *co* and *fd* procedures were recorded (figure 3b). DSC trace of vitamin B6 presents a sharp endothermic melting peak at 212°C, followed by an exothermic peak, possibly melt degradation. The thermogram of the *β*-CD showed an endothermic transition between 70-110°C, due to the loss of water molecules and another endothermic peak at around of 300°C indicating the degradation of cyclodextrin. By comparing the thermograms of the pure compounds with the thermograms of products the decreasing of dehydration endothermic peak of cyclodextrin was observed, as well as a disappearance of the melting peak of the vitamin B6 in the inclusion compounds, but the exothermic peak of melt degradation remains for all inclusion compounds.

### *3.4. $^1$H NMR*

The preliminary NMR measurements are presented in the figure 4 where we can see the $^1$H NMR spectra of both molecules, B6 vitamin and *β*-CD, separately, and of the 1:1 mixture of the two components, all as aqueous solutions in $D_2O$. The aim of these measurements was to notice if the mixture spectrum is or not an algebraic sum of the components spectra. This fact can give us information about the complex formation during the preparation procedure. A very careful

Journal of Physics: Conference Series **182** (2009) 012003 doi:10.1088/1742-6596/182/1/012003

examination of the mixture spectrum' points out some chemical shifts variation for any protons belonging both to B6 vitamin and β-CD. For instance, we can notice a modification of the chemical shift corresponding to H5 and H3 protons of guest and of the B6 molecule ethyl protons.

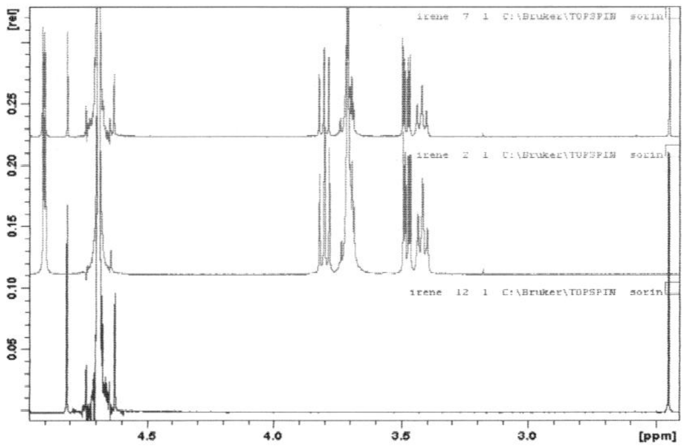


**Figure 4.** [1]H NMR spectra of the inclusion complex of vitamin B6 and β-CD. Bottom: vitamin B6; middle: β-CD; top: inclusion complex of vitamin B6:β-CD (1:1).

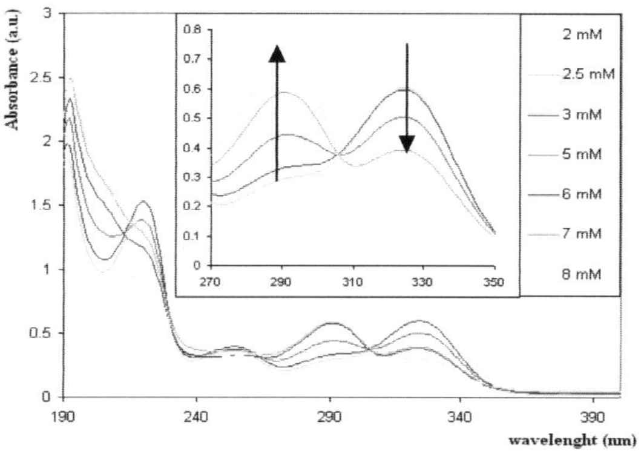


**Figure 5.** UV-Vis spectra of the inclusion complex of vitamin B6 (pyridoxine) with β-CD ([B6] = 0.1 mM in $H_2O$, [β-CD]=variable).

All these allowed us to conclude that the inclusion complex of these two molecules is formed, and the NMR measurements are in good accordance with the other measurements presented in this paper. Further NMR measurements connected with the possibilities to determine the stoichiometry and the complexation constant of the complex are in progress.

### *3.5. UV-Vis*

The vitamin B6 has two absorption maxima, at 292 and 324 nm (figure 5: the arrows suggest the absorption maxima variations with the increase of the β-CD concentration). With increasing concentration of β-CD, the absorption intensity at 292 nm gradually increased, whereas the absorption intensity at 324 nm decreased gradually. The isosbestic point was observed at 307 nm. This can confirm the supramolecular inclusion complex formation between vitamin B6 and β-CD.

## 4. Conclusions

Based on FTIR, X-ray powder diffraction and DSC data it was evidenced that an inclusion compound between vitamin B6 and β-CD was formed. The intensity diminishing or the disappearance of the melting endothermic peak characteristic to pure substance and the diminishing of the endothermic peak characteristic to bonded and non bonded cyclodextrin water elimination show the inclusion complex formation, also. All the preparation methods employed were successfully in obtaining these supramolecular assemblies. The changes in the chemical shift corresponding to H5 and H3 protons of guest and of the B6 ethyl protons demonstrate the formation of B6–β-CD inclusion complex in aqueous solution. The isosbestic point of the UV-Vis spectrum demonstrates also the inclusion complex formation in aqueous solution.

**Acknowledgments**

This work received financial support from the Romanian Research and Education Ministry under the Core Project PN-09-44 02 01.

## References

[1] Leucuta S E 1989 *Farmacocinetica şi biodisponibilitatea în terapia medicamentoasă* (Bucureşti: Editura Medicală)
[2] Albers E and Muller B W 1992 *J. Pharm. Sci.* **81** 756-61
[3] Ficarra R, Ficarra P and Rameri D 2000 *J. Pharm. Biomed. Anal.* **23** 231-6
[4] Acerbi D, Bovis G, Carli F, Pasini M, Pavesi L and Peveri T 1990 *Drug Investigation* **2**, 29-36
[5] Lopata A, Darvas F, Stadler-Szóke A and Szejtli J *J. Pharm. Sci.* 1995 **74** 211-3.

Journal of Physics: Conference Series **182** (2009) 012004 doi:10.1088/1742-6596/182/1/012004

# Spectroscopic investigation of the interaction between β-cyclodextrin and ascorbic acid

**Ioan Bratu[1], Marieta Muresan-Pop[2], Irina Kacso[1] and Sorin I Fărcaş**

[1] National Institute for Research and Development of Isotopic and Molecular Technologies, 65-103 Donath, 400293 Cluj-Napoca, Romania

[2] Babeş-Bolyai University, Faculty of Physics, 1 Mihail Kogălniceanu 400084 Cluj-Napoca, Romania

E-mail: ioan.bratu@itim-cj.ro

**Abstract**. Inclusion compound of vitamin C (ascorbic acid) with β-cyclodextrin (β-CD), prepared by different methods (kneading, co-precipitation and freeze-drying) has been caracterized by several spectroscopic techniques (FTIR, $^1$H NMR, UV-Vis), powder X-ray diffraction and DSC technique. Based on the chemical shifts observed in the $^1$H-NMR and on FTIR spectra the tentative conclusion is that vitamin C probably enters the cyclodextrin torus forming the inclusion complex.

## 1. Introduction

Cyclodextrins and their derivatives are used to increase the solubility and to improve the drug delivery rate of the bioactive substances. Cyclodextrins are cyclic oligosaccharides having a hydrophobic external surface and an inner hydrophilic surface (see figure 1a). A bioactive compound act orally, parentherally or locally depending on the administration form (solid or liquid); in function of the auxiliary substances, its biodisponibility is influenced [1]. In order to solve this problem, the most employed method is the complexation with cyclodextrins (CD) [2-4].

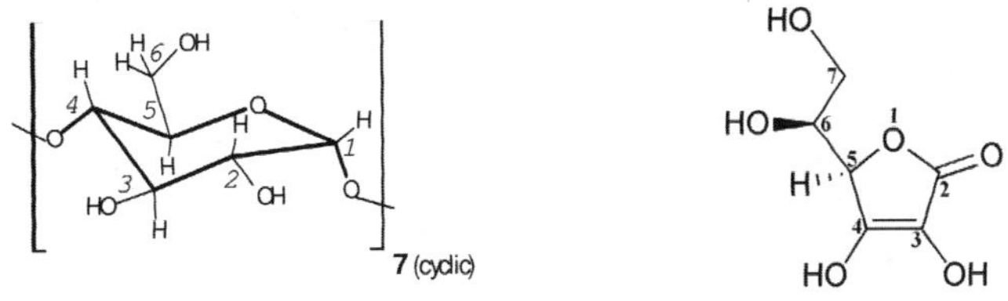


(a) (b)

**Figure 1.** (a) β-Cyclodextrin molecule. (b) Ascorbic acid molecule.

They are capable to form more soluble guest-host systems through non-covalent bonds, Van der Waals interactions, hydrophobic effect, solvent molecules reorganization and hydrogen bonds [5].

Vitamin C (ascorbic acid) (figure 1b), known as the antiscorbutic vitamin, essential nutrient with attributes in cellular and tissular oxido-reduction processes, is implied in the glucose and collagen metabolism, of the folic acid or some amino acids in the neutralization of the free radicals, in immunological reactions.

Based on calorimetric and spectrophotometric investigation of the interaction between HP-β-CD and ascorbic acid, the authors demonstrate an 1:1 molecular complex formation [6]. The effects of degree of substitution and of pH on complex formation between HP-α-CD and HP-β-CD with ascorbic acid have been also investigated [7]. The interaction of ascorbic acid with HP-β-CD was also investigated by FTIR, $^1$H NMR and UV spectroscopy [8]. $^1$H NMR study of α- and β-CD complexation with ascorbic acid was performed to establish the protons of α- or β-CD involved in the inclusion process [9].

The aim of this paper was to investigate the supramolecular assembly of the ascorbic acid with β-CD through several spectroscopic methods (FTIR, UV-Vis and $^1$H NMR), DSC and powder X-ray diffraction.

## 2. Experimental

The inclusion compounds were prepared by kneading (*kn*), co-precipitation (*co*) and freeze-drying (*fd*) methods. Also, physical mixtures (*pm*) were prepared in order to be compared with the inclusion compounds. FTIR, DSC and X-ray powder diffraction (XRD) were employed to evidence the inclusion compounds formation. FTIR measurements were performed with JASCO 6100 spectrometer. DSC thermograms were obtained with DSC 60 Shimadzu calorimeter and X-ray diffraction patterns were obtained with a Bruker D8 Advance diffractometer. The inclusion complexes of vitamin C with *β*-CD were prepared in aqueous solutions starting from milimolar "mother solutions" in $D_2O$ for $^1$H NMR and in $H_2O$ for UV-Vis, respectively. $^1$H NMR spectra were collected with Bruker Avance 500 spectrometer. UV-Vis spectra were obtained with JASCO V-550 spectrometer.

## 3. Results and discussion

### *3.1. FTIR*

The O–H stretching frequency (see figure 2) remains constant for the *pm* and *kn* products, whereas for *co* and *fd* inclusion compounds is shifted to 3414 and 3408 cm$^{-1}$, respectively. The vitamin C's peak at 3410 cm$^{-1}$ can be identified in the *pm* and *kn* products, so the inclusion process is not so efficient in the case of kneaded product. As concerned the functional C=O groups, they were located at 1673 cm$^{-1}$ and 1754 cm$^{-1}$ for pure vitamin C, *pm* and *kn* products, whereas they are shifted to 1694 and 1763 cm$^{-1}$, respectively for *co* and *fd* products.

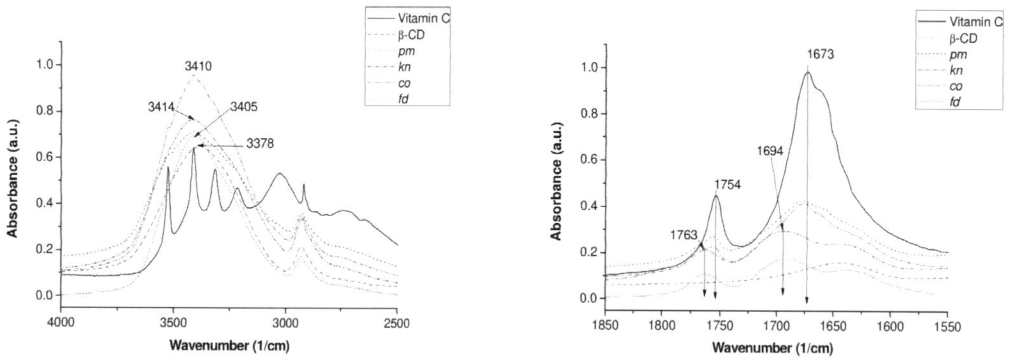


**Figure 2.** FTIR spectra of: vitamin C, *β*-CD, their 1:1 *pm* and the products obtained by *kn*, *co*, and *fd* procedures, 4000-2500 cm$^{-1}$ (left) and 1850-1550 cm$^{-1}$ (right) spectral ranges.

The frequency shift toward higher frequencies demonstrates the breaking of the hydrogen bonds during the inclusion process.

### *3.2. X-ray powder diffraction*

From powder diffraction patterns (see figure 3a) one can see that inclusion compounds are formed in a different degree. Both crystalline and amorphous phases are present in the so obtained compounds.

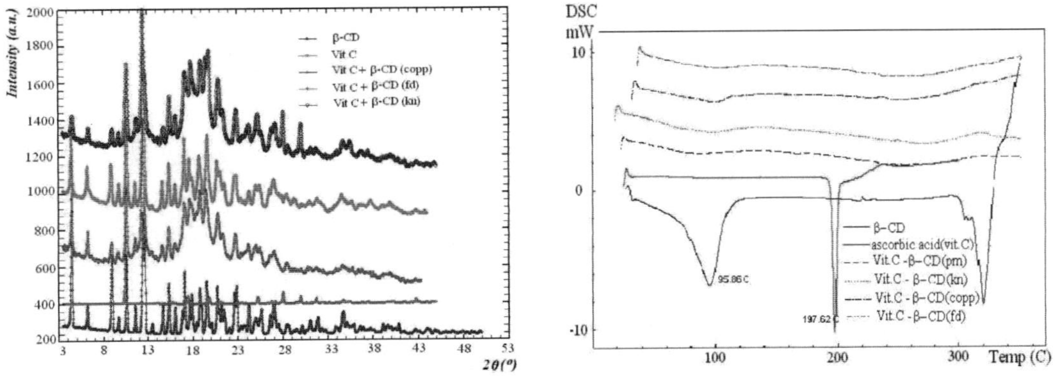


(a) (b)

**Figure 3.** (a) X-ray patterns of: vitamin C, β-CD, and the products obtained by *kn*, *co* and *fd* procedures; (b) DSC thermograms of the vitamin C, β-CD and of the obtained inclusion compounds.

The difference between these patterns is due to different ratios between crystalline and amorphous phases.

### *3.3. DSC*

DSC trace of vitamin C (figure 3b) presents a sharp endothermic melting peak at 197°C, followed by an exothermic peak at around 240°C, possible the start of degradation. The thermogram of the *β*-CD shows an endothermic transition between 70-110°C and an endothermic peak at around of 300°C. By comparing the thermograms of the pure compounds with those of the products, the decreasing of dehydration endothermic peak of cyclodextrin, as well as a disappearance of the melting peak of the vitamin C in the inclusion compounds were observed.

### *3.4. $^1$H NMR*

The $^1$H NMR measurements are used to prove the inclusion phenomena of C vitamin in β-CD identifying the protons implied in the complexation process and belonging both to the guest and to the host molecules, respectively. The inclusion of a guest molecule into a host one is shown by changes in chemical shifts of some of the guest and host protons in comparison with the chemical shifts of the same protons in the free compounds (figure 4). The complexation is a dynamic process and the exchange between the free and bound molecules is very rapid. In any case, it was observed a variation of the chemical shifts of the same protons in the case of NMR spectrum of the free and bound molecules. This was the aim of the preliminary measurements, and the results are in good accordance with the other measurements presented in this paper.

### *3.5. UV-Vis*

One can observe a hypsochromic shift of the maximum (see figure 5 – the arrows suggest the absorption maxima shifts with the increase of the β-CD concentration) in comparison with the corresponding one for the pure substance (265 nm), with the increase of the β-cyclodextrin excess.

This maximum frequency shift demonstrates the inclusion complex formation and can be used to determine its stoichiometry and the stability constant.

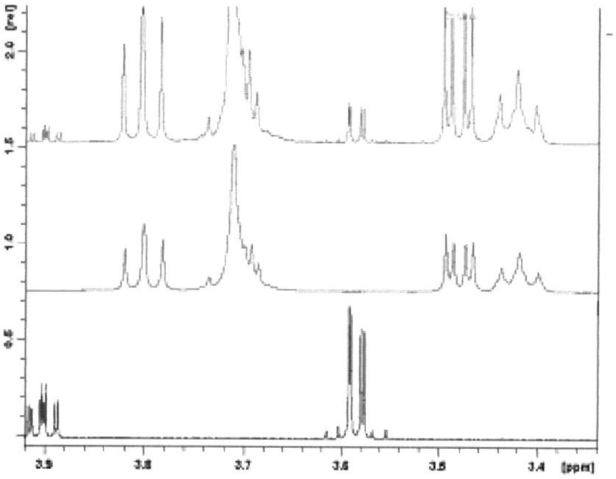

**Figure 4.** $^1$H NMR spectra of the inclusion complex of vitamin C and β-CD. Bottom: vitamin C; middle: β-CD; top: inclusion complex of vitamin C:β-CD (1:1).

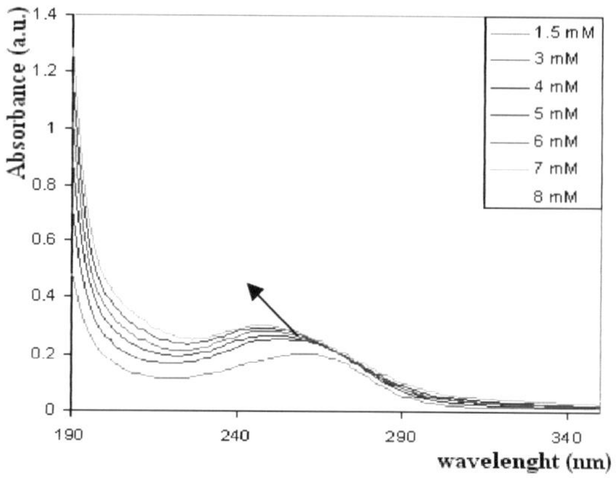


**Figure 5.** UV-Vis spectra of the inclusion complex of vitamin C (ascorbic acid) with β-CD ([C]=0.08 mM in $H_2O$, [β-CD] = variable).

## 4. Conclusions

FTIR, X-ray powder diffraction and DSC data reveal that an inclusion compound between ascorbic acid and β-CD was formed. The changes appearing in the DSC data of the pure substance as compared with those of the kneaded, freeze-dried and co-precipitated products confirm the inclusion compound formation, all employed preparation methods being successfully in obtaining these supramolecular assemblies. The $^1$H NMR and UV-Vis data demonstrates also the inclusion complex formation in aqueous solution.

## Acknowledgments

This work received financial support from the Romanian Research and Education Ministry under the Core Project PN-09-44 02 01.

## References

[1] Leucuta S E 1989 *Farmacocinetica şi biodisponibilitatea în terapia medicamentoasă* (Bucureşti: Editura Medicală)
[2] Albers E and Muller B W 1992 *J. Pharm. Sci.* **81** 756-61
[3] Ficarra R, Ficarra P and Rameri D 2000 *J. Pharm. Biomed. Anal.* **23** 231-6
[4] Acerbi D, Bovis G, Carli F, Pasini M, Pavesi L and Peveri T 1990 *Drug Investigation* **2**, 29-36
[5] Lopata A, Darvas F, Stadler-Szóke A and Szejtli J 1995 *J. Pharm. Sci.* **74** 211-3
[6] Terekhova I V and Obukhova **N A** 2005 *Mendeleev Commun.* **15** 38-40
[7] Terekhova I V, Obukhova N A, Agafonov A V, Kurochkina G I, Syrtsev A N and Gratckev M K 2005 *Russ. Chem. Bull. Int. Ed.* **54** 1883-6
[8] Garnero G and Longhi M 2007 *J. Pharm. Biomed. Anal.* **45** 536-45
[9] Terekhova I V, Kulikov O V, Kumeev R S, Nikiforov M Yu and Al'per G A 2005 *Russ. J. Coord. Chem.* **31** 218-20.

# Short range correlation of the erythrocyte membrane fluctuations

**Luiza Buimagă-Iarinca, Vasile V Morariu**

National Institute for Research and Development of Isotopic and Molecular Technologies, 65-103 Donath, 400293 Cluj-Napoca, Romania

E-mail: luiza.iarinca@itim-cj.ro

**Abstract**. The erythrocyte membrane fluctuations analysis was performed for two suspension media, plasma and phosphate buffered saline (PBS) respectively. The investigation methods consist of detrended fluctuation analysis (DFA) and spectral analysis applied on data series formed by successive values of cellular area which were obtained by managing the sequential images set for each cell. We have shown that the suspension media influences significantly the membrane fluctuation characteristics. Detrended fluctuation analysis revealed two short range correlations both for cells suspended in their natural medium and artificial medium. Moreover, we found out the strength on interaction between terms in series by using spectral analysis and autoregressive modeling. The correlation between parameters obtained by the above mentioned methods was evidenced by theoretical models and certified by our experiments.

## 1. Introduction

The human red blood cells (RBC) are characterized by vibratory motions, referred to as flickering [1]. The first quantitative studies [2] revealed that flickering under normal conditions does not display any significant isolated frequency, but rather is made of a mixture of a large number of low frequencies varying from about 40 Hz to zero. The interest for this phenomenon appeared to be constant and significant [3]. The character of the cell fluctuation was basically done by spectral and fluctuation analysis respectively of the time series describing the fluctuation of various geometrical parameters of the cells.

The aim of this paper was to study the influence of the liquid (*i.e.* plasma or PBS) on RBSc membrane fluctuations. The reason is that the natural environment of the erythrocytes is the plasma which has a complex composition whereas all the *in vitro* experiments are performed in an artificial environment. This usually consists of isotonic buffered saline. Indeed we found significant differences in the flickering of the two different environments. Laboratory experiments confirmed that the mathematical model is suitable for describing the membrane flickering phenomenon for both suspension media.

## 2. Experimental

The fluctuations of the erythrocyte membrane were analyzed on diluted cell suspension, obtained from human blood. RBCs were prepared as described in [4]. Investigations were performed for a set of 110 cells. In our experiments we prepared 20 μl sample volumes, using Rosenthal counting cell with 0.2 mm optical depth. We recorded 1000 sequential images for each one RBCs. The images set were captured at 0.2 s interval using a CCD camera (Optikam PRO3) mounted on an inversed optical microscope (OPTIKA XDS-2). The cell images were processed using Image J [5] software in order to obtain the area of red blood cells. This software reduces the red cell shape to a simple form and extracts the area values, returning a numerical data series.

## 3. Detrended fluctuation analysis

The method of detrended fluctuation analysis has proven useful in revealing the extent of long-range correlations in time series [6-8].

In our case the log-log plot is not linear and, instead, the relationship between *logF(n)* and *log(n)* appears to have two distinct linear segments, the slopes of both lines are calculated separately and termed $\alpha_1$ and $\alpha_2$, respectively. The value of $\alpha_1$ proved to be superior to $\alpha_2$ in terms of prognostic ability.

For erythrocytes suspended in plasma we have recorded closer values for $\alpha_1$ an $\alpha_2$. Performing *Two Sample Independent t-Test* we have obtained the P value being 0.439. We could suppose that the phenomenon is governed by a 1/f power law while the median correlation exponent value is 1.02 ± 0.01.

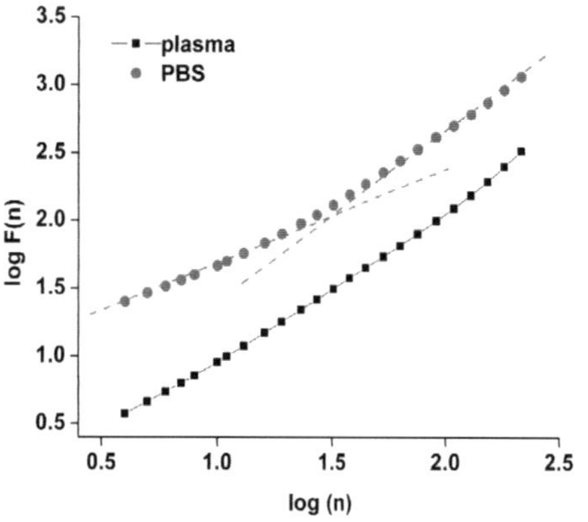


**Figure 1.** DFA slope for cellular area fluctuations when suspended in plasma (black) and PBS (red). Two slopes can be observed in PBS case (short range correlation) while plasma case could be approximate with 1/f slope.

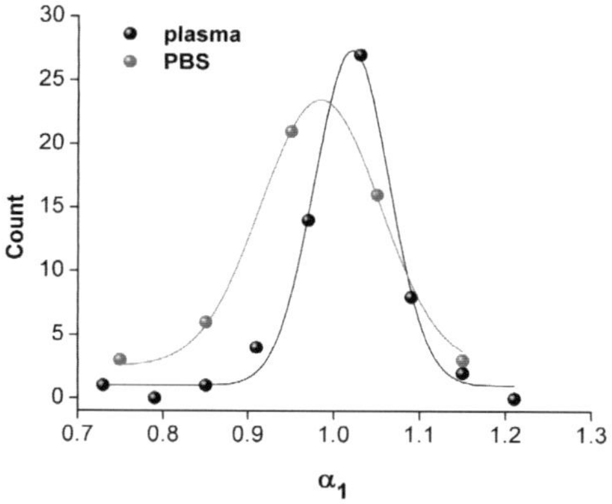


**Figure 2.** Gaussian fit for the first correlation segment $\alpha_{1plasma}$ (black) and $\alpha_{1PBS}$ (red) distributions.

We have considered $\alpha_1$ parameter (correlation between closer terms) both for erythrocytes suspended in plasma and PBS (figure 2) to evidence the different effect of the suspension media on the fluctuation characteristics. The P value was 0.00896 ($\alpha_{1plasma} = 1.00 \pm 0.00$, $\alpha_{1PBS} = 0.96 \pm 0.01$).

## 4. Autoregressive model (AR1)

AR1 model may apply to temporal or spatial processes and the significance of interaction factor $\varphi$ can be better understood when applied to such particular processes. For example in the case of a temporal process the parameter $\varphi$ can be understood in terms of a relaxation time $\tau = -1/\log(\varphi)$ (for details see [9]).

Parameter $\varphi$ can alternatively be regarded as the strength of interaction among the terms. Obviously the more distant the terms of the series, the lower is the correlation. Regardless of a temporal or spatial process, the parameter $\varphi$ can be understood as the scale of short range memory of the system.

In order to perform the spectral analysis, we have detrended the series by substrating a $10^{th}$ order polynomial which approximate the local trend. Moreover, the spectrum was averaged at 21 terms to indentify the spectrum appearance [10]. An immediate consequence of the averaging procedure is that deviation from a power-law description of the spectra may be better disclosed. Sometimes, such a deviation can be obvious even if the spectrum is not averaged.

The differences for spectrum appearance are evidenced in figure 3, where shown the averaged spectrum for two RBCs, suspended in plasma and PBS. It could be observed that none of those spectra can be fitted by a straight line. They are fitted better with a first order autoregressive model. The Gaussians for interaction factors computed with AR1 (figure 4) confirms results obtained with DFA, while medians are $\varphi_{plasma} = 0.78 \pm 0.00$ and $\varphi_{PBS} = 0.73 \pm 0.01$. P value is 0.056.

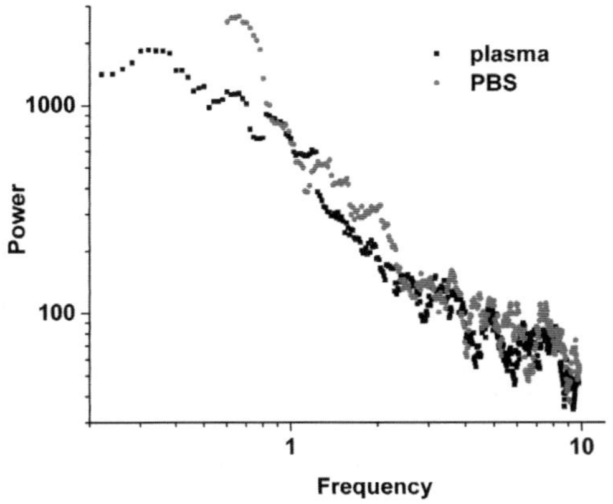


**Figure 3.** Averaged spectrum for RBC's area when suspended in plasma (black) and PBS (red). None of them can be fitted linear.

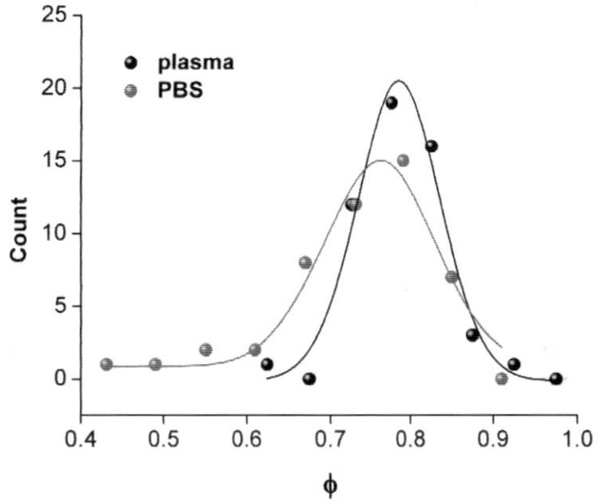


**Figure 4.** Gaussian fit for the interaction factor $\varphi_{plasma}$ (black) and $\varphi_{PBS}$ (red).

## 5. Theoretical model

We computed DFA on series governed by autoregressive law generated by using a software developed by us in MATLAB [11], with $\varphi \in [0.1, 0.9]$ [12]. Two slopes for the DFA plot were observed. We represented $\alpha_1$ function of $\varphi$ (figure 5) and found out an exponential dependency.

Experimental data are in good agreement with the theoretical model. The black line (figure 5) is for median value (Gaussian distribution) for $\alpha_{1\ plasma}$ and $\varphi_{plasma}$ and the red line for $\alpha_{1PBS}$ and $\varphi_{PBS}$. We can consider that the law which governs the fluctuation phenomena in plasma case only seeming 1/f. When phenomenon is rigorously verified, it appears to be short range correlation, and moreover, being fitted proper by a first order autoregressive model.

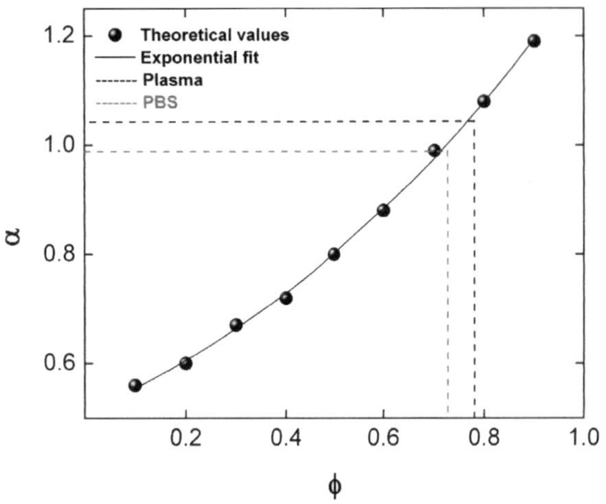


**Figure 5.** The experimental data fitted by the theoretical model $\alpha_{1plasma}$ and $\varphi_{plasma}$ (black) and $\alpha_{1PBS}$ and $\varphi_{PBS}$ (red).

## 6. Conclusions

The suspension media influences the characteristics of the membrane fluctuations. Both DFA and spectral analysis evidenced the stress hereupon the cell is submitted when is removed by its natural suspension medium (plasma) and resuspended into an artificial medium (PBS).

The matematical metods used in our study to investigate the RBCs membrane fluctuations are complementary and returns information about how strong is the correlation among terms in series and about the mathematical law that governs this fluctuations. It has been shown that those phenomena present short range correlations, even if the RBCs are suspended in plasma or PBS. The differences between suspension media are revealed both by DFA and spectral analyses. Moreover, the DFA parameters are strongly correlated with AR1 parameters, the theoretical model confirming the experimental results.

## References

[1] Blowers R, Clarkson E M and Maizels M 1951 *J. Physiol.* **113** 228-39
[2] Brochard F and Lennon J F 1975 *J. Phys.* **36** 1035-47
[3] Costa M, Ghiran I, Peng C K, Nicholson-Weller A and Goldberger A L 2008 *Phys. Rev.* E: *Stat., Nonlinear, Soft Matter Phys.* **78** 020901
[4] Buimaga-Iarinca L 2008 *Rom. J. Biophys.* **18** 67-72
[5] Rasband W S 1997-2006, ImageJ, U S National Institutes of Health, Bethesda, Maryland, USA, http://rsb.info.nih.gov/ij
[6] Peng C K, Buldyrev S V, Havlin S, Simons M, Stanley H E and Goldberger A L 1994 *Phys. Rev.* E **49** 1685-9.
[7] Hu K, Ivanov P Ch, Chen Z, Carpena P and Stanley E H 2001 *Phys. Rev.* E **64** 0111114-1
[8] Chen Z, Ivanov P C, Hu K and Stanley H E 2002 *Phys. Rev.* E **65** 041107-1
[9] Vamos C, Soltuz S M and Craciun M 2007 *Rev. d'Analyse Numerique et de Theorie de l'Approximation* **36** 201-16
[10] Mandelbrot B B 1998 *Multifractals and 1/f noise* (New York: Springer)
[11] MATLAB R2008a © 1994-2009 The MathWorks, Inc.
[12] Morariu V V, Buimaga-Iarinca L, Vamos C and Soltuz S M 2007 *Fluctuations and Noise Lett.* **7** 249-55.

# Determination of cobalt in human biological liquids from electrothermal atomic absorption spectrometry

**Felicia Drăgan[1], Lucian Hîncu[2] and Ioan Bratu[3]**

[1] University of Oradea, Faculty of Medicine and Pharmacy, 29 N Jiga, 410028 Oradea, Romania

[2] University of Medicine and Pharmacy "Carol Davila", Faculty of Pharmacy, 6 Traian Vuia, 020956 Bucuresti, Romania

[3] National Institute for Research and Development of Isotopic and Molecular Technologies, 65-103 Donath, 400293 Cluj-Napoca, Romania

Email: fdragan@uoradea.ro

**Abstract**. Problems and possibilities of the determination of Co in serum and urine samples by electrothermal atomic absorption spectrometry (ETAAS) are described. Optimal instrumental parameters as well as a suitable atomizer, calibration procedure and hydrogen peroxide as modifier are proposed for direct ETAAS measurement of Co in serum and urine. The detection limit achieved was 0.1 µg $L^{-1}$ for both matrices and relative standard deviations varied in the range 5-20% depending on the Co concentration in the sample. The validity of the method was verified by the analyses of standard reference materials. For serum samples with Co content lower than the detection limit, a separation and preconcentration procedure based on liquid/liquid extraction is suggested prior to determination of Co in the organic phase by ETAAS. This procedure permits determination of 0.02 µg $L^{-1}$ Co in serum samples with a relative standard deviation of 10-18%.

## 1. Introduction

Current development of human health related studies requires a growing number of elements to be monitored in biological matrices, including all essential and well-recognized toxic metals. Cobalt is an essential element in the human body as a component of vitamin B12 and it is clear that monitoring of body fluids for cobalt is essential for the control of nutritional deficiencies and perhaps prevention of its toxic effect in cases of occupational exposure. The most serious problems of cobalt determination in serum and urine samples are: (*i*) the extremely low concentration of this element in body fluids (normal serum and urine levels are in the range 0.1-0.5 µg $L^{-1}$); (*ii*) possible sample contamination during blood collection; (*iii*) high content of organic matrix causing numerous physical and chemical interferences. Double focusing sector field inductively coupled plasma mass spectrometry is a very useful and superior technique for this analysis. However, these instruments are not available in all laboratories. Another option is to use electrothermal atomic absorption spectrometry. Therefore an accurate measurement of cobalt at concentrations 0.5-1 µg $L^{-1}$ is still a challenge.

In the present study, the possibilities of electrothermal atomic absorption spectrometry (ETAAS) are investigated for Co determination in serum and urine samples from healthy persons and patients on haemo dialysis. The originality of the paper is that procedures based only on direct determination of Co without any preliminary sample pretreatment are proposed. This makes it possible to achieve very low detection limits due to the very low blank sample level. The methods of direct determination and the procedures of extractive separation of cobalt from serum have been optimized (type of graphite tube, modifier, temperature programs and calibration procedure). Analytical quality assurance of the method was carried out by analyzing standard reference materials.

## 2. Experimental

### *2.1. Instrumentation*

The atomic absorption spectrometer Varian Spectra AA 640Z Zeeman AAS equipped with GTA100 graphite furnace (Varian, USA) and PSD-100 autosampler (Varian) was used. Pyrocoated tubes and tubes with centre fixed platforms were used as atomizers. A Varian cobalt hollow cathode lamp was used and the measurements were performed at 242.5 nm. Argon was applied as protective gas and 10μl serum or urine samples were injected into the graphite furnace (GF). The graphite furnace operating parameters are presented in table 1. Only integrated absorbance values (peak area) were used for quantification.

### *2.2. Reagents and samples*

The working standard solutions were prepared by dissolving the Merck (Germany) stock solution containing 1 g $L^{-1}$ cobalt in nitrate form. Hydrogen peroxide produced in the Laboratory for High Purity Substances was additionally purified by ion exchange. Mg $(NO_3)_2$ *p.a.* (Merck), $NH_4F$ and $NH_4SCN$ *p.a.* (Fluka, Switzerland) were additionally purified by recrystallization. Ammonium pyrrolidine dithiocarbamate, APDC *p.a.* (Merck) and isobutylmethyl ketone, IBMK *p.a.* (Merck) were used for the extraction procedures. Lypochek Urine Metals Control Level 1 (Lot 69031, Bio Rad, Germany) and Clin Rep Level 1 serum (Recipe, Germany) samples were used as reference materials for method validation. All disposable devices were rigorously cleaned shortly before use by immersion in hot concentrated nitric acid and rinsing with doubly distilled water.

Serum and urine samples were obtained from 20 presumably healthy volunteers and additional serum samples were obtained from 15 dialyzing patients and were transferred into plastic tubes. Serum samples were collected with a plastic iv canula No. 24 with injection valve. Urine samples were taken as spot samples. The samples were acidified up to pH 2 and were kept frozen (–18°C) until analysis.

All patients signed agreement for cobalt testing of their serum during a systematic medical checkup.

### *2.3. Procedures*

*Co direct ETAAS determination in serum and urine.* The serum or urine samples of 10 μl were directly introduced into the graphite furnace together with equal volume of modifiers. Calibration curves (1-5 μg $L^{-1}$ Co) were prepared by using aqueous standard solutions of Co.

### *2.4. Extraction-ETAAS determination of Co in serum*

*Extraction of Co as chelate complex with APDC.* Serum sample (3.0 ml) was placed in the extraction tube, 1 ml 1% solution of APDC was added and extraction was performed for 3 min in 500 μl IBMK. Cobalt was measured in the organic phase by ETAAS.
*Extraction of Co as ion associate complex with $NH_4SCN$.* Serum sample (5.0 ml) was placed in the extraction tube, 0.2 g $NH_4SCN$ and 0.05 g $NH_4F$ were added and extraction was performed for 3 min in 500 μl IBMK. Cobalt was measured in the organic phase by ETAAS.

Recoveries for both procedures were calculated against organic standard solutions of Co.

*Data analysis.* Detection limit and determination limit were calculated based on the variability of the blank (3 $SD_b$ and 6 $SD_b$, respectively). Common statistical tests (ANOVA, Student's *t*-test, 95% confidence level) were used.

## 3. Results and discussion

### *3.1. ETAAS determination of Co in serum and urine*

*GFAAS program optimization.* Serum and urine samples were spiked with 4 μg $L^{-1}$ Co, ashing and atomization curves were constructed for both atomizers investigated in the presence of the modifiers studied. The results obtained are presented in figure 1. It was found that ash temperature of 1200°C for pyrocoated tubes and 1300°C for tubes with centre fixed platforms can be used without any loss of Co even in the absence of a modifier. These temperatures are high enough to ensure complete matrix decomposition and removal during the pyrolysis step. The modifiers used do not influence the thermal behaviour of Co. The only advantage achieved with $H_2O_2$ as modifier is the elimination of carbon residue formed at the end of the atomization cycle. It is very important to keep the atomization temperature as low as possible, because a rise in the atomization temperature leads to higher values of nonspecific absorption: for pyrocoated tubes the optimal atomization temperature is around 2100°C and for tubes with centre fixed platform around 2300°C. The optimal instrumental parameters for Co determination in serum and urine are summarized in table 1.

**Table 1.** Instrumental parameters for determination of Co in serum and urine samples by ETAAS.

| Step | Temperature (°C) | | Time (s) | Argon flow ($L\ min^{-1}$) |
|---|---|---|---|---|
| | Pyrocoated tubes | Tubes with centre fixed platforms | | |
| Drying | | | | |
| 1 | 85 | 85 | 5 | 3 |
| 2 | 95 | 95 | 40 | 3 |
| 3 | 120 | 120 | 10 | 3 |
| Pyrolysis | | | | |
| 4 | 1200 | 1300 | 5 | 3 |
| 5 | 1200 | 1300 | 5 | 3 |
| 6 | 1200 | 1300 | 2 | 0 |
| Atomization | | | | |
| 7 | 2100 | 2300 | 2 | 0 |
| 8 | 2100 | 2300 | 2 | 0 |
| Cleaning | | | | |
| 9 | 2300 | 2400 | 2 | 3 |

*Calibration.* Matrix interferences were evaluated according to the slopes of calibration curves obtained in the presence of the matrix (serum or urine), and for aqueous standard solutions. The results obtained are recorded in table 2. As can be seen, tubes with centre fixed platforms are perfect atomizers for Co determination in urine – the slope of aqueous standard calibration curve (3.9 ± 1.2%) and the slope of the calibration curve in presence of urine (3.8 ± 1.4%) are practically equal. This means that calibration against aqueous standard curves could be used, simplifying in this way the analytical procedure. Pyrocoated tubes are preferable atomizers for serum samples, but in this case significant differences were observed between the slopes of the calibration curves for aqueous standards (3.8 ± 1.2%) and in the presence of serum (2.5 ± 2.0%). The standard addition method was applied for calibration. However, further experiments performed showed that no significant differences were found between the slopes of calibration curves for different serum samples; consequently, the analytical procedure requires at least a single standard addition to one of the serum samples analyzed. It should be mentioned that statistically different slopes of the standard addition curves were obtained when standard solution was added beforehand to the sample and then injected into the furnace and for the standard additions made directly into the graphite furnace by using the facilities of the auto sampler. In the former case, the slope was remarkably higher. Depression of Co atomization was probably due to the formation of some complexes with serum components; this requires standard addition before sample injection.

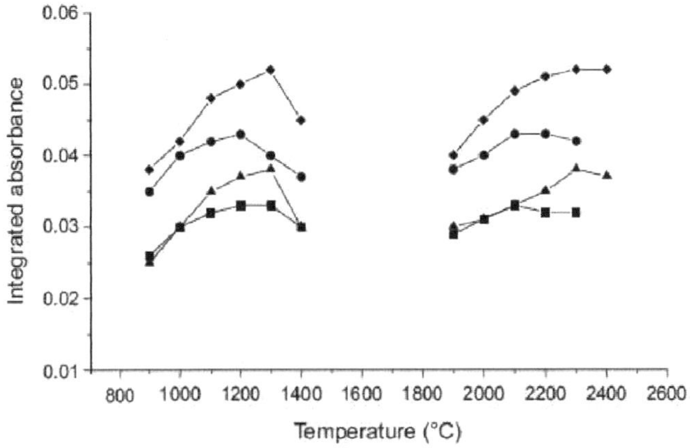


Figure 1. Ashing and atomization curves for Co: in serum: (●) – pyrocoated tube, (▲) – tube with centre fixed platform, and in urine (■) – pyrocoated tube, (♦) – tube with centre fixed platform.

**Table 2.** Slopes (A $\mu g^{-1}$) of the best fit linear regression models for calibration curves.

| Sample | Slopes of calibration curves | | | |
|---|---|---|---|---|
| | Pyrocoated tube | | Tube with center fixed platform | |
| | Mean value | RSD (%)[a] | Mean value | RSD (%)[a] |
| Aqueous standard solutions | 3.8 | 1.2 | 3.9 | 1.2 |
| Serum | 2.5 | 2.1 | 2.1 | 2.3 |
| Serum in the presence of H2O2 | 2.7 | 2.0 | 2.2 | 2.1 |
| Urine | 1.9 | 2.4 | 3.8 | 1.4 |
| Urine in the presence of H2O2 | 2.0 | 2.3 | 3.9 | 1.3 |

*Analytical figures of merit.* The linearity range for Co determination is 0.2-30 $\mu g\ L^{-1}$ for both matrices. The detection limit for cobalt in serum and urine was 0.1 $\mu g\ L^{-1}$ and the determination limit was 0.2 $\mu g\ L^{-1}$. Within-batch precision strongly depends on analyte concentration in the measuring solution. For direct ETAAS, for Co in the range 0.1-0.15 $\mu g\ L^{-1}$, within-batch precision is 16-20% while for Co in the range 0.15-0.20 $\mu g\ L^{-1}$ it is 5-8%. Between-batch precision (calculated as the standard deviation for results obtained for parallel samples analyzed on different days) is 12-17%.

**Table 3.** Analysis of Co in reference materials by direct and extraction ETAAS methods.

| Standard reference material | Certified value ($\mu g\ L^{-1}$) | Determined by ETAAS[a] | |
|---|---|---|---|
| | | Direct ETAAS ($\mu g\ L^{-1}$) | Extraction ETAAS ($\mu g\ L^{-1}$) |
| Lypocheck Urine Metals Control Level 1 | 3.7 (2.9 ± 4.4) | 3.9 ± 0.3 | 3.8 ± 0.4 |
| Clin Rep Level 1 serum | 5.0 (3.7 ± 6.7) | 5.1 ± 0.4 | 5.2 ± 0.5 |

[a] Mean ±SD, $n = 5$

The accuracy of the proposed procedure for ETAAS determination of Co in serum and urine samples was checked by spike recovery experiments. Serum and urine samples were spiked with Co in the concentration range 0.5-10 μg $L^{-1}$ Co. The recovery values obtained ranged between 96.0 and 98.0% for all spikes.

The accuracy of the analytical methods developed was also checked by analyzing certified reference materials for urine and serum. A comparison of the results given in table 3 shows good agreement with the certified values, *i.e.* the validity and versatility of the analytical procedure is confirmed.

### *3.2. Extraction-ETAAS determination of Co in serum*

Since cobalt content in some serum samples is lower than the determination limit achieved by ETAAS, the separation and preconcentration procedures should be proposed for these samples. Liquid/liquid extraction was tested as an effective analytical procedure for Co preconcentration directly from undigested serum samples. Two extraction reagents were studied for this purpose.
*Ammonium pyrrolidinedithiocarbamate (APDC).* This reagent forms chelate complexes with Co and ensures quantitative extraction of Co in a wide pH range from 4 to 7. Recoveries achieved for extraction of Co with APDC from 3 ml serum samples into 500 μl IBMK are in the range 95-97%, which means that this extraction system could be used for Co separation and preconcentration from serum samples. However, in some cases relatively high values of non-specific absorption were observed during the ETAAS measurement due to the high iron content co-extracted in the organic phase. To eliminate these interferences another extraction procedure was suggested.
*$NH_4SCN$.* Quantitative extraction of Co could be achieved from 0.5 mol $L^{-1}$ solution of $NH_4SCN$ into IBMK. Iron as $Fe^{2+}$ or $Fe^{3+}$ strongly depressed the degree of extraction.

Therefore, for serum samples, $NH_4F$ is proposed as a reagent forming stable complexes with iron ions and thus eliminating this interference. ETAAS measurement of Co in the organic phase is free of matrix interferences because Fe is blocked in the aqueous phase. This extraction system ensures high recoveries (96-98%) and interference-free determination of Co. Hence, extraction of Co from a 5.0 ml serum sample in the presence of 0.2 g $NH_4SCN$ and 0.05 g $NH_4F$ into 500 μl IBMK is proposed for Co separation and preconcentration.

*Analytical figures of merit.* The extraction analytical procedure with $NH_4SCN$ allows detection of 0.01 μg $L^{-1}$ Co and determination of 0.02 μg $L^{-1}$ Co in serum samples.

Within-batch precision for Co in the range 0.02-0.1 μg $L^{-1}$ is 10-18%.

The accuracy of the results obtained by the proposed extraction procedures was verified by comparative analysis of serum samples with a relatively high content of Co by direct ETAAS and by the suggested extraction ETAAS procedure. Very good agreement between the results obtained (table 3) for reference standard samples confirmed the accuracy of the proposed preconcentration method.

### *3.3. Analytical application*

The analytical methods developed were applied to the determination of Co in serum and urine (direct method and extraction method with $NH_4SCN$ for a low concentration of Co) samples obtained from 20 presumably healthy volunteers and in serum samples from 10 patients on dialysis. The results obtained revealed that the Co content in serum of healthy volunteers varied in the range 0.04 to 0.09 μg $L^{-1}$ and in urine in the range 0.11 to 1.22 μg $L^{-1}$. As could be expected, the Co content in the serum samples of persons on dialysis was significantly higher than in healthy persons, with serum Co ranging from 1.70 to 4.70 μg $L^{-1}$.

## 4. Conclusions

The results of this study indicate that direct ETAAS determination of cobalt in serum and urine samples is possible in the range 0.2-30 μg $L^{-1}$ by using graphite tubes with a centre fixed platform for urine samples and pyrolytically coated graphite tubes for serum samples. The suitable modifier proposed is $H_2O_2$, which prevents carbon residue formation in the graphite tube. Pretreatment temperature of 1200°C (1300°C) and atomization temperature of 2100°C (2300°C), respectively, for pyrolytically coated graphite tubes and graphite tubes with centre fixed platforms are recommended.

Journal of Physics: Conference Series **182** (2009) 012006 doi:10.1088/1742-6596/182/1/012006

For serum samples with a cobalt content below 0.2 μg $L^{-1}$, the extraction procedure with $NH_4CSN$, in the presence of $NH_4F$ as a masking reagent for iron, into IMBK was developed. ETAAS measurements were carried out using the same temperature program.

## References

[1] Lyon T D B, Fell G S, Hutton R C and Eaton A N 1998 *J. Anal. At. Spectrom.* **3** 265-71
[2] *Trace Element Analysis in Biological Specimens* 1994ed R F M Herber and M Stoeppler (Amsterdam: Elsevier)
[3] Begerow J, Turfeld M and Dunemann L 1997 *Fresenius J. Anal. Chem.* **359** 427-9
[4] Alimonti A, Petrucci F, Fioravanti S, Laurenti F and Caroli S 1997 *Anal. Chim. Acta* **342** 75-81
[5] Barany E, Bergdahl I A, Schutz A, Skerfving S and Oskarsson A 1997 *J. Anal. At. Spectrom.* **12** 1005-9
[6] Schramel P and Wendler I 1998 *Fresenius J. Anal. Chem.* **361** 487-91
[7] Krachler M, Alimonti A, Petrucci F, Forastiere F and Caroli S 1998 *J. Anal. At. Spectrom.* **13** 701-5
[8] Begerow J, Turfeld M and Dunemann L 2000 *J. Anal. At. Spectrom.* **15** 347-52
[9] Rodushkin I, Odman F, Olofsson R and Axelesson M 2000 *J. Anal. At. Spectrom.* **15** 937-44
[10] Bouman A, Platenkamp A and Posma F 1986 *Ann. Clin. Biochem.* **23** 346-50
[11] Sampson B 1988 *J. Anal. At. Spectrom.* **3** 465-9
[12] White M A 1999 *J. Trace Elem. Med. Biol.* **13** 93-101
[13] Branchflower J W, Cannavan A and Kennedy G D 1990 *Analyst* **115** 1323-5.

# Structural investigation of Lisinopril by powder X-ray diffraction and solid-state NMR

**Xenia Filip[1], Carmen Tripon[1], Gheorghe Borodi[1], Luminiţa Oprean[2] and Claudiu Filip[1]**

[1] National Institute for Research and Development of Isotopic and Molecular Technologies, 65-103 Donath, 400293 Cluj-Napoca, Romania

[2] University of Medicine and Farmacy Iuliu Haţieganu, 8 Victor Babeş, 400012 Cluj-Napoca, Romania

E-mail: xenia.filip@itim-cj.ro

**Abstract**. Structural studies on polycrystalline *Lisinopril* (N-N-[(s)-1-carboxy-3-phenylpropyl] -L-lysil-L-proline) are performed by combined powder X-Ray diffraction and $^{13}C$ solid-state nuclear magnetic resonance (NMR). The crystal structure of this drug, used primarily for the treatment of hypertension, has not yet been determined due to the impossibility of synthesizing single crystals of sufficient quality. It is shown here that valuable insights into the crystal and molecular structure of *Lisinopril* can be obtained on polycrystalline powder based on the complementary character of the information provided by the two techniques.

## 1. Introduction

*Lisinopril* is a drug of the angiotensin converting enzyme (ACE) inhibitor class that is primarily used in treatment of hypertension, congestive heart failure, heart attacks and also in preventing renal and retinal complication of diabetes [1]. *Lisinopril* has a number of properties that distinguishes it from other ACE inhibitors: it is hydrophilic, has long half-life and tissue penetration and is not metabolized by the liver.

The preferred conformation of *Lisinopril* in solution was investigated by liquid-state NMR [2], and the result was correlated with its pharmacological activity. However, crystal structure of *Lisinopril* is not yet reported in the Cambridge Structural Database (CSD) most probably because it was not possible to obtain single crystals of sufficiently good quality. Therefore, useful structural features of *Lisinopril* in the solid form are determined here based on powder X-Ray diffraction and $^{13}C$ solid-state NMR. These methods are often used as complementary techniques for structural investigation in such cases when single crystal X-Ray diffraction cannot be applied.

## 2. Experimental

Experimental powder diffraction pattern was obtained with Pananalytical X-ray diffractometer using Cu $K\alpha_1$ radiation. X-ray powder diffraction pattern (figure 1) was indexed using *Dicvol 91* and *Ito* programs, whereas the lattice parameters were refined using the *Checkcell* program. The solid-state $^{13}C$ NMR spectrum of *Lisinopril* (figure 2) was recorded at room temperature on a Bruker AVANCE-400 spectrometer (at 100 MHz $^{13}C$ Larmor frequency) by using the Cross-Polarization under Magic

Angle spinning (CP/MAS) technique [3]. Quantum mechanical computation of $^{13}C$ chemical shifts of *Lisinopril* in solid was performed by using the GIPAW module within the QUANTUM ESPRESSO program [4] based on the structural model provided from the powder X-Ray data. *G*auge *I*ncluding *P*rojector *A*ugmented-*W*ave method (GIPAW) was introduced as an extensions to the Projector Augmented-Wave method, being valid for systems in non-zero uniform magnetic fields [5]. For chemical shift calculations we use norm-conserving GIPAW-pseudopotentials, doing first a geometry optimization for the H atoms positions. The proposed spectral assignment in figure 3 is based on comparing the measured $^{13}C$ chemical shifts with those obtained in solution and computed by the GIPAW method.

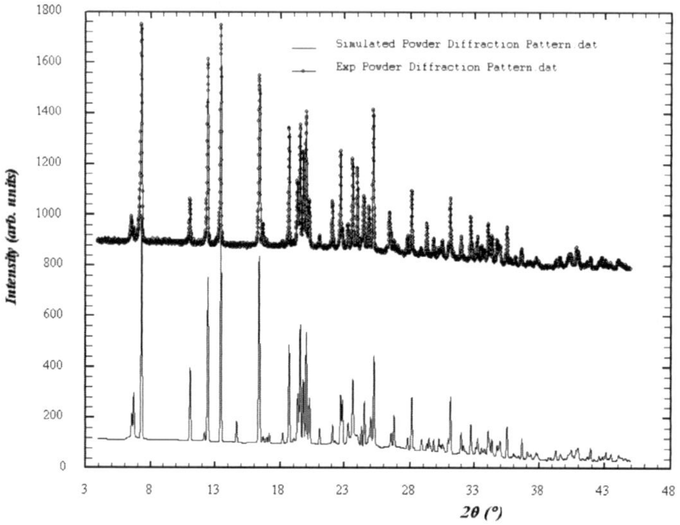


**Figure 1.** Experimental (top) and simulated (bottom) powder X-Ray diffraction pattern of *Lisinopril.*

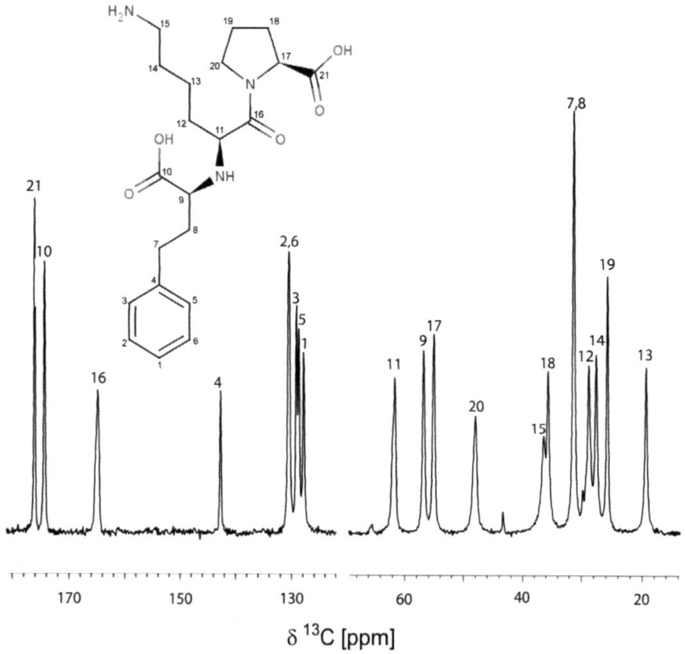


**Figure 2.** The $^{13}C$ solid-state NMR spectrum of *Lisinopril* and the proposed spectral assignment.

## 3. Results and discussion

From the indexing procedure we obtained that *Lisinopril* crystallizes in the monoclinic system having the following lattice parameters: a = 14.6632Å, b = 5.9097Å, c = 14.3140Å, $\beta$ = 113.140$^0$ and the volume of the unit cell, V = 1137.7 Å$^3$. Taking into account the molecular weight of *Lisinopril*, M = 405.488, and the forbidden reflexions, we conclude that the most probable space group is $P2_1$, which has two molecules per unit cell. The calculated density of 1.18 g/cm$^3$ is consistent with the typical values for organic molecules containing only carbon, oxygen, nitrogen and hydrogen atoms.

The problems associated with obtaining single crystals are being increasingly overcome by full structural determination from powder diffraction patterns [6]. However, the success rate in such applications essentially depends on the ability to reach a reliable initial structural model by the existing search algorithms. The difficulty in determining such models largely increases with the number of degrees of freedom (DOF) to be explored for the investigated molecule. From this point of view *Lisinopril*, which has 19 flexible torsion angles that can be modified during the search procedure is approaching the upper limit of structural complexity that can be reliably solved from X-Ray powder diffraction data [6]. The crystal structure analysis was performed based on the direct space search method implemented in the *Fox* program [7]. The initial configuration of *Lisinopril* was given using the *z* matrix with 29 non-hydrogen atoms. For obtaining a structural model both, simulated annealing and parallel tempering algorithms were employed. Multiples runs were performed in order to find the best solution: the structural model selected in the end corresponds to the lowest value of the $R_{pw}$ parameter [7].

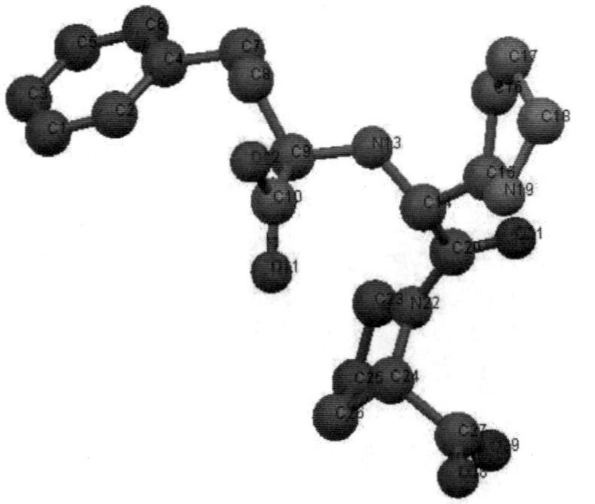

**Figure 3.** The molecular structure of *Lisinopril.*

In the molecular conformation obtained from analyzing the X-Ray data (see figure 3) the three branches of the molecule make approximately the same angles around the central C14 atom, namely 115.5$^0$ (N13, C14, C20), 110.2$^0$ (N13, C14, C15), and 112.3$^0$ (C15, C14, C20). Another distinctive structural feature is that the C14 - C19 carbons, which form a straight chain in the starting configuration of the isolated molecule, are found to be bent to a ring-like conformation due to the crystal packing. The arrangement of the two *Lisinopril* molecules in the unit cell is shown in figure 4: according to this structural model, crystal packing appears to be favored by intermolecular hydrogen bonding between N13-H...O11, and O21-H...O28, respectively.

The analysis of the chemical shifts obtained from MAS NMR, together with their values computed by explicitly considering the crystal lattice, often provides valuable local structural details that can be used for increasing the reliability of the structural model derived from the powder X-Ray data. For instance, large differences between the chemical shift values extracted from the $^{13}$C CP/MAS spectrum shown in figure 2 and those reported for *Lisinopril* in solution [2], are obtained for the C24 (~ 9 ppm), C9 (~ 8 ppm) sites, the C14-C19 carbon chain (~ 4-5 ppm), the C20 carbonyl (~ 5 ppm), and the C27 carboxyl site (~ 4 ppm), respectively. They are indicative for the conformational changes in solid, and generally consistent with the structural model described above. At the other extreme, there are little changes associated with the phenyl ring, the carbon sites connecting it with the rest of the molecule (C7, C8), and the central C14 carbon, for which chemical shift variations were found only between 0.5 and ~ 2 ppm: these small variations are most likely determined by crystal packing effects, and not by significant changes of the molecular conformation. Also, within the accepted error limits of the GIPAW method [5], the computed $^{13}C$ chemical shifts are generally in good agreement with the measured values. Relatively large deviations of up to10 ppm were found though in the case of the C9, C24, carboxyl and carbonyl sites: however, this result is not completely surprising since the structural model used in calculations is not yet refined, and these particular carbons are known to be very sensitive even to small deviations from the real local geometry.

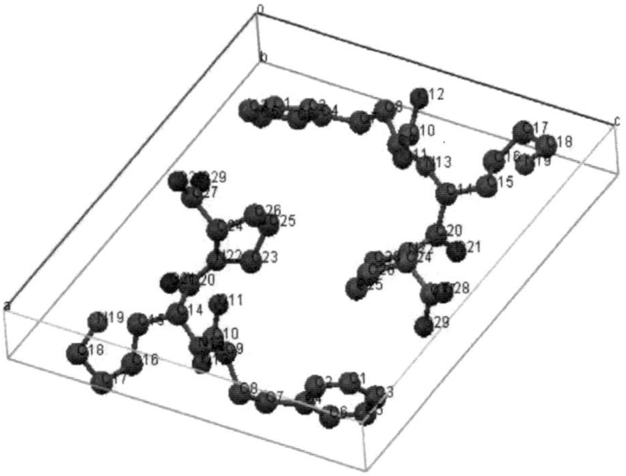

**Figure 4.** The unit cell of *Lisinopril.*

## 4. Conclusions

In summary, structural investigations on polycrystalline *Lisinopril* (N-N-[(s)-1-carboxy-3-phenyl-propyl]-L-lysil-L-proline), a drug of the angiotensin converting enzyme (ACE) inhibitor class that cannot be synthesized as single crystal, have been carried out by powder X-Ray diffraction and $^{13}C$ solid-state NMR. The lattice parameters were extracted with high accuracy, and a structural model has been obtained through direct space search by employing the simulated annealing and parallel tempering algorithms. This model shows a series of important conformational changes compared to the preferred molecular conformation of *Lisinopril* in solution. Most of these structural features are consistent with the experimentally measured $^{13}C$ chemical shifts in solid, and also with the values computed through the GIPAW method. In conclusion, the derived model constitutes a reliable starting structure for the next stage of crystal structure refinement.

**Acknowledgments**

Financial support from the ANCS (project 61-003) is gratefully acknowledged.

**References**

[1] Armayor G M and Lopez L M 1988 *Drug. Intell. Clin. Pharm.* **22** 365
[2] Sakamoto Y and Ishi T 1993 *J. Molec. Struct.* **298** 129
[3] Stejskal E O, Schaefer J and Waugh J S 1977 *J. Magn. Reson.* **56** 354
[4] Ceresoli D, Seitsonen A P, Gerstmann U and Mauri F, GIPAW module within the Quantum Espresso package (http://www.Quantum-Espresso.org/)
[5] Pickard C J and Mauri F 2001 *Phys. Rev.* **B63** 245101
[6] Harris K D M, Treymayne M and Kariuki B M 2001 *Angew. Chem. Int. Ed. Engl.* **40** 1626
[7] Radovan C and Favre-Nicolin F 2007 *Z. Krisstallogr.* **222** 105.

# Analysis of the secondary structure of a protein's N-terminal

**C G Floare[1], M Bogdan[1], O Horovitz[2], A Mocanu[2] and M Tomoaia-Cotisel[2]**

[1] National Institute for R & D of Isotopic and Molecular Technologies, 65-103 Donath, 400293 Cluj-Napoca, Romania

[2] Babes-Bolyai University, Faculty of Chemistry and Chemical Engineering, Department of Physical Chemistry, 11 Arany Janos, 400028 Cluj-Napoca, Romania

E-mail: `calin.floare@itim-cj.ro`

**Abstract.** The major protein component from aleurone cells of barley (Hordeum vulgare L.), PACB, is related to 7S globulins present in other cereals and to the vicilin-type 7S globulins of legumes and cotton seed. It contains 4 subunits of about 20, 25, 40 and 50 kDa molecular weights. The N-terminal sequence of 16 amino acids (over 260 atoms) in the protein was previously determined, and our aim is the prediction of its secondary structure. The empirical Chou-Fasman method was applied in an improved version as well as the empirical DSC method (discrimination of protein secondary structure class) with quite similar results. A molecular dynamics simulation was also performed, using the FF99SB forcefield within AMBER version 9.0. Solvation effects were incorporated using the Born model. The results are compared and a 3D model is proposed.

## 1. Introduction

Proteins adopt 3-dimensional structures, which enable them to function in specific ways, for example as biologically active enzymes. Shorter polypeptides generally adopt less compact and less stable structures in solution, making the experimental characterization of their preferred folds a non-trivial task. The end of the polypeptide chain with an unbound amine group, the N-terminus or N-terminal, is usually the first part of the protein that exits the ribosome during protein biosynthesis. It often contains sequences that act as targeting signals, basically intracellular zip codes that allow for the targeting of the protein to its designated location within the cell. Therefore, there is a growing interest in the characterization of these N-terminal signal sequences and of their role [1, 2, 3]. Plants store amino acids for longer periods in the form of specific storage proteins, deposited in seeds, in root and shoot tubers or in other vegetative organs [4]. The latest advances in computer technology and modelling techniques facilitate simulations of small peptide folding in atomic detail. In this work, we are trying to predict the secondary structure for the N-terminal of the major globular storage protein from aleurone cells of barley (*Hordeum vulgare L.*). This protein is related to 7S globulins present in other cereals and to the vicilin-type 7S globulins of legumes and cotton seed. It contains 4 subunits of about 20, 25, 40 and 50 kDa molecular weights. Its N-terminal sequence of 16 amino acids in the protein is given as follows: $^{1}$X$^{2}$Glu$^{3}$Gln$^{4}$Gly$^{5}$Asp$^{6}$Ser$^{7}$Arg$^{8}$Arg$^{9}$Pro$^{10}$Tyr$^{11}$Val$^{12}$Phe$^{13}$Gly$^{14}$Pro$^{15}$Arg$^{16}$(Ser or His)$^{17}$Phe, or with the one-letter notations of amino-acids $^{1}$XEQGDSRRPYVFGPRS(H)F$^{17}$, where X stands for the first amino acid of the N-terminal of aleurone protein which was not identified [5]. The secondary structure of this protein was investigated by advanced spectroscopies [6, 7, 8, 9] and its adsorption at various solid and liquid interfaces and on nanoparticles was studied [10, 11].

## 2. Methodology

We started by using two empirical techniques. The *Chou-Fasman* method is used to predict secondary structure in proteins and depends on assigning a set of prediction values to a residue and then applying a simple algorithm to those numbers [12]. The method is based on analyses of the relative frequencies of each amino acid in $\alpha$-helices, $\beta$-sheets, and turns based on known protein structures solved with X-ray crystallography. From these frequencies a set of probability parameters were derived for the occurrence of each amino acid in each secondary structure type, and these parameters are used to predict the probability that a given sequence of amino acids would give a certain conformation in a protein. The *Chou-Fasman* parameters found some strong tendencies among individual amino acids to prefer one type of secondary structure over others. Alanine, glutamate, leucine, and methionine were identified as helix formers, while proline and glycine commonly end a helix. The original *Chou-Fasman* parameters were derived from a very small and non-representative sample of protein structures due to the small number of such structures that were known at that time. These original parameters have since been shown to be unreliable and have been updated from a current dataset, along with modifications to the initial algorithm [13]. The *DSC* Method (Discrimination of protein Secondary structure Class), fully described by King & Sternberg [14], is accessible as 'Secondary Structure Prediction' (SSP) option from *Accelrys Discovery Visualizer v1.7.*

On the other hand, molecular dynamics (MD) is a method in which Newton's equations of motion are integrated to reproduce the time evolution (trajectory) of the atoms in a system and is widely used to study dynamic processes in biomolecular systems at atomic resolution. MD simulations have been shown to be a reliable instrument to investigate the process of reversible folding of various peptides [15, 16]. Encouraging agreement with experiment has been obtained, showing that the biomolecular forcefields mimic microscopic processes very well. On a classical MD simulation of a biomolecular system some important points must be emphasized. Firstly, the dynamics of individual atoms during the process of folding cannot yet be directly observed experimentally. Experimental data at atomic resolution is only available for equilibrium distributions of conformations under specific conditions. Secondly, the time scale on which the relevant biomolecular systems can be simulated in atomic detail is currently on the order of picoseconds to nanoseconds, while the lower limit for proteins as well as most peptides to fold is believed to be microseconds. Thirdly, the process of peptide folding involves many (peptide and solvent) degrees of freedom and is governed by small (free) energy differences, on the order of a few multiples of kT (k - Boltzmann's constant), which puts very high demands on the accuracy of the atomic interaction function and model that is used [17]. We specify also that when an experimentally determined structure is not available it is difficult to evaluate the quality of conformations sampled on the trajectory. Consequently, there is some inherent uncertainty in this kind of study at which we add the fact that we do not consider the interactions with the body of the protein (which hasn't been determined).

## 3. Computational Details

The on-line *Chou-Fasman* predictor CHOFAS version 2.0u66 September 1998 was used for the two N-terminals of barley aleurone, one containing serine (S) and the other histidine (H) residue. SSP option in Accelrys Discovery Visualizer v1.7 (based on DSC method) was also used. The MD simulations have been performed using AMBER version 9.0 [17], starting with the linear structure: GLU GLN GLY ASP SER ARG ARG PRO TYR VAL PHE GLY PRO ARG SER

PHE, built using *xLEaP* tool. Initially a short minimization was performed (1000 steps - 500 using steepest descent followed by 500 of conjugate gradient algorithms) to cleanup the structure and fix up hydrogen positions, such that the MD be stable. We ran a non-periodic simulation and, since this system has only 16 residues, we could afford to use a large non-bonded cutoff of 999 Å and to include all atom pairs in the effective Born radii calculation. The MD was started using *sander* executable: the system was slowly heated up to 325K over 50 ps in a total of 7 stages. In this way, the chances that the system will blow up are reduced, by allowing it to equilibrate at each temperature. The temperature, slightly higher than the normal 300K, was chosen in order to avoid the system being kinetically trapped in local minima. The simulation was performed in the weak-coupling canonical ensemble, using Berendsen thermostat [18] with a time constant for heat bath coupling of 1 ps, the Stony Brook parameterization of the FF99 forcefield and the implicit Born solvation model [19], which includes the solvent effects within the forcefield equations. To allow the initial system to relax in a controlled fashion, a very short time step of 0.5 fs was used for the heating stage and then it was switched to 2 fs for the equilibration. On this stage, where there is an increased chance of system instability, we also set the coordinate write frequency on the trajectory to 50, in order to permit the identification of possible problems occurring. On the production MD, a more appropriate value of 500 was used. With this equilibrated structure, a long production run at 325 K was started: a 200 ns simulation using the same forcefield and the time step 2 fs. On account of this large time step, all bonds involving hydrogen had to be constrained. The SHAKE algorithm was used on the whole time. The Berendsen thermostat was used for temperature control, more closely coupled (system time constant for heat bath coupling of 0.5 ps) to keep the system close to 325 K.

## 4. Results and Discussion

The on-line Chou-Fasman predictor and the DSC method gave the following results for the N-terminal of barley aleurone containing S (serine) residue (the analysis of the sequence containing histidine residue gave essentially similar results):

| Chou-Fasman Predictor | | DSC Method | |
|---|---|---|---|
| | EQGDSRRPYVFGPRSF | | EQGDSRRPYV FGPRSF |
| helix | | DSC-SEC | CCCCCCCEEE ECCCCC |
| sheet | EE | PROB-H | 0001001000 011000 |
| turns | T TT | PROB-E | 1111114688 742111 |
| | | PROB-C | 9998995422 357999 |

**Table 1.** The results from Chou-Fasman predictor are straightforward. There are turns at prolyl residues and sheet between them. The first section of the report for the DSC method shows 5 lines for each sequence: (1) The sequence; (2) The predicted secondary structure type (DSC_SEC); (3, 4, 5) The probability of being helix (PROB_H), a beta strand (PROB_E) or a coil (PROB_C). [Values of PROB_H, PROB_E, and PROB_C are from 0 to 9, with 9 being the highest probability.]

From Table 1 we can infer that we have a coil structure with turns where the prolyl residues are present and sheet between them. A database analysis of X-ray diffraction structures of 7S seed storage proteins reveals [20, 21] that these proteins are typically trimeric, lacking cysteine residues (it cannot form disulfide bonds). The $\beta$-sheet conformation predominates and the N- and C-terminals are generally coils. The MD computed trajectory for the N-terminal containing

Journal of Physics: Conference Series **182** (2009) 012008 doi:10.1088/1742-6596/182/1/012008

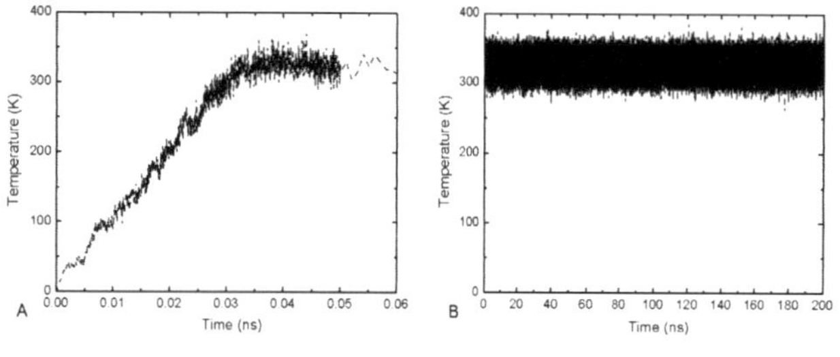


**Figure 1.** Temperature evolution during the first 60 ps of the simulation (A) and on the entire trajectory (B).

the serine residue is analyzed by plotting temperature and potential energy from the output file. From figure 1 it is evident that the temperature rose smoothly and equilibrated at 325 K. Similarly, changes in the kinetic and potential energy should be smooth and should level off at equilibrated values. Any sudden spikes in the temperature or energy plots are indicative of a problem with our simulation protocol (bad starting structure, too long time step, inappropriate parameters etc.).

In figure 2A is shown the potential energy as a function of time during this simulation. The highest energy difference between conformations is 115.9 kcal/mol, but overall a decrease of around 50 kcal/mol can be observed. The potential energy plot helps to locate the structure with the lowest energy, considered to be the folded state. In this case it's showing an initial important decrease over the course of the first 10 ns and then, in steps, the system explore different conformations until the moment when, before 160 ns, it reaches the minimum.

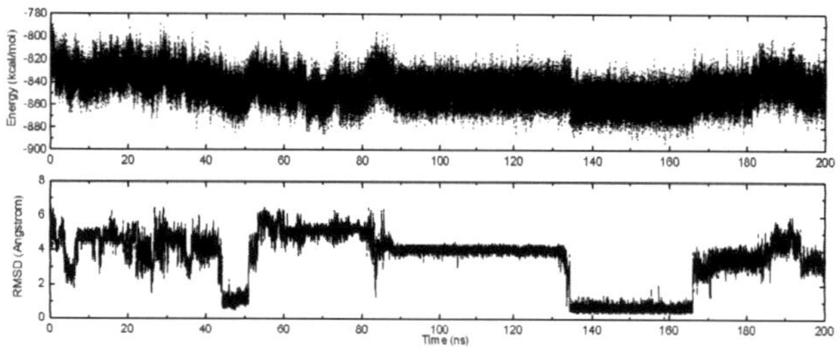


**Figure 2.** (A) Potential energy of PACB as a function of time during the MD. (B) Backbone RMSd during the same MD, compared to the lowest energy conformation.

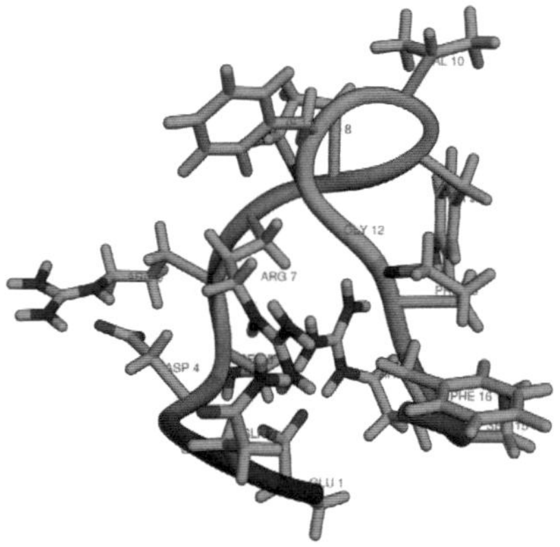


**Figure 3.** The PACB N-terminal structure corresponding to the lowest potential energy.

Using the *ptraj* program, the backbone root mean square deviation (RMSd) of structures relative to the lowest-energy conformation has been calculated and is represented in figure 2B. From the analysis of this representation and of the trajectory, using VMD and Sirius programs, it seems that the N-terminal of PACB stays most of the time in alternative structures and has not folded up to the native structure. Even if the trajectory is fairly long, the system is still exploring the potential energy surface, since the end point of the trajectory shows that the system tends to unfold and fold up differently. Here we must also keep in mind the fact, mentioned before, that shorter polypeptides generally adopt less compact and less stable structures in solution. This simulation should be continued, but we can consider the conformation having the minimum potential energy as an intermediary folded state.

The N-terminal structure corresponding to the lowest potential energy is given in Figure 3. It potential energy is -894.6279 kcal/mol, and the model is already well folded and presents hydrogen bonds between aminoacids. In the model there are turns at prolyl residues, a fact which is consistent with the previous analysis, with the empirical methods. The Molecular graphics image was produced using the UCSF Chimera package [22].

## 5. Conclusions

In order to determine the secondary structure of the N-terminal of the major protein component present in embryos and aleurone layers of barley, we used two different approaches. Initially we analyzed this system using two empirical prediction methods. Additionally, we studied the folding of this peptide by performing a 200 ns all-atom fully unrestrained molecular dynamics simulation in implicit solvent. Both empirical methods predicted a coil conformation with two turns where the prolyl residues are present, and sheet conformation between these 2 turns. For the structure containing the serine residue, analyzed by molecular dynamics, the conformation found to have the lowest potential energy has a coil structure and presents turns at prolyl residues, i.e. consistent with the previous empirical results. Actually, in known structures of other 7S seed storage proteins, determined by X-ray diffraction, the $\beta$-sheet conformation predominates and the N- and C-terminals are generally coils. From the RMSd between the trajectory structures and lowest energy structure found in the simulation, it seems evident that the N-terminal of PACB spends most of time in alternative structures and has not yet folded up to the native structure, making further analysis necessary. This could be achieved in several ways. A longer trajectory could be run (possibly with the use of explicit solvent). Alternatively, an annealing approach or a replica exchange could be simulated.

## References

[1] Meijer A B, Spruijt R B, Wolfs C J A M and Hemminga M A 2001 *Biochemistry* **40** 5081
[2] Hegde R S and Bernstein H D 2006 *Trends in Biochemical Sciences* **31** 563
[3] Meinnel T and Giglione C 2008 *Proteomics* **8** 626
[4] Müntz K 1998 *Plant Molecular Biology* **38** 77
[5] Yupsanis T, Burgess S R, Jackson P J and Shewry P R 1990 *Journal of Experimental Botany* **41** 385
[6] Tomoaia-Cotisel M 2006 in *Convergence of Micro-Nano-Biotechnologies, Series in Micro and Nanoengineering* ed M Zaharescu, E Burzo *et al* (Bucharest: Romanian Academy Press) **9** 147
[7] Tomoaia-Cotisel M, Tomoaia-Cotisel A, Yupsanis T, Tomoaia G, Balea I, Mocanu A and Racz C 2006 *Revue Roumaine de Chimie* **51** 1181
[8] Tomoaia-Cotisel M, Mocanu A, Leopold N, Vasilescu M, Chis V and Cozar O 2007 *JOAM* **9** 637
[9] Bratu I, Tomoaia-Cotisel M, Damian G and Mocanu A 2007 *J. Optoelec. Adv. Mat.* **9** 672
[10] Horovitz O, Tomoaia G, Mocanu A, Yupsanis T and Tomoaia-Cotisel M 2007 *Gold Bulletin* **40** 213 & 295
[11] Tomoaia-Cotisel M, Horovitz O, Borostean O, Bobos L D, Tomoaia G, Mocanu A and Yupsanis T 2008 *Studia Universitatis Babes-Bolyai, Chemia* **53** 143
[12] Chou P Y and Fasman G D 1974 *Biochemistry* **13** 211 & 222
[13] Chen H, Gu F and Huang Z 2006 *BMC Bioinformatics* **7** (Suppl 4) S14
[14] King R D and Sternberg M J E 1996 *Protein Science* **5** 2298
[15] Shea J E and Brooks C L 2001 *Annual Reviews of Physical Chemistry* **52** 499
[16] Onuchic J N and Wolynes P G 2004 *Current Opinion in Structural Biology* **14** 70
[17] Case D A, Cheatham T E, III, Darden T, Gohlke H, Luo R, Merz K M, Jr, Onufriev A, Simmerling C, Wang B and Woods R 2005 *Journal of Computational Chemistry* **26** 1668
[18] Berendsen H J C, Postma J P M, Gunsteren van W F, DiNola A and Haak J R 1984 *Journal of Chemical Physics* **81** 3684
[19] Tsui V and Case D A 2001 *Biopolymers (Nucleic Acid Sciences)* **56** 275
[20] Ko T P, Ng J D, Greenwood A and McPherson A 1993 *Acta Cryst. D. Biol. Cryst.* **49** 478
[21] Maruyama N, Adachi M, Takahashi K, Yagasaki K, Kohno M, Takenaka Y, Okuda E, Nakagawa S, Mikami B and Utsumi S 2001 *European Journal of Biochemistry* **268** 3595
[22] Pettersen E F, Goddard T D, Huang C C, Couch G S, Greenblatt D M, Meng E C and Ferrin T E 2004 *Journal of Computational Chemistry* **56** 1605

Journal of Physics: Conference Series **182** (2009) 012009 doi:10.1088/1742-6596/182/1/012009

# Inclusion compound of vitamin B13 in β-Cyclodextrin. Structural investigations

**Irina Kacso, Gheorghe Borodi, Sorin I Fărcaş and Ioan Bratu**

National Institute for Research and Development of Isotopic and Molecular Technologies, 65-103 Donath, 400293 Cluj-Napoca, Romania

Email: ioan.bratu@itim-cj.ro

**Abstract.** Structural characterization of inclusion compound of vitamin B13 (orotic acid) with *β*-cyclodextrin (*β*-CD) prepared by different methods (kneading, co-precipitation and freeze-drying) has been performed by using FTIR spectroscopy, powder X-ray diffraction and DSC, FTIR data being compared with molecular modeling of these supramolecular architectures.

## 1. Introduction

A bioactive compound acts depending on the administration form (solid or liquid); in consequence, its biodisponibility can be influenced [1]. In order to solve this problem, the most employed method is the complexation with cyclodextrins (CD) [2-4]. Cyclodextrins are cyclic oligosaccharides having a hydrophobic external surface and an inner hydrophilic surface. There are different CD types depending on the glucopyranose unit number [*α*-six; *β*-seven, *γ*-eight, etc.] (see figure 1a). They are capable to form more soluble guest-host systems through non-covalent bonds, Van der Waals interactions, hydrophobic effect, solvent molecules reorganization and hydrogen bonds [5].

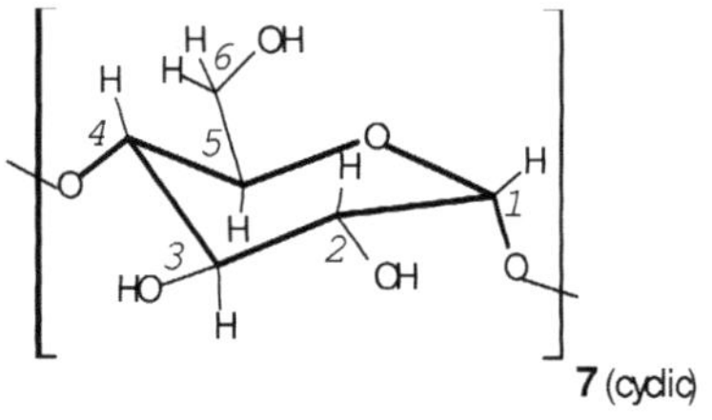


(a)

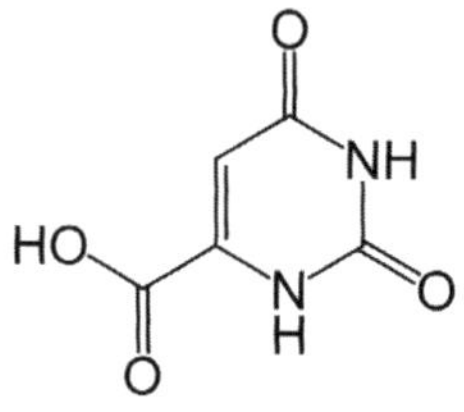

(b)

**Figure 1.** (a) *β*-Cyclodextrin molecule. (b) Orotic acid molecule.

Vitamin B13 (orotic acid) (figure 1b) is not really recognized as a vitamin, since it is manufactured by the body intestinal flora. The orotic acid plays a central role in the metabolism of folic acid and vitamin B12. It assists and may enhance the transportation and absorption of essential nutrients, especially calcium and magnesium, across cell membranes and helps the production of genetic material. It may be beneficial after a heart attack and has been used in conditions such as multiple sclerosis and chronic hepatitis. It is also reported to prevent liver-related complications and premature aging. Found naturally in whey and root vegetables, such as beets, turnips, and carrots. Orotic acid is easily destroyed by water and sunlight. Deficiency of vitamin B13 may be cause of liver disorders, cell degeneration, premature aging.

The investigation of the molecular encapsulation of the vitamin B13 in *β*-CD through different techniques (FTIR spectroscopy, DSC and X-ray diffraction) and the comparison of this information with that obtained by molecular modeling is the purpose of this paper.

## 2. Experimental

The inclusion compounds were prepared by kneading, co-precipitation and freeze-drying methods. Also, physical mixtures were prepared in order to be compared with the so prepared inclusion compounds.

### *2.1. Experimental investigation methods*

To evidence the inclusion compounds formation, several techniques were employed. FTIR measurements were performed with JASCO 6100 spectrometer in the 4000 to 400 $cm^{-1}$ spectral region with a resolution of 4 $cm^{-1}$ using KBr pellet technique. DSC thermograms were obtained with DSC 60 Shimadzu calorimeter and X-ray diffraction patterns were obtained with a Bruker D8 Advance diffractometer in the $2\theta = 2\text{-}50°$ angular domain using Cu $K\alpha_1$ radiation.

### *2.2. Molecular modeling*

Guest molecules can have interaction with the hosting CD via the latters OH groups, by Van der Waals forces, electrostatic forces and hydrogen bonding [6]. The interaction may be external but also internal. In the latter case, the guest enters the CD cavity, thus forming an inclusion complex. To assess the precise type of complex, a starting model was built by positioning the orotic acid molecule at the larger side of the *β*-CD cavity. The geometry of the complex was optimized in vacuum using the molecular- mechanics algorithm of the HyperChem software [7]; the details are given elsewhere [8]. The well-known MM+ (with the Polak-Ribière conjugate gradient) method was used to minimize the energy of the structures until a RMS gradient lower than 0.015 kcal $mol^{-1}$ Å$^{-1}$ was obtained.

## 3. Results and discussion

### *3.1. FTIR*

As concerned the C=O stretching band (see figure 2), its maximum located at ~1709 $cm^{-1}$ is diminished strongly in intensity after complexation with *β*-CD, the same occurs with C=C stretching band at 1674 $cm^{-1}$ [9].

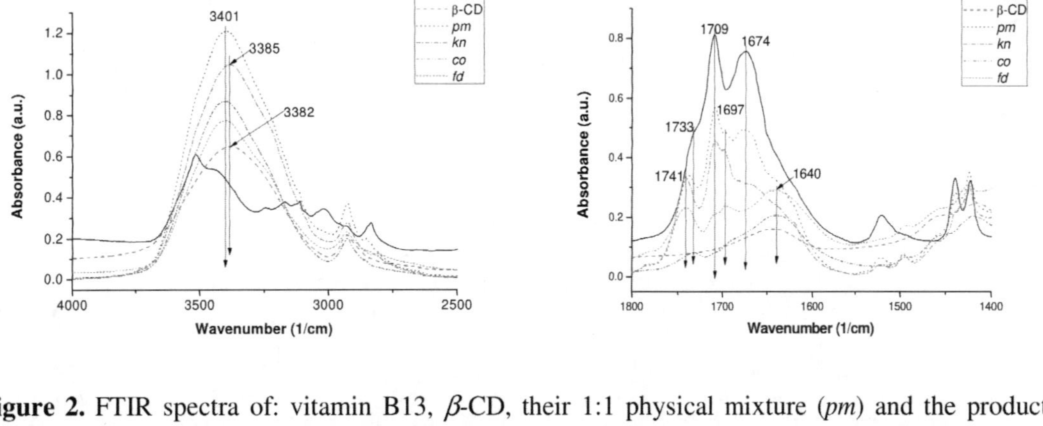


**Figure 2.** FTIR spectra of: vitamin B13, *β*-CD, their 1:1 physical mixture (*pm*) and the products obtained by kneading (*kn*), co-precipitation (*co*), and freeze-drying (*fd*) procedures, 4000-2500 $cm^{-1}$ (left) and 1900-1400 $cm^{-1}$ (right)spectral ranges.

The intensity diminishing can be related with the mobility hindering of the ring, an indirect argument of the inclusion process. This frequency shift can be related to the appearance of hydrogen bonds during the complexation process, especially for the *fd* product. The frequency of the O-H bending vibration, located at 1640 cm$^{-1}$ in pure *β*-CD is not shifted for *kn*, *co* and *fd* products. This already mentioned complexation mechanism must be taken into account for the *kn* and *co* products, also.

### *3.2. X-ray powder diffraction*

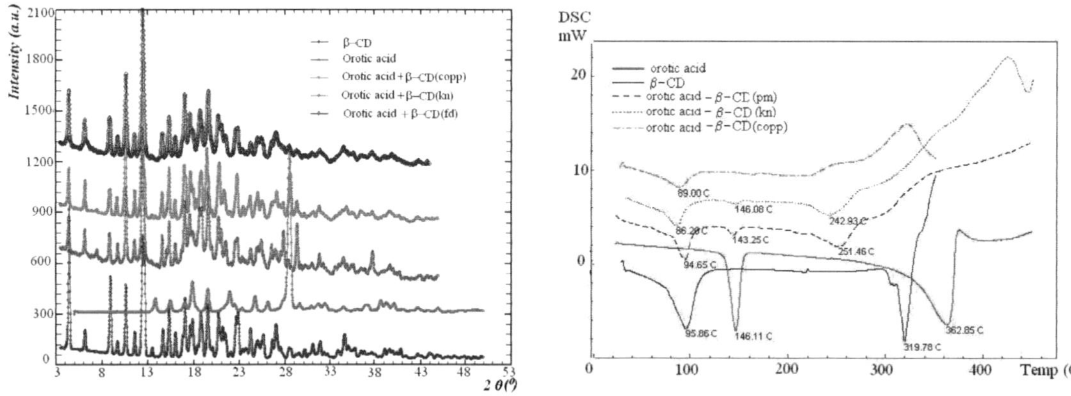


(a) (b)

**Figure 3.** (a) X-ray patterns of: vitamin B13, *β*-CD, and the products obtained by *kn*, *co* and *fd* procedures. (b) DSC thermograms of vitamin B13, *β*-CD and of the obtained inclusion compounds.

From powder diffraction patterns (see figure 3a) one can see that inclusion compounds are formed in different degree depending on the preparation method. The difference between these patterns is due to different crystalline /amorphous ratios. The highest degree of formation is obtained for freeze-drying procedure.

### *3.3. DSC*

The thermogram of the *β*-CD showed an endothermic transition between 70-110°C, due to the loss of water molecules and other endothermic peak at around 300°C indicating the degradation of cyclodextrin. DSC trace of vitamin B13 presents a sharp endothermic melting peak at 146°C. By comparing the thermograms of the pure compounds with that of the obtained products, the decreasing of dehydration endothermic peak of cyclodextrin was observed, as well as a decreasing until disappearance of the melting peak of the vitamin B13 in the inclusion compounds obtained by different methods (see figure 3b). One can conclude that the inclusion compound formation depends of the preparation method employed.

### *3.4. Molecular modeling*

An analysis of the vibrational frequency shifts based on FTIR spectra (C=O and C=C stretching modes) led to the tentative conclusion that the orotic acid probably enters the cyclodextrin torus when forming the β-CD - orotic acid inclusion compound. Molecular mechanics calculations are consistent with this model, but a more detailed interpretation is not possible due to the MM+ approximations used.

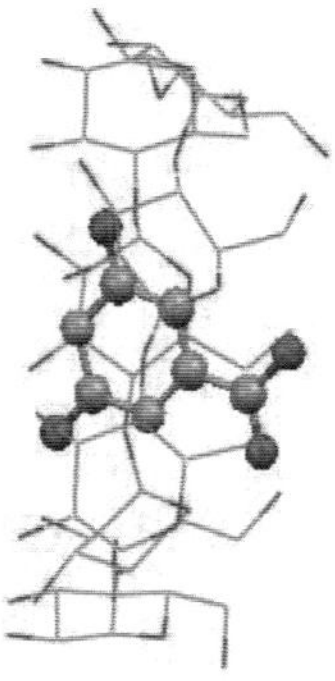

**Figure 4.** Molecular modeling (MM+) of the vitamin B13–β-CD supramolecular assembly.

More work needs to be done to establish the nature [6] of the forces implied in the inclusion process and in particular the expulsion processes of the intracavity water molecules [10] must be taken into account.

## 4. Conclusions

The inclusion compound between vitamin B13 and β-CD was confirmed based on FTIR, X-ray powder diffraction and DSC data. All preparation methods employed were successfully in obtaining these supramolecular assemblies. Molecular modeling with Hyperchem software based on MM+ molecular mechanics method was successfully used to determine the geometry of this supramolecular assembly: the orotic acid molecule enters the cyclodextrin torus when forming the inclusion compound.

**Acknowledgments**

This work received financial support from the Romanian Research and Education Ministry under the Core Project PN-09-44 02 01.

**References**

[1] Leucuta S E 1989 *Farmacocinetica şi biodisponibilitatea în terapia medicamentoasă* (Bucureşti: Editura Medicală)
[2] Albers E and Muller B W 1992 *J. Pharm. Sci.* **81** 756-61
[3] Ficarra R, Ficarra P and Rameri D 2000 *J. Pharm. Biomed. Anal.* **23** 231-6
[4] Acerbi D, Bovis G, Carli F, Pasini M, Pavesi L and Peveri T 1990 *Drug Investigation* **2**, 29-36
[5] Lopata A, Darvas F, Stadler-Szóke A and Szejtli J 1995 *J. Pharm. Sci.* **74** 211-3
[6] Liu L and Guo Q-X 2002 *J. Incl. Phenom. and Macrocyclic Chem.* **42** 1-14
[7] HyperChem™ Release 4 1994 (Waterloo: Hypercube Inc)
[8] *Computational Chemistry* 1994 (Waterloo: Hypercube Inc)
[9] Hernanz A, Billes F, Bratu I and Navarro Raquel 2000 *Biopolymers* **57** 187-98
[10] Gavira J M, Hernanz A and Bratu I 2003 *Vib. Spectrosc.* **32** 137-46.

# New ionic liquids as innovative reaction media

**J Liebscher, J Shah, Z Yacob, S Sadiq, S Hanelt and H Blumenthal**

*Institute of Chemistry, Humboldt-University Berlin, Brook-Taylor Str. 2, 12489 Berlin, Germany*

*E-mail: liebscher@chemie.hu-berlin.de*

**Abstract.** New ionic liquids were developed based on triazolium salts and guanidinium salts. By introduction of functions, such as organocatalytic units, into these units task-specific ionic liquids could be obtained. These products can be used as efficient catalysts in asymmetric syntheses.

## 1. Introduction

Ionic liquids (IL) are salts that are liquid at low temperature (< 100 °C). They present a new class of solvents of nonmolecular, ionic character. As such they are almost non-volatile, can easily be recycled in many cases and reused and thus are often being termed as "green" solvents. Furthermore, they can show specific solvent effects on reactions. Their polarity and miscibility with other solvents and water can be tailored by the choice of anions and cations.

Even though the first examples of ionic liquids can be traced back to 1876 the general interest in this class of compounds has developed in the last 20 years. Nowadays, ionic liquids such as imidazolium salts **1** or quaternary ammonium salts **2** are widely used in different fields, such as chemical synthesis, electrolysis, as sealing materials, lubrication, heat storage, as liquid crystals, as electroelastic materials and in separation techniques. They have been virtually applied in all types of organic reactions, polymerisation and biological reactions included.

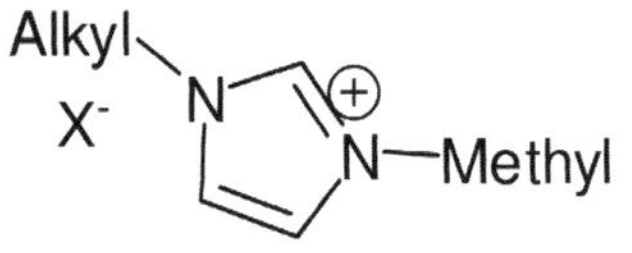


$R_4N^+ X^-$

**1** **2**

They are widely used in catalysis, where better performance and easier recycling could be achieved in a number of cases. Because IL have practically no vapour pressure or can be sparingly soluble in other solvents, easy product separation is possible by distillation or extraction, respectively.

As a recent development so-called task specific ionic liquids (TSIL) have to be mentioned. Here, a function (often a catalytic function) is covalently attached to the ionic liquid and thus such TSIL can be used as solvent and catalyst at the same time or can be dissolved in another IL by its IL-substructure-tag. If the function is a catalyst the assemblies are also called IL-tagged catalysts.

Journal of Physics: Conference Series **182** (2009) 012010 doi:10.1088/1742-6596/182/1/012010

## 2. Results and discussion

We developed new ionic liquids and TSIL based on guanidinium salts **3** and 1,2,3-triazolium salts **4**.

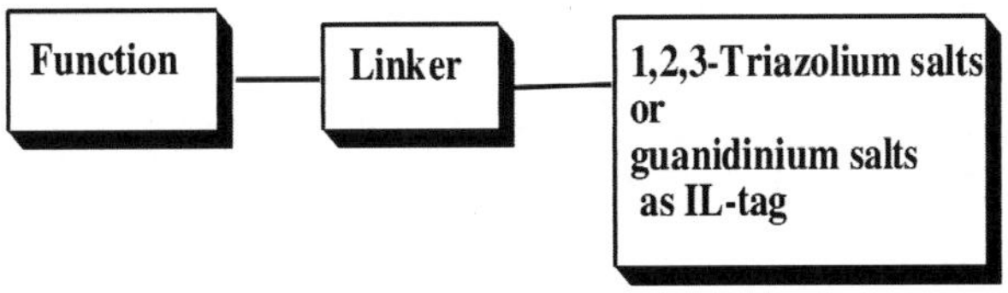


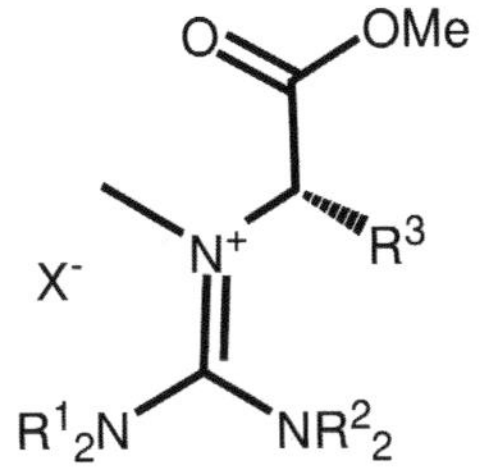


**3**

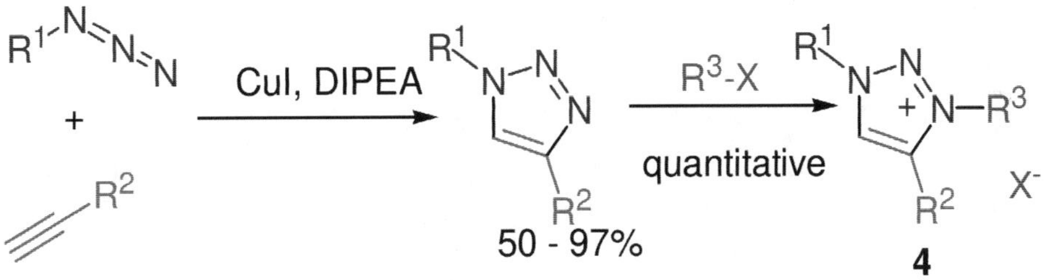


**4**

$R^1$, $R^2$, $R^3$ = alkyl, benzyl, (S)-pyrrolidin-2-methyl, 4-(1-pyrenyl)-butanoylaminomethyl, 3-phthalimidopropyl, (S)-prolinolcarbonyloxymethyl

Both classes can be synthesized in straight forward and versatile routes. Thus the guanidinium salts **3** were obtained from α-amino esters and isothiuronium salts or chloroamidinium salts and N-methylation of the resulting guanidines. Triazolium salts **4** were synthesized from organic azides and terminal alkynes by Cu-catalyzed Huisgen-Meldal-Sharpless cycloaddition ("click-reaction") followed by N-alkylation. Both steps occur in high yields. Anions $X^-$ can be further exchanged by salt metathesis. In this way halides were replaced by tetrafluoroborate or dicyanoamide. All salts **3** and **4** obtained melt below 100 °C, i. e. they are ionic liquids. Most of them are liquid at room temperature and thus represent room temperature ionic liquids. Several functions, such as organocatalysts, ligands, fluorescent markers or reactive groups can be introduced into the triazolium moiety **4**. First applications of these new TSIL in asymmetric synthesis will be reported. Thus the guanidinium salts **3** as well as their precursors could be successfully applied in asymmetric Baylis-Hilman-reaction.

Triazolium salts **4** equipped with (*S*)-proline or proline derivatives were used as IL-tagged organocatalysts in aldol and Michael reactions. High enantioselectivities could be obtained and the organocatalysts could be recycled and reused several times. The reactions could be performed in traditional organic solvents or in ionic liquids. Non-functionalized guanidinium salts were found to improve the performance of (*S*)-proline as organocatalyst in aldol reactions.

**Acknowledgements**
Financial support by the Bundesministerium für Bildung und Forschung (BMBF) of Germany and Deutscher Akademischer Austauschdienst (DAAD) is acknowledged.

**References**

[1] Wasserscheid P and Welton T *Ionic Liquids in Synthesis* 2; ed. Wiley-VCH: Weinheim, 2008.
[2] Hanelt S and Liebscher J 2008 *Synlett* 1058
[3] Shah J, Blumenthal H, Yacob Z and Liebscher J 2008 *Adv. Synth. Catal.* **350** 1267
[4] Yacob Z, Shah J, Leistner J and Liebscher J 2008 *Synlett* 2342
[5] Shah J and Liebscher J 2008 *Synthesis* 917
[6] Tornoe C W, Christensen C and Meldal M 2002 *J. Org. Chem.* **67**, 3057
[7] Rostovtsev V V, Green L G, Fokin V V and Sharpless K B 2002 *Angew. Chem. Int. Ed.* **41** 2596

# Assessment of molecular interaction in a cycluron-cyclodextrin inclusion complex

**Mihaela Mic[1], Dyanne Cruickshank[2] and Ioan Turcu[1]**

[1] National Institute for Research and Development of Isotopic and Molecular Technologies, 65-103 Donath, 400293 Cluj-Napoca, Romania

[2] Department of Chemistry, University of Cape Town, Rondebosch 7701, South Africa

Email: mihaela.hossu@itim-cj.ro

**Abstract**. Thermodynamic parameters and the host-guest stoichiometry of inclusion complex cycluron with $\beta$-cyclodextrin in aqueous solution have been determined. The supramolecular structure has been investigated by isothermal titration nanocalorimetry and $^1$H NMR spectroscopy at 298.15 K.

## 1. Introduction

Cycluron is a urea herbicide used to control broadleaf and grassy weeds in cereals and other crops. The inclusion of agrochemicals into cyclodextrins is used to improve the physical properties of many agrochemicals. The inclusion of the herbicides into cyclodextrins is an attempt to improve several physical properties: the low solubility of agrochemicals in water, reduced stability and keeping their efficiency given by the high toxicity.

Cyclodextrins (CDs), are cyclic oligosaccharides having hydrophilic out surfaces and a hydrophobic cavity. They are very good candidates as host molecules because of their ability to form inclusion compounds with a large variety of molecules by selectively incorporating such molecules on the basis of size considerations and selective interactions [1,2]. Accordingly, cyclodextrins are good candidates able to enhance the aqueous solubility, chemical reactivity and spectral properties of cycluron. Among this class of "host" molecules, $\beta$-cyclodextrin ($\beta$-CD) with seven glucose units in its molecule has been proved to be the optimal molecular partner for cycluron.

The purpose of our research is to investigate and characterize the complexation thermodynamics of $\beta$-CD with cycluron using isothermal titration calorimetry (ITC) and to relate the thermodynamic parameters with the structure, by measuring $^1$H NMR spectra of the inclusion complex.

## 2. Materials and methods

### *2.1. Materials*

The solutions were prepared as follows: 15 mM $\beta$-CD by dissolving 191.85 mg of $\beta$-CD in 10 mL of water and 0.75 mM cycluron by dissolving 7.44 mg in 50 mL of water. Both solutions were prepared using degassed water in a thermal bath at 35-40 °C. The weight of each sample has been measured with a precision of ± 10 μg.

### 2.2. *Isothermal titration calorimetry*

ITC is an extremely powerful, highly sensitive technique, able to characterize the weak interaction associated with $\beta$-cyclodextrin–cycluron complexation. An ITC experiment provides the complete thermodynamics of a binding reaction, including $K$, $\Delta G$, $\Delta H$, $\Delta S$ and reaction stoichiometry ($n$).

The experiments were performed on a VP-ITC (2G) MicroCal (New Castle, Delaware, USA) calorimeter at 25°C. From a computer-controlled 250-µL Hamilton syringe a series of 40 injections into the 1.4 mL calorimeter cell has been performed. Each injection leads to heater compensation peak, which is integrated to obtain the reaction heat response. The 15 mM $\beta$-CD was injected into the stirred (150 rpm) sample cell containing 0.75 mM cycluron in aliquots of 5 µL at 300 s intervals. As compared to reaction heat the dilution contribution was found to be negligible.

### 2.3. *$^1H$ NMR*

$^1$H NMR experiments were performed on a Brucker Avance III 500. The outcome used to characterize the $\beta$-cyclodextrin–cycluron complex were the chemical shifts of all the protons from the investigated macromolecular structure. The experimental data allows us to determine the reaction stoichiometry and to calculate the association constant $K$ of the guest-host complex.
The stoichiometry of the complex was determined by the continuous variation method (Job plot).

### 2.4. *Methods*

The equilibrium in guest–host reaction is described by the following hypothetical scheme:

$$A + nB \leftrightarrows AB_n \equiv C;\ K=C/(AB^n)$$

where $A$, $B$, and $C$ stand for the concentrations of cycluron, $\beta$-cyclodextrin and $\beta$-cyclodextrin-cycluron complex respectively. The implicit guess is that a cycluron molecule can bind one or two ($n$ = 1 or 2) $\beta$-cyclodextrin molecules.

#### 2.4.1. *ITC*

In an ITC experiment the reaction heat per mole of injected molecules after the $i^{th}$ injection is given by:

$$\frac{q_i}{vB_0} = \frac{\Delta H}{2}\left[1 + \frac{1 - r_i/n - (r_0 + r_i)/nKB_0}{\sqrt{(1 + r_i/n + (r_0 + r_i)/nKB_0)^2 - 4r_i/n}}\right] \quad (1)$$

where $r_0 = \dfrac{B_0}{A_0}$; $r_i = \dfrac{B_i}{A_i} = r_0\dfrac{iv}{V}$, $A_0$ and $B_0$ are the initial concentrations of cycluron and of the $\beta$- CD respectively, $\Delta H$ is the reaction enthalpy, $V$ and $v$ standing for the cell and injection volumes. Equation (1) will be used as fit function for the ITC data. The ITC experiments alowd us to obtain the the stoichiometry $n$, the reaction enthalpy $\Delta H$ and the echilibrium constant $K$.

#### 2.4.2. *$^1H$ NMR*

In the $^1$H NMR experiments, the overall concentration of the two species was kept constant ($A_t + B_t = M$) and the ratio $\rho = B_t / M$ was varied form 0 to 1. $A_t$ and $B_t$ are the total concentrations of the host and guest respectively.This was accomplished by using equimolar solutions of A and B and mixing them to constant volume to the desired ratio $\rho$. Taking into account that: $A_t = A + C$, $B_t = B + nC$ we have: $C = K(\rho M - C)(M - \rho M - nC)^n$.

The absolute concentration of C will reach a maximum at $dC/d\rho = 0$, at $\rho = 1/(1+n)$. For 1:1 stoichiometry ($n$ = 1) we have $\rho = 0.5$.

The absence of new peaks that could be assigned to the complex suggested that complexation is a dynamic process, the included guest being in a fast exchange between the free and bound states.

In this case the observed chemical shift $\delta_o$ for a certain proton ($X = A$ or $B$) is given by:

$$\delta_o = \frac{X\delta_f + C\delta_c}{X_t} \Rightarrow X_t \Delta\delta_o = C\Delta\delta_c \tag{2}$$

where $\Delta\delta_o = \delta_f - \delta_o$, $\Delta\delta_c = \delta_f - \delta_c$, $\delta_f$ and $\delta_c$ standing the chemical shifts corresponding to free $X$ and to the $\beta$-cyclodextrin–cycluron complex ($C$) respectively. The quantity $\Delta\delta_o X_t$, given by the changes in the chemical shift of H-3 and H-5 protons of the $\beta$-cyclodextrin and also of the protons from the methylene an methyl groups of cycluron, was plotted against ρ (figure 4). As fitting function the following ρ - dependence of the complex concentration:

$$C = \frac{M}{2}\left\{1 + \frac{1}{KM} - \left[\left(1 + \frac{1}{KM}\right)^2 - 4\rho(1-\rho)\right]^{1/2}\right\} \tag{3}$$

has been used. All the curves presents a maximum at $\rho = 0.5$ indicating the existence of a 1:1 stoichiometry within the range of the investigated concentrations. Knowing the association constant K from the ITC measurements, the four data sets have been used for finding the chemical shifts of the investigated protons in the $\beta$-cyclodextrin-cycluron complex.

## 3. Results

The equilibrium constant obtained from ITC measurement was $K = 4753\ M^{-1}$ which corresponds to a Gibbs free energy $\Delta G = -21$ KJ/mol. The reaction enthalpy of the host-guest inclusion complex is $\Delta H = -8.92$ kJ/mol. This negative value indicates that the formation of host-guest inclusion complex is a weak exothermic process. The entropic effect of the host-guest interaction ($\Delta S = 40.5\ JK^{-1}/mol$) is positive being due to the balance between the negative contribution brought by the complex formation, and positive contribution induced by the release of water molecules from the $\beta$-cyclodextrin cavity.

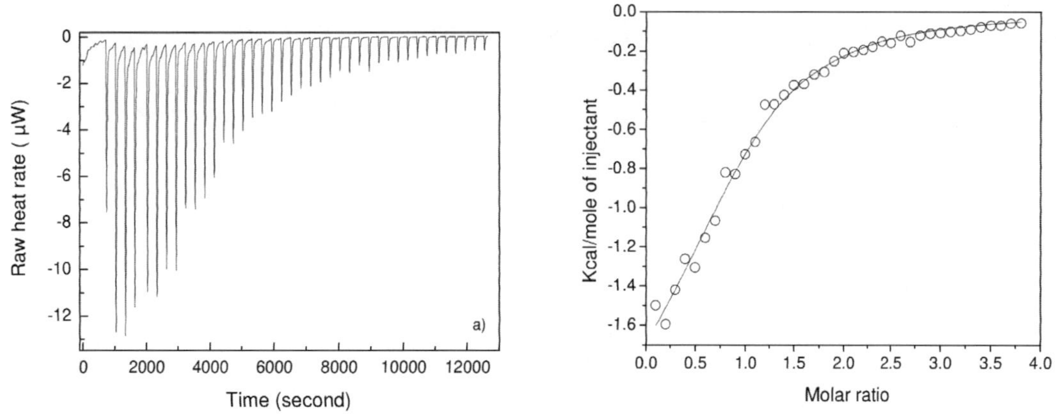


(a) (b)

**Figure 2.** Calorimetric titration of β-CD solution in cycluron solution. (a) Raw data obtained from 40 automatic injection of 5 μL aliquots of 15 mM $\beta$-CD solution into 0.75 mM cycluron solution 25°C. The heat of dilution has been subtracted. (b) Integration plot of the data calculated from the raw data. The solid line correspond the best fit curve.

The main NMR results are given in table 1 and consist in the predicted chemical shifts for several protons from the host-guest inclusion complex.

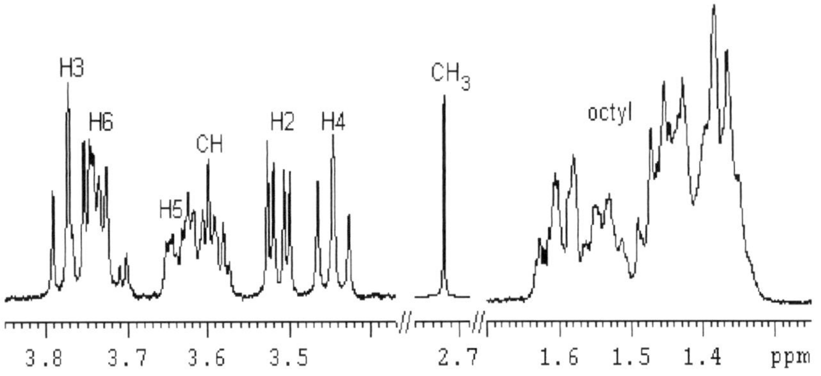


**Figure 3.** $^1$H NMR spectrum of $\beta$-CD–cycluron system.

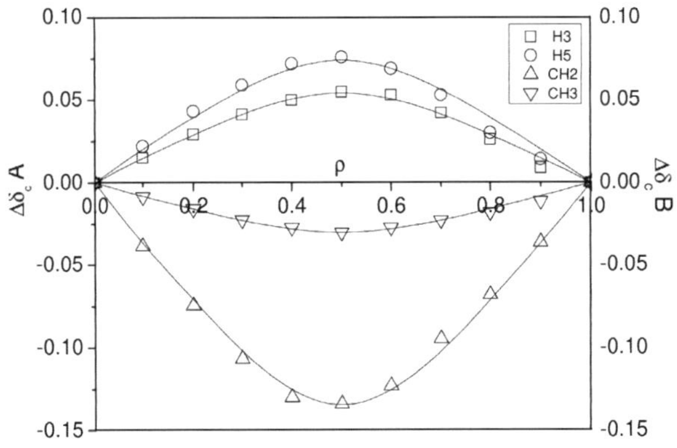


**Figure 4.** Job plot for H-3, H-5 protons of $\beta$-CD and also of the protons from the methylene and methyl groups of cycluron.

**Table 1.** Chemical shifts of the investigated protons in the $\beta$-cyclodextrin-cycluron complex.

| Chemical shifts | H3 | H5 | H ($CH_2$) | H ($CH_3$) |
|---|---|---|---|---|
| $\delta_c$ | 3.783 | 3.631 | 1.749 | 2.749 |
| $\Delta\delta_c$ | 0.0652 | 0.0891 | -0.1615 | -0.036 |

## 4. Conclusions

The inclusion process between cycluron and $\beta$-CD has been investigated systematically by isothermal calorimetry and $^1$H NMR in aqueous solution at 298.15 K. The microcalorimetry allowed us to obtain the reaction enthalpy, stoichiometry, association constant, Gibbs free energy changes and the entropy effect of the inclusion process based on the directly calorimetric data utilizing non-linear fitting procedure. $^1$H NMR spectra showed that the inclusion phenomenon take place since the most marked modifications of the chemical shift parameters occur for protons that were oriented toward the cavity. The stoichiometry of inclusion complex of $\beta$-CD with cycluron is 1:1 revealing the recognition function of the host and guest molecules.

## References

[1] Bria M, Cooke G, Cooper A, Delattre F, Hewage S G, Rabani G, Nutley M and Woisel P 2007 *Tetrahedron Letters* **48** 8430

[2] Sun De-Zhi, Li Ling, Qu Xiu-Kui and Yin Bao-Lin 2006 *Acta Chimica Sinica* **17** 1765.

# Nuclear magnetic resonance studies of liquids morphology inside partially saturated porous media

**Ruben Nechifor, Codruta Badea and Ioan Ardelean**

Technical University of Cluj-Napoca, Physics Department, 15 Constantin Daicoviciu, 400020 Cluj-Napoca, Romania

E-mail: ioan.ardelean@phys.utcluj.ro

**Abstract**. In the present contribution we are investigating the relationship between liquid morphology and relaxation time distribution in partially saturated pores. The filling fluids are acetone (polar) and tetradecane (nonpolar). The porous sample is a silica glass (Vitrapor#5) with the nominal mean pore size is d = 1 μm (± 0.6 μm). All nuclear magnetic resonance relaxation experiments are performed at 20C using a NMR instrument operable at 20 MHz proton resonance frequency. The experimental results are compared with a two phase exchange model providing us information on the strength of surface relaxation.

## 1. Introduction

The behaviour of fluids in systems with restricted geometry is known to be very different from that in bulk. The physical properties of molecules adsorbed on the surface or confined in small pores may substantially differ from their properties in bulk. These deviations may be very important for technical applications (separation, catalysis, storage) of such host-guest systems. Consequently a complete understanding of the dynamical effects carried out by the interaction of molecules with the surface in porous media is very important.

Owing to its completely non-invasive character, nuclear magnetic resonance (NMR) is widely used to investigate the dynamics of molecules confined in porous media. NMR is the most versatile tool to investigate molecules dynamics. An NMR experiment is able to provide much more information than the simple resonance frequency of nuclear spins (*i.e.* their chemical shift) or the associated line width. NMR measurements of diffusion coefficients and relaxation times render quantitative data on the dynamics of confined molecules and the restrictions the confinement imposes on their translational and rotational mobility [1]. NMR relaxation experiments are suitable in studying liquids in porous media mainly due to an observed enhanced relaxation rate under confinement conditions. This behaviour is generated by interactions of the probed molecule with the surface depending on porous sample composition, morphology, the nature of the filling fluid and its distribution [2].

Most of the studies reported in the literature about relaxation distribution in porous media refer to saturated porous samples. Here we will investigate the partial saturated pores. Thus the relationship between relaxation time distribution liquid morphology under partially saturated conditions and the polar character of the confined liquid will be investigated.

## 2. The theory of NMR relaxation in partially saturated pores

In the case of porous media saturated with liquids it is often considered that the observed relaxation rate is a weighted average between the relaxation rate of the bulk-like liquid and the surface relaxation rate of molecules that are confined in a tiny layer [2]. It is assumed that the confined molecules experience short range interactions with the surface and a fast exchange process between the molecules in the surface region and the remaining liquid in saturated pores (bulk-like region) is present. Consequently the induced nuclear magnetization relaxes mono-exponentially in time, with a transverse relaxation rate given as the weighted average between the surface relaxation rate and the bulk-like rate [2].

If we adapt this model to the partially saturated porous media then the observed transverse relaxation rate can analogously be considered the result of an exchange process between the molecules in the surface region and the remaining liquid, partially filling the sample (bulk-like region). Consequently the measured relaxation rate can be expressed as:

$$\frac{1}{T_2}=\frac{\lambda S\rho_s}{\lambda S\rho_s+V_b\rho_b}\frac{1}{T_{2s}}+\left(1-\frac{\lambda S\rho_s}{\lambda S\rho_s+V_b\rho_b}\right)\frac{1}{T_{2b}}. \tag{1}$$

Here $T_{2b}$ and $T_{2s}$ refer to the bulk-like and surface relaxation times, respectively. $S$ represents the surface area visited by molecules during the NMR relaxation experiment. It depends on the effective diffusion coefficient of the liquid inside the porous structure. In the case of partially saturated samples this diffusion coefficient can be enhanced significantly due to the vapour phase contribution [3,4]. $V_b$ is the bulk like liquid volume bounded by the S surface. $\lambda$ represents the distance from the surface where the enhanced relaxation takes place (surface region). This distance is generally assumed to be of the order of one intermolecular distance [5]. $\rho_s$ and $\rho_b$ are the mass densities inside the surface region and bulk like region respectively. Note that equation (1) is valid only at filling factors corresponding to multilayer surface coverage. For lower filling factors, when one monolayer is approached, all molecules are interacting with the surface. Consequently the measured relaxation time coincides with the surface value.

The above equation can be expressed as a function of the filling factor $f$ defined by the ratio between the mass of confined molecules in the case of partially filled and saturated sample

$$f=\frac{\lambda S\rho_s+V_b\rho_b}{\lambda S\rho_s+(V_0-\lambda S)\rho_b}. \tag{2}$$

$V_0$ is the volume of the void space bounded by the $S$ surface. With the filling factor given above and if we assume that the densities inside the relaxation layer and the bulk-like region are the same $(\rho_s=\rho_b)$ then equation (1) can be rewritten as:

$$\frac{1}{T_2}=\frac{1}{T_{2b}}+\frac{\lambda S}{V_0}\frac{1}{f}\left(\frac{1}{T_{2s}}-\frac{1}{T_{2b}}\right)=\frac{1}{T_{2b}}+\xi\frac{S}{V_0}\frac{1}{f}. \tag{3}$$

Here $\xi=\lambda(1/T_{2s}-1/T_{2b})$, the surface relaxation strength, represents a phenomenological physical constant characteristic of the liquid-solid interaction [2].

If the local relaxation time varies throughout the sample due to different local surface to volume ratios $(S/V_0)$, different surface relaxation strengths or different filling factors then a multi-exponential attenuation of the magnetization can be observed:

$$A(t)=\int_0^\infty P(T_2)\exp(-t/T_2)\,dT_2\,. \tag{4}$$

An analysis of the experimental data using a regularised numerical Laplace inversion algorithm (CONTIN) [5] will allow us extraction of relaxation times distribution $P(T_2)$. This distribution may provide us in turn the liquid distribution inside pores assuming certain pores shape.

## 3. Experimental

As porous sample a silica glass (Vitrapor#5) was used. It was purchased from ROBU Glasfilter-Geräte, Germany. The nominal mean pore size is $d$ = 1 μm (± 0.6 μm) as indicated by the manufacturer. Before use the samples were cleaned by 30 minutes of boiling in 30 % $H_2O_2$ followed by several rinses with distilled water. After that they were left in vacuum for 24 hours at 95 °C. At the end, the samples are considered to be dry with a nominal filling factor $f$ = 0. The NMR measurements done on the dry sample have indicated no signal.

The solvents were filled into the porous glass with the aid of the bulk-to-bulk method resulting in a filling factor 1 for the corresponding solvent. The solvents examined in this study were tetradecane (non-polar; non-wetting) and acetone (polar; wetting). These solvents were purchased from Merck, Germany. Partial evaporation was then used for the adjustment of the desired filling factor. The solvent content reached by this procedure was determined by weighing. After that the samples were placed in a sealed container with practically no empty space that would allow for further evaporation. All experiments were done only after about one hour permitting the fluid to equilibrate with respect to phase distributions. After this annealing period no changes in the NMR signal were perceptible any longer.

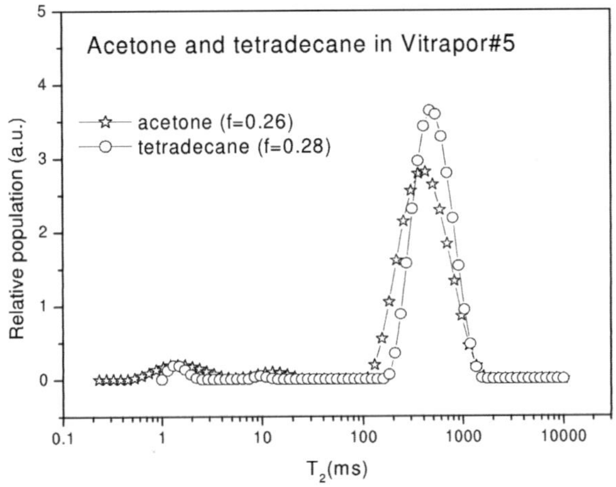


**Figure 1.** Relaxation time distribution of acetone and tetradecane in partially saturated Vitrapor#5 for a filling factor as indicated in the legend.

All relaxation experiments were performed on a Bruker MINISPEC MQ20 spectrometer operating at a proton resonance frequency of 20 MHz. The data were recorded at 20 °C using the standard Carr-Purcell-Meiboom-Gill (CPMG) technique [1]. The length of one hard 90° pulse was 2.5 μs. The echo time of 200 μs has prevented for the appearance of diffusion effects on echo decay [6].

## 4. Results and conclusions

The echo decay curves (not shown here) have indicated a non exponential dependence on evolution time both in the case of acetone and tetradecane filled samples. This dependence has generated the relaxation times distributions as those depicted in Figure 1 for a chosen filling factor. One can observe here three peaks which can be interpreted in the frame of the liquid distribution proposed by the puddle model [7]. Thus, the first peak (from the left) corresponds to the molecules forming one monolayer or two on the surface of porous media (surface region). The second peak may be associated with the molecules in the in the silica interstices (puddles) and the third peak with the molecules in the

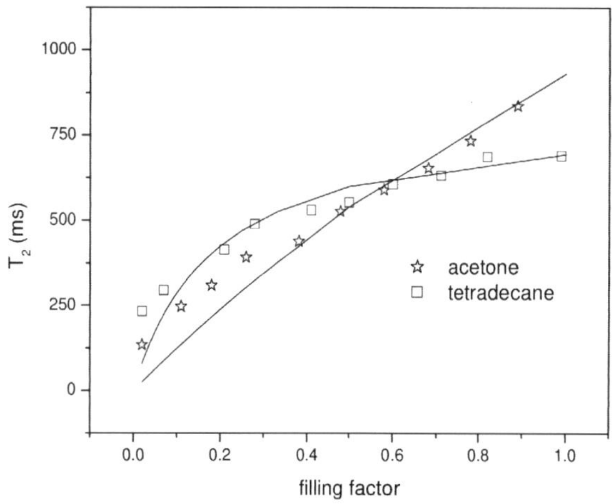


**Figure 2.** The transverse relaxation time versus filling factor for acetone (stars) and tetradecane (squares). The theoretical lines are fits to equation (3) with only one parameter ($\xi$).

bulk-like region. Note that for high filling factors only the third peak is visible due to its dominant contribution.

The relaxation times extracted from the position of the third peak versus filling factor are represented in Figure 2. Assuming that the pores region visited by molecules during the time of the experiment has a cylindrical shape with a radius of 0.5 μm one can compare the extracted relaxation times with the theoretical curves generated on the basis of equation (3). The best fits were obtained for the surface relaxation strength of $1.95{\cdot}10^{-7}$ m/s in the case of acetone and $0.57{\cdot}10^{-7}$ m/s in the case of tetradecane. As can be observed the value is smaller in the case of nonpolar tetradecane as compared with the polar acetone indicating stronger surface effects on relaxivity for polar molecules. Note however that only one parameter $(\xi)$ was fitted in equation (3). The bulk values of relaxation time used in our fitting were those obtained experimentally ($T_{2,b-acetone} = 3400\,ms$ ; $T_{2,b-tetradecane} = 825\,ms$ ).

In our knowledge there is no unitary view about distribution of molecules on the pores surface at low saturation degrees. Some experimental results and the simulation have led to two representations of liquid distributions: the puddle model [7] or the plugs model [8]. Our results could be well explained in the frame of a puddle model. However more investigations are necessary in order to fully elucidate the liquid morphology under saturated conditions.

**Acknowledgments**
Financial support by the Romanian CNCSIS and European Social Fund (project POSDRU/6/1.5/S/5) is gratefully acknowledged.

**References**

[1] Kimmich R 1997 *NMR Tomography, Diffusometry, Relaxometry* (Berlin: Springer)
[2] D'Orazio F, Bhattacharja S and Halperin W P 1990 *Phys. Rev.* **B42** 9810
[3] Farrher G, Ardelean I and Kimmich R 2007 *Magn. Reson. Imag.* **25** 453
[4] Valiullin R, Naumov S, Galvosas P, Kärger J, Woo H J, Porcheron F and Monson PA 2006 *Nature* **443** 965
[5] Provencher S W 1982 *Comp. Phys. Comm.* **27** 229
[6] Casieri C, De Luca F and Fantazzini P 2005 *J. Appl. Phys.* **97** 043901
[7] Allen S G, Stephenson P C L and Strange J H 1998 *J. Chem. Phys.* **108** 8195
[8] Troyer W E, Holly R, Peemoeller H and Pintar M M 2005 *Solid State Nucl. Magn. Reson.* **28** 238.

Journal of Physics: Conference Series **182** (2009) 012013 doi:10.1088/1742-6596/182/1/012013

# NMR Spectroscopic characterization of β-cyclodextrin inclusion complex with vanillin

**Adrian Pîrnău, Mircea Bogdan and Călin G Floare**

National Institute for Research and Development of Isotopic and Molecular Technologies, 65-103 Donath, 400293 Cluj-Napoca, Romania

E-mail: adrian.pirnau@itim-cj.ro

**Abstract.** The inclusion of vanillin by β-cyclodextrin was investigated by $^1$H NMR. The continuous variation technique was used to evidence the formation of soluble 1:1 complex in aqueous solution. The association constant of vanillin with β-cyclodextrin has been obtained at 298 K by fitting the experimental chemical shifts differences, $\Delta\delta_{obs} = \delta_{free} - \delta_{obs}$ of the observed guest and host protons, with a non-linear regression method. Besides the effective association constant, the fitting procedure allows a precise determination of all chemical shift parameters characterizing the pure complex. They can by used for an analysis of the geometry of the molecular complex in solution.

## 1. Introduction

Vanillin, on 4-hydroxy-3-methoxybenzaldehide is the primary component of the extract of the vanilla bean. Natural "vanilla extract" is a mixture of several hundred different compounds in addition to vanillin. Artificial vanilla is a solution of pure vanillin, usually of synthetic origin. Because of the scarcity and expense of natural vanilla extract, there has long been interest in the synthetic preparation of its predominant component. Today, most artificial vanillin is produced from the petrochemical raw material guaiacol. The largest use of vanillin is as flavouring, usually in sweet foods. The ice cream and chocolate industries together comprises 75% of the market of vanillin as flavouring. Vanillin displays antioxidant and antimicrobial properties and may be used as a food preservative. It is also used in the fragrance industry, in perfumes, and the mask unpleasant odours or tastes of drugs. Vanillin has been encapsulated in oxidized starches and in β-cyclodextrin in order to enhance the protection towards oxidation.

Cyclodextrins (CDs) are cyclic oligosaccharides containing six (α-CD), seven (β-CD) or eight (γ-CD) α-1,4-linked glucopyranose units with a hydrophilic hydroxyl group on their outer surface and a hydrophobic cavity in the center. The hydrophilic exterior of the CD molecules can make them water soluble, but the hydrophobic cavity provides an environment for appropriate sized non-polar molecules. In aqueous solution CDs are capable of forming inclusion complex with many molecules by taking up a whole molecule or some part or it, into the cavity. These non-covalent complexes offer a variety of physicochemical advantages over uncomplexed molecules including increased water solubility and stability. In this paper, we report a $^1$H NMR study of the inclusion complex formed between vanillin (figure 1a) and β-cyclodextrin (figure 1b), in aqueous medium. Analysis of our data by the continuous variation method confirms that the inclusion occurs and the complex has 1:1 stoichiometry. The association constant was calculated by a non-linear least squares regression analysis of the observed chemical shift changes of the vanillin and β-CD protons as a function of β-CD concentration.

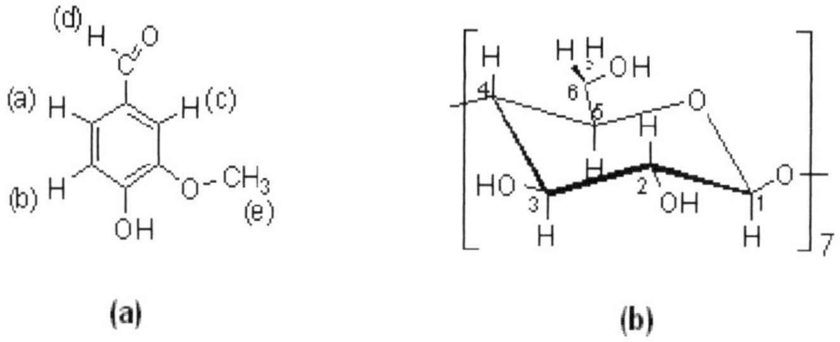


**Figure 1.** Chemical structure of (a) vanillin and (b) β-CD.

## 2. Experimental

### *2.1. Materials*
β-cyclodextrin (water content 8 mol/mol) was purchased from Sigma Chemie GmbH Germany and vanillin from Alfa Aesar GmbH&CoKG, Germany. Both chemicals were used without any further purification. Deuterium oxide (99.7% D) was obtained from was obtained from Heavy Water Plant Romag - Prod, Romania.

### *2.2. Apparatus*
The $^1$H NMR experiments were performed at 500 MHz on a Bruker Avance III, spectrometer. The NMR spectra were recorded in $D_2O$ solution at 298 K and all chemical shifts were measured relative to TMS. Typical conditions were as follows: 32 K data points, sweep width 3500 Hz giving a digital resolution of 0.1 Hz/point. The $90^0$ pulse width was 10 μs and the spectra were collected by co-addition of 16 to 32 scans.

### *2.3. Procedure*
In order to study the complexation process between vanillin and β-CD in solution, two stock solutions in $D_2O$, both having 10 mM were prepared. Based on these two equimolar solutions, a series of nine samples (i = 1 - 9) containing both the vanillin and β-CD molecules were prepared. This was accomplished by mixing the two solutions to constant volume at varying proportions, so that a complete range ($0 < r < 1$) of the ratio $r = [X] / ([H] + [G])$ was sampled. X = G or H and [H] and [G] are the total concentrations of the host (β-CD) and guest (vanillin), respectively. Thus the total concentration $[H] + [G] = [M] = 10$ mM was kept constant for each solution. The same set of samples was used both for the determination of stoichiometry and association constant.

## 3. Results and discussion
NMR is a technique which provides the most evidence for the inclusion of a guest molecule into the hydrophobic CD cavity in solution. The inclusion of vanillin in β-CD is shown by the change in the chemical shift of some guest vanillin and host (β-CD) protons in the complex, in comparison with the chemical shifts of the same protons in the free components. Partial $^1$H NMR spectra of pure components and vanillin: β-CD mixture in a 1:1 molar ratio is presented in figures 2 and 3.

The absence of new peaks that could be assigned to the complex suggested that complexation is a dynamic process, the included vanillin being in a fast exchange between the free and bound states.

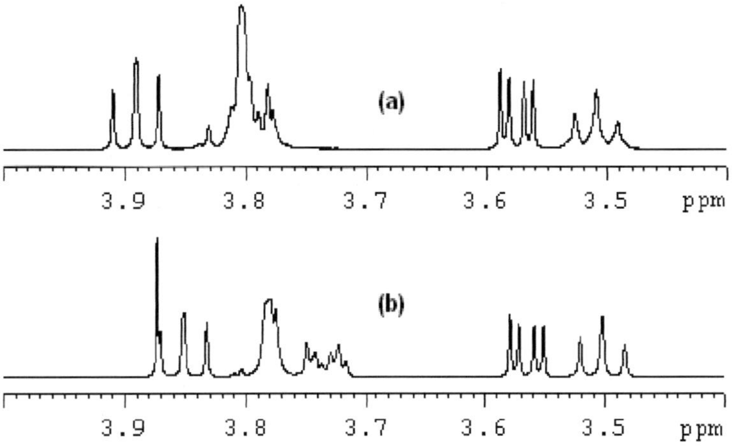


**Figure 2.** Partial $^{1}$H NMR spectrum of (a) 10 mM β-CD and (b) 5 mM β-CD and 5 mM of vanillin.

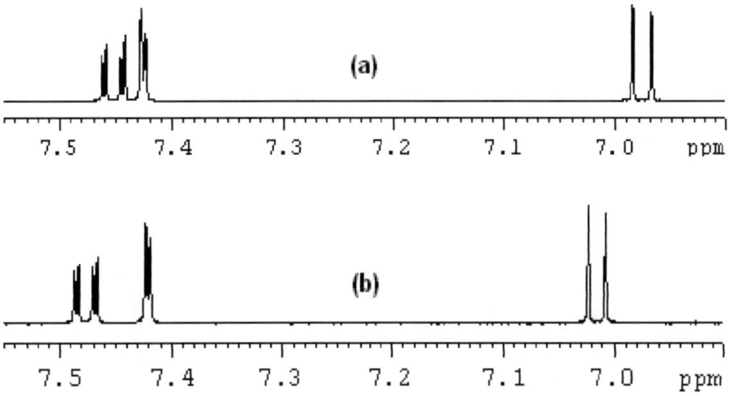


**Figure 3.** Partial $^{1}$H NMR spectrum of (a) 10 mM vanillin and (b) 5mM vanillin and 5 mM β-CD.

### *3.1. Determination of the stoichiometry*

The stoichiometry of the complex must be calculated before proceeding with any association constant calculations. Determination of the stoichiometry of the vanillin-β-CD complex, by the continuous variation method was based on $^{1}$H NMR spectra obtained for vanillin and β-CD mixtures in which the initial concentrations of the two species were maintained constant and that ratio r varied between 0 and 1 (see section *2.3. Procedure*). A physical parameter directly related to the concentration of the complex (in our case the chemical shift, δ) is measured under these conditions and is plotted as a function of r. The maximum value for this parameter well occur at $r = n / (m + n)$, where m and n are, respectively, the proportions of vanillin and β-CD in the complex $(\text{vanillin})_m$:$(\beta\text{-CD})_n$. The calculated quantities Δδ[βCD] are proportional to the concentration of the complex [1], and can be plotted against r. The resulting continuous variation plots demonstrate that because r has a maximum value of 0.5 and highly symmetrical shapes (figure 4), the complex has 1:1 stoichiometry.

The induced shift, $\Delta\delta$ is defined as the diference in chemical shifts in the absence and in the presence of the other reactant for a given ratio r.

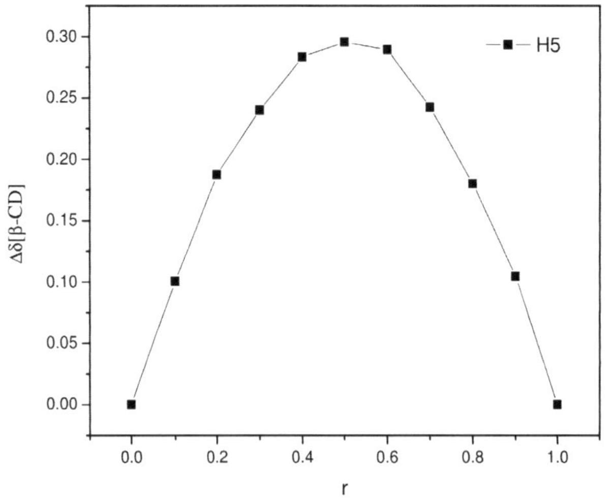


**Figure 4.** Concentrations variation plot for H-5 proton of β-CD.

### *3.2. Determination of the association constant*

In order to determine the extent of the intermolecular binding between vanillin and β-CD, the association constant has been evaluated. The association constant, K for an 1:1 complex can be determined according to the following equation [2]:

$$\Delta\delta^{(i,j)} = \frac{\Delta\delta_c^{(j)}}{2[X]}\left\{[M]+\frac{1}{K}-\left[\left([M]+\frac{1}{K}\right)^2-4[H]^{(i)}[G]^{(i)}\right]^{1/2}\right\} \tag{1}$$

where *i* counts the sample number and j the studied proton. If the studied proton belongs to the guest or host molecule, X = G or H respectively. $\Delta\delta_c^{(i)}$ represents the chemical shift deference (for a given proton) between the free component and the pure inclusion complex. Equation (1) involves no approximations and correlates the total concentrations of the guest and host molecules with the observed deference in the chemical shift:

$$\Delta\delta^{(i,j)} = \delta_{free}^{(i,j)} - \delta_{obs}^{(i,j)} \tag{2}$$

We used a computer programme [3] based on an iteration procedure following specific algorithms in order to fit the experimental values of $\Delta\delta^{(i,j)}$ to the appropriate equation. Each iteration sets up a quadratic programme to determine the direction of search and the loss function,

$$E = \sum_{i,j}(\Delta\delta^{(i,j)} - \Delta\delta_{calc}^{(i,j)})^2 \tag{3}$$

until the search converges.

The treatment of the whole set of protons studied yields one single K value characterizing the inclusion process and a set of calculated $\Delta\delta_c^{(i,j)}$ values. The programme is quite flexible since it allows observing the chemical shift variation of the host, guest or both molecules as a function of variable guest or host concentrations.

In our case, we applied equation (1) for a set of protons consisting in H3 and H5 of β-CD and Ha, Hb and He of vanillin.

The association constant obtained using the above described procedure is K = 170.2 $M^{-1}$ with E = $8.29 \cdot 10^{-5}$ and a correlation factor r = 0.9995.

The complete set of chemical shifts in the free state and in pure complex is reported in table 1.

**Table 1.** Chemical shifts of the proton of vanillin and β-CD in the free and complexed states.

| Proton | $\delta_{free}$ (ppm) | *$\delta_c$ (ppm) |
|---|---|---|
| H3 | 3.8907 | 3.7774 |
| H5 | 3.7819 | 3.6125 |
| Ha | 7.4522 | 7.5179 |
| Hb | 6.9767 | 7.0817 |
| He | 3.8459 | 3.9215 |

*$\delta_c = \delta_{free} - \Delta\delta_c$

## 4. Conclusion

The vanillin:β-CD system in aqueous solution has been studies by $^1$H NMR. Analysis of our data by the continuous variation method indicates that the inclusion occurs and the complex has 1:1 stoichiometry.

The complexation-induced chemical shift of H3 and H5 protons of β-CD are those expected as a result of interaction with those protons of the vanillin (Ha, Hb and He) which are oriented towards the β-cyclodextrin cavity, thus confirming a vanillin / β-CD interaction.

The association constant K for the inclusion complex was evaluated from the observed difference in chemical shifts for vanillin and β-CD protons.

## References

[1] Loukas Y L 1997 *J. Pharm. Pharmacol.* **49** 944
[2] Bogdan M, Caira M R and Farcas S I 2002 *Supramol. Chem.* **14** 427
[3] Floare C, Balibanu F and Bogdan M 2005 *Studia Univ. Babes-Bolyai Physica* **L4a** 451.

Journal of Physics: Conference Series **182** (2009) 012014 doi:10.1088/1742-6596/182/1/012014

# Preparation of 1D nanostructures using biomolecules

**Stela Pruneanu[1], Liliana Olenic[1], Lucian Barbu Tudoran[2], Irina Kacso[1], Said A Farha Al-Said[3], Reda Hassanien[3], Andrew Houlton[3] and Benjamin R Horrocks[3]**

[1] National Institute for Research and Development of Isotopic and Molecular Technologies, 65-103 Donath, 400293 Cluj-Napoca, Romania

[2] Babes-Bolyai University, Electron Microscopy Center, 1 Mihail Kogălniceanu, 400006 Cluj-Napoca, Romania

[3] School of Chemistry, Bedson Building, Newcastle University, Newcastle upon Tyne, NE1 7RU, UK

E-mail: stela.pruneanu@itim-cj.ro

**Abstract**. In this paper we have shown that one-dimensional (1D) particle arrays can be obtained using biomolecules, like DNA or amino-acids. Nano-arrays of silver and gold were prepared in a single-step synthesis, by exploiting the binding abilities of λ-DNA and L-Arginine. The morphology and optical properties of these nanostructures were investigated using AFM, TEM and UV-Vis absorption spectroscopy.

## 1. Introduction

The controlled organization of metallic nanoparticles into one-dimensional (1D) arrays has attracted an increased interest due to their possible application in nanoelectronics, optoelectronics and biosensor devices [1]. Template synthesis (*e.g.* filling the pores of an alumina membrane with the desired material) has proved to be one of the most successful methods which can provide anisotropic nanostructures, such as nanorods and nanowires [2]. However, due to the inherent difficulties related with the selection and usage of a single nanowire, other methods have been developed, by employing supramolecular templates like DNA, peptide nanofibrils or phospholipids [3-5].

Besides these methods, particular attention has been given to preparation of 'template-free' 1D arrays. This new approach is based on coupling gold nanoparticles by using amino-acids or by partial removal of the organic stabilization layer which increases dipole-dipole interactions, leading to anisotropic orientation of nanoparticles [6].

In this work we report the preparation of extensive 1D arrays of metallic nanoparticles: (*i*) by using DNA as a template and (*ii*) by exploiting the binding ability of amino-acids (L-Arginine). The evolution in time of morphological and optical properties of metallic nanoparticles attached to biomolecules, were investigated using AFM, TEM and UV-Vis absorption spectroscopy.

## 2. Experimental

λ-DNA stock solution (500 µg/mL in 10 mM Tris-HCl pH 8 + 1 mM EDTA) was purchased from New England Biolabs, United Kingdom. Chloroauric acid ($HAuCl_4$), sodium citrate, sodium borohydride and L-arginine were obtained from Aldrich Chemicals and used as received. Colloidal gold (GNP) was prepared by reducing $HAuCl_4$ with sodium citrate and $NaBH_4$. Arginine-capped gold nanoparticles (Arg-GNP) were obtained by reducing $HAuCl_4$ with $NaBH_4$, in the presence of L-Arginine (L-Arg). Preparation of DNA templated silver nanostructures (Ag-DNA) is given in reference [7].

### *2.1. UV-Vis Spectroscopy*

Optical properties of Ag-DNA solution, colloidal gold and arginine-capped gold nanoparticles were monitored using an UV-Vis spectrophotometer (JASCO V-570).

### *2.2. TEM measurements*

For TEM analysis drops of colloidal gold or arginine-capped gold nanoparticles were placed on carbon-coated copper grids. The grids were allowed to dry prior to TEM measurements (JEOL-JEM 1010 instrument).

### *2.3. AFM*

AFM imaging was performed in air on a Dimension Nanoscope V system (Veeco Inc., Metrology group) using Veeco NanoProbe tips. All of the AFM images are height images taken in tapping mode.

## 3. Results and discussion

### *3.1. Silver 1D nanostructures templated on DNA*

DNA's rich functionality (*e.g.* negatively charged phosphate groups and nucleobases) allows it to interact with different metals, forming 1D particle arrays or continuous wires (depending on the reaction time). We have investigated the interaction between silver and DNA because the high conductivity of silver makes it ideal for integration in electrical circuits. The templating reaction is efficient and after 24 hours, silver nano crystals can be visualized strongly attached to DNA. Figures 1 a,b shows typical examples of such 1D nanostructures (beads-on-a string appearance) along with randomly dispersed silver nanocrystals, on the $SiO_2$/Si surface. Some strands show DNA regions smoothly covered by silver (figure 1a) while others are densely packed with nanoparticles (figure 1b).

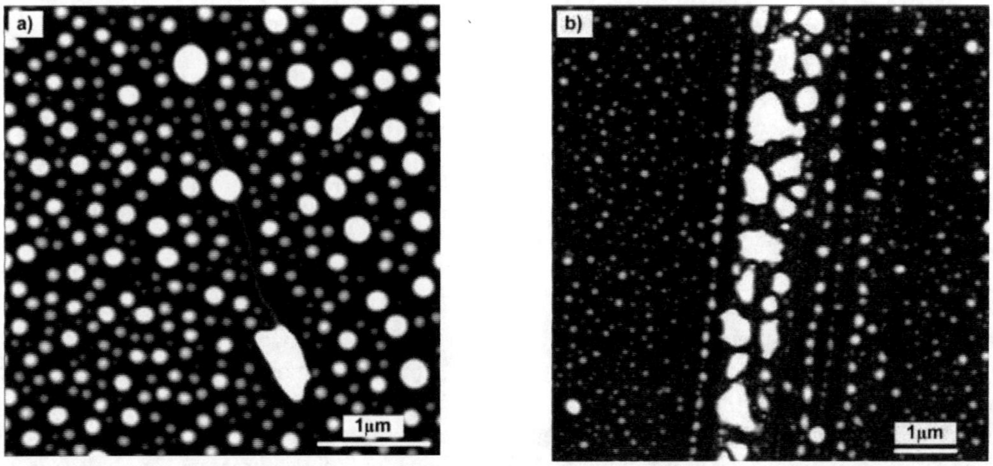


**Figure 1.** a, b: AFM images of silver crystals arranged as 1D nanostructures along DNA strands; the grayscale corresponds to a height range of 30 respectively 35 nm.

The mean diameter of silver nanoparticles attached to DNA is around 50 nm (measured as height by AFM tip). Besides these, one can notice randomly dispersed nanoparticles on $SiO_2/Si$ surface, heaving a mean diameter of 20 nm (figure 1a) respectively 10 nm (figure 1b). It is important to emphasize that nanoparticles attached to DNA are bigger than those randomly dispersed in solution, which can suggests that the reduction of silver ions is catalyzed by DNA.

In order to better understand the interaction between silver and DNA, we have also monitored the evolution of the Ag-DNA mixture by UV-Vis spectroscopy. Metallic nanostructures (nanoparticles, nanorods) can exhibit strong local resonances of light-induced electron plasma oscillations [8]. The electric field of an incoming light wave induces a polarization of the free electrons of metallic nanoparticle, with respect to its ionic core. A charge difference is created at the nanoparticle surface, which in turn acts as a restoring force (Coulomb attraction). In this way, a dipolar oscillation of the free electrons results, whose frequency is determined by the electron density, effective mass and particle shape. This electronic motion is specific for nanoparticles due to the geometrical confinement effects of free electrons. The resonance of the free electronic cloud with the incident light wave generates an absorption band, also known as surface plasmon resonance band (SPB). This band is observed in the UV-visible spectrum and its position is characteristic for each metal (silver, gold). This is the reason which supports the study of silver-DNA interaction, by UV-Vis spectroscopy (figure 2).

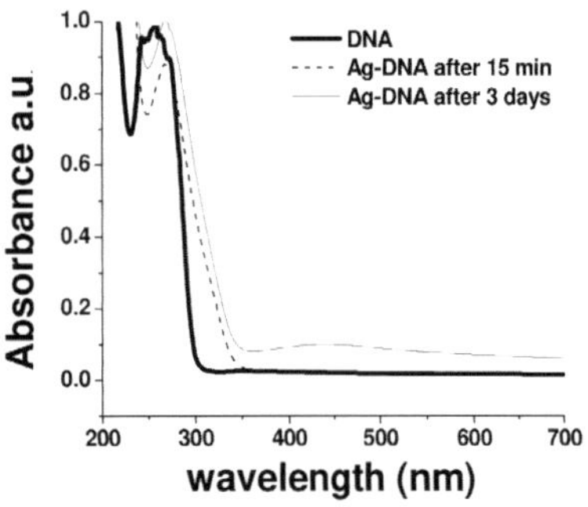


**Figure 2.** Evolution in time of DNA solution after mixing with silver' reagent.

The first spectrum is that of stock DNA solution which exhibits the characteristic nucleic acid absorption band at 260 nm. The next spectrum was recorded after incubating DNA with silver' reagent, for 15 minutes. A decrease of the 260 nm absorption peak was detected together with a shift towards higher wavelengths (271 nm). The evolution of the mixture was then closely monitored, but no further change was observed within 2 days. The spectrum recorded on the third day revealed a new maximum developed around 440 nm, due to surface plasmon bands of silver nanoparticles. The strong absorption band below 300 nm can be attributed to non-metallic silver clusters ($Ag_4^{2+}$) formed in the initial stage of silver reduction. This type of clusters can be stabilized by sodium dodecyl sulphate [9] or sodium poly(phosphate) [10]. In our case, the phosphate groups on DNA strands can act as stabilizing group, which explains the increase in band intensity with time. The transition from non-metallic to metallic particle is confirmed by the appearance of the 440 nm surface plasmon band. This takes place due to agglomeration of silver clusters which generates nanometer-sized metal particles.

### *3.2. One-dimensional nanostructures prepared from arginine-capped gold nanoparticles*

Alternatively, 1D metallic nanostructures can be obtained in the absence of a template, by exploiting the binding ability of amino-acids (L-Arg). In order to show the influence of L-Arg on nanoparticles morphology, we have also prepared citrate-capped gold nanoparticles. Figure 3a shows a typical TEM image of citrate-capped gold nanoparticles, one hour after preparation. One can notice the sphere shape of particles as well as the uniform distribution of diameters (mean diameter 4 nm).

In contrast with this, arginine-capped gold nanoparticles agglomerate into larger spherical or elliptical aggregates (diameter around 20 nm, figure 3b). Within 3 hours since the reducing agent was added to the solution, the appearance of Arg-GNP has radically changed and nanoparticles are no longer forming spherical aggregates. They have connected with each other, forming nanochains of various lengths, from tens to hundreds of nm (figure 4 a,b). In particular, figure 4b shows that the gold nanoparticles are well fused into each other (the length of the branch is around 450 nm). With time, the chain like nanostructures increases in length and develops additional branches and closed loops. These 1D structures remains stable in solution for weeks, although they can be destroyed by ultrasound.

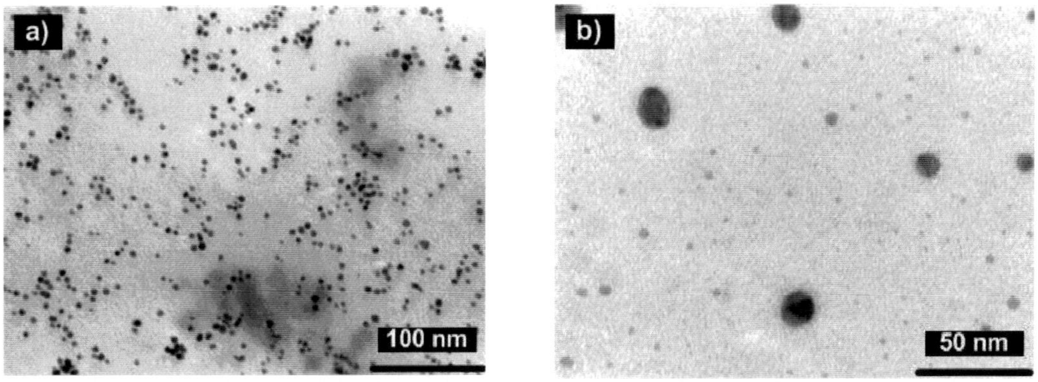


**Figure 3.** TEM images: citrate-capped gold nanoparticles (a); Arg-GNP immediately after preparation (b).

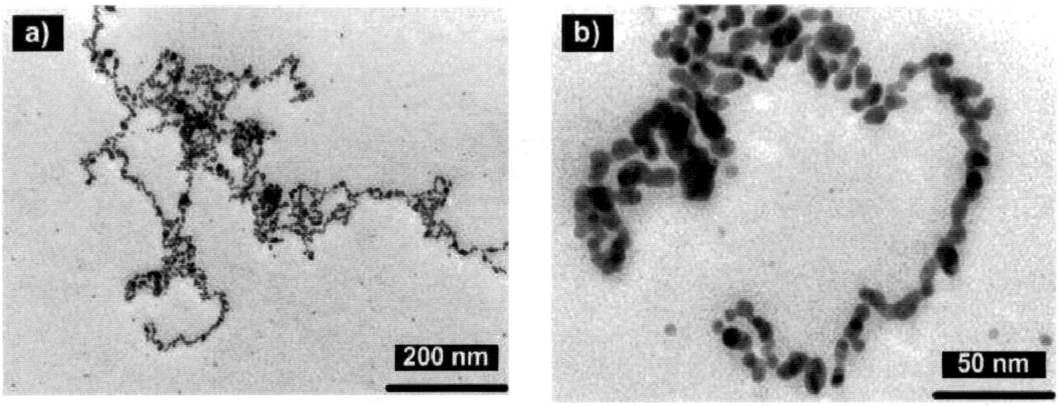


**Figure 4.** a, b: TEM images of Arg-GNP, 3 hours after preparation.

The evolution of UV-Vis spectra (figure 5) is in good agreement with TEM images, confirming the change in time of nanoparticles morphology. This is clearly shown by the change in surface plasmon resonance band, which depends on the dielectric properties of metal and surrounding medium as well as on the dimension of nanoparticles [11]. For a small sphere, Mie theory shows that the light absorption is given by a single SPB, which is associated with a dipolar charge distribution on the surface (transverse surface plasmon resonance) [12]. As the particles become less symmetric (*e.g.* by assembling into nanowires) the charge distribution induced on the surface can result in dipolar modes with different resonant frequencies. Consequently, an additional SPB band is developed, generally red shifted (longitudinal surface plasmon resonance).

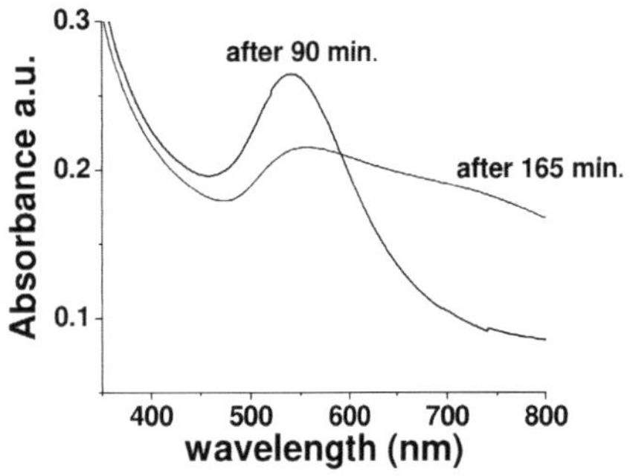


**Figure 5.** Evolution in time of UV-Vis spectra for arginine-capped gold nanoparticles.

In the case of arginine-capped gold nanoparticles, the transition from dispersed nanoparticles to one-dimensional nanochains is clearly evidenced by the absorption spectra. After preparation (90 min.) the spectrum of Arg-GNP is that characteristic for dispersed nano spheres with a sharp peak at 530 nm (see also the TEM image, fig.3b). This is assigned to the dipole resonance, which gives rise to the transverse surface plasmon band. Within 3 hours, the plasmon peak broadened and decreased in intensity. A new broad peak developed around 750 nm, due to the linear arrangement of nanoparticles, which generates the longitudinal surface plasmon band. The linear self-assembly of nanoparticles is favored by arginine binding ability, which interacts not only with gold but also with adjacent molecules, forming 1D nanostructures.

## 4. Conclusions

In conclusion we have shown that 1D nanoparticles arrays can be obtained either by using DNA as a template or by spontaneously assembly of arginine-capped colloidal gold. AFM, TEM and UV-Vis spectroscopy have confirmed the transition from dispersed nanoparticles to linear assembled nanostructures (silver and gold).

## References

[1] Tang Z Y and Kotov N A 2005 *Adv. Mater* **17** 951
[2] Nagle L, Ryan D, Cobbe S and Fitzmaurice D 2003 *Nano Lett.* **3** 51
[3] Braun E, Eichen Y, Sivan U and Ben-Yoseph G 1998 *Nature* **391** 775
[4] Djalali R, Chen Y and Matsui H 2002 *J. Am. Chem. Soc.* **124** 13660
[5] Zhang D B, Qi L M, Ma J M and Cheng H M 2001 *Chem. Mater.* **13** 2753
[6] Lin S, Li M, Dujardin E, Girard C and Mann S 2005 *Adv. Mater.* **17** 2553
[7] Farha Al-Said S A, Hassanien R, Hannant J, Galindo M A, Pruneanu S, Pike A R, Houlton A and Horrocks B R 2009 *Electrochem. Commun.* **11** 550
[8] Gonzalez A L and Noguez C 2007 *Phys. Stat. Sol.* (c) 4 **11** 4118
[9] Henglein A 1979 *J. Phys. Chem.* **83** 2209
[10] Henglein A 1989 *Chem. Phys. Lett.* **154** 473
[11] Moores A and Goettmann F 2006 *New J. Chem.* **30** 1121
[12] Mie G 1908 *Ann. Phys. Leipz.* **25** 377.

Journal of Physics: Conference Series **182** (2009) 012015 doi:10.1088/1742-6596/182/1/012015

# Atomic force microscopy study of nanocrystalline ceria thin films

**T Ristoiu, T Petrisor Jr, M S Gabor, M Nasui, B Mos, L Ciontea and T Petrisor**

Materials Science Laboratory, Technical University of Cluj-Napoca, 15 C Daicoviciu, 400020 Cluj-Napoca, Romania

E-mail: tristoiu@phys.utcluj.ro

**Abstract**. This work reports on the structural and morphological characterization of ceria thin films grown by chemical solution deposition (CSD) techniques. The coating solution was prepared starting from cerium 2,4-pentadionate, $Ce(CH_3COCHCOCH_3)_3$, and propionic acid, $C_2H_5COOH$, further concentrated by distillation. The as-prepared solution was spin coated on single crystalline (100) $SrTiO_3$ (STO) substrates. The precursor films were heat treated in air at 900°C both by a rapid heating and quenching to room temperature (A) and annealing (B). The X-ray structural analysis and the SEM and AFM morphological analysis for the $CeO_2$/STO thin film have revealed the influence of the thermal treatment on the growth mechanism. The as-obtained polycrystalline nanostructured ceria thin films present an average grain size of 40-50 nm and a roughness of 1.5-5.4 nm, depending on the thermal treatment. The annealing renders the ceria thin film a smoother surface with respect to the rapid heat treatment followed by quenching, performed both at the same temperature.

## 1. Introduction

Cerium oxide thin films have a large range of applications in electronics, optics, catalysis, wear resistance, corrosion protection and superconductivity as buffer layer in the coated conductor architecture [1]. Lately, doped ceria is being extensively studied as a potential solid electrolyte for fuel cells [2].

Chemical solution deposition (CSD) has emerged as a highly attractive, non-vacuum method for producing oxide thin films. Metal 2,4-pentadionate (β-diketonates, acetylacetonates), due to their high volatility, low decomposition temperature, ease of use, commercial availability and relatively low cost have become adequate precursors for the chemical solution deposition of metal oxides thin films. The metal acetylacetonates easily dissolve in propionic acid up to high concentrations. The structure and the morphology of the thin films is strongly influenced by the thermal treatment performed in order to transform the precursor thin film into a crystalline one, the growth mechanism being still under investigation [3].

In this work we report on the fabrication and characterization of cerium oxide thin films deposited on single crystalline (100)STO substrates starting from cerium 2,4-pentadionate and propionic acid. The influence of the thermal treatment on the structural properties and surface morphology of the film were investigated by X-ray diffraction (XRD), scanning electron microscopy (SEM) and atomic force microscopy (AFM), respectively.

Journal of Physics: Conference Series **182** (2009) 012015 doi:10.1088/1742-6596/182/1/012015

## 2. Experimental

The cerium acetylacetonate, $Ce(C_5H_7O_2)_3.xH_2O$ and the propionic acid, $CH_3CH_2COOH$ with a purity of 99% used in this work have been purchased from Alfa Aesar. The coating solution has been prepared by dissolving the $Ce(C_5H_7O_2)_3.xH_2O$ in an excess of propionic acid under stirring to form a clear yellow solution. The as-obtained solution was concentrated by distillation (43 mbarr and 75°C bath temperature) resulting in a change of colour to red-brown.

The thin films were deposited by spinning on 10x10 mm$^2$ (100)STO substrates at 4000 rpm for 60s. Prior to the deposition, the substrates were ultrasonically cleaned with isopropanol.

The precursor films have been thermally treated in air at 900°C for 15 minutes by: (A) introducing the film directly in the heated furnace at 900°C, followed by quenching down to room temperature-named quenching hereafter, and (B) by annealing, with a heating rate of 10°C/min. These conditions have been proposed in agreement with the TG-DTA analyses of the precursor powdered solution [4]. The structural characterization of the thin films was performed using a X-ray Bruker D8 Discover diffractometer with $CuK_\alpha$ radiation, equipped with a graphite monochromator on the diffracted beam to suppress the $K_\beta$ component and the parasitic scattering. The morphology was investigated by Scanning Electron Microscopy (SEM) using a LEO 1525 field emission-high resolution scanning electron microscope and a Veeco D3100 AFM in contact mode with a SNL-10 Veeco contact tip, having a nominal tip radius of 2 nm. The AFM images were analyzed using the WSxM software [5].

## 3. Results and discussion

In order to study the influence of the heat treatment on the growth behavior of the ceria thin films, the thin films were subsequently characterized by X-ray θ-2θ scans. The indexing of the reflections was based on the fluorite lattice cell of $CeO_2$ with a lattice parameter of $a_{CeO2}$ = 5.4Å. Figure 1, registered between 27-34°, significant for the main reflections, shows that both ceria films are polycrystalline, but no secondary phases have been identified. Thus, beside the (111) reflection at 2θ = 28.58°, the (200) reflection at 2θ = 33.02° is present in an almost equal intensity in the annealed film (B), while in the quenched film (A) it is undetectable. This suggests that the annealed film is partially [100](100) textured.

The SEM analysis of the thin films have revealed homogeneous, continuous, crack free $CeO_2$ surfaces for both thermal treatments, as shown in figure 2. It can be seen that they have very fine grains. Nevertheless, the quenched film (A) contains more dark zones, which can be attributed to bigger gaps between the grains.

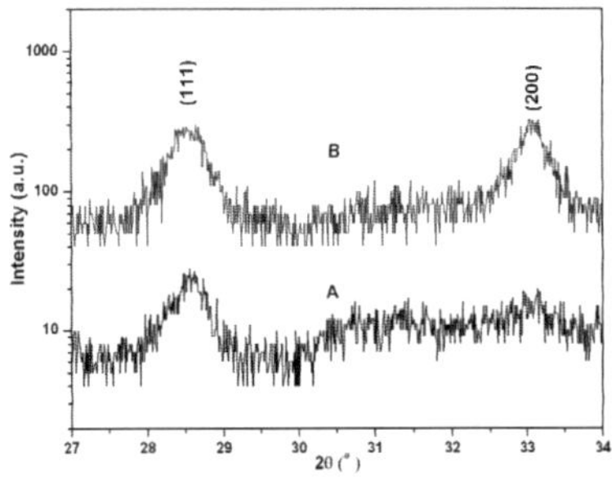


**Figure 1.** X-ray diffraction pattern of the quenched film (A) and of the annealed film (B).

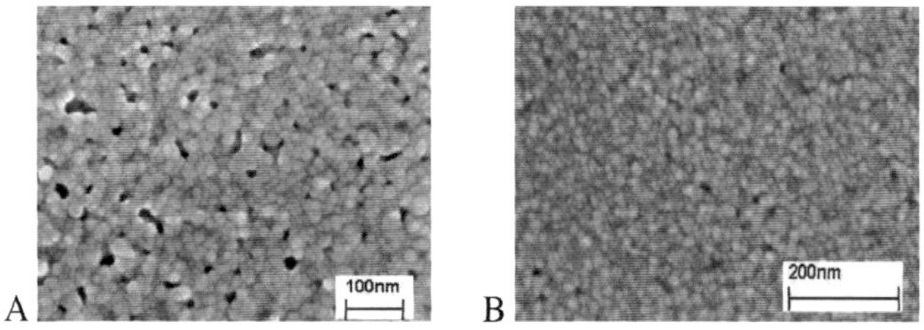


**Figure 2.** SEM images (200.00 kx) of the quenched film (A) and of the annealed film (B).

From the AFM images, figure 3 and figure 4, of the as-grown $CeO_2$ films it is found that the surface has a granular structure, similar to the one presented in the SEM analyses. However, the crystallites are larger in size in the AFM images due to tip convolution. For the same reason, the narrow gaps between the grains, visible in the SEM images, appear to have less sharper borders. The root mean square roughness (RMS) of the ceria thin films, maximum value and average height, calculated on an area of 1 µm x 1 µm and 10 µm x 10 µm, respectively, are synthetically presented in table 1. The values indicate that the annealed thin films are smoother than the quenched ones. The grain size estimated from the accompanying line profiles are 35 nm for the annealed film (B) and 45 nm for the quenched film (A) and have a more uniform distribution, as shown in the topographic representation, figure 3. The observed surface grain structure did not allow any conclusions to be drawn concerning the structure within the volume of the film.

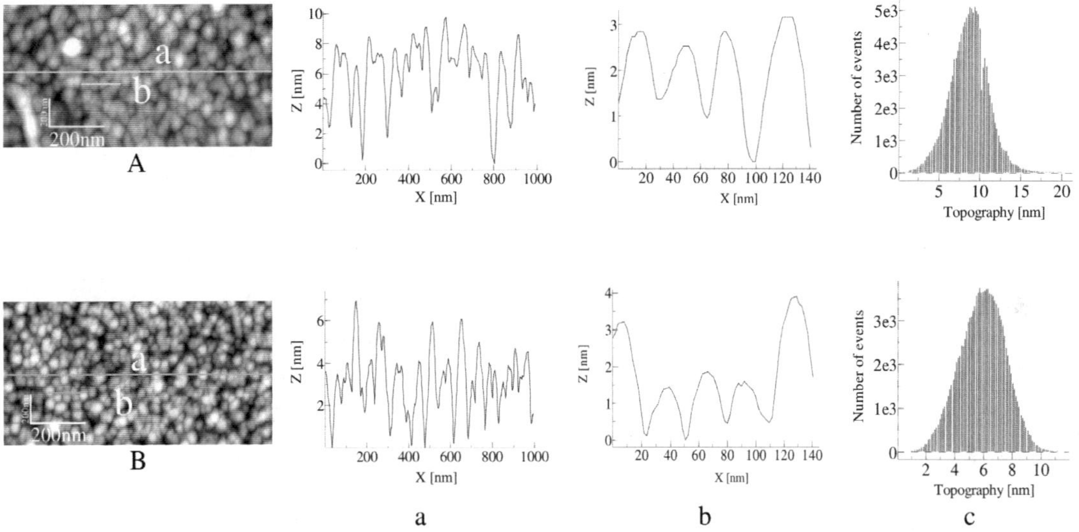


**Figure 3.** 2D AFM images, the accompanying line profiles (a, b) and the topographic profile (c) of the quenched film (A) and of the annealed film (B) for 1 µm x 1 µm area.

**Table 1.** Principal data obtained from the AFM analysis.

| Area | RMS | | Maximum value | | Average height | |
|---|---|---|---|---|---|---|
| | A | B | A | B | A | B |
| 1 µm x 1 µm | 2.3 nm | 1.5 nm | 20.9 nm | 11.5 nm | 8.8 nm | 5.9 nm |
| 10 µm x 10 µm | 5.4 nm | 3.7 nm | 73.5 nm | 98.6 nm | 19.9 nm | 10.4 nm |

The 3D representation, figure 4, reveals the slight wavy nature of the film surface due to thermal stress induced by the rapid heat treatment (A).

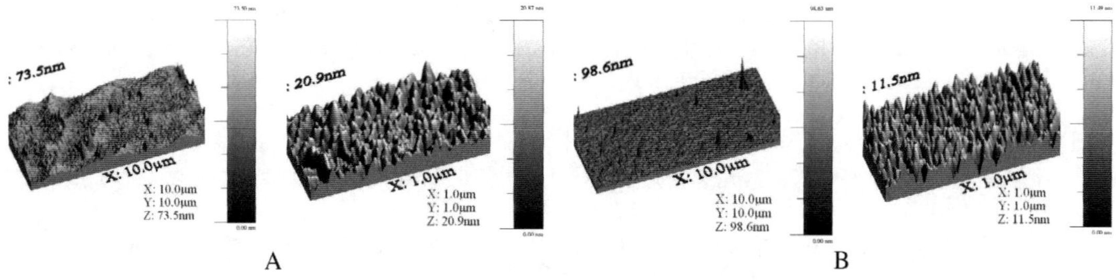


**Figure 4.** 3D AFM images, for different surface areas, of the quenched film (A) and the annealed film (B).

The interpretation of the experimental results can be done taking into account the two types of thermal treatments: a one step and two step crystallization processes. In the case of a one step process the pyrolysis and crystallization are concurrently and take place simultaneously at an upper temperature; this corresponds to the quenching thermal treatment (A). Due to the fast heating, the decomposition of the precursor is not complete, although the temperatures are sufficiently high to initiate crystallization. This contributes to grain growth and results in a larger average grain size film. In the two step process, pyrolysis and crystallization take place successively: pyrolysis corresponds to the burn-off of the organic moieties, resulting in an amorphous film, and crystallization which takes place at a higher temperature. In our case, since the heating rate is relatively low (10°C/min) and the temperature range for the precursor powder decomposition being 300-500°C [4], the annealing of the film (B) can be treated as a two step process, in spite of the lack of an intermediate thermal treatment.

From a thermodynamic point of view it has been demonstrated that the driving force that governs the transformation from the amorphous film to into a crystalline oxide film can play a significant role in defining the active nucleation events, and thereby the film microstructure [1]. It is known that the crystal growth from solution derived films is more complicated and difficult than in vacuum deposition techniques, because the nuclei form not only at the interface between the film and the substrate, but also at the film surface and/or on impurity particles in the film [6]. In our case, in spite of the fact that the films are not epitaxially grown, we assume that during the quenching process (A) the nucleation takes place in the film volume, while during the annealing process (B) the nucleation takes place rather at the interface between the film and the substrate. This assumption is sustained by the oriented character of the annealed films and the smaller roughness with respect to the quenched ones.

## 4. Conclusions

Ceria thin films have been prepared by CSD using cerium 2,4-pentadionate and propionic acid as reactants. The structural and morphological analyses have demonstrated that the annealing thermal treatment at 900°C is more adequate for the growth of nanocrystalline, smooth, quasi-oriented ceria thin films. The film growth has been interpreted in terms of volume and interface nucleation for the applied thermal treatments. In case of volume nucleation the film presents larger grain size and higher surface roughness, while the interface nucleated film exhibits smaller grain sizes, lower roughness and a [100](100) partial texture. The results of this study are promising for the use of ceria as buffer layers in the coated conductor architecture. Higher temperatures are necessary for the epitaxial growth of ceria thin films.

## Acknowledgments

The authors thank ENEA-Frascati, Italy, Superconductivity Division for the X-ray and SEM analysis. The financial support from PN II under contract number 71-045 was highly appreciated.

## References

[1] Bhuiyan M S, Paranthaman M and Salama K 2006 *Supercond. Sci. Technol.* **19** R1
[2] Chockalingam R, Amarakoon V R W and Giesche H 2008 *J. Eur. Ceram. Soc.* **28** 959
[3] Coll M, Gazquez J, Huhne R, Holzapfel B, Morilla Y, Garcia-Lopez J, Pomar A, Sandiumenge F, Puig T and Obradors X 2009 *J. Mater. Res.* **24** 1446
[4] Ristoiu T, Ciontea L, Petrisor Jr T, Gabor M S, Thalmaier Gy and Petrisor T 2009 submitted to *J. Optoelectron. Adv. Mat.*
[5] Horcas I, Fernández R, Gómez-Rodríguez J M and Colchero J 2007 *Rev. Sci. Instrum.* **78** 013705
[6] Schwartz R W, Schneller T and Waser R 2004 *C. R. Chimie* **7** 433.

# The extraction and chromatographic determination of the essentials oils from *Ocimum basilicum* L. by different techniques

**Maria Loredana Soran[1], Simona Codruta Cobzac[2], Codruta Varodi[1], Ildiko Lung[1], Emanoil Surducan[1] and Vasile Surducan[1]**

[1] National Institute for Research and Development of Isotopic and Molecular Technologies, 65-103 Donath, 400293 Cluj-Napoca, Romania

[2] Babes-Bolyai University, Faculty of Chemistry and Chemical Engineering, 11 Arany Janos, 400028 Cluj-Napoca, Romania

E-mail: loredana.soran@itim-cj.ro

**Abstract**. Three different techniques (maceration, sonication and extraction in microwave field) were used for extraction of essential oils from *Ocimum basilicum* L. The extracts were analyzed by TLC/HPTLC technique and the fingerprint informations were obtained. The GC-FID was used to characterized the extraction efficiency and for identify the terpenic bioactive compounds. The most efficient extraction technique was maceration followed by microwave and ultrasound. The best extraction solvent system was ethyl ether + ethanol (1:1, v/v). The main compounds identified in *Ocimum basilicum* L. extracts were: α and β-pinene (mixture), limonene, citronellol, and geraniol.

## 1. Introduction

Basil (*Ocimum basilicum* L.) is an aromatic herb that is used extensively to add a distinctive aroma and flavor to food. The leaves can be used fresh or dried for use as a spice. Essential oils extracted from fresh leaves and flowers can be used as aroma additives in food, pharmaceuticals, and cosmetics [1]. The interest in medicinal plants and their biologically active derivatives has increased in recent years, in relation to the possible development of novel potential drugs [2]. Traditionally, basil has been used as a medicinal plant in the treatment of headaches, coughs, diarrhea, constipation, warts, worms, and kidney malfunction. Major aroma compounds from volatile extracts of basil present anti-oxidative activity [1].

Any changes of metabolism equilibrium cause the alteration of volatile oil extracts compositions, which might be called 'aroma profile characteristics' by analogy with the fingerprint [3].

Conventional essential oil extraction techniques include maceration (passive extraction) [4], steam distillation (SD) [5, 6], simultaneous distillation extraction (SDE) [7], purge and trap (P&T) [8], static (S-HS) and dynamic headspace [9], and head-space solid-phase microextraction (HS-SPME)[10, 11]. Steam distillation is the routine method recommended by pharmacopoeias for controlling the quality of plant materials as essential oil sources [12]. These classical methods require long extraction time, large amounts of solvents and multiple steps. Moreover, many unstable volatiles compounds would be thermally decomposed and degraded during thermal extraction. Due to the relative simplicity SD and

SDE are still extensively used for essential oils extraction. In recent years, some advanced extraction techniques, such as headspace solvent drop microextraction (HSME) [13], pressurized liquid extraction (PLE) [14], supercritical fluid extraction (SFE) [15], solvent free microwave extraction [16], microwave assisted hydrodistillation extraction and ultrasound-assisted extraction [17] were used.

Among these extraction techniques, high-temperature water extraction of herb like basil is of particular interest because the water extraction is performed around to 100°C and therefore may mimic the cooking process in the kitchen. The extraction in the microwave field has the same practical importance, as well.

The essential oils extracts are analyzed by various chromatographic techniques such as: high - performance liquid chromatography (HPLC) [18, 19] and thin layer chromatography (TLC) [19, 20]. Due to the high volatility of the analytes, the specific technique is GC. The more precisely information in qualitative analysis are obtained by gas-chromatography coupled with mass spectrometry (GC-MS) [21]. For quantitative determination gas-chromatography with flame ionization detector (GC-FID) and GC-MS are preferred [1, 2, 20, 22].

The main goal of our investigations was to evaluate the extraction efficiency of essential oils from basil using various techniques and solvent systems. The chromatographic determinations were performed by TLC and GC-FID. The resulting chromatograms were analyzed. Some essential oils were identified using standard solutions.

## 2. Experimental data

### *2.1. Materials*

The plant material was commercially purchased. The essential oils standards were obtained from Fluka (Germany). The chromatographic plates were from Merck (Germany). All the solvents were from Chimopar (Bucharest, Romania). All chemicals were of analytical grade. Stock solutions were prepared in ethanol at 100 $\mu g\ ml^{-1}$.

### *2.2. Extraction procedure*

The vegetal material of *Ocimum basilicum* L. for culinary purpose was purchase from Kotany, Austria as dried leaves. After grinding with a hand mill (grinder), the powder was exactly weighed in portions of 0.5 g and subjected to solvent extraction with different systems and techniques. Following solvent were chose to perform the extraction: E1 – hexane; E2 – ethyl ether; E3 – ethanol; E4 – hexane + ethyl ether (1:1, v/v); E5 – ethyl ether + ethanol (1:1, v/v) and E6 – hexane + ethyl ether + ethanol (1:1:1, v/v). Each extraction procedure was optimized with respect the principal factors.

*Maceration* (M) was performed 14 days at room temperature with 15 ml extraction solvents E1-E6. After filtration and washing the final volume was adjusted at 25 ml.

*Ultrasound solvent assisted extraction* (UAE) was performed in two steps using a Transsonic T 310 bath at 35 kHz and an installed power of 95 W. In the first step sample was soaked 10 min with 10 ml extraction solvent (E1-E6). After 15 min of sonication the extract was separated (by decantation) and the sample was once again subjected for other 15 min sonication with 10 ml solvent (E1-E6). The sample was finally filtered and the residuum washed. The extracts were reunited and than the final volume were adjusted at 25 ml. For avoiding solvent leaks the extraction temperature was established at 4°C (ice bath) and the extraction vessels were tightly closed.

*Microwave solvent assisted extraction* (MAE) was performed using a home made apparatus [23]. The device has the possibilities to control the operation time, temperature and duty cycle. Sample (0.5g) together with extraction solvent (20 ml) was placed into the extraction cell. In concordance with sonication the extraction procedure consist in two steps, 10 min soaking followed by microwave extraction. Taking in account the specificity of plant material the following parameter were selected: maximum temperature 30°C, action time 1 min and duty coefficient of 40% at an installed power of 900 W. Depending of the absorption capacity of the solvent, the entire extraction time takes more than

1 min because the cell needs to cool down bellow 30°C. Because of the low operation temperature, the solvent systems used do not boiled, so the extraction can be conducted at atmospheric pressure. After filtration and washing the final volume was adjusted at 25 ml.

### *2.3. TLC analysis*
TLC analyses were performed on two kinds of plates: TLC Sil G $F_{254}$ and HPTLC Sil G $F_{254}$ pre-coated plates. Prior using the TLC plates were conditioned with methanol and dried at 110°C for 3 h. The samples were applied with a Linomat 5 device as 5 mm bands, 20 µl for plant extracts and 7 µl for standards. In the case of HPTLC plates the applied volume was decreased at 10 µL for extracts and at 5 µl for standards. Every time a mixture of toluene-ethyl acetate (93:7, v/v) was used as mobile phase. The developed plates were sprayed with anisaldehyde and than heated 3 min at 110°C when red-bluish bands appear. The plates were inspected in daylight and also at 366 nm in UV range [24].

### *2.4. GC-FID analysis*
GC analysis was performed with a Shimadzu GC-2010 gas chromatograph with flame- ionization detection (FID). Compounds were separated on a methyl silicone column OV-17 (2m x 3.16 mm, 80-100 Mesh). Helium was used as carrier gas at 15 ml/min flow rate. The oven temperature was programmed 2 min at 80°C increased to 200°C with 4°C/min, maintained 1 min and then with 20°C/min to 260°C and held for 35 min. The injection port and detector temperature were 260°C and 240°C respectively. The injection volume was 2 µl for extract samples and standards (100 mg/ml).

## 3. Results and discussions
In the case of plant extracts TLC/HPTLC is used to provide fingerprint information [25]. Only when a very good resolution and no doubt about the identity of compounds are achieved a quantitative analyze is possible.

For identification of some essential oils, HPTLC was used. The E6 extraction solvent was chose because it has the capacity to extract the compounds with different polarities. After separation the plates were pulverized with anisaldehyde, heated for spots coloring and inspected in visible range (figure 1).

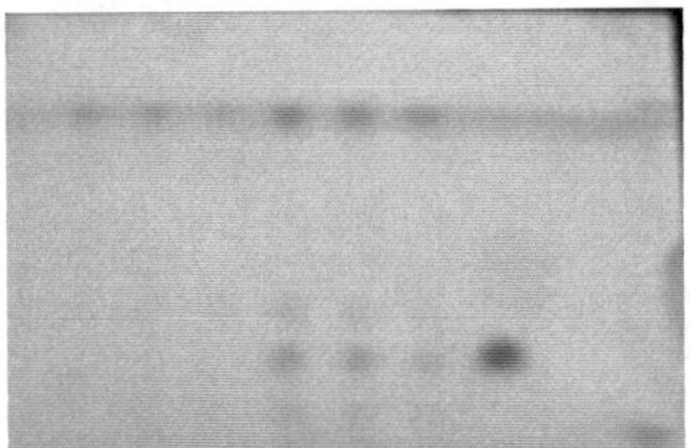


**Figure 1.** Identification chromatograms of thymol, linalool and cineole in *Ocimum basilicum* L. extracts E6.

The $R_f$ values were determined and compared with those existent in literature. Based on this it is possible that the *Ocimum basilicum* L. extracts to contain cineole and linalool. The presence of linalool is confirmed by [24, 25]. The TLC fingerprint of extracts obtained with different techniques – M, MAE and UAE showed to be similar, that means that no degradations processes happens or no new compounds were extracted (figure 2).

Journal of Physics: Conference Series **182** (2009) 012016 doi:10.1088/1742-6596/182/1/012016

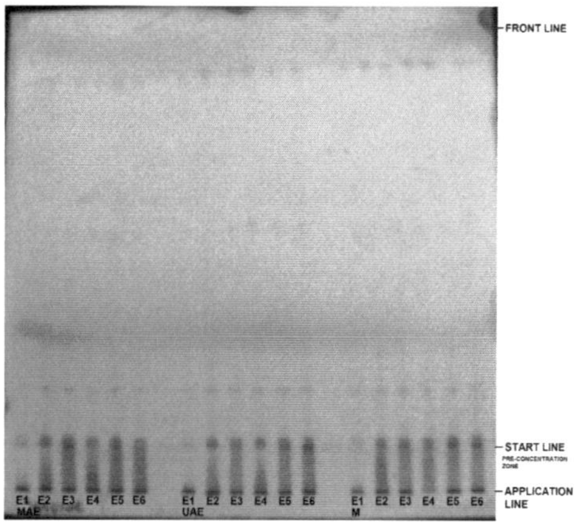


**Figure 2**. The chromatogram of *Ocimum basilicum* L. extracts obtained with solvent system E1-E6 using M, MAE and UAE techniques.

Better results were obtained with E3, E4 and E6, the spots being more intense. Even so we cannot choose the best extraction condition (solvent and technique), because that together with essential oils, some other compounds were extracted. In this case is indicated to be employed GC as analytical technique.

The chromatograms of extracts E1-E6 obtained by MAE and some standards were registered by GC-FID. The maximum intensity of peaks was obtained using the mixture E5 as extraction solvent.

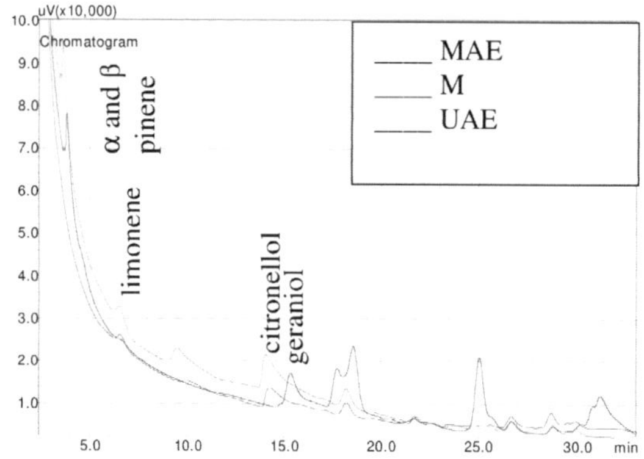


**Figure 4.** The chromatograms of *Ocimum basilicum* L. extracts E5 obtained by MAE and M, extracts E6 obtained by UAE.

The chromatograms for extracts E5 obtained with the studied techniques were compared (figure 4).

Good results were obtained by MAE and M techniques. A great advantage of extraction in microwave field was the short time for extraction comparing with maceration or sonication.

By overlayering the standards and extracts chromatograms some essential oils were identified: α and β-pinene (mixture), limonene, citronellol and geraniol.

There may be observed from other papers as well, that the essential oils fingerprint depend on extraction technique, solvent system and operation conditions. Of course modern techniques are advantageous, but they can change the composition of extract, enhancing some compounds and depleting other [26, 27].

## 4. Conclusions

The extracts were analyzed by TLC/HPTLC technique and the fingerprint information was obtained. The GC-FID was used for to establish the best conditions for extractions and for identifying the essential oils. The best extraction of essential oils was obtained by maceration using the mixture ethyl ether + ethanol (1:1, v/v) as extraction solvent. In *Ocimum basilicum* L. extracts were identified: α and β-pinene (mixture), limonene, citronellol, and geraniol.

**References**

[1] Lee S J, Umano K, Shibamoto T and Lee K G 2005 *Food Chem.* **91** 131
[2] Lampronti I, Saab A M and Gambari R 2006 *Int. J. Oncol.* **29** 989
[3] Klimánková E, Holadová K, Hajšlová J, Čajka T, Poustka J and Koudela M 2008 *Food Chem.* **107** 464
[4] Yang Y, Kayan B, Bozer N, Pate B, Baker C and Gizir A M 2007 *J. Chromatogr.* A **1152** 262
[5] Zhang Z M and Li G K 2007 *Microchem. J.* **86** 29
[6] Quinn B P, Bernier U R and Booth M M 2007 *J.Chromatogr.* A **1160** 306
[7] Siani A C, Garido I S, Monteiro S S, Carvalho E S and Ramos M F S 2004 *Biochem. Syst. Ecol.* **32** 477
[8] Sinyinda S and Gramshaw J W 1998 *Food Chem.* **62** 483
[9] Texteira S, Mendes A, Alves A and Santos L 2007 *Anal. Chim. Acta* **584** 435
[10] Stashenko E, Jaramillo B E and Martinez J R 2004 *J. Chromatogr.* A **1025** 93
[11] Rosillo L, Salinas M R, JGarijo and G L Alonso 1999 *J. Chromatogr.* A **847** 155
[12] *** 1997 *European Pharmacopoeia*
[13] Seidani A B, Jabbari A and Yamini Y 2005 *Anal. Chim. Acta* **530** 155
[14] Dawidowicz A L, Rado E, Wianowska D, Mardarowicz M and Gawdzik J 2008 *Talanta* **76** 878
[15] Pourmoriazavi S M and Hajimirsadeghi S S 2007 *J. Chromatogr.* A **1163** 2
[16] Lucchesi M E, Chemat F and Smadja J 2004 *J. Chromatogr.* A **1043** 323
[17] Kimbaris A C, Siatis N G, Daferera D J, Tarantilis P A, Pappas C S and Polissiou M G 2006 *Ultrason. Sonochem.* **13** 54
[18] Rauber C S, Guterres S S and Schapoval E E S 2005 *J. Pharm. Biomed. Anal.* **37** 587
[19] Buiarelli F, Cartoni G P, Coccioli F and Ravazzi E 1991 *Chromatographia* **31** 489
[20] Háznagy-Radnal E, Czigle S and Máthé I 2007 *J. Planar. Chromatogr.* **20** 189
[21] Cong Z, Meiling Q, Qinglong S, Shan Z and Ruonong F 2007 *J. Pharm. Biomed. Anal.* **44** 464
[22] Wang S Y, Wu C L, Chu F H, Chien S C, Kuo Y H, Shyur L F and Chang S T 2005 *Holzforschung* **59** 295
[23] Surducan E and Surducan V 2008 *Procedure and device for dynamic processing of materials* Romanian Patent, RO-00112063 B1
[24] Wagner H, Bladt S and Zgainski E M 1983 *Drogen Analyse, Dunnsichtchromatographische Analyse von Arzneidrogen* (Berlin: Springer) pp 5-26
[25] Wicht M 1994 *Herbal Drugs and phytopharmaceuticals* (CRC Press)
[26] Stashenko E E, Jaramillo BE and Martinez J R 2004 *J. Chromatogr.* A **1025** 93
[27] Vinatoru M 2001 *Ultrason. Sonochem.* **8** 303.

# Dielectric properties of *Zea mays* kernels – studies for microwave power processing applications

**Emanoil Surducan, Camelia Neamtu and Vasile Surducan**

National Institute for Research and Development of Isotopic and Molecular Technologies, 65-103 Donath, 400293 Cluj-Napoca, Romania

E-mail: emanoil.surducan@itim-cj.ro

**Abstract**. Microwaves absorption in biological samples can be predicted by their specific dielectrical properties. In this paper, the dielectric properties ($\varepsilon'$ and $\varepsilon''$) of corn (*Zea mays*) kernels in the 500 MHz - 20 GHz frequencies range are presented. A short analysis of the microwaves absorption process is also presented, in correlation with the specific thermal properties of the samples, measured by simultaneous TGA-DSC method.

## 1. Introduction

Dielectric properties are characterized by the dielectric constant or permittivity, $\varepsilon^*$, and are defined as the correlation factor between the electric induction *D* and the electrical field *E* applied to the material:

$$D = \varepsilon^* E \tag{1}$$

For a variable electric field, $E = E_0 \exp^{-j\omega t}$, the permittivity has the expression [1, 2]:

$$\varepsilon^* = \varepsilon_o \left(\varepsilon' - j\varepsilon''\right) \text{ (F/m)} \tag{2}$$

where:

$\varepsilon_0 = 8.856 \times 10^{-12}$ F/m represents the dielectric properties of free space,

$\varepsilon'$ is the degree of the induced dipolar variable-order (dispersion) by the electric field *E* and

$\varepsilon''$ represents the loss of electric energy (absorption) by normal conduction (electronic or ionic) and dipolar displacement or dielectric loss:

$$\varepsilon'' = \frac{\sigma}{\omega \varepsilon_o} + \varepsilon''_{die} \tag{3}$$

In equation (1) σ is the material's conductivity and $\omega = 2\pi f$, with *f* the frequency of the electric field variations.

Usually the relative permittivity $\varepsilon_r$, a non-dimensional quantity, is used:

$$\varepsilon_r = \frac{\varepsilon^*}{\varepsilon_o} = \varepsilon' - j\,\varepsilon'' = \varepsilon' - j\left(\varepsilon''_{die} + \frac{\sigma}{\omega\varepsilon_o}\right) \tag{4}$$

together with the loss factor:

$$\tan\delta = \frac{\varepsilon''}{\varepsilon'} \tag{5}$$

The paper presents measurements of the ε' and ε" values for corn (*Zea mays*) kernels in the 500 MHz to 20 GHz frequency range for the microwave field.

The material's absorption ε" or, more precisely, the loss factor, $\tan\delta$, is the most important parameter in the microwave power processing of materials. The microwave power absorbed into the material is defined as [3]:

$$P_a = \omega\,\varepsilon_0\,\varepsilon'\tan\delta\,E_i^2 = \omega\,\varepsilon_0\,\varepsilon''\,E_i^2 \;\;(\mathrm{W/m^3}) \tag{6}$$

where $E_i$ is the sample's internal electric field and it depends on the incident microwave power $P_{oi}$ through the probe impedance $Z_p$ in the microwave field, $E_i^2 = P_{oi}\cdot Z_p$.

If we consider that all the microwave power absorbed in the material during the period $t$ of treatment is transformed into heat, $P_a\cdot t = Q$, it is possible to evaluate the material's temperature rising rate during microwave power thermal treatments:

$$\frac{\Delta T}{\Delta t} = \frac{P_a}{m\,c} \tag{7}$$

where $m$ is the sample's mass and $c$ is the specific heat.

The corn kernel's mass and heat flow variations *versus* temperature have been also investigated by TGA-DSC measurements, in order to determine the volatile content and the popping temperature of the seed.

## 2. Dielectric properties measurements

Dielectric properties measurements were performed with an Agilent N 5230A network analyzer and related accessories for dielectric properties measurements (Agilent 85070E Dielectric Probe Kit 200 MHz to 50 GHz). Two experimental configurations in the microwave field have been used, based on the reflected wave and cavity perturbation methods [2, 4].

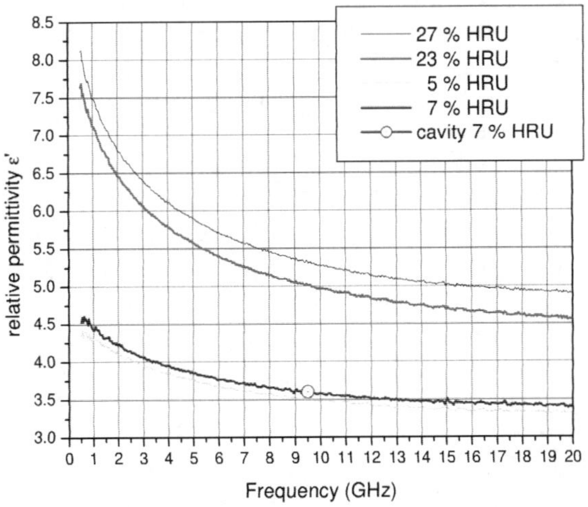


**Figure 1.** Dielectric relative permittivity ε' versus frequency, for various water contents in the sample (relative humidity, % HRU).

The reflected wave technique is suitable for wide frequencies band measurements of both ε' and ε" values of the permittivity, but it is sensitive to the initial calibration procedures and to the experimental setup. In the cavity perturbation method, the results depend on the probe's volume and shape and on the precision in the determination of the probe's volume; however, no initial calibrations are necessary. This method is very precise, even if it can be applied only at the resonance frequency of the cavity. The results obtained with the reflected wave method (figure 1, figure 2) have been verified by the perturbation method, at one frequency.

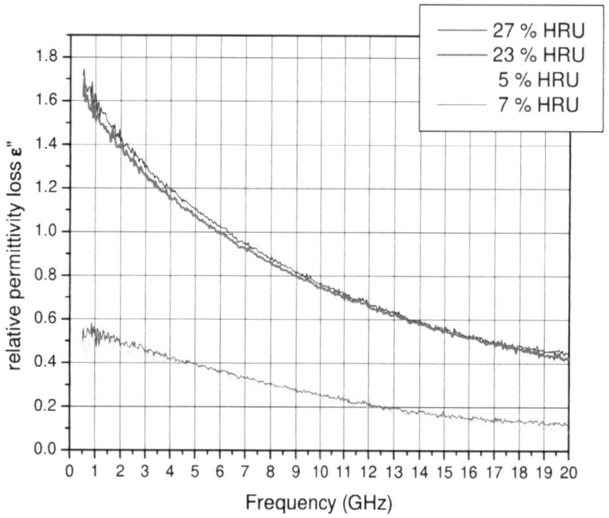


**Figure 2.** Dielectric relative permittivity $\varepsilon$" (loss) versus frequency, for various water contents in the sample (relative humidity, % HRU).

For the reflected wave method, a single corn kernel was polished to obtain a minimum of 6 $mm^2$ of flat surface, where the coaxial microwave probe was applied. In the case of cavity perturbation method, a cylindrical shape having a volume of 5 $mm^3$ was cut out from the corn kernel and was introduced in the centre of a cylindrical cavity resonating at $f_{probe}$ = 9.5207 GHz, in the $TM_{010}$ mode.

## 3. Thermal properties measurements on corn kernel

The samples (a regional variety of flint corn) were investigated by simultaneous TGA-DSC measurements, using a SDT Q600 analyzer (TA Instruments). The purge gas was argon, at a rate of 100 ml/min. The temperature and heat flow of the instrument were calibrated using alumel and nickel (Curie transition temperatures) and sapphire, respectively. The corn samples (about 5 mg) were weighed in open alumina pans. A single slice, cut from the kernel's horny endosperm and placed at the bottom of the pan, was used for each measurement. An empty pan was used as reference. The experiments were carried out in a temperature range of 25-350°C at a heating rate of 10°C/min. The obtained data were analysed using the TA Universal Analysis Software Version 4.5A.

Figure 3 shows the TGA curve obtained for a corn kernel specimen. The mass loss starts very close to room temperature, and amounts to 13-14% weight, but the main thermal decomposition of the sample is initiated at around 270°C. The derivative weight loss curve indicates that the point at which the transition (popping) occurs is in very good agreement with the DSC results, as shown in figure 4. An endothermic peak was obtained, corresponding to the phase transition of the sample. As expected, the onset temperature at 225.09°C is higher than the one reported for popping corn (popcorn) variety [5], due to the specific experimental conditions (flint corn sample, measurement performed in inert gas flow, only a piece of the kernel, not the whole, was used).

## 4. Microwave absorption analysis in corn kernels

Microwave power absorption $P_a$ is difficult to be evaluated because of the probe impedance $Z_p$ in the microwave field, which depends not only on the material's dielectric properties, but also on the sample's volume and orientation in the incident wave or microwave coupling effect. Equation (6) is used in probing dielectric properties optimisation by adding supplementary moisture or other absorbent on the surface of the sample.

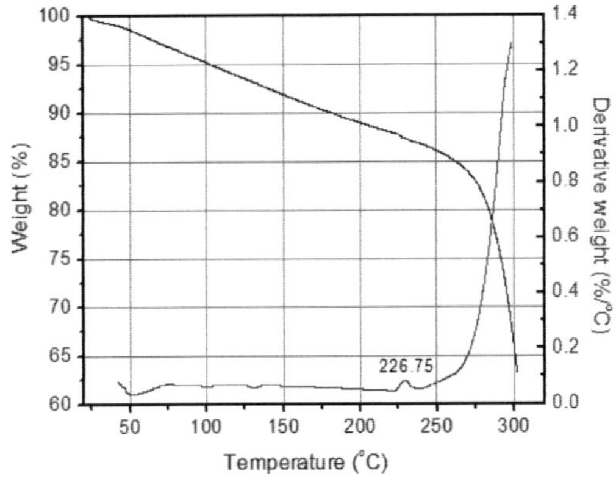


**Figure 3**. TGA curve recorded for a flint corn kernel sample, together with its calculated first derivative (in temperature).

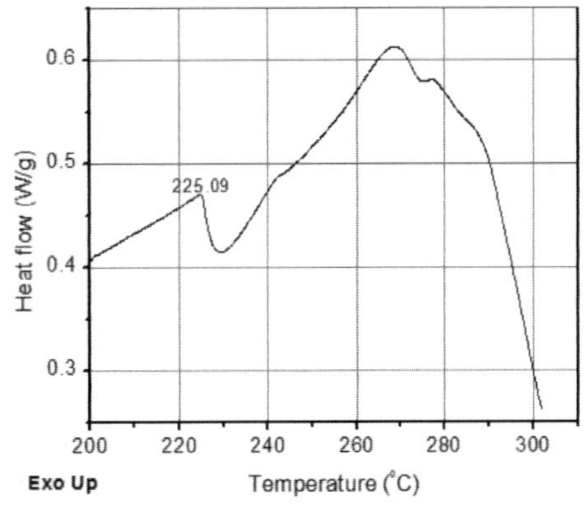


**Figure 4.** Heat flow thermogram of flint corn sample (cutted from the horny endosperm) in inert atmosphere.

Considering a corn kernel mass of $m = 0.152$ g, a specific heat of $c = 2000$ J/Kg K and an effective power of $P_a = 2$ W transformed in heat, a rate of the temperature rising of $\Delta T/\Delta t = 5.02$ K/s is found. To obtain the popping transformation of the corn kernel, only a time interval $\Delta t = 40.8$ s (starting from room temperature) is necessary. The most important technological problem is to transfer the microwave power in the seed with a high rate.

## 5. Conclusions

The dielectric properties investigations indicate that the moisture content of the corn kernels has an important role in their dielectric properties in the frequencies range of 0.5 - 3 GHz. An additional 20% relative moisture increases with 200% the dispersion value $\varepsilon'$ and with 300% the absorption loss $\varepsilon''$. This result suggests the possibility of optimisation of the corn kernel treated in microwaves, by analysing it as a metamaterial with a high absorption in a specific frequency band [6].

Thermal analysis measurements confirmed that, for the investigated samples, the water content was of about 14% and the popping temperature is in the 150-250°C range [7, 8].

## References

[1] Pozar D M 1990 *Microwave Engineering* 3rd Edition (New York: Addison-Wesley)
[2] Musil J and Zacek F 1986 *Microwave Measurements of Complex Permittivity by Free Space Methods and Their Applications* (Amsterdam; New York: Elsevier)
[3] Surducan E 1997 *Microwaves absorption measurements for substance characterization* PhD Thesis (Cluj-Napoca, Romania: Babes-Bolyai University)
[4] Agilent Application Note 5989-2589EN Basics of measuring the dielectric properties of materials
[5] Roshdy T H, Hayakawa K and Daun H 1984 *J. Food Sci.* **49** 1412-4
[6] Caloz C and Tatsuo I 2006 *Electromagnetic Metamaterials: Transmission Line Theory and Microwave Applications* (Hoboken, New Jersey: Wiley Interscience)
[7] Byrd J E and Perona M J 2005 *Cereal Chem.* **82** 53-9
[8] Vozárová V 2005 *Int. Agrophysics* **19** 187-92.

Journal of Physics: Conference Series **182** (2009) 012018 doi:10.1088/1742-6596/182/1/012018

# The effect of coloured light on *Ipomoea purpurea* growth

**Vasile Surducan, Ildiko Lung and Emanoil Surducan**

National Institute for Research and Development of Isotopic and Molecular Technologies, 65-103 Donath, 400293 Cluj-Napoca, Romania

E-mail: vasile.surducan@itim-cj.ro

**Abstract.** *Ipomoea purpurea* is a climbing ornamental plant native to Mexico. The paper is describing the experimental setup and results for indoor growing plants exposed to white LED light (inside a reference chamber) and four different wavelength LED lights (inside a measure chamber). Four growing experiments of 12-15 days, took place in identical environmental conditions (identical temperature and relative humidity inside the reference and measure chambers, similar lighting conditions and soil moisture). At the end of the experiments, the plant chlorophyll and xanthophylls content have been measured and the plant aspect (vegetal mass, leaves colour and robustness) has been observed. The smallest content in chlorophyll (a and b) was developed by the plants growth in blue light (480 nm), however those plants where 10% taller than plants growth in white light, but less robust. The higher content in carotenoids and xanthophylls was observed in plants which growth in white and red light.

## 1. Introduction

A few US patents [1, 2] including NASA research, claim the combination between red and blue LED light effect for plant growing acceleration. The experiment presented in our work, is searching for the optimal wavelength combination for *Ipomoea purpurea* growth.

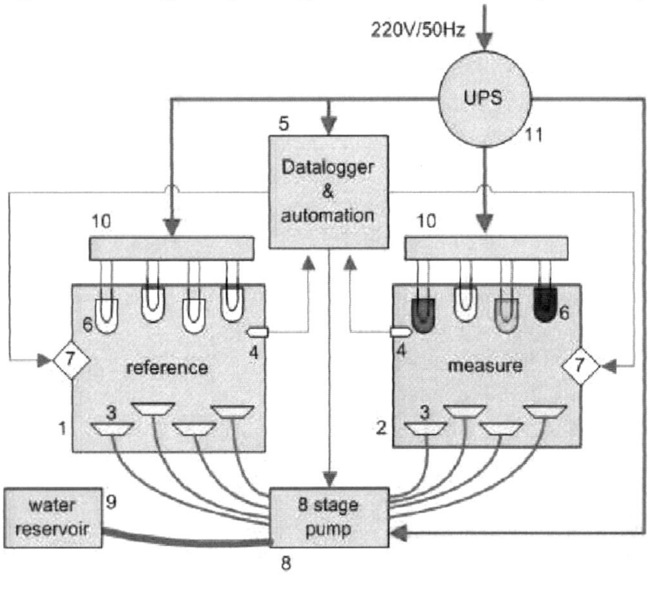


**Figure 1.** The experimental setup.

## 2. Experimental setup

The experimental setup (figure 1) consists of two identical anechoic chambers (1, 2), both with door access, four LED light sources (6), temperature and relative humidity sensors (4) and fans (7). The reference chamber (1) has four white LED-group sources while the measure chamber has four different LED sources: red 650 nm, yellow 590 nm, green 550 nm and blue 480 nm. The chambers are anechoic, against any external radiofrequency waves which can accelerate the growing plants process [3]. The humidity / temperature sensors of both chambers are connected to a microcontroller datalloger and automation device (5) which is storing to own memory the temperature and humidity samples taken with a programmed sampling time. For this set of experiments, the sampling time was set to 5 minutes over the entire experiment period of 12-15 days. The microcontroller is also driving the fans (7) which keep the same humidity and temperature in both chambers within ± 3% RH and ± 0.5°C. Each of the LED lamps (6) are supplied from a controlled current power supply (10) to achieve a constant illuminating intensity of 300 lux measured at the soil level in both chambers. For light intensity measurement an Iso-tech 1332A Digital Illuminance Meter with a known amplitude to wavelength sensitivity has been used. To avoid any power supply malfunction, the datalogger (5) has its own lithium accumulator and the whole experiment is supplied from a 2KVA UPS unit (11) with an authonomy of 24 hrs in absence of the main power supply.

Seeds with almost similar shape, size and colour have been selected at the beginning of each experiment. Five seeds laid in every four transparent plastic vessels (3) placed below the LED light spots (6). Seeds have been moistured initially with 2.5 ml of tap water and kept exposed to light for 24 hrs (figure 2).

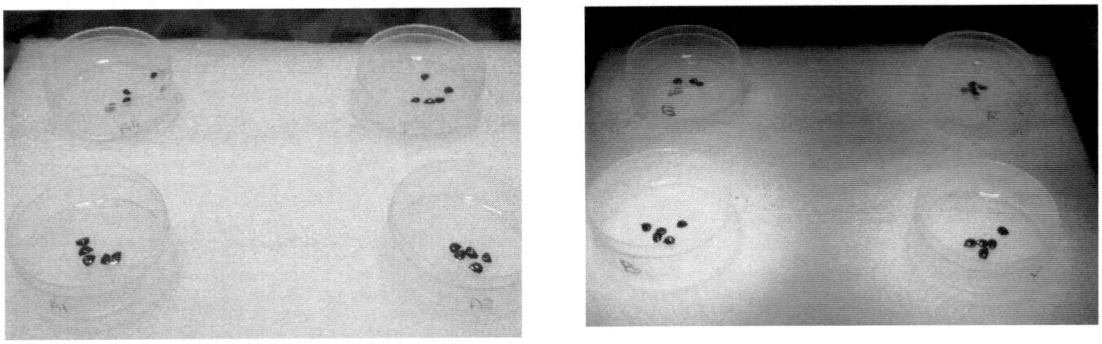

**Figure 2.** Seeds in moisture under white light. Seeds in moisture under coloured light.

On the second day of each experiment, seeds have been buried below a 5cm$^3$ of uncompressed Terra Dena soil manufactured by Matecsa Ker. Es Kert. KFT Hungary and moistured with 5 ml of tap water. Twice per day over the entire experiment period, the soil was moistured with the same amount of 2.5 ml of water using an 8 stage modified peristaltic pump (8) model Cole Palmer Masterflex 7568-10, driven from the same datalogger/automation device (5). During the experiment, the growing ratio of the plants from various plastic vessels has been observed, but not all five seeds have always germinated.

**Figure 3.** Plants after 12 days in white light. Plants after 12 days in coloured light.

When the smallest plant from the measuring chamber reached a size of 5cm tall and leaves expanded (figure 3), the taller plant from each vessel has been removed and the vegetal content was analyzed (figure 4) and leaf pigments have been measured.

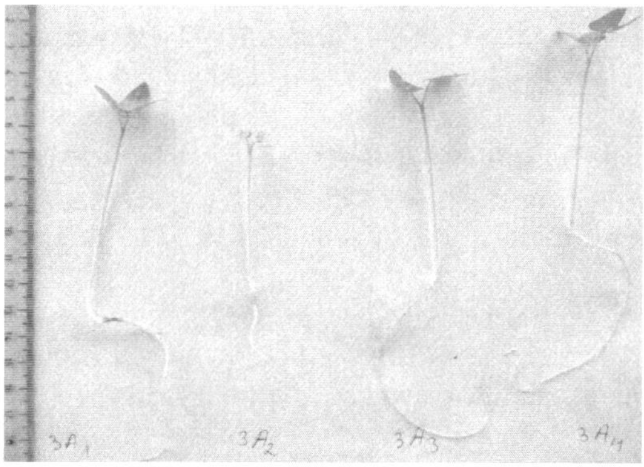

**Figure 4.** Plants length grow in white light (reference chamber, A1…A4= white).

Plants length grow in coloured light (measure chamber, G=green, R=red, Y=yellow, B=blue).

The remaining plants have been moved in flower pots using the same soil; they were placed in the room and the growing speed and robustness (alternating moisture to dryness periods) were observed. Four experiments took place between february 2009 to april 2009 (the experiment is still in progress).

## 3. Measurements

### *3.1. Humidity and temperature measurements inside the reference and measuring chambers*

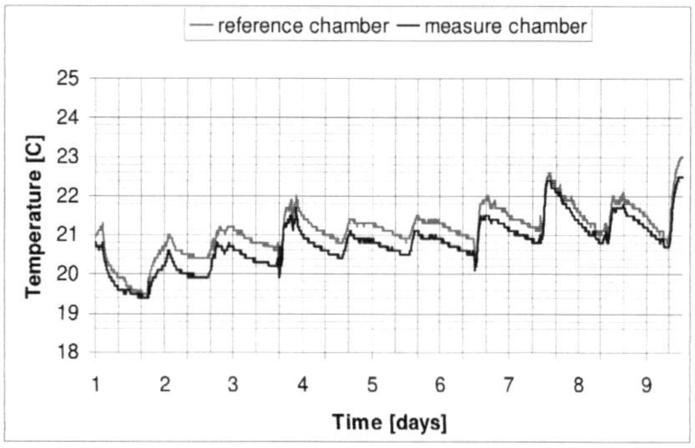


**Figure 5.** Temperature variation in chambers over 9 days (less than 0.5°C difference between chambers).

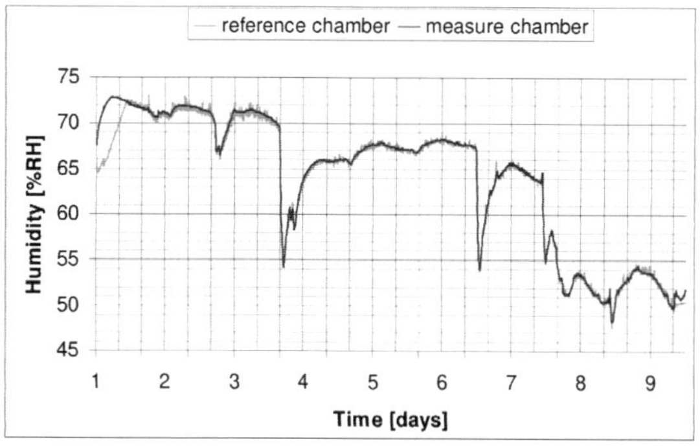


**Figure 6**. Humidity variation in chambers over 9 days (less than 3% RH difference between chambers).

At the end of each experiment, datalogger memory is transferred to a PC using the RS232 interface. The temperature and humidity variation between chambers over 9 days of experiment is represented in figures 5 and figure 6. The same environmental conditions have been kept inside of both chambers over the entire experiment.

### *3.2. Pigments determination*

The leaf pigments were colorimetric determined in 80% acetone extract, at wavelengths of 663 nm, 646 nm and 470 nm. 0.1 g plant material was ground in the presence of quartz sand and washed several times with 100% acetone, filtered in vacuum and passed quantitatively into a 10 ml quoted flask containing 2 ml distilled water. The acetone extract of plant was spectrophotometrically dosed and compared to a blank of 80% acetone. Results were calculated based on formulas developed by Mckiney [4] and values were expressed in mg/100g plant material (equation1 to equation3 and table1)

$$Clorophyll\ a = (12.21 * A_{663} - 2.81 * A_{646}) * 5 \tag{1}$$

$$Clorophyll\ \mathrm{b} = (20.13 * A_{646} - 5.03 * A_{663}) * 5 \tag{2}$$

$$carotenoids\ \&\ xanthophylls = \frac{1000 * A_{470} - 3.27 * Clorophyll\ \mathrm{a} - 1.04 * Clorophyll\ \mathrm{b}}{229} * 5 \tag{3}$$

## 4. Results and discussion

**Table 1.** Pigment quantities in plants growing under different LED light.

| **Pigment** [mg/100 g plant] | White light | Blue light (480 nm) | Green light (550 nm) | Red light (650 nm) | Yellow light (590 nm) |
|---|---|---|---|---|---|
| Chlorophyll a | 96.6375 | 81.0385 | 90.2607 | 85.6420 | 78.0534 |
| Chlorophyll b | 89.0604 | 69.1358 | 84.9778 | 94.2400 | 73.2494 |
| Carotenoids and xanthophylls | 34.5147 | 28.1003 | 33.1352 | 34.1374 | 32.8025 |

The following plant behaviour have been observed:

- in each experiment, plants growing in red light (650 nm) have used less water for photosynthesis than other plants;
- on each experiment, plants growing in red, yellow and green light are growing facing to the blue light, while plants growing in blue light are growing straight upwards (figure 3);
- plants growing in white light developed the largest chlorophyll (chlorophyll a + chlorophyll b) content (table 1) followed by the plants growing in green and red light;
- on most experiments, plants growing in blue light where 10% taller than other plants but they have the smallest robustness to dryness once they are moved in natural light;
- there is no major difference in total chlorophyll content between plants growing in red or green light (table 1).

## References

[1] Thomas C, Robert M, Jeffrey E and Marty G 2009 http://www.freepatentsonline.com/7473008.html *Marine LED lighting system and method* US patent 7473008 B2

[2] William A Jr and Larry C 2005 http://www.freepatentsonline.com/6921182.html *Efficient LED lamp for enhancing commercial and home plant growth*, US patent 6921182 B2

[3] Surducan E, Surducan V and Halmagyi A 2008 OSIM Bucuresti *Plant growing stimulation in microwave field* Romanian patent pending: A/00429

[4] McKinney G 1941 *J. Biol. Chem* **140** 315-22.

# Infrared and UV-Vis spectroscopic study of 3,7,10-substituted-phenothiazine derivatives adsorbed on gold nanoparticles

**N Tosa[1], L Olenic[1], I Bratu[1], R Turdeanu[2] and I Turcu[1]**

[1] National Institute for Research and Development of Isotopic and Molecular Technologies, 65-103 Donath, 400293 Cluj-Napoca, Romania

[2]Babes-Bolyai University, Faculty of Chemistry and Chemical Engineering, Organic Chemistry Department, 11 Arany Janos, 400028 Cluj-Napoca, Romania

E-mail: nicoleta.tosa@itim-cj.ro

**Abstract**. Absorption of 3,7,10-substituted-phenothiazine derivatives on gold nanoparticles was investigated by optical spectroscopic methods. Gold colloids provide active surfaces for the self-assembly of redox-active thiolate and disulfide phenotiazine compounds monolayers by forming one Au-S bond. The coverage of their surfaces is a long term process being accompanied by stability of the functionalized gold nanoparticles.

## 1. Introduction

Several studies have been reported dealing with preparation and characterization of self-assembled monolayer on various type of substrates due to their applications ranging from nanotechnology to fundamental surface science [1,2]. Particularly, gold nanoparticles (GNP) have been the subject of intense research due to their unique properties: *i*) gold can be functionalized through tiol functional groups[2,3], *ii*) the synthesis of size controlled GNP is quite easy, following the reduction of a gold(III) compound by a reducing agent [4,5], *iii*) their plasmon characteristics can be tuned from visible domain to the near infrared one, making them ideal candidates for sensing and biolabelling [6].

Obtaining of modified nanoparticle surfaces by organic molecules is a topic of great interest. Thus, heterocyclic compounds such phenotiazine derivatives, which present remarkable pharmaceutical applications[7], represent a challenging option from this point of view. Moreover exhibiting low oxidation potential, phenotiazines can generate radical cations playing important roles in physiology [8].

In this work we report on the interaction of 3,7,10-substituted phenotiazine derivatives[9], bearing tiol and disulfide functional groups, with gold nanoparticles in colloidal organic solutions, by using optical spectroscopic (UV-Vis and FTIR) methods. These results represent primary steps for applications in chemical and biological sensing.

## 2. Experimental

### *2.1. Chemicals*
The tetrachloroauric (III) acid ($HAuCl_4$, 99.5%) and tetrahydrofuran (THF, spectroscopic grade) were purchased from Merck. The trisodium citrate dihydrate ($Na_3C_6H_5O_7 \cdot 2H_2O$, 99%) and sodium borohydride ($NaBH_4$, 98%) were purchased from Sigma Aldrich. Deionized water (18 $M\Omega \cdot cm^{-1}$) was added to prepare all aqueous solutions and THF to prepare all organic solutions.

### *2.2. Preparation of gold nanoparticles (GNP)*
GNPs were prepared by direct reduction of tetrachloroauric (III) acid with sodium borohydride in the presence of trisodium citrate[5], in order to limit the size of colloidal gold nanoparticles. Then they were subjected to repeated centrifugal purification to obtain a concentrated colloidal solution.

### *2.3. Preparation of functionalized GNPs*
3,7,10-substituted-phenothiazine derivatives, denoted Tiol8CH2 and BTiol8CH2 respectively, have been synthesized according to the protocols previously described[8]. Each compound was dissolved in THF to achieve $10^{-3}$ M homogeneous solutions, and then was added in 3:1 v/v ratio to the colloidal concentrated aqueous solution. The samples were kept at room temperature in dark sealed vessels and investigated by optical spectroscopic methods after 24 h and seven months, respectively.

### *2.4. Instrumentation*
The UV-Vis absorption spectra were recorded on a Jasco V550 UV-Vis spectrophotometer in the 190-900 nm wavelengths range with a 2 nm resolution, using 10 mm length quartz cells. FTIR measurements were performed in the range 4000 to 350 $cm^{-1}$ on a JASCO 6100 spectrometer (single beam) with a 2 $cm^{-1}$ resolution, and 512 scans were performed to collect each spectrum. KBr pellet technique for measurements in absorbance mode was employed.

## 3. Results and discussion
Obtaining of functionalized gold nanoparticles with 3,7,10-substituted-phenothiazine derivatives consisted of two main steps: the first, GNPs synthesis and centrifugal cleaning, the second, adsorption of organic molecules onto the obtained gold nanoparticles.

### *3.1. UV-Vis characterization*
The bare and the functionalized GNPs, in their appropriate solutions, were characterized by UV-Vis spectroscopy. The visible absorption spectra of the gold solution (figures 1 and 2) presents a well-defined peak with maximum at $\lambda_{max}$ = 528 nm, characteristic for plasmon (SPR) absorbance of nanometric Au particles ascribed to an average size bellow 10 nm.

The interaction of GNPs with compound Tiol8CH2 (figure 1) bring about a significant decrease of the (SPR) peak intensity, as well as a red-shift of 15 nm ($\lambda_{max}$ = 543 nm). This effect can be assigned to the chemical adsorption of organic molecules on the gold surface through the sulfur atom, followed by their aggregation. Two additional peaks of low intensity at 686 nm and 731 nm come to support this tendency.

In case of BTiol8CH2 interaction with GNPs, there are noticeable two maxima of absorption, at 532 nm and at 743 nm (figure 2). Taking into account that both experiments were carried out in the same conditions and the spectra are comparable on the absorbance scale, one can consider that compound BTiol8CH2 was chemisorbed on the gold surface, but the aggregation tendency of capped nanoparticles is bigger- the red-shift being considerable. The UV-Vis spectra of the considered phenotiazine compounds exhibit no absorption bands in the visible range of wavelengths, their maxima lying in the 200-350 nm UV range.

The S–H bond of the tiol and S-S bonds of the disulfide will be broken during adsorption onto the gold surface resulting in individual fragments stabilizing the metal particles. The generated nanoparticles were highly stable both in THF solution as well as in the solid state. Solubility characteristics do not change over 7 month, but UV-Vis spectra recorded after 7 months present a dramatically decrease of absorbance and the disappearance of the peaks around 700 nm, ascribed to the higher level of aggregation. However, the main absorption maxima (543 nm and 532 nm) do no shift once the deposition is finished, revealing that the aggregation or decomposition does not occur.

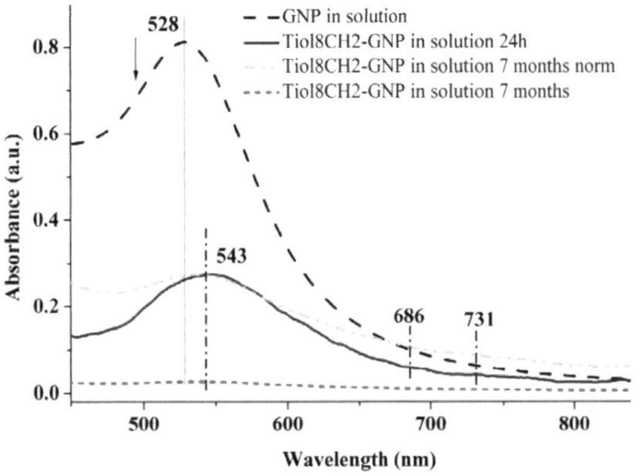


**Figure 1.** UV-Vis spectra of Au colloidal solution and of Tiol8CH2 adsorbed on GNPs in THF solution after 24 h and 7 months, respectively.

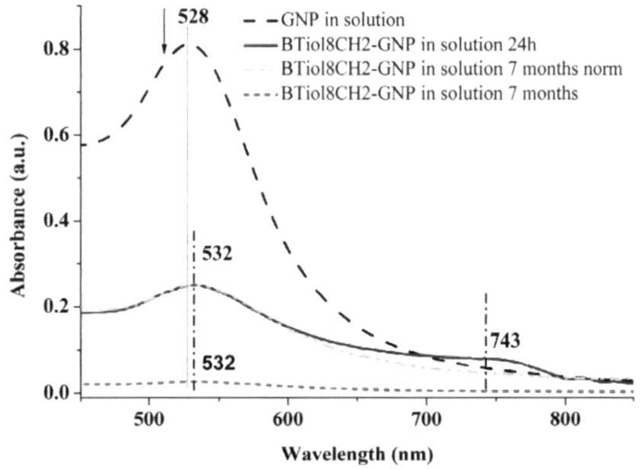


**Figure 2.** UV-Vis spectra of Au colloidal solution and of BTiol8CH2 adsorbed on GNPs in THF solution after 24 h and 7 months, respectively.

*3.2. FTIR characterization*

The FTIR spectra of Tiol8CH2 (figure 3) and BTiol8CH2 (figure 4), both neat and chemisorbed on GNPs, were recorded in transmission, in KBr matrix.

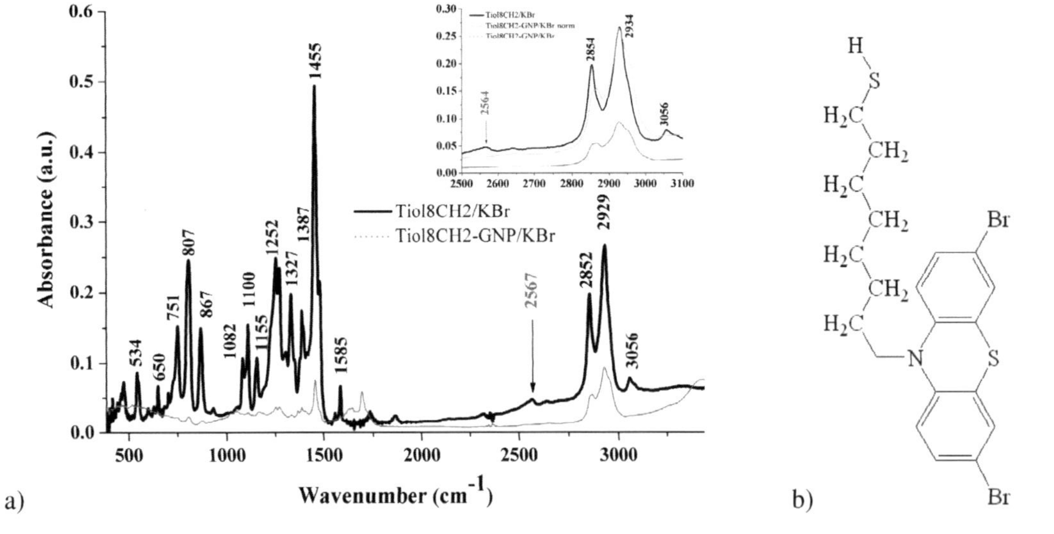


**Figure 3.** a) FTIR spectra of Tiol8CH2, neat and adsorbed on GNPs, in KBr matrix, b) Tiol8CH2.

The FTIR spectrum of neat Tiol8CH2 (figure 3) indicates the presence of three C–H stretching modes: two for the $-CH_2$ group, consisting of one symmetric at 2852 $cm^{-1}$ and one antisymmetric at 2929 $cm^{-1}$ and one for the –CH aromatic group at 3056 $cm^{-1}$. The stretching vibration characteristic of tiol group is located at 2567 $cm^{-1}$ and those of C=C aromatic is located at 1585 $cm^{-1}$.

The region 1500-1100 $cm^{-1}$ exhibits medium intensity peaks assigned to the deformation frequencies, characteristic of aromatic and saturated aliphatic compounds. Thus, the 1446 $cm^{-1}$ peak corresponds to the asymmetric bending frequency of $–CH_2$ group and the 1342 $cm^{-1}$ peak corresponds to the symmetric one. A sharp band is observed at 1252 $cm^{-1}$ is assigned to a skeleton vibration. The intense band at 1110 $cm^{-1}$ corresponds to a skeleton motion, as well as the highly coupled skeletal vibrations in the region 800-900 $cm^{-1}$. The C-S stretching vibration is identified at 650 $cm^{-1}$ as a weak intensity peak.

In FTIR spectrum of Tiol8CH2 adsorbed on GNPs (insert of figure 3) no peak was seen around 2567 $cm^{-1}$ where the S-H stretching vibration normally occurs. This observation confirms that the thiol proton is lost upon surface adsorption and the resulting anion is in direct interaction with the gold atom of the GNPs surface via the sulfur atom. To avoid any doubt concerning the low intensity of the absorbance peaks in the 7 months after FTIR spectrum, we rescaled it at the intensity of the free molecule FTIR spectrum. In this respect, we have considered as reference the absorbance of the $–CH_2$ asymmetric stretching vibration peak. Similar procedure was applied for spectrum of BTiol8CH2 adsorbed on GNPs (insert of figure 4).

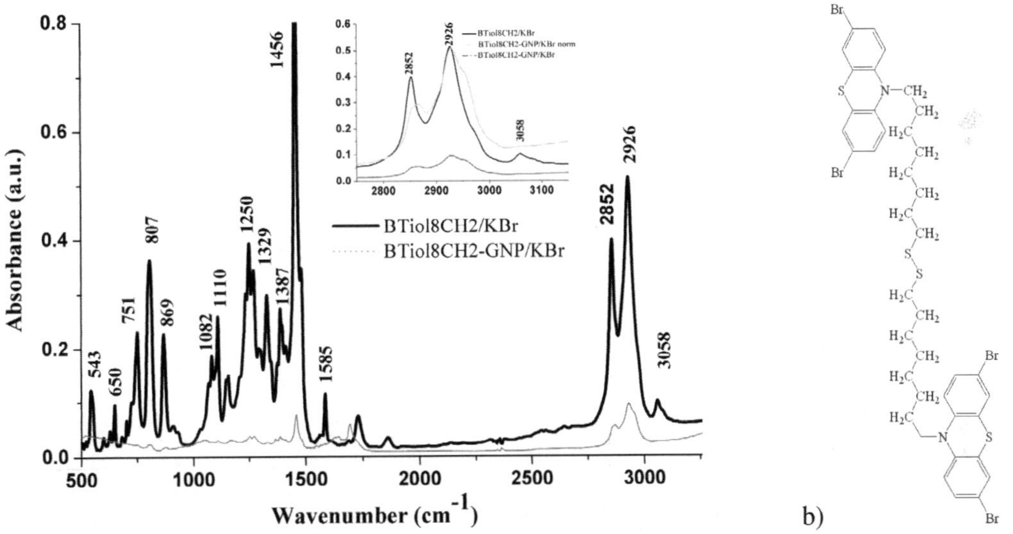


**Figure 4.** a) FTIR spectra of BTiol8CH2, neat and adsorbed on GNPs, in KBr matrix, b) BTiol8CH2.

The FTIR spectrum of neat BTiol8CH2 (figure 4) reveals very similar assignments with those of Tiol8CH2. The major difference consists of the lack of –SH stretching vibration. Unfortunately, the S-S vibration, located in the 450-550 $cm^{-1}$ region, exhibits a very weak peak without relevance in infrared spectra.

FTIR spectrum of BTiol8CH2 adsorbed on GNPs (figure 4) is similar to that of Tiol8CH2 adsorbed on GNPs. It is worth mentioning that BTiol8CH2 molecule represents in fact a linear dimer of Tiol8CH2 molecule, connected through a disulfide bridge. During the adsorption of BTiol8CH2 onto the gold surface the S–S bonds are broken [10] and generate thiolate (Tiol8CH2) fragments that will bind with the gold atoms from the GNP surface. In this situation, we have the same functionalized gold nanopaticles as if we should deposited Tiol8CH2 on GNPs. However, one cannot support the chemisorption by the missing of S–S vibration but one can consider the $CH_2$ group peak as an internal standard to validate the functionalization process comparison to Tiol8CH2, when identical sample quantities were taken into account.

To avoid any doubt concerning the low intensity of the absorbance peaks in the 7 months after UV-Vis spectra, we rescaled them at the 24h molecule UV-Vis spectra, using as reference the intensity of the absorbance peaks with maxima at 543 nm and 532nm, respectively.

## 4. Conclusions

In conclusion, a combined UV-Vis and FTIR spectroscopic study converge in the evidencing that phenotiazine derivatives Tiol8CH2 and BTiol8CH2 chemisorb on the surface of gold nanoparticles by forming Au-S bonds. The chemical binding of the considered organic molecules on the gold surface is accompanied by a capped-GNPs aggregation, supported by a significant red-shift of SPR absorption peak relatively to the bare GNPs. Gold colloids functionalization is a long term process, which evolves at room temperature long time after the 24 h "well-established" period for deposition. Tiol and disulfide group bearing phenotiazine derivatives in conjunction with gold nanoparticles offer a good approach for building molecular defined "blocks" useful to design chemical and biological sensing devices.

## Acknowledgements

The financial support of this work by CEEX program (grant 2Cex06-11-93), TD 401 and PN2 61-003 is acknowledged.

## References

[1] Ulman A 1996 *Chem. Rev.* **96** 1533-54
[2] Love J C, Estroff L A, Kriebel J N, Nuzzo R G and Whitesides G M 2005 *Chem. Rev.* **105** 1103-69
[3] Gooding J J, Mearns F, Yang W and Liu J 2003 *Electroanal.* **15** 81-96
[4] Frens G 1972 *Nature* **241** 20-2
Patil V, Malvankar R B and Sastry M 1999 *Langmuir* **15** 8197-206
Premkumar T, Kim D, Lee K and Geckeler K E 2007 *Gold Bull.* **40** 321-7
[5] Brown K R, Lyon L A, Fox A P, Reiss B D and Natan M J 2000 *Chem. Mater.* **12** 314-23
[6] Barbillon G, Bijeon J-L, Plain J, Chapelle M L, Adam P-M and Royer P 2007 *Gold Bull.* **40** 240-44
[7] Silberg I A, Cormos G and Oniciu D C 2006 *Adv. Heterocycl. Chem.* **90** 205-37
[8] Elisei F, Aloisi G G, Latterini L, Mazzucato U, Viola G, Mioio G, Vedaldi D and Dall'Acqua F 2002 *Photochem. Photobiol.* **75** 11-21
[9] Turdean R, Bogdan E, Terec A, Petran A, Vlase L, Turcu I and Grosu I 2009 *Cent. Eur. J. Chem.* **7** 111-7
[10] Gropeanu R A, Tintas M, Pilon C, Morin M, Breau L, Turdean R and Grosu I 2007 *J. Heterocycl. Chem.* **44** 521-7.

# The skin's electrical asymmetry

**N M Bîrlea[1], S I Bîrlea[2] and V Toşa[3]**

[1] Technical University of Cluj-Napoca, Physics Department, 15 Constantin Daicoviciu, 400020 Cluj-Napoca, Romania

[2] National University of Ireland, Electrical and Electronic Engineering Department, Galway, Ireland

[3] National Institute for Research and Development of Isotopic and Molecular Technologies, 65-103 Donath, 400293 Cluj-Napoca, Romania

E-mail: mbirlea@phys.utcluj.ro

**Abstract**. We studied the asymmetry of skin response to a constant current pulse (4mA/0.1ms) of positive or negative polarity, followed by a free decay of skin's potential. Negative polarity pulse is related with a lower skin resistance and higher electrical capacity. This behavior corresponds to 2 of the 3 relaxation times found in our skin measurements.

## 1. Introduction

In order to find the relevant electrical parameters describing the skin condition, we studied the relaxation times of the skin potential in response to a constant current impulse of positive or negative polarity. We believe that the relaxation times are markers of skin's structure or processes.

Bioimpedance spectroscopy has many applications [1] in skin assessment from measuring skin moisture [2], to detect skin cancer [3]. Usually, bioelectrical measurements are performed using sinusoidal currents at several frequencies and the experimental data are fitted with electrical models or empirical formulas. An alternative approach uses square-wave electrical pulses [4] to measure the skin's electrical properties. The method has the advantage of measuring electrical response in a broad range of frequencies and allows a more direct and explicit modality to measure skin characteristics [5].

## 2. Experimental Method

The electrical properties of human skin were measured using constant current electrical impulses. The measuring system comprises of a signal generator (made in the laboratory) with a 9 V battery as power supply, a bipolar transistor as an open collector current drive, and an oscilloscope (Picoscope 2203) connected to a laptop, which visualizes the potential difference between the electrodes and the current through the electrodes, as shown in figure 1(a). The signal generator supplies an electrical current impulse (width T = 0.1 ms, the "ON" period) followed by a pause (1 s, the "OFF" period) when the electrodes can not receive or transmit current (high impedance status).

The oscilloscope (Picoscope 2203, maximum sampling rate 20 MS/s) displays the potential difference between the two electrodes on channel A, the oscilloscope's "ground" being connected to the common point of the electrode with the 100 Ohm resistance, used for measuring the current on channel B. The signal on the channel B triggers the data acquisition and the display. Two metallic electrodes (Ag), covered by wet cotton (saline solution), deliver the electrical signal to the skin. The reference electrode has a greater surface (25 cm$^2$) than the active electrode (2 cm$^2$). To avoid influences on the results from sweat, the skin is wiped using ethanol before the measurement, and after 5 minutes the electrodes are attached on the desired place using elastic ribbons. The reference electrode was placed on the middle of the ventral side of the forearm (hairless skin), and the measuring electrode on the forearm close to the wrist. A constant current I = 4 mA is injected with the described device for a period of T = 0.1 ms, then the skin's potential is recorded. Figure 1(b) shows the oscilloscope traces for the potential difference between the electrodes (upper and middle traces, channel A) and the potential difference on the 100 Ω resistor (lower trace, channel B), generated by an electrical current of 4 mA.

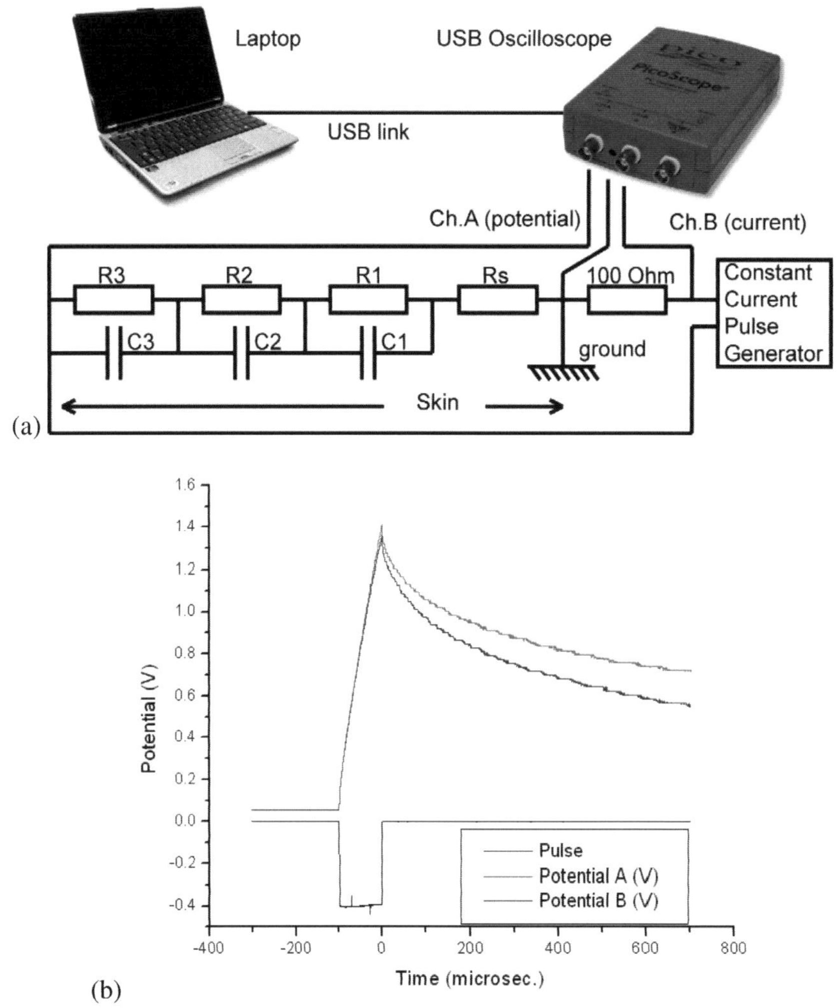


**Figure 1.** (a) Experimental arrangement and (b) oscilloscope traces for the potential difference between the electrodes (upper trace for positive impulse and middle trace for negative impulse) and the potential difference on the 100 Ω resistor (lower trace), generated by an electrical current of 4 mA. Potential difference between electrodes is 10 times greater than that on the scale (scope probe 1/10).

## 3. Model and data analysis

The electrical signal applied to the skin is an impulse of constant current with the width T, the "ON" period, followed by a pause when the skin can not receive or transmit current, the "OFF" period. The electrical model for the skin we have chosen consists of 3 parallel $R_i$ $C_i$ configurations (I = 1, 2, 3) that are all connected in series and a series resistance $R_s$ ($Z = R_s+\Sigma R_i||C_i$), as shown in figure 1(a). The solutions of the Kirchhoff equations system for $t\in[0,T]$ are:

$$I_i = I\,[1-\exp(-t/\tau_i)] \tag{1}$$

$$U= IR_s + IR_1[1-\exp(-t/\tau_1)] + IR_2[1-\exp(-t/\tau_2)] + IR_3[1-\exp(-t/\tau_3)] \tag{2}$$

where $\tau_i= R_iC_i$ is the time constant of the i-th loop;
I = constant current through circuit for the time period [0,T];
$I_i$ = current trough parallel resistance $R_i$;
U = the potential difference between the two electrodes;
$U_i$ = the potential difference on $R_i$ ($I_i$ $R_i$);
$U_s$ = the potential difference on $R_s$ (I $R_s$).

For t > T, the OFF period, the skin does not receive any electrical current so the potential drop on the series resistance is equal to 0 and we are left with a simple exponential decay of the potential for each $R_iC_i$ configuration. For sake of simplicity we adopted the time value t = 0 at the beginning of the decay, thus the potential difference between the electrodes will be:

$$U = \Sigma_i U_i = I\cdot\Sigma_i\, R_i[1-\exp(-T/\tau_i)]\exp(-t/\tau_i) \tag{3}$$

Dividing the potential difference by the excitatory current we obtain the skin response function (dimensionally an impedance)

$$U/I = \Sigma_i\, R_i[1-\exp(-T/\tau_i)]\exp(-t/\tau_i) \tag{4}$$

We extracted from the measured data only the points corresponding to the free exponential decay of the electrical potential of the skin (t > 0). Each time series of the potential difference between the electrodes was divided by the injected current used for measurement. The resulting response function, dimensionally an impedance, presented in figure 2, was fitted with a third order exponential decay function using the Microcal OriginPro8 software package. The results of the numerical analysis are given in table 1.

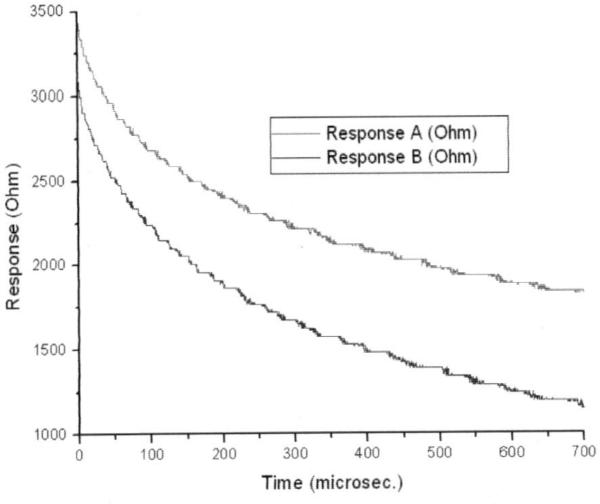


**Figure 2.** Response function (free decaying skin potential divided by pulse current 4 mA, an impedance) for positive pulse (upper curve) and negative pulse (lower curve).

Fitting to multiple exponentials is more difficult than fitting to a single exponential. Because the fitting function is:

$$y = y_0 + \Sigma_i A_i \exp(-t/t_i) \quad (5)$$

the parameter $y_0$ must be zero, as imposed by the physics of our problem (at very long time the skin potential vanishes). To avoid the numerical problem instability we begun the nonlinear fitting session by fixing only $A_1 = 0$ and $t_1$. When the program attains a stable value for fitting parameters we permit $A_1$ and $t_1$ to vary. The chi-square values were monitored all the time, in order to see if it is necessary to increase or decrease the number of time constants used.

Comparing equation (4) for the response function and the fitting function (5) we can calculate the resistance $R_i$ and the capacitance $C_i$ as below and the values are given in table 1:

$$R_i = A_i/[1-\exp(-T/t_i)] \text{ and } C_i = t_i/R_i = t_i\,[1-\exp(-T/t_i)]/A_i \quad (6)$$

**Table 1.** Values of skin resistances and capacitances corresponding to the amplitudes and time constants resulted from the fitting procedure.

| I (mA) | $A_1$ (Ω) | $t_1$ (μs) | $R_1$ (Ω) | $C_1$ (nF) | $A_2$ (Ω) | $t_2$ (μs) | $R_2$ (Ω) | $C_2$ (nF) | $A_3$ (Ω) | $t_3$ (μs) | $R_3$ (Ω) | $C_3$ (nF) |
|---|---|---|---|---|---|---|---|---|---|---|---|---|
| +4 | 296 | 28.0 | 304.6 | 91.9 | 800 | 177.9 | 1403 | 126.7 | 2303 | 2828 | 66287 | 42.7 |
| –4 | 238 | 26.9 | 243.9 | 110.3 | 832 | 136 | 1598 | 85.1 | 1967 | 1309 | 26744 | 73.5 |

## 4. Discussion and conclusion

Biological materials have unusual dielectric spectra with large values of time constants and static polarization. These complex effects result from physical processes at the solid-liquid interface of a porous system [6]. Identifying the characteristic electrical signatures of these physical mechanisms is important for the correct interpretation of experimental data. The chosen theoretical model has 3 time constants, electrical equivalent of the three RC groups connected in series. This choice was made because the electrical current flows successively across the different layers of the skin. This specific type of measurement, with constant current excitation, offers the advantage of a simple link between the measured data and the model parameters.

From table 1 we can see that the polarity of the excitatory current has a marked effect on the skin's electrical response. Because the epidermal stratum corneum is credited with the largest value of the resistance among skin layers, it is natural to consider that parameters $t_3$, $R_3$, $C_3$ belong to it. Here the parallel resistance $R_3$ for negative polarity pulse is 2.48 times smaller than that for positive polarity pulse, but capacity for negative pulse is larger than that for positive pulse.

Interestingly, the second set of parameters $t_2$, $R_2$, $C_2$ has an opposite behavior, the resistance increases and the capacitance decreases for negative pulse. We believe that this set of parameters belongs to a granular layer of keratinizated cells (stratum granulosum, stratum lucidum) because of the large value of resistance. For the shortest relaxation time $t_1$, parameters $R_1$, $C_1$ have a similar behavior with those in stratum corneum, resistance decreases and capacitance increases for negative pulse, but with much lower amplitude.

Our measurements show the presence of 3 relaxation times in the free decay of skin electrical potential after a constant current pulse. The electrical discharge of skin charge is faster after a negative pulse than after a positive one.

## References

[1] Grimnes S and Martinsen Ø G 2000 *Bioimpedance & Bioelectricity Basics* (London: Academic Press)
[2] Martinsen Ø G, Grimnes S and Karlsen J 1995 *Skin Pharmacol.* **8** 237-45
[3] Åberg P 2004 *Skin cancer as seen by electrical impedance* (PhD thesis) (Stockholm: Karolinska Institutet)
[4] van Boxtel A 1977 *Med. & Biol. Eng. & Comput.* **15** 679-57
[5] Bârlea N M, Bârlea S I and Culea E 2008 *Rom. J. Biophys.* **18** 87-98
[6] Chelidze T L and Gueguen Y 1999 *Geophys. J. Int.* **137** 1-15.

Journal of Physics: Conference Series **182** (2009) 012021 doi:10.1088/1742-6596/182/1/012021

# Liquid as template for next generation micro devices

**Jérôme Charmet[1], Henri Haquette[1], Edith Laux[1], Ganna Gorodyska[2], Marcus Textor[2], Guido Spinola Durante[3], Erwin Portuondo-Campa[3], Helmut Knapp[3], Roland Bitterli[4], Wilfried Noell[4] and Herbert Keppner[1]**

[1] HES-SO ARC, Institut des Microtechnologies Appliquées, La Chaux de Fonds, Switzerland

[2] ETHZ, BioInterfaceGroup, Zürich, Switzerland

[3] CSEM Centre Suisse d'Electronique et de Microtechnique SA, Alpnach, Switzerland

[4] Ecole Polytechnique Fédérale de Lausanne, Institute of Microengineering, Sensors, Actuators and Microsystems laboratory, Rue Jaquet Droz 1, 2000 Neuchâtel, Switzerland

E-mail: Jerome.Charmet@he-arc.ch

**Abstract:** Liquids have fascinated generations of scientists and engineers. Since ancient Greece, the perfect natural shape of liquids has been used to create optical systems. Nowadays, the natural shape of liquid is used in the fabrication of microlens arrays that rely on the melting of glass or photoresist to generate high quality lenses. However shrinkage normally associated to the liquid to solid phase transition will affect the initial shape and quality of the liquid structure. In this contribution, a novel fabrication technique that enables the encapsulation and replication of liquid templates without affecting their natural shape is presented. The SOLID (SOlid on LIquid Deposition) process [1] allows for a transparent solid film to be deposited and grown onto a liquid template (droplet, film, line) in a way that the liquid shapes the overgrowing solid layer. The resulting configuration of the SOLID devices is chemically and mechanically stable and is the base of a huge variety of new micro-nano systems in the field of microfluidics, biomedical devices and micro-optics among others. The SOLID process enables in a one step process the encapsulation of liquid microlenses, fluidics channels, drug reservoir or any naturally driven liquid structure. The phenomenon and solid-liquid interface resulting from the SOLID process is new and still unexploited. The solid layer used for the SOLID process chosen in this paper is poly-para-xylylene called Parylene, a transparent biocompatible polymer with excellent mechanical and chemical properties. Moreover, as the solid layer is growing over a liquid template, atomically smooth surfaces channels can be obtained [2]. The polymerization of Parylene does not exert stress and does not change the shape of the liquid; this latter aspect is particularly interesting for manufacturing naturally driven liquid structures. In this paper the authors explore the limits of this new method by testing different designs of SOLID encapsulated structures and their potential to deliver next generation micro devices.

## 1. Parylene deposition on liquids

Poly(chloro-para-xylylene) is deposited using a downstream LPCVD process known as Gorham process [3] that is carried out at typically 1-7 Pa. The dimer dichloro-para-xylylene is evaporated at typically 120 °C and cracked in a pyrolysis station at 650°C into reactive monomers. The deposition occurs in a two step reaction mechanism in a deposition chamber at room temperature. The low pressure of the standard Gorham process let water and other liquids disappear due to evaporation. The authors use in this work several liquids that have very low vapor pressures in a way that no liquid loss occurs at low pressure during deposition. Suitable liquids are *e.g* glycerol, silicone oils, adipates and phthalates offering different viscosities. Other liquids such as water and aqueous solutions, have been meanwhile successfully coated by the authors [2], using APCVD parylene deposition (inert carrier gas at atmospheric pressure). For the purpose of this work a 2 µm Parylene layer was deposited over the liquid loaded substrates, following the conventional Gorham process.

## 2. Trapping scenarios

The first approach consisted in the encapsulation of liquids with naturally driven shapes. Drops with contact angle as high as 150° were successfully encapsulated by the SOLID process as can be seen in figure 1 below. This paper will also describe the study of the encapsulation of meniscus in capillaries and other naturally driven structures.

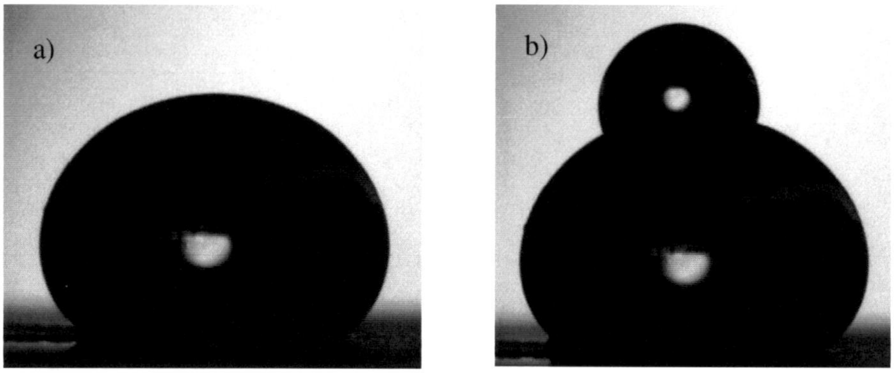


**Figure 1.** a) SOLID encapsulated drop of glycerol on fluorinated oleophobic surface. The parylene layer of about 2 microns was deposited using the conventional Gorham process. The structure is mechanically, chemically and thermally stable (given by the intrinsic properties of the parylene film). b) Deposition of another drop on the SOLID encapsulated structure as a proof that the drop in indeed encapsulated in a stable membrane.

As shown in figure 1, the shape of the liquid features encapsulated by the SOLID process depends solely on the surface energies at the liquid – solid and gas interfaces. Therefore in order to fabricate structures for specific purposes, trapping scenarios be they physical, chemical or a combination of both were investigated. Figure 2 gives a schematic overview of the trapping scenarios.

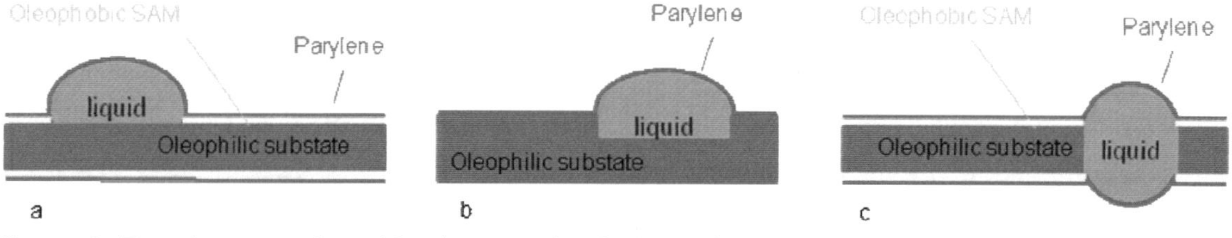


**Figure 2.** Trapping scenarios, side view: a) chemical trapping; b) physical trapping; c) chemico-physical trapping.

Microcontact printing (μCP) or direct dispensing of self-assembled monolayers (SAMs) are good chemical traps transfer techniques for manufacturing and fast prototyping respectively. However, in order to test the limits of SOLID process in terms of feature size and feature architecture, the molecular-assembly patterning by lift-off (MAPL) process [4] was investigated thoroughly (figure 3).

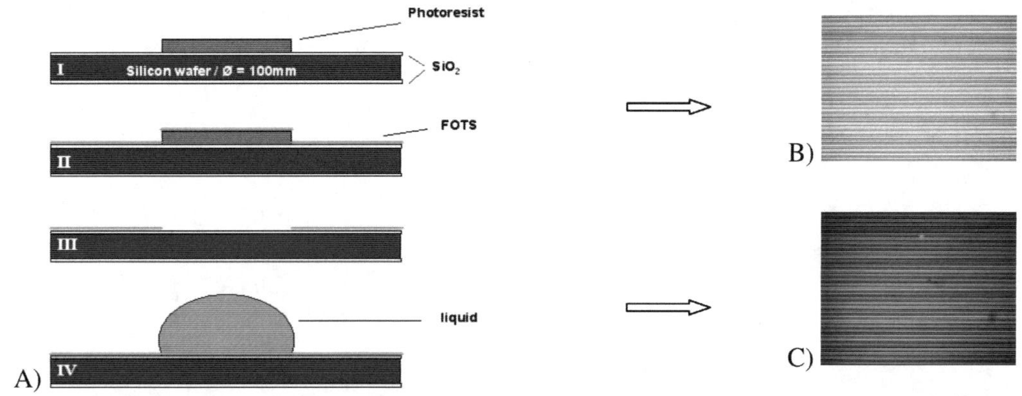


**Figure 3**. A) Schematics of Molecular Assembly Patterning by Lift-off (MAPL). A photoresist is pre-patterned on top of $SiO_2$ grown on a silicon wafer (I); FOTS is deposited onto the surface from vapor phase (II); lift-off of the residual resist in an organic solvents (III), and finally filling of bare areas with driver liquid (IV). B) Optical microscopy of pre-patterned photoresist stripes about 4 μm in width and periodicity 10 μm. C) Optical microscopy of the same stripes on the stage IV filled with bis(2-ethylhexyl)adipate.

Moreover a series of experiments were performed using laser microstructures to create physical, chemical and physico-chemical patterns. The physical traps were obtained using UV laser bulk micromachining to generate channels on a Silicon substrate. The chemical traps were obtained by first evaporating oleophobic SAMs on quartz slides that were selectively removed by laser micro-machining to create patterns.

Test structure including lines, circles, squares and more complex forms at dimensions from a few microns to several tens of microns were designed for that purpose (figure 4).

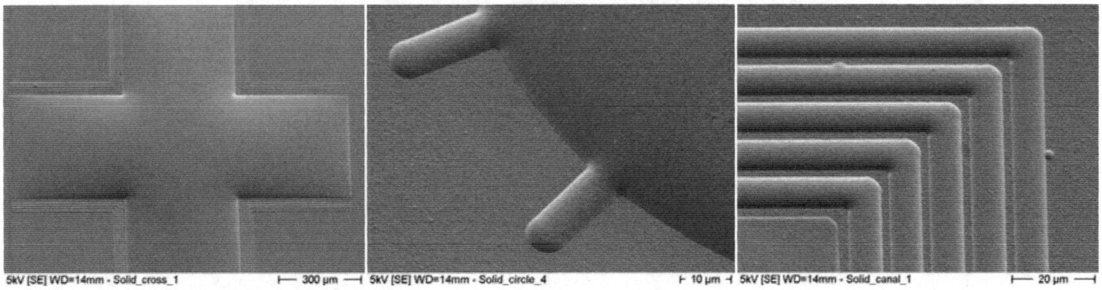


**Figure 4.** SEM micrographs of complex SOLID coated structures with a 2 microns parylene layer.

## 3. Loading scenarios

In the case of chemical trapping, after the substrate is functionalized, liquids can be loaded by dip-coating: as the substrate is withdrawn, the liquid remains confined to the wettable areas, reproducing the structures transferred in the previous step.

This method has obvious advantages for a quick patterning of liquid networks on a flat surface. However, it provides little control for the amount of liquid loaded and the neatness of the unwettable surface, which is often spotted by satellite droplets. Both, the liquid load, and the presence of residual droplets on the functionalized surface, are complex functions of the physicochemical properties of the liquid and of the dip-coating parameters, namely: the liquid density, viscosity and surface tension, the wetting angles on the substrate, the shape of the structures, the withdraw velocity and the tilt of the substrate. All these parameters determine the outcome of the dip-coating process in such a tangled way, that it is impossible to predict it, unless detailed numerical simulations are performed. With this aim we have carried out numerical simulations using CFD-ACE+ to test different dip-coating scenarios with different liquid properties. The results have proven their reliability and constitute an important tool for setting up new protocols of liquid patterning for SOLID applications.

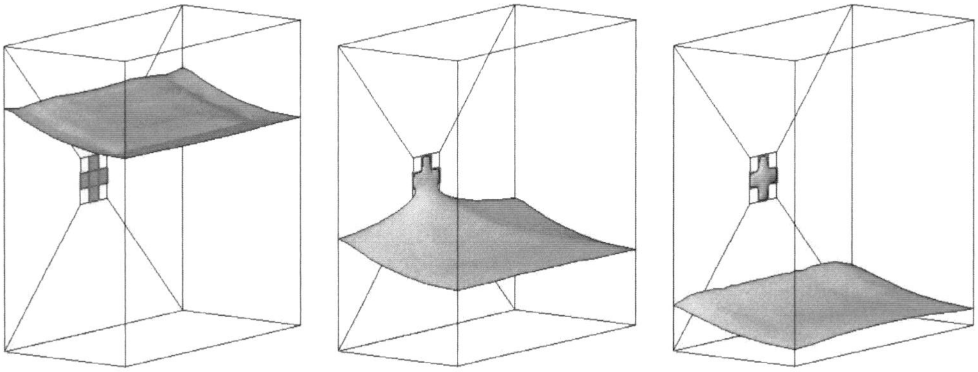

**Figure 5.** Sequence of snapshots from a dip-coating simulation of a cross shaped wettable patch. The global borders of the simulation cell are represented as a box. The substrate is the back side of the box, and the colored cross is the wettable area on it. The liquid-air interface is represented in green and is gradually brought down with every time step, as the cell is drained from below. As the liquid surface passes beyond the cross, part of the liquid detaches from the main volume, thus loading the wettable structure.

An alternative to the dip-coating method for loading the substrates is by direct dispensing at predefined positions.

The advantages of this system with respect to the dip-coating process are the better control of the dispensed volume and the absence of residual droplets outside the wettable pattern. Besides, this method is better suited for filling 3D structures, since there is no risk of bubble trapping during the load. The drawbacks are the increased cost of the technology and loading time. Also, for 2D structures smaller than 100 μm, dip-coating offers a more accurate filling than dispensing. The choice of one technique or the other depends ultimately on the specific requirements of each application.

## 4. Further results

The study of the SOLID encapsulation of dynamic structures will also be presented in this publication. In this approach, standing waves are generated on a liquid during the SOLID encapsulation.

## 5. Conclusions

The fact that a solid (polymer) layer can be deposited onto a liquid substrate has the potential to contribute to next generation micro devices. The paper presents different scenarios to fabricate complex naturally driven and tailored structures using the novel SOLID technology Physical and chemical liquid patterning scenarios were investigated to test the limits of the fabrication technology with respect to feature size and layout. The possibility of dip-coating allows loading complex trap structures at high throughput and paves the way to low-cost scenarios, whereas a more expensive dispensing alternative allows a better control of the volume dispensed. A problem-adapted simulation tool allows accurate predictions for loading of liquids on various trapping structures, whereby the quantity of trapped liquids is (predicted and experimentally confirmed) self-stabilizing.

Parylene has shown to be a suitable polymer with respect of conformal shape-replication, pinhole freeness, shape conservation due to stress-less overgrowing of the liquid template. Looking at later biomedical devices or drug delivering scenarios, parylene is particularly attractive due to its biocompatibility. In particular, the combination of SOLID technology together with MAPL technology allows complex parallel processing proved to be an extremely reliable self-stabilizing trap loading at laboratory level.

Some layouts using liquid structures are surprising indeed, and, in some cases, strictly reserved to the use of liquid templates. Generally spoken such Solid on Liquid templates contribute in a substantial way to push the potential of next generation micro and nano systems far beyond the state of the art.

## References

[1] Keppner H and Benkhaïra M 2006 *Method for producing a plastic membrane device and the thus obtained device*, patent WO/2006/063955
[2] Paper submitted
[3] Gorham W F 1966 *J. Polym. Sci. Al* **4** 3027-39
[4] Falconnet D, Koenig A, Assi F and Textor M 2004 *Adv. Funct. Mater.* **14** 749-56.

# Ion mobility analyzer - quadrupole mass spectrometer system design

**C Cuna[1], M Leuca[1], N Lupsa[1], V Mirel[1], V Bocos-Bintintan[2], Stela Cuna[1], V Cosma[1] and Florina Tusa[1]**

[1] National Institute for Research and Development of Isotopic and Molecular Technologies, 65-103 Donath, 400293 Cluj-Napoca, Romania

[2] Babes-Bolyai University, Faculty of Environmental Sciences, 3 Fantanele, 400294 Cluj Napoca, Romania

E-mail: cornel.cuna@itim-cj.ro

**Abstract**. Because of their extremely high sensitivity for chemicals with elevated electronegativity or high proton affinity the ion mobility analysers are ideal for the ultra-trace detection of toxic or explosive chemicals, most of these situated often at concentration levels of sub-ppb (parts-per-billion). Ion mobility spectrometers (IMS) can be used to identify illicit drugs or environmental pollutants. Since resolution of an IMS is relatively low, to achieve an accurate identification of target analyte it is recommended to couple the IMS with a quadrupole mass spectrometer (QMS) or a time of flight mass spectrometer, acquiring in this way confirmatory information. This coupling is made through a specific interface. In this paper, an experimental model of such a tandem instrument, IMS-QMS is described. Accomplishment of this general purpose will be done, overcoming a series of specific issues. This implies the solving, using innovative solutions, of a series of complex issues: ensuring the stability of the ions beam generated by ion source; transfer with a good efficiency of the ionic current from IMS analyser to QMS; and realization of a special electronic circuitry which will be able to detect both positive and negative ions.

## 1. General scheme

The general scheme of the IMS-QMS tandem is given in figure 1. We have proposed us to build a complex detector of tandem type, IMS-QMS, ultra-sensitive and having a fast response, to identify dangerous substances (organophosphoric chemicals).This tandem can be used also for the detection of other compounds, as environmental pollutants.

Positive and negative ions are generated by a corona discharge ionisation source, and are then transported by the electric field and by the carrier gas towards the gap between the two cylinders of the IMS analyser. Depending on the voltages that are applied on those two cylinders (direct compensation voltage and asymmetric wave alternative voltage, respectively), the ions are separated and sent to the interface. From this point, the ions are introduced into the quadrupole analyser, where they are mass analysed. Finally, the ionic current is measured using an electrometer and is further picked up by either an oscilloscope or an analogue-to-digital interface.

Others groups reported results obtained with similar combination of analysers (the level of perchlorate in water matrices) /1/, (the explosive detection) /2/, (isomeric dihalogenated benzenes) /3/.

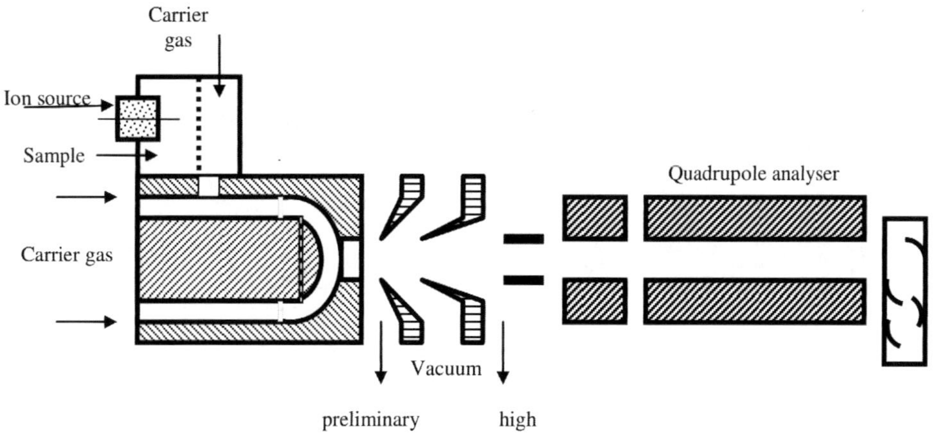


**Figure 1.** The general scheme of tandem IMS-QMS.

## 2. Ion source

The ion source we have designed, manufactured and tested was a corona discharge ionisation source. The well-known corona discharge is, by itself, a gas electrical discharge that reunites the ionisation processes of this gas in the proximity of the active electrode /4/, /5/, where the electric field is strong. Corona discharge implies usually the existence of two electrodes – one having a strong curvature (e.g., a very small diameter needle or wire), and the other having a small curvature (like a plane); therefore, a so-called "point-to-plane" configuration of the corona discharge is used. The strong curvature will ensure a very high gradient of potential around the tip electrode, which shall be enough to generate the necessary plasma.

If the tip electrode is polarised positive, relative to the flat electrode, we will have a positive corona discharge, while by reversing this polarity one shall have a negative corona discharge. The unipolar current density for positive and negative corona discharge is distributed on the plane in accordance with the Waiburg law:

$$j(\theta) = j_0 \cos^5 \theta \qquad (1)$$

$$j_0 \sim I/(2d^2) \qquad (2)$$

where $tg\theta = r/d$, $I$ is the total corona current and $d$ is the distance from point to plane. Field lines and ion trajectories are somewhat shifted by the repelling force due to the spatial charge. The unipolar current $I$ at a certain value $U_0$ cannot be larger than the saturation limit.

$$I_{sat} \sim 2\mu\varepsilon_0 U_0^2 / d \qquad (3)$$

where $\mu$ is ionic mobility and $\varepsilon_0$ is permittivity.

One can demonstrate that more than 95% of the input electrical energy for an unipolar corona discharge goes to the neutral gas, and only several percent go toward the producing a directional movement of the gas. Thus, the largest part of the energy is used to heating the gas, and the positive ions can strongly excite the atoms and the neutral molecules present in the drift region with low intensity of the field.

## 3. Cylindrical FAIMS (high-field asymmetric waveform ion mobility spectrometry) analyser

The cylindrical FAIMS /6/ analyser consists of two cylinders, which are denoted electrode I and electrode II. These cylinder electrodes have been manufactured using stainless steel (type

Journal of Physics: Conference Series **182** (2009) 012022 doi:10.1088/1742-6596/182/1/012022

10TiNiCr180) and are concentric, having a gap of 0.5 mm between them. The asymmetric waveform is applied on the external cylinder (5), while on the internal cylinder (6) the direct voltage (compensation voltage) has been applied. This entire system was mounted on a flange made of Teflon PTFE (3). The ions generated by the corona discharge enter the analytical gap between the two cylinders through four holes, and the analysed ions, which were travelling through the gap, exit toward detector through a small hole with 0.5-2 mm diameter (figure 2).

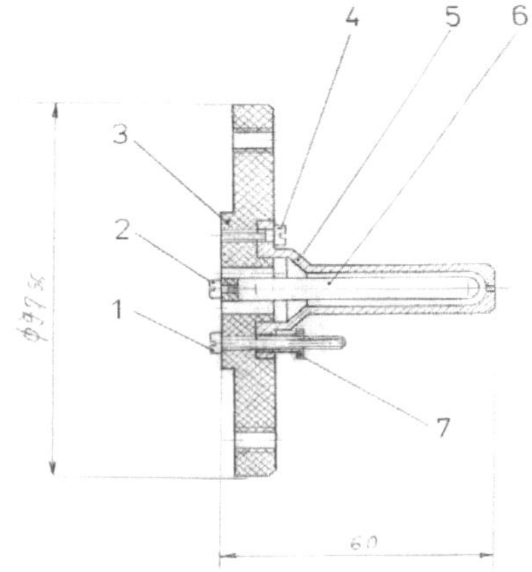


**Figure 2.** Cylindrical analyzer: 1-screw, 2-screw, 3-flange, 4-screw, 5-outer electrode, 6-inner electrode; 7-v. bucea.

## 4. Interface IMS – quadrupole MS

Figure 3 presents the above mentioned interface, together with the mounting system for Einzel electrostatic lenses and for the hexapole. Here, one may observe the bodies I (1) and II (17). The Einzel lens is mounted in the body I by using two separator pieces (9) and 10). The lenses (16) are made of non-magnetic stainless steel discs; they are being secured by using a set of bushed bearings (11) and insulating washers (12). Electrical connections are made between the bushed bearing IV (11) to bushed bearing III (7), and from here to the special screw (4), which has a pin to electrical contact. Those pins are also connected to the electrical wires for hexapole's polarisation. Ion beam is coming from the ion mobility analyser, enters the repelling hole (15) and is accelerated towards the superior hole of body II. From this point, the ion beam is directed toward the Einzel lens, where it is focused, then it is transported through nozzle (19) to the hexapole electrostatic lens mounted in body I (1). This hexapole eliminates the second order image aberration of the ionic beam and sends this beam to be analysed in the quadrupole filter.

According to the theory /7/ a series of relations between the basic parameters of a QMS, which may be used to get a first estimation of their values, have been chosen. Thus, the maximum mass of the ions that could pass through a quadrupole filter is given by:

$$\mathrm{m_m} = \frac{7x10^6 V_m}{f^2 r_0^2} \tag{4}$$

where $V_m cos2\pi ft$ represents the radiofrequency voltage applied between the adjacent rods, $r_0$ (in meters) is the radius of the rods, and $m_m$ the maximum mass of the ion (in a.m.u.). For the vast majority of the instruments, choosing of $V_m$ and $r_0$ is relatively limited. Thus, $V_m$ is on the order of 3,000 V, and $r_0$ in the range 3-10 mm. Another useful relation is:

$$\frac{m}{\Delta m} = \frac{1}{K} N^n \tag{5}$$

where $N$ is the number of *rf* cycles of the field applied to the ion, and $\Delta m$ is the width of the peak having the mass $m$. It was found that *"n"* has the value n=2, and K=20 /7/.

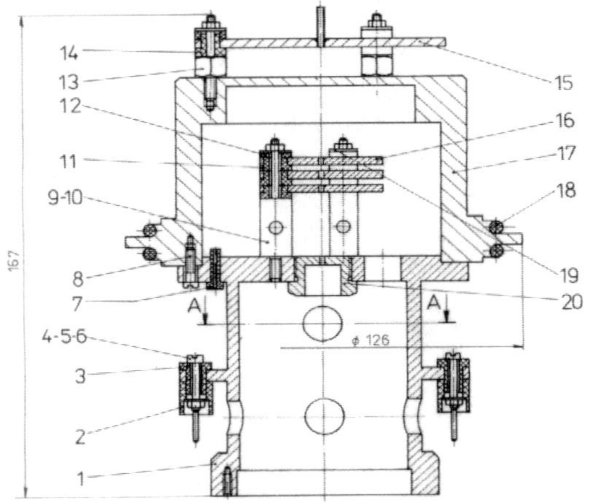


**Figure 3.** The assembly of the Einzel lenses within the interface: 1-body I, 2-v. bucea I, 3-v. bucea II, 4-special screw I, 5-washer, 6-screw nut, 7-v. bucea III, 8-screw, 9-distance piece, 10-washer, 11-v. bucea IV, 12-insulating washer, 13-special screw, 14-v. bucea V, 15-ion repeller, 16-disk, 17-body II, 18-O-ring, 19-washer, 20-nozzle.

Equation (5) may be written as:

$$\frac{m}{\Delta m} = 0{,}05 \left\{ fL\sqrt{\frac{m}{2eV_z}} \right\}^2 \tag{6}$$

where *"e"* is the charge of the electron (C), *m* the molecular mass and *Δm* mass difference (Kg), K = 20 and n = 2. From this we will obtain the expression for the maximum resolution:

$$\frac{m_m}{\Delta m} = \frac{L^2 V_m}{570 V_z r_0^2} \tag{7}$$

Using cylindrical rods for the electrodes, instead of hyperbolic profile electrodes – as an approximation of the hyperbolic field – represents a factor that degrades the instrument performances, especially its mass resolution. By exploiting the (the above mentioned relation), the following main parameters of the quadrupole analyser were calculated: $r_0$ = 4 mm; L = 200 mm; $V_m$ = 1250 V; F = 2,082 $MH_z$; $m_m$ = 300 a.m.u.; R = 300.

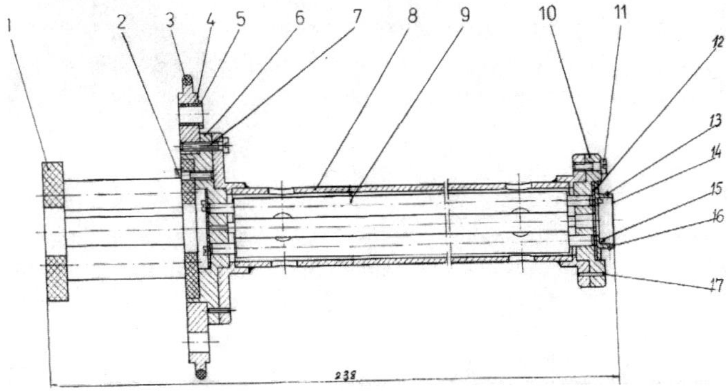


**Figure 4.** The scheme of the quadrupole analyser: 1-hexapole, 2-screw, 3-O-ring, 5-v. bucea, 6-rod support I, 7-screw, 8-analyser's housing, 9-rods, 10-rod support II, 11-screw, 12-lid, 13-safety lock, 14-collector grid, 15-screw, 16-grid support, 17-peg.

The length of the rods is 200 mm, and diameter of the rods is $\phi$ = 8mm. The alignment of the quadrupole rods has been achieved with an accuracy of the $10^{-5}$ m, being an essential step in generating a symmetric and uniform electric field inside the quadrupole analyser. In figure 4 can be observed the rods (9), analyser's housing (8), and the system for securing the rods in place (6). A hexapole electrostatic lens (1) has been mounted and aligned inside a dedicated room within the analyser's flange, having the role of eliminating the second-order optical aberrations of the ionic beam coming through the interface. From the hexapole the ions enter the quadrupole analyser where they are separated after their m/q ratio and directed to the collector.

## 5. Vacuum system used for testing the tandem IMS – QMS

This system should ensure a continuous pressure drop, from atmospheric pressure (existent inside the IMS) to a vacuum of at least $10^{-6}$ mbar (inside the quadrupole analyser). Achievement of this goal, combined with the appropriate electrical polarisation, will assure the transport of ions from the IMS to the quadrupole mass analyser, through the interface.

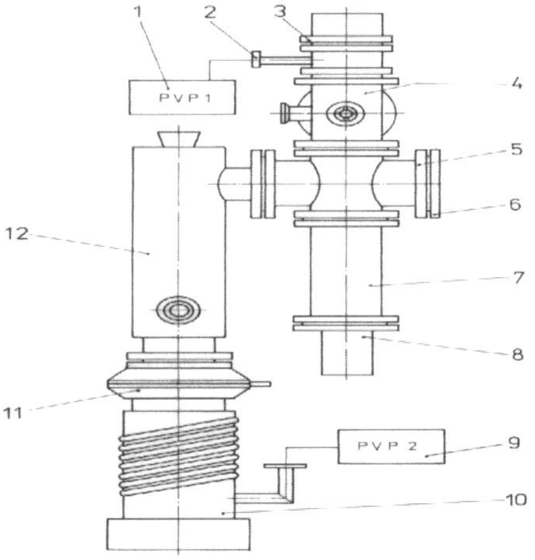


**Figure 5.** Diagram of the whole system: 1 - preliminary vacuum pump, 2 - intermediary body, 3 - ion source, 4 - body I, 5 - body II, 6 - blind flange,7 - body IV, 8 - ion detection system, 9 - preliminary vacuum pump, 10 - diffusion pump, 11 - water trap, 12 - liquid nitrogen trap.

As depicted in figure 5, the IMS has been mounted in section (2) and (3), the IMS-QMS interface in sections (4) and (5), while the quadrupole filter and its associated ion detector were mounted in sections (7) and (8). The preliminary vacuum pump PVP-1 (1), with a flow rate of 24 $m^3$ $h^{-1}$, will create a vacuum of $10^{-3}$ mbar inside the interface, while the vacuum system (composed of diffusion pump with oil (10) with its preliminary vacuum pump PVP-2 (9) and nitrogen trap (11)) will generate inside the quadrupole filter a vacuum of $10^{-5}$-$10^{-6}$ mbar.

## 6. Detection and processing of ionic current

Detection and processing of ionic current will be done by using a home-made electrometer, designed and manufactured by our institute, which may be coupled to a fast oscilloscope (TEKTRONIX MSO-4104) or to a National Instruments card (NI SCC-68) interfaced to a personal computer.

## 7. Conclusions

We succeeded in designing and realising the mechanic and electronic projects of a complex tandem composed of an asymmetric waveform IMS and a QMS. This hyphenated instrument will be used to detect ultra-traces of toxic, pollutants and explosive chemicals. Our solutions for the main components of this tandem were aimed to obtaining the highest possible ionic current (about $10^{-10}$ A), a stable ionic beam and a mass resolution of at least 300.

## References

[1] Ells B, Burnett A D, Purves W R and Guevremont R 2000 *J. Environ. Monit.* **2** 393
[2] Hunka E D and Austin E D 2006 Sandia Report SAND 2005-6908 1
[3] Borsdorf H, Nazarov E G and Eiceman G A 2004 *Int. J. of Mass Spectrom. Ion Proc.* **232** 117
[4] Eiceman G A and Karpas Z 2005 *Ion Mobility Spectrometry* (Boca Raton, Fl: CRC Press)
[5] Goldman M, Goldman A and Sigmund R S 1985 *Pure and Appl. Chem.* **57** 1353
[6] Purves R W and Guevremont R 1998 *Rev. Sci. Instrum.* **69** 4094
[7] Dawson P H 1976 *Quadrupole mass spectrometry and its applications* (Amsterdam: Elsevier).

# On the sensitivity of FPPE – TWRC method in thermal effusivity investigations of solids

**Dorin Dadarlat, Mihaela Streza, Mircea N Pop and Valer Tosa**

National Institute for Research and Development of Isotopic and Molecular Technologies, 65-103 Donath, 400293 Cluj-Napoca, Romania

E-mail: nicolae.dadarlat@itim-cj.ro

**Abstract**. The front detection configuration (FPPE) together with the thermal-wave-resonator-cavity (TWRC) method was used for direct measurement of the thermal effusivity of solid materials inserted as backings in the FPPE detection cell. It was demonstrated that the normalized phase of the FPPE signal has an oscillating dependence as a function of sample's thickness. The paper presents experimental results on solid materials, with various values of thermal effusivity (Cu, brass, steel, bakelite, wood). A study of the sensitivity of the technique for different liquid/backing effusivity ratios is performed. The highest sensitivity was obtained when investigating solids with values of thermal effusivity not far from the effusivity of the liquid layer of the detection cell.

## 1. Introduction

During the last decades, the photopyroelectric (PPE) calorimetry developed many ways in order to obtain the values of the thermal parameters of condensed matter samples [1, 2]. In principle, one may combine detection configurations (back or front), sources of information (PPE amplitude or phase) and scanning parameters (chopping frequency or sample's thickness), in order to obtain one dynamic thermal parameter (usually, thermal diffusivity or effusivity) [3, 4].

It is well known that, when investigating solid samples, it is not easy to obtain accurate quantitative results, due to the influence of the coupling (sample-sensor) fluid. The PPE calorimetry is a contact technique and the coupling fluid, always necessary when investigating solids, leads to uncontrolled errors.

If we focus our attention on the front photopyroelectric (FPPE) configuration, it was largely used in the past to obtain thermal parameters of liquid and solid samples [1, 2, 5, 6]. Mandelis and co-workers [1, 7, 8] proposed, during the last decade, the so called thermal-wave resonator cavity (TWRC) method, which, in the back (BPPE) configuration, showed to be very accurate for investigating thermal properties (especially thermal diffusivity) of liquids [5, 6, 9, 10].

Recently, we proposed an alternative technique that combines the FPPE configuration with the TWRC method [11]; in this configuration, by monitoring the thickness of a fluid sample, the thermal effusivity of a solid material used as backing in the detection cell can be measured. In such a way the coupling fluid is not anymore a disturbing factor, but its properties can be used in order to obtain the value of the thermal effusivity of a solid.

In this paper we perform an investigation on the sensitivity of the coupled FPPE-TWRC technique in measuring the thermal effusivity of solids, as a function of liquid/backing effusivity ratio.

## 2. Theory

In the FPPE configuration, the radiation impinges on the front surface of the pyroelectric sensor, and the sample, in good thermal contact with its rear side, acts as a heat sink. The detection cell is composed by 4 layers: air, (directly irradiated) sensor, liquid sample and solid backing.

In the approximation of the one-directional heat propagation, and thermally thin limit for the sensor ($exp(\pm\sigma_p L_p)=1\pm\sigma_p L_p$), the normalized complex PPE signal is given by [11]:

$$V_n = \frac{\sigma_p L_p + b_{sp}}{(\sigma_p L_p) + b_{sp}\left[\frac{1 + R_{bs}\exp(-2\sigma_s L_s)}{1 - R_{bs}\exp(-2\sigma_s L_s)}\right]} \tag{1}$$

where

$$S=exp(\sigma_s L_s),\ P=exp(\sigma_p L_p),\ \sigma_j = (1+i)a_j;\ \mu=(2\alpha/\omega)^{1/2},\ b_{ij} = e_i/e_j \tag{2}$$

In equations (1) and (2), $\omega$ is the angular chopping frequency of radiation, σ and $a$ are the complex thermal diffusion coefficient and the reciprocal of the thermal diffusion length ($a = 1/\mu$), respectively. Symbols $p$, $s$ and $b$ refer to pyroelectric sensor, liquid sample and backing material, respectively. $R_{ij} = (b_{ij}-1)/(b_{ij}+1)$ represents the reflection coefficient of the thermal wave at the $ij$ interface.

Equation (1) was obtained after a double normalization: first with the signal obtained with empty sensor directly irradiated and then, with the signal obtained with very thick liquid sample ($exp(-\sigma_s L_s) = 0$). Equation (1) indicates that the normalized PPE signal depends on the thermal effusivity of the backing material and a thickness scan of the amplitude and/or phase of the PPE signal (at constant chopping frequency) can lead to its direct measurement.

Mathematical simulations of the phase of the FPPE signal as a function of liquid sample's thickness are displayed in figure 1.

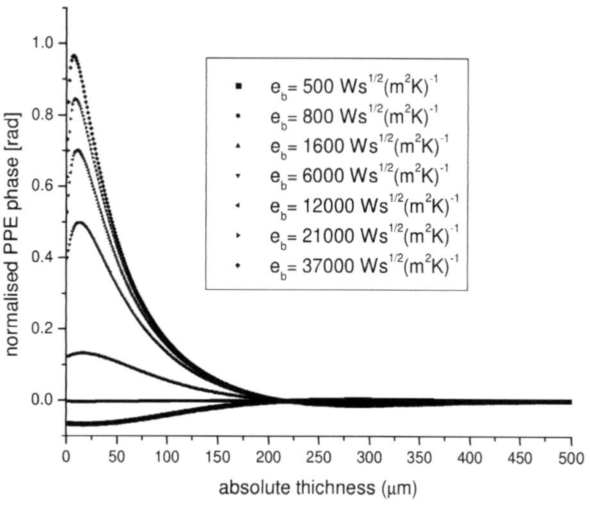


**Figure 1.** Mathematical simulations (equation (1)) of the behaviour of the normalized PPE phase as a function of the absolute thickness of the sample.

## 3. Experimental results

The experimental set-up was largely presented elsewhere [4, 6, 11]. We will give here only some particular details of the experiment. The modulation frequency was 1 Hz, the thickness variation step was 0.9μm and ethylene glycol was used as liquid layer. Solid samples (copper, glass, bakelite, wood), with values of thermal effusivity covering a large range (350-37000 $Ws^{1/2}/m^2K$) were inserted in the detection cell as backing materials. The values of the thermal effusivity were obtained by optimizing the fit performed with equation (1), with the effusivity of the backing as fitting parameter.

The results obtained for the behaviour of the phase of the FPPE signal as a function of liquid's thickness, together with the values obtained for the thermal effusivity are displayed in figure 2.

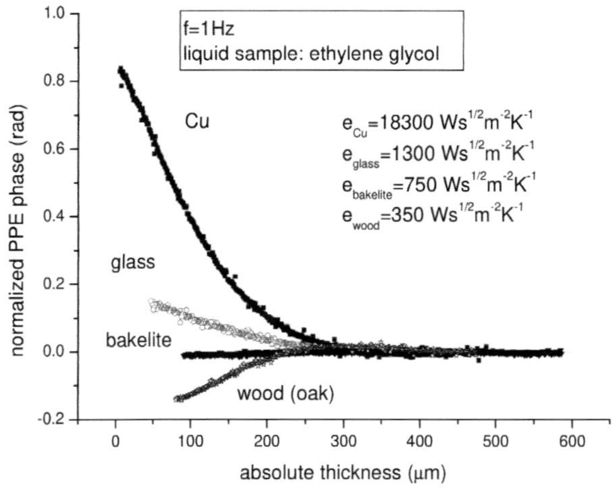


**Figure 2.** Normalized PPE phase as a function of sample's thickness for four solid backing materials with different values of thermal effusivity: Cu, glass, bakelite, wood.

Figures 3 and 4 represent contour maps of the precision of the fits for two materials with very different values of thermal effusivity.

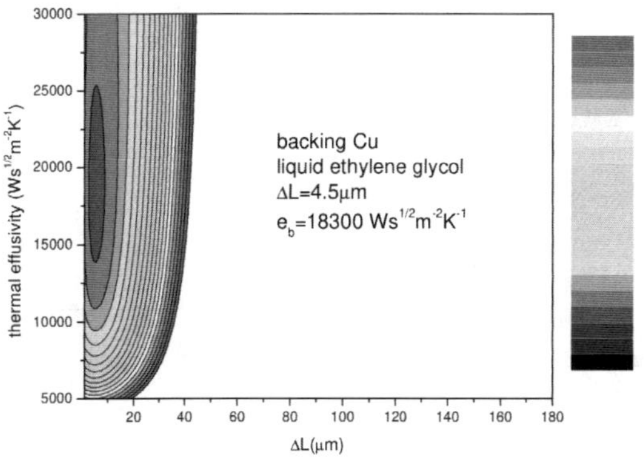


**Figure 3.** Contour map of the precision of the fit performed with equation (1) on the experimental data obtained on Cu backing. X-axis represents the error in the measurement of the absolute liquid's thickness. Conclusion: good localization of the position of the backing and rather low precision (relative error ± 30%) in the value of thermal effusivity.

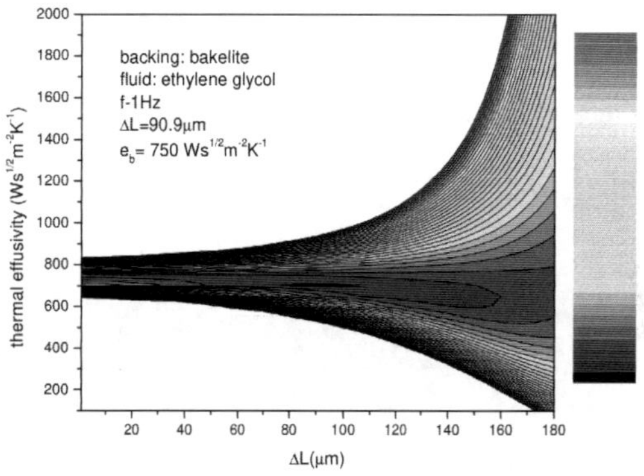


**Figure 4.** Contour map of the precision of the fit performed with equation (1) on the experimental data obtained on bakelite backing. X-axis represents the error in the measurement of the absolute liquid's thickness. Conclusion: bad localization of the position of the backing and rather good precision (relative error ± 5%) in the value of thermal effusivity.

## 4. Concluding remarks

The front PPE configuration, coupled with the TWRC method, represents a suitable alternative for the direct measurement of thermal effusivity of solid materials. The value of the thermal effusivity of the solid material used as backing in the detection cell was obtained by performing a sample's (liquid) thickness scan of the phase of the complex FPPE signal. The accuracy of the measurement (compared with results reported in the literature) is dependent on the signal-to-noise ratio (about 1000 in our experiment), on the thickness variation step (0.9 μm), but also on the ratio of the liquid/solid effusivities. As revealed in figures 3 and 4, the best accuracy was obtained when the $e_s/e_b \approx 1$ and, at this stage of investigation, it is of about 90%. For backings with large values of thermal effusivity (metals), the method shows accuracies of only about 50%-70%, due to the saturation of the signal for thermally thin regime of the liquid (see equation (1)).

## References

[1] Mandelis A and Matvienko A 2007 *Pyroelectric Materials and Sensors* (Kerala: D. Remiens)
[2] Delenclos S, Dadarlat D, Houriez N, Longuemart S, Kolinsky C and Hadj Sahraoui A 2007 *Rev. Sci. Instrum.* **78** 024902
[3] Mandelis A and Zver M M 1985 *J. Appl. Phys.* **57** 4421
[4] Dadarlat D, Bicanic D, Visser H, Mercuri F and Frandas A 1995 *J.Amer. Oil Chem. Soc.* **72** 273
[5] Dadarlat D, Chirtoc M, Neamtu C, Candea R and Bicanic D 1990 *Phys. Stat. Sol.* **(a)121** K231
[6] Dadarlat D and Neamtu C 2009 *Acta. Chim. Slovenica* **56** 225
[7] Shen J and Mandelis A 1995 *Rev. Sci. Instrum.* **66** 4999
[8] Shen J, Mandelis A and Tsai H 1998 *Rev. Sci. Instrum.* **69** 197
[9] Pittois S, Chirtoc M, Glorieux C, van den Bril W and Thoen J 2001 *Analytical Sci. (Japan)* **17** S110
[10] Balderas-Lopez L A, Mandelis A and Garcia J A 2000 *Rev. Sci. Instrum.*, **71**, 2933
[11] Streza M, Pop M N, Kovacs K, Simon V, Longuemart S and Dadarlat D 2009 *Laser Phys.* **19** 1340.

Journal of Physics: Conference Series **182** (2009) 012024 doi:10.1088/1742-6596/182/1/012024

# Mass spectrometry of toluene clusters using a laser produced plasma for tunable single photon ionization

**Tonia M Di Palma and Antonio Borghese**

Consiglio Nazionale delle Ricerche, Istituto Motori, Via Marconi 8, 80125 Napoli, Italy

E-mail: t.dipalma@im.cnr.it

**Abstract.** Here we report the first operation of a new apparatus of time of flight mass spectrometry using the tunable Vacuum-UV radiation emitted by a laser produced plasma as ionising source. Toluene cluster mass spectra acquired at different ionising wavelength are presented. Due to massive fragmentation the mass spectra change significantly by increasing the photon energy above the toluene ionisation potential. We found that the more reliable information on the cluster distribution can be gained by tuning the photon energy at value slightly higher than the toluene ionisation energy. The results demonstrate that the developed tunable vacuum-UV source can be successfully used as a soft ionisation method in the study of weakly bound complexes as well as of real complex gaseous mixture.

## 1. Introduction

The chemical analysis of gaseous mixtures of complex composition with high selectivity and sensitivity is a topic of relevance in many branches of science and technology. The impact of the combustion by-products emitted from vehicles, industries and incinerators, on the human health and environment, both on local and global scale, is beyond any doubt. There is therefore a clear need for developing high-throughput methods for comprehensive analysis of a large number of compounds, including the case of low concentrations. The key-concern here is the simultaneous multi-species detection of compounds in complex mixtures with both high sensitivity and selectivity to give a comprehensive picture of the mixture composition.

The requirements of sensitive, selective and simultaneous detection of species in complex mixtures can be achieved by combining the features of the Time of Flight (TOF) Mass Spectrometry with those of the Vacuum-UV (VUV) photoionization. Compared to other ionization methods used in combination with TOF spectrometry [1], the VUV photoinization is not dissociative, giving less congested mass spectra and enabling more affordable analysis. Moreover, the use of VUV photons continuously tunable across and above the ionization threshold adds a spectroscopic tool to the mass analysis and gives access to the details of the molecular structures, particularly if supported by quantum-mechanical modelling. The method has been demonstrated in many experiments using tunable VUV synchrotron light sources at large-scale facilities [2], which are obviously unfit to be integrated in compact instruments for widespread use.

Recently, we have carried out characterization studies on Laser-Produced Plasmas (LPPs) [3] and have proposed an LPP-based VUV light source, suitable for application of tunable Single-Photon Ionization (SPI) to TOF mass spectrometry [4]. LPP-based VUV light sources provide at once: *i*)

single-photon ionization, as they emit in the useful spectral range 6-12 eV, *ii*) spectral tunability, as their emission spectrum is almost continuum in that band, *iii*) photon fluxes much higher than those achievable with glow-discharge lamps.

Here we report preliminary results using a set-up based on TOF mass spectrometry coupled with a LPP as VUV photoionization source. As a test case, we chose to study the toluene clustering in a supersonic jet. The toluene clusters are van der Waals (vdW) aggregates with bonds far weaker than the chemical bond. The dynamics of the vdW cluster formation in the supersonic jet has been extensively studied [5] by means of mass spectrometry coupled with laser ionisation. The fragmentation induced by the multiphoton ionization was the main drawback of the laser ionization. Great efforts have been devoted to infer reliable information on the neutral cluster distribution (i.e. the real distribution of the clusters formed in the adiabatic expansion of the gas jet) from the analysis of the ion distribution. More information can be gained by using soft and tunable ionisation sources.

The analysis of the toluene clusters formed in a supersonic jet constitutes a severe test for photoionization experiments devoted to detect molecular compounds without fragmentation, aiming eventually towards a comprehensive chemical analysis of complex gaseous mixtures.

## 2. Experimental apparatus

The SPI-TOF apparatus is housed in two vacuum chambers, separated by a skimmer and evacuated by two 1000 L/s turbomolecular pumps (see figure1).

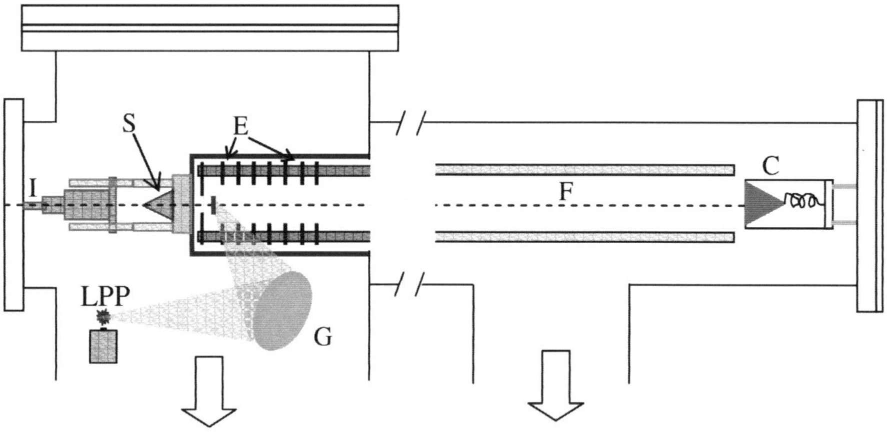


**Figure 1**. The SPI-TOF set-up. I = gas sample inlet, S = skimmer, E = extraction and focusing electrodes, F = 900 mm free flight region, C = channeltron, G = grating.

A first chamber includes the VUV source based on a Laser Produced Plasma on a Xe supersonic jet. A detailed description of the LPP source set-up has been given in ref [3] and[4] therefore here is only mentioned. Briefly, a Nd-Yag laser beam (1064 nm, 10 ns, 200 mJ) is focused on the axis of a supersonic Xe jet generated by a solenoid pulsed valve (0.8 mm nozzle) at about 1 mm away from the nozzle. The stagnation pressure is 10 bar and the valve aperture time is 400 □s. The radiation emitted by the laser spark is collected, spectrally dispersed and focused by a concave flat field grating (70 mm diameter, 210 mm focal length, 8 nm/mm dispersion) in a line shape volume at the centre of the vacuum chamber. The emission spectrum of the xenon plasma is mostly continuum in the 100-200 nm wavelength range and VUV photon fluxes exceeding $10^{12}/cm^2$.nm.pulse (with 2% BW @150 nm) have been measured on the grating focus in the first 20 ns from the laser onset. More details on the spectroscopic characterization of the LPP source have been reported elsewhere [3,4]. In the same chamber, a second pulsed valve is used for the gas sample intake. We used a gaseous mixture of 2 bar of He and vapour of liquid toluene (Aldrich HPLC 99.9%) at 25°C. A supersonic jet is formed in the expansion of the gas and a molecular beam is selected by a 0.4 mm conical skimmer, coaxially to the TOF tube and crossing the VUV optical axis at 90°. The skimmer is hold directly on the wall of an inner chamber (*i.d.* 60 mm) housing the extraction fields of the TOF, so that the distance between the valve nozzle and the ionization region is minimized to 60 mm. The inner chamber, connected to the flight tube, is provided with 10×30 $mm^2$ $MgF_2$ windows which allow the VUV photons to reach the molecular beam while preventing most of the valve-injected gases from reaching the ionization and flight zone. The mass spectrometer is housed in a second vacuum chamber. It is a home-made linear TOF in a Wiley Mc-Laren geometry, equipped with the extraction/acceleration electrodes, seven cascaded focusing electrodes and a 90cm-long free flight region. The ion detector is a channeltron. With a 200 V/cm continuum extraction field, the resulting mass resolution is m/□m = 200 @92 a.m.u.. A stepping motor is used for the wavelength scanning and each mass spectrum is acquired by accumulating 300 shots by means of a home made LabVIEW® program driving a DAQ card (National Instrument PCI-5152).

## 3. Results and discussion

Mass spectra of toluene clusters have been acquired by scanning the photon energy from 8 eV to 12.5 eV with a step 0,3 eV. Figure 2 shows the mass spectra of the toluene clusters at three photon energies (8.8, 10.4, 11.5 eV).

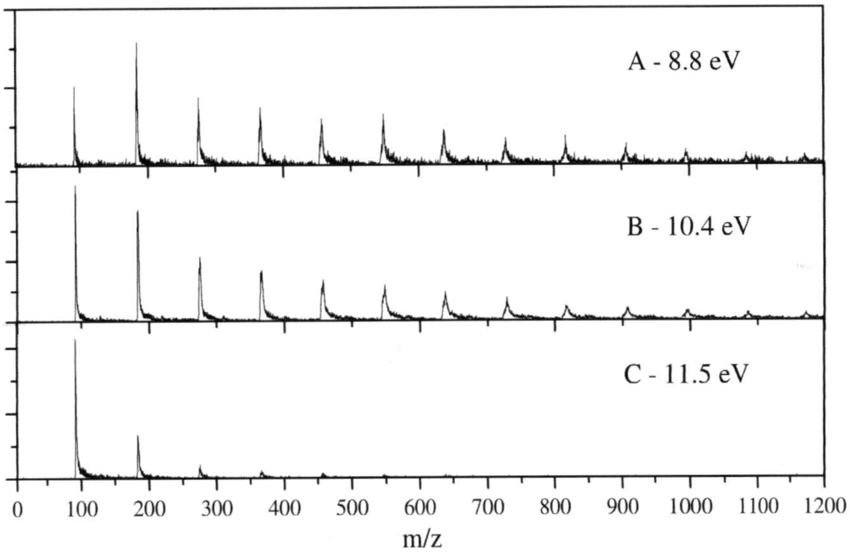


**Figure 2.** Cluster distribution at three photon energies. The leftmost peak correspond to the toluene monomer, the following peaks to clusters of 2, 3, … molecules.

The cluster ion distributions reported in figure 2A-C show different shapes at increasing photon energy, since the ionization efficiency depends, besides instrumental response, on the neutral cluster concentration as well as on the photoionization cross-section of each cluster, the latter being a function of the photon energy. We measured values of the ionisation potentials of the toluene clusters falling in the range 8.7-8.5 eV, down-shifted with respect to that of monomer (8.8 eV). By using up 9.3 eV photon energy the mass spectra appear like that reported in figure 2.A in which the dimer's peak is higher than the monomer's. By increasing the photon energy (up to 10.7 eV), the monomer's peak increases and the mass distribution show clusters with up to 13 molecules (figure 2 B). At even higher energy the cluster ion distribution changes significantly (see tha case of 11.5 eV, figure 3C), and heavier clusters disappear from the mass spectra as a result of a massive fragmentation.

We conclude that reliable information on the toluene clustering in the supersonic jet can be gained at near-threshold photoionization (≤ 9 eV). Furthermore, scanning across the ionization threshold can provide additional information about the size distribution of the neutral cluster. Finer details are achievable by increasing the spectral resolution of the ionizing photons.

The results reported above on toluene clusters demonstrate that the developed tunable VUV ionizing source, based on LPP, can be successfully applied to the study of complex gaseous mixture as well as of weakly bound complexes.

**Acknowledgments**
T.M.D.P. acknowledges the financial support from "Ricerca Spontanea a Tema Libero" (*Curiosity Driven Research*) Grant no. 326 of the Italian National Research Council.

**References**

[1] Muhlberger F, Zimmermann R and Kettrup A 2001 *Anal. Chem.* **73** 3590
[2] Yang Z, Zhang T, Pan J, Hong X, Tang Z and Qi F 2009 *J. Am. Soc. Mass Spectrom.* **20** 430
[3] Di Palma T M and Borghese A 2007 *Nucl. Instrum. Methods Phys. Res.* B **254** 193
[4] Borghese A and Di Palma T M 2007 *Appl. Opt.* **46** 4948
[5] Squire D W and Bernstein R B **1984** *J. Phys. Chem.* **88** 4944.

Journal of Physics: Conference Series **182** (2009) 012025 doi:10.1088/1742-6596/182/1/012025

# Increasing the accuracy of structural investigations by MAS spin-echo solid-state NMR experiments

**Xenia Filip[1], Carmen Tripon[1], Steven P Brown[2] and Claudiu Filip[1]**

[1] National Institute for Research and Development of Isotopic and Molecular Technologies, 65-103 Donath, 400293 Cluj-Napoca, Romania

[2] Department of Physics, University of Warwick, Coventry CV4 7AL, UK

E-mail: claudiu.filip@itim-cj.ro

**Abstract**. The use of rotor-synchronized Hahn-echo pulse train (RS-HEPT) $^1$H decoupling, in a MAS (Magic Angle Spinning) $^{15}$N spin-echo NMR experiment allows the use of longer the spin-echo durations compared to the case when conventional decoupling is applied, hence enabling the accurate determination of very small *J* couplings down to a few Hz. Here, a thorough numerical analysis of the $^{15}$N spin-echo under RS-HEPT decoupling is performed with the purpose of characterizing the factors that still limit its efficiency. The obtained results are promising with respect to further improving the accuracy in the measured *J* couplings.

## 1. Introduction

To characterize the hydrogen bonding network formed in molecular superstructures, solid-state nuclear magnetic resonance (SS-NMR) spectroscopy emerges as a powerful technique capable to provide structural details with atomic resolution. In particular, the modulating effect of *J* couplings upon spin echoes is a key property [1], which enables the measurement of very weak coupling constants in SS-NMR experiments designed to provide through-bond structural characterization. Since *J* couplings are not normally resolved in ordinary one-dimensional SS-NMR spectra, a $\tau$-$\pi$-$\tau$ spin-echo magic angle spinning (MAS) experiment is needed to refocus evolution due to magnetic field inhomogeneity or disorder. The use of efficient heteronuclear decoupling gives transverse dephasing times, $T_2$', which are usually much longer than the decay constants of the non-refocused NMR signal, $T_2$*, that characterize the free-induction decay. Hence it is possible to obtain refocused line widths ($\Delta' = 1/\pi T_2'$) which are sufficiently narrow so as to observe splitting due to *J* couplings [2].

Previous work [3,4] has been demonstrated the use of cross-polarization (CP) spin-echo MAS experiments employing a low duty-cycle $^1$H decoupling scheme based upon rotor-synchronized $\pi$ pulses, namely the RS-HEPT sequence [5]. This enabled the accurate measurement of small *J* couplings for two synthetic deoxyguanosine derivatives by $^{15}$N MAS NMR spin-echo experiments [3]. Here, we provide a thorough numerical analysis of the $^{15}$N spin-echo under RS-HEPT decoupling with the purpose of characterizing the factors that still limit the efficiency of the proposed technique. The derived conclusions are expected to constitute the theoretical basis for further improvements in the applied heteronuclear decoupling and, consequently, reduce the magnitude of the smallest possible measured *J* couplings.

## 2. Experimental setup

Figure 1 shows the MAS spin-echo pulse sequence. $^{15}$N transverse magnetization is created using cross-polarization from $^{1}$H. The rotor-synchronized Hahn-echo pulse train (RS-HEPT) decoupling sequence [5], applied during the spin-echo evolution period of $2\tau$ consists of a $^{1}$H $\pi$ pulse every two rotor periods. This has been applied previously in one-dimensional $^{13}$C CP MAS experiments on adamantane. For each resonance in the $^{15}$N NMR spectrum acquired by the MAS spin-echo experiment, corresponding spin-echo curves have been measured by incrementing the echo time with integer multiples of the rotor periods, $\tau_R$, and observing the dependence of the corresponding line intensity as function of $2\tau$. An exponential decay was found in all cases with decay rates, $T_2'$.

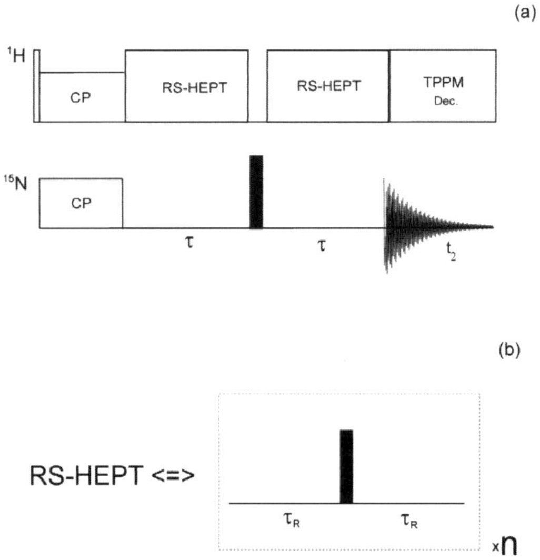


**Figure 1.** (a) The pulse sequence employed for the $^{15}$N spin-echo experiment. (b) The basic block in the Rotor-Synchronize Hahn-Echo Pulse train (RS-HEPT).

## 3. Numerical simulations

The use of RS-HEPT decoupling in the spin-echo experiment is a key element for measuring weak *J* coupling constants with relevance in identifying structural motifs induced by hydrogen bonding in supra-molecular systems. The decoupling performance of the sequence yields sufficiently long $T_2'$ dephasing times to allow the determination of the *J* couplings for the non-protonated $^{15}$N resonances. It is to be noted that in the absence of $^{1}$H decoupling, $T_2'$ values were two to three times shorter as compared to the values determined with RS-HEPT decoupling. While more efficient than no decoupling at all, the RS-HEPT decoupling is less efficient than current high-power decoupling techniques, such as TPPM decoupling. However, the use of high-power continuous $^{1}$H heteronuclear decoupling, is unpractical for measuring very weak *J* couplings because the long spin-echo duration required to identify their modulation effect upon the spin-echo decay curve can cause serious damage to the *r.f.* electronics.

By employing the experimental setup in figure 1, spin-echo decay times up to 200 ms could be obtained [3], which made it possible to measure weak *J* couplings, down to 3.8 Hz (corresponding to intra-molecular *J* couplings over two bonds). However, the measurement of even smaller *J* coupling constants would be beneficial with respect to better structural characterization of supramolecular aggregates. To accomplish this, however, the heteronuclear decoupling efficiency during the spin-echo evolution by RS-HEPT has to be further improved. For this purpose, in the present work a thorough numerical analysis has been performed, which is aimed at identifying the main mechanisms that still limit the decoupling performances of the RS-HEPT sequence. In this study, numerical simulations of the $^{15}$N spin-echo experiment in figure 1 on extended NH7 spin systems have been performed using the SPINEVOLUTION program [6], by considering a spatial arrangement that is representative for organic compounds.

The results of the numerically simulated $^{15}$N spin-echo decay curves obtained by incrementing the spin-echo time $2\tau$ are shown in figure 2. First, the experimental results obtained in [3] are compared with the simulated spin-echo curve (represented by thin continuous line in figure 2a), where the employed isotropic, and anisotropic, $^{1}$H chemical shift parameters are typical for the involved protons in deoxyguanosine, and a spinning frequency of 22.5 kHz has been considered. As can be seen, spin-echo durations up to 200 ms have been also obtained in our theoretical investigation, which agrees very well with the experimental values. One can thus conclude that the investigated spin system is sufficiently large to correctly reproduce the main experimental findings.

Next, we investigate two major mechanisms under pulsed-decoupling by RS-HEPT that prevent us from completely eliminating the influence of protons upon the evolution of low-$\gamma$ nuclear spins. In ref. [5] they have been identified as third order effects due to mixing terms between the isotropic $^{1}$H chemical shift, homo- and heteronuclear dipolar interaction Hamiltonians, and the interference of the latter with $^{1}$H chemical shift anisotropy, respectively. To obtain more insights into these effects, the spin-echo curve in the absence of proton chemical shift interactions (both, isotropic, and anisotropic) has been simulated and considered as a reference curve for ideal decoupling conditions (thick continuous line in figure 2a). As can be seen, there is almost no decay under this particular conditions, which reveals the strong influence of the $^{1}$H chemical shift upon the achievable decoupling efficiency. Furthermore, to identify the separate effect of the isotropic and anisotropic parts of the chemical shift interaction, two distinct computations have been performed for the spin-echo curve under the hypothetical conditions that only the isotropic chemical shifts (dashed line in figure 2a), or only the $^{1}$H chemical shift anisotropies (dotted line in figure 2a), are active during the spin-echo evolution time. As can be seen comparing these two spin-echo curves, their effect upon the ideal behavior are of comparable magnitudes, and manifests themselves by a strong reduction of the decoupling efficiency under RS-HEPT. Obviously, the thin continuous line in figure 2a represents the combined contribution of both, isotropic, and anisotropic, $^{1}$H chemical shift interactions, and this results in a further reduction of the practical decoupling performances.

Finally, practical solutions for reducing these detrimental effects of the $^{1}$H chemical shift interactions are analyzed in the reminder of the present work based again on numerical simulations. First, we have studied the decoupling efficiency, measured through the decay rate of the acquired $^{15}$N spin-echo curve, by using a phase alternation in the RS-HEPT sequence, by which the relative phases of consecutive pulses are shifted by an amount, $\Delta\varphi$. Considering first the case of the spin-echo recorded at 22.5 kHz MAS frequency that was discussed above, the best performance was obtained for $\Delta\varphi = 30^0$. The corresponding spin-echo decay curve is represented in figure 2b by dashed line. For comparison, also the spin-echo curve obtained without phase-alternation is shown (thin line). Next, the effect of increasing spinning frequency to 50 kHz was investigated. This has become possible in practice due to the recent progresses in the ultra-fast MAS probehead technology. The spin-echo curve obtained at 50 kHz MAS frequency is drawn in figure 2b with dotted line in the case when no phase-shifting of the RS-HEPT pulses was applied. A further improvement in the decoupling efficiency is obtained in this case compared to the characteristic efficiency under 22.5 kHz spinning frequency and phase-shifting. Interestingly, if the both changes are considered simultaneously, *i.e.*, 50 kHz MAS frequency and phase-shifted pulses by $\Delta\varphi = 30^0$ in the RS-HEPT sequence, one obtains a significant increase in the spin-echo decay time, of up to 800 ms (thick line in figure 2b), which is much larger than the value that would be obtained by simply adding the separate effects. This result is very promising with respect to improving the practical performances of the proposed SS-NMR spin-echo technique. For instance, the result obtained in the present study indicates the possibility of measuring very weak hydrogen-bond mediated $J$ coupling constants, of less than 1 Hz, and thus identify more subtle structural details in supramolecular aggregates of current research interest.

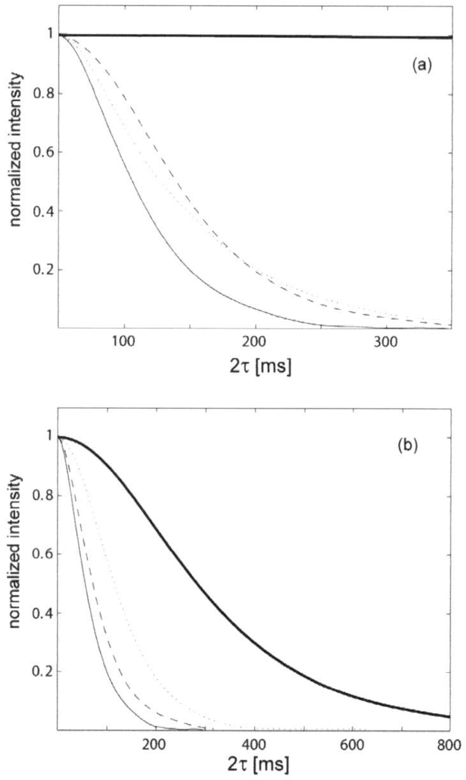


**Figure 2.** Numerical simulations of the $^{15}N$ spin-echo curves obtained on a 8-spin system (NH7) by using different simulation parameters, described in the text.

## 4. Conclusions

To further increase the accuracy of the spin-echo experiment, a numerical analysis has been performed with respect to improving the decoupling efficiency by RS-HEPT. It was found that using higher sample spinning frequencies (up to 50 kHz) combined with a suitable phase modulation of the $\pi$-pulses in the RS-HEPT sequence, could lead to almost a fourfold improvement in the characteristic spin-echo decay time. This is expected to lead to a substantial increase in the practical performances of the $^{15}N$ spin-echo SS-NMR experiment.

## Acknowledgments

Financial support from ANCS (project IDEI) and the Royal Society is gratefully acknowledged.

## References

[1] Dingley A J and Grzesiek S 1998, *J. Am. Chem. Soc.* **120** 8293
[2] Lesage A, Bardet M and Emsley L 1999 *J. Am. Chem. Soc.* **121** 10987
[3] Pham T N, Griffin J M, Masiero S, Lena S, Gottareli G, Hodgkinson P, Filip C and Brown S P 2007 *Phys. Chem. Chem. Phys.* **9** 3416
[4] Griffin J M, Tripon C, Samoson A, Filip C and Brown S P 2007 *Magn. Reson. Chem.* **45** S198
[5] Filip X. Tripon C and Filip C 2005 *J. Magn. Reson.* **176** 239
[6] Veshtort M and Griffin R G 2006 *J. Magn. Reson.* **178** 248.

Processes in Isotopes and Molecules IOP Publishing
Journal of Physics: Conference Series **182** (2009) 012026 doi:10.1088/1742-6596/182/1/012026

# Unexpected synchronization

**Szabolcs Horvát,[1,2] E. Ágnes Horváth,[1,3] Gabriell Máté,[1,3] Erna Káptalan[1] and Zoltán Néda[1]**

[1] Faculty of Physics, Babeş–Bolyai University, Kogălniceanu 1, 400084 Cluj-Napoca, Romania
[2] Institute of Physics, University of Bergen, Allégaten 55, 5007 Bergen, Norway
[3] Pázmány Péter Catholic Univ., Dept. of Information Technology, 1083 Budapest, Hungary

E-mail: szhorvat@ift.uib.no

**Abstract.** An ensemble of oscillators capable of a highly nontrivial synchronization is investigated. The oscillators are stochastic units which can emit a signal and detect signals emitted by others. They have two operational modes characterized by two different oscillation periods. Switching between the modes is governed by a simple rule: the average output intensity is kept around a prescribed threshold. This simple dynamical rule realizes the coupling of the units and leads to a complex collective behaviour. Computer simulations proved that for a given interval of the threshold parameter, partial synchronization of the units occurs. The appearance and disappearance of this synchronization as a function of the threshold parameter indicates a phase-transition type behaviour.

## 1. Introduction

Spontaneous synchronization occurs in diverse systems in nature. Examples include biological systems such as fireflies flashing in unison or crickets chirping together, rhythmic applause, mechanically coupled metronomes, the menstrual cycles of women living together, pacemaker cells in the heart, etc. [1]

Several mathematical models have been put forward to explain how a large group of coupled oscillators with different frequencies can synchronize spontaneously, without an external driving force. Most of these models fall into two broad categories: those describing phase-coupled oscillators, and those which use pulse-coupling between the units. In the present work we shall take a look at a totally different type of synchronization mechanism: one in which the interaction between the oscillating units is not chosen explicitly to induce synchronization. Instead of being the result of an evident phase-difference minimizing force, synchronization arises as a side effect of a simple optimization rule. Similar models are described in [2, 3, 4].

The classical phase-coupled model of synchronization was introduced by Kuramoto and Nishikawa [5]. In their model, each oscillator has an associated phase between 0 and $2\pi$. The oscillators evolve according to a set of coupled first order differential equations, with a coupling that minimizes the phase difference between them. The form of the equations was chosen to allow for an analytic solution. In the thermodynamic limit, this model shows a second-order phase transition as a function of the coupling strength. The critical coupling depends on the variance of the oscillators' frequencies.

Pulse-coupling is used in integrate-and-fire type models [6]. Each oscillator has a state variable, which increases monotonically until it reaches a given value. At this point the oscillator emits a pulse (it *fires*) and resets its state. The state variable of all those other oscillators that can detect the pulse increases instantly by a fixed value. Under the right conditions, a single pulse can trigger an avalanche of pulses in the system, thus causing a high proportion of the units to fire at the same time.

## 2. The Model

The model consists of an assembly of $N$ similar oscillators, each of which can either be active at a particular instant in time, outputting a signal of strength $1/N$, or inactive, outputting no signal at all. Thus the total output signal of the assembly can vary between 0 and 1. It is easy to picture these oscillators as flashing units, so the active oscillators will be referred to as *lit*, while the inactive ones will be called *unlit*. Correspondingly, the total output level of the system can be thought of as the total light intensity.

The units are two-mode stochastic oscillators, i.e. they can oscillate with a longer or a shorter period, and these periods are random variables. At the end of a period, oscillators stay lit for an amount of time.

Each oscillator cycles between three states, which will be denoted by $A$, $B$ and $C$ (see Figure 1a) [2, 3, 4]. The length of time a unit stays in each of the three states will be denoted by $\tau_A$, $\tau_B$ and $\tau_C$.

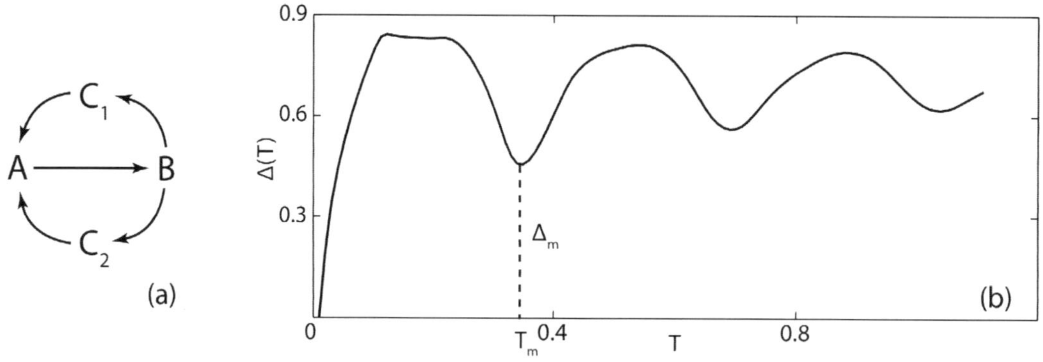


**Figure 1. (a)** Dynamics of the oscillators. Each oscillator cycles between three states, $A$, $B$ and $C$. The duration of phase $C$ can be long or short ($C_1$ and $C_2$). **(b)** Shape of the $\Delta(T)$ function for parameters $f^* = 0.2$, $\tau^* = 0.05$ and $N = 2000$.

State $A$ is the stochastic phase of the oscillators. Its length, $\tau_A$, is a random variable following an exponential distribution:

$$P(\tau_A) = \frac{1}{\tau^*} e^{-(\tau_A/\tau^*)} \tag{1}$$

The average duration of state $A$ is $\langle \tau_A \rangle = \tau^*$.

Phase $B$ has a fixed length $\tau_B$. This ensures that the oscillator stays unlit for at least a time of $\tau_B$.

State $C$ is the lit state of the units, and can have two durations, a longer one, $\tau_{C1}$, and a shorter one, $\tau_{C2}$. These correspond to oscillating modes 1 and 2, respectively. When entering state $C$, the oscillators decide how long to stay in that state based on the total output of the system. If the output $f$ is less than a prescribed value $f^*$, the oscillator will choose mode 1. If $f > f^*$, it will choose mode 2.

By choosing mode 1 when $f < f^*$ and choosing mode 2 when $f > f^*$, the units try to keep the output close to $f^*$. As a side effect of this optimization, for certain values of the model parameters the oscillators will flash synchronously, and the total output of the system becomes approximately periodic in time.

This model is a variant of the models presented in [2, 3, 4]. In the earlier models, instead of phase C, it was the length of phase $B$ that was different between the two modes. Note that unlike in the earlier versions, in the present model the length of a full oscillation period is lengthened rather than shortened by a low total output.

To back up the claim that the units are capable of flashing in unison, we need an objective way to detect synchronization in the system. For this, a $\Delta(T)$ function that characterizes the periodicity level of the total $f(t)$ output was chosen [2, 3]:

$$\Delta(T) = \frac{1}{2M} \lim_{x\to\infty} \frac{1}{x} \int_0^x |f(t) - f(T)| \, dt \tag{2}$$

where

$$M = \lim_{x\to\infty} \frac{1}{x} \int_0^x |f(t) - \langle f(t)\rangle| \, dt \quad and \quad \langle f(t)\rangle = \lim_{x\to\infty} \frac{1}{x} \int_0^x f(t) \, dt \tag{3}$$

The shape of this function is sketched on Figure 1b. The more periodic the output signal is (assuming a period $T$), the smaller the value of $\Delta(T)$ will be. Thus the period of the approximately periodic function $f(t)$ can be considered to be $T_m$, where $\Delta_m = \Delta(T_m)$ is the deepest minimum of $\Delta(T)$ (excluding the obvious minimum at $T = 0$). It is easy to see that for a perfectly periodic function $\Delta_m = 0$, and that $\Delta_m$ does not depend on the amplitude of the signal. We can define the *periodicity level* of the output as $p = 1/\Delta_m$.

The quantity $p/p_1$ was chosen as the order parameter characterizing the synchronization level of the assembly of oscillators. Here, $p_1$ is the periodicity level of a single unit oscillating in the long mode ($\tau_{C1}$), hence $p/p_1$ is the increase in the periodicity level of the output due to the coupling between the oscillators.

## 3. Results

The model was implemented on a computer, and studied numerically using simulations. The level of synchronization as a function of $\tau^*$ and $f^*$ was studied. The other parameters of the model were fixed at the following values: $\tau_B = 0.20$, $\tau_{C1} = 0.15$ and $\tau_{C2} = 0.10$. (Other choices lead to qualitatively similar results.)

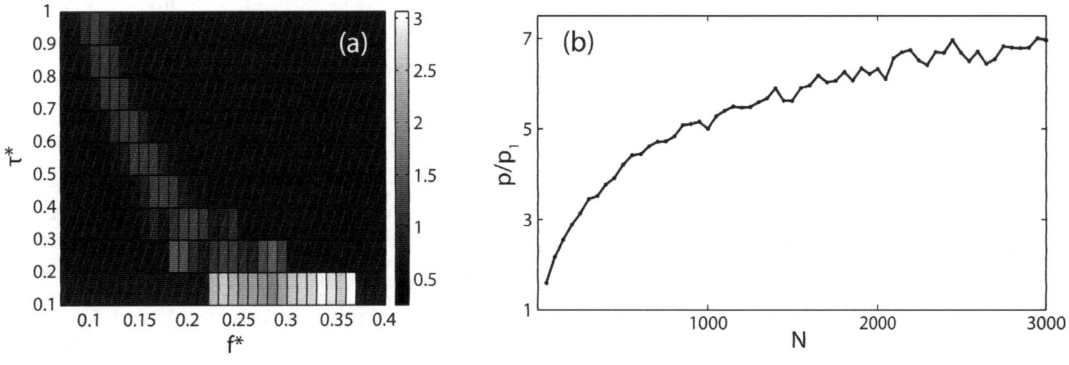


**Figure 2. (a)** Synchronization level as a function of $\tau^*$ and $f^*$. Lighter shades of grey indicate a higher $p/p_1$ value. Synchronization occurs only in a certain region of the $f^*$-$\tau^*$ space.
**(b)** Synchronization level versus the number of oscillators.

Similarly to earlier models, where the modes differed in the value of $\tau_B$, the present model exhibits synchronization in an island-like region in $\tau^*$-$f^*$ space (Figure 2a). It was also found that in this region, the synchronization level, as measured by $p/p_1$, increases monotonically with the number of units in the system (see Figure 2b).

For a fixed $\tau^*$, synchronization appears and disappears abruptly as $f^*$ is varied (see Figure 3a). The transition gets sharper as $N$ is increased, suggesting a phase-transition like phenomenon.

The distribution of the followed oscillation modes is plotted on Figure 3b for a few values of $f^*$. As expected, for those values of $f^*$ where synchronization occurs, both modes are present. For low or high $f^*$, solely one mode occurs, barring any possibility of non-random shift in the length of oscillation cycles and thus synchronization.

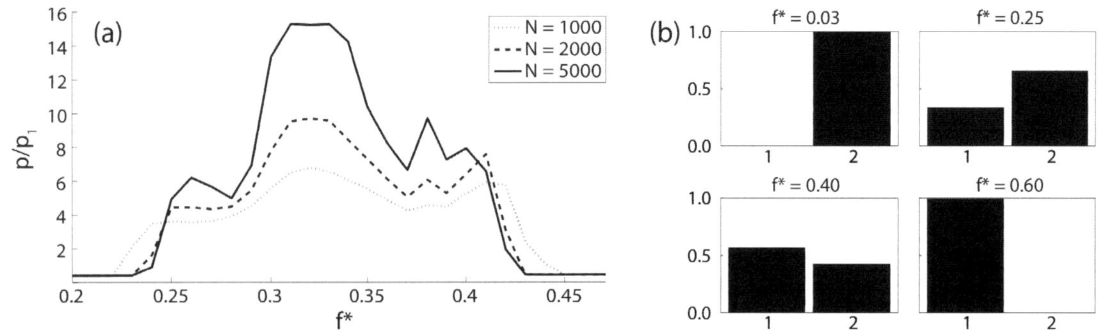


**Figure 3. (a)** Synchronization level as a function of $f^*$. $\tau^*$ was fixed at 0.05, and the number of units, $N$, is indicated in the legend. **(b)** Distribution of the oscillation modes for various values of $f^*$. Results obtained for $\tau^* = 0.05$ and $N = 2000$.

## 4. Conclusions

An assembly of two-mode stochastic oscillators were investigated by computer simulations. At the end of its oscillation period, each unit emits a signal of either long or short duration (corresponding to the two modes), depending on the total output intensity in the system. When the light intensity is less than a threshold value, $f^*$, the system emits a long signal, otherwise a short one. This dynamics tries to keep the average output around $f^*$. As an unexpected side effect, for certain values of $f^*$ synchronization occurs. The abrupt appearance and disappearance of this synchronization resembles a phase transition. Synchronization is improved by increasing the number of units in the system.

The synchronization mechanism found in this study is interesting and puzzling because no phase-difference minimizing interactions are involved in the dynamics.

## Acknowledgments

The work of Sz. H. is supported by the European Social Fund through a POSDRU 2007–2013 PhD grant. Supported by grant PNCDII/ElBioArch under Contract No. 12121.

## References

[1] Strogatz S 2003 *SYNC: The Emerging Science of Spontaneous Order* (Hyperion, New York)
[2] Nikitin A, Néda Z and Vicsek T 2001 *Phys. Rev. Lett.* **87** 024101
[3] Néda Z, Nikitin A and Vicsek T 2003 *Physica A* **321** 238–247
[4] Sumi R, Néda Z, Tunyagi A, Boda Sz and Szász Cs 2009 *Phys. Rev. E* **79** 056205
[5] Kuramoto Y and Nishikawa I 1987 *J. Stat. Phys.* **49** 569–605
[6] FitzHugh R 1955 *Bull. Math. Biol.* **17** 257–278

# Characterization of some extracts for therapeutic use by GC/MS

**Andreea Maria Iordache, Monica Culea and Onuc Cozar**

Babes-Bolyai University, Faculty of Physics, 1 Mihail Kogalniceanu, 400084 Cluj-Napoca, Romania

E-mail: mculea@phys.ubbcluj.ro

**Abstract.** A liquid-liquid extraction method (LLE), compared with a solid-phase extraction method (SPE), was used for characterizing some plant extracts for therapeutic use. Precision of the extraction methods gave relative standard deviation lower than 3%. The extracts were analyzed by GC-FID and GC-MS. A HP-5 capillary column, 30 m x 0.32 mm, 0.25 µm film thickness was used. By using an appropriate mixture of solvents, good recovery mean values were obtained by LLE (81%) comparable with the recoveries of other extraction methods as SPE (92%). The study was applied to characterize the compounds of therapeutic value extracted from herb plants.

## 1. Introduction

In the last years the interest for the study of the organic compounds from plants and their activity has increased. A lot of extraction methods and analytical methods as spectrophotometry, high performance liquid chromatography, gas chromatography-mass spectrometry (GC/MS) were developed for plant active compounds study [1-7].

The aim of the present work is the characterization of the compounds extracted from some herb plants.

The methods were applied to characterize "Floratonic", a "swedish" tincture, used internal or external, in traditional medicine in our country. It is prepared from some herbs growing in Romania as hydroalcoholic extract which contains 12 plant species: *Aloe* (folium), *Angelica arhangelica* (radix), *Juniperus communis*, *Fraxinus excelsior* (fructus) *Gentiana lutea* (radix), *Iris palida* (radix), *Robinia pseudacacia* (flos), *Polygonum aviculare* (herba), *Arctium lappa* (radix), *Laurus nobilis* (folium), *Abies alba.*

## 2. Experimental

### *2.1. Apparatus*

A Hewlett Packard GC 5890 coupled with a MS engine 5989B in the EI mode was used for compounds identification. The GC was equipped with a HP-5MS capillary column 30 m x 0.25 mm diameter, 0.25 µm film thickness, in the temperature program: 50°C for 2 min, to 250°C with a rate of 8°C/min, then increased with 30°C/min to 310°C and kept 10 min. The GC/MS interface line and the ion source were maintained at 280°C and respective 200°C, and quadrupole analyser at 100°C. Electron energy was 70 eV and electron emission 300 µA.

### 2.2. *LLE extraction*
A mixture of three solvents, named *solvent A*, was prepared as follow: ethyl acetate: hexane: methylen chloride (5:1:1, v/v/v). The LLE extraction procedure was: 3ml hydroalcoholic extract (or 10 µl *mixture* (stock solution) in 3 ml solution distilled water: ethanol (1:1, v/v)), 3 ml distilled water and 1ml *solvent A* (3:3:1, v/v/v), 3 g NaCl were mixed 5 min and then centrifuged 2 min. For recovery study 1µl methyl myristate was added to the supernatant and then 1 µl was injected two times by using the autosampler injector. For method validation, 20 ng/ml or 2 µg/g methyl myristate, the internal standard (IS), was added before extraction.

### 2.3. *SPE extraction*
The solid phase was conditioned with 3 ml methanol and 3 ml distilled water. After sample application, washing and drying 10 min at vacuum, the sample was eluted with 3 x 0.3 ml *solvent A*. After the addition of 1 µl external standard to the extract, 1 µl was injected by using the autosampler injector. Each sample was injected twice.

### 2.4. *Standard solutions*
A stock solution was obtained by diluting 100 µl of each compound.

### 2.5. *Method validation*
Linearity was studied in the range 0-24ng. The regression curves obtained for the standards gave good correlation coefficients higher than 0.99 for each standard: 1) 3-hepten-2-one y = 0.03 x – 0.03, r = 0.994; 2) fenchone y = 0.04 x – 0.01, r = 0.998; 3) phenyl ethyl alcohol y = 0.05 x – 0.04, r = 0.995; 4) terpinen-4-ol y = 0.05 x – 0.002, r = 0.997; 5) citronellol y = 0.04 x – 0.03, r = 0.997; 6) carvone y = 0.05 x – 0.01, r = 0.997; 7) anisaldehyde y = 0.04 x – 0.01, r = 0.998.

Precision for LLE extraction gave relative standard deviation (RSD) lower than 3% (n = 4) and for SPE, RSD was lower than 3% (n = 5). Recovery for LLE was 81% (n = 3) and for SPE 92% (n = 4). Accuracy for 20 ng shows a mean value for RSD of 5% and for 24 ng of 6%. The sensitivity was lower than 100 pg at a ratio S/N = 10 and LOD was 10 pg, S/N = 10 for each standard.

## 3. Results and discussion
The separation chromatogram of the standards studied for the two extraction modes is presented in figure 1. Figure 2 shows the regression curves obtained for the standards studied.

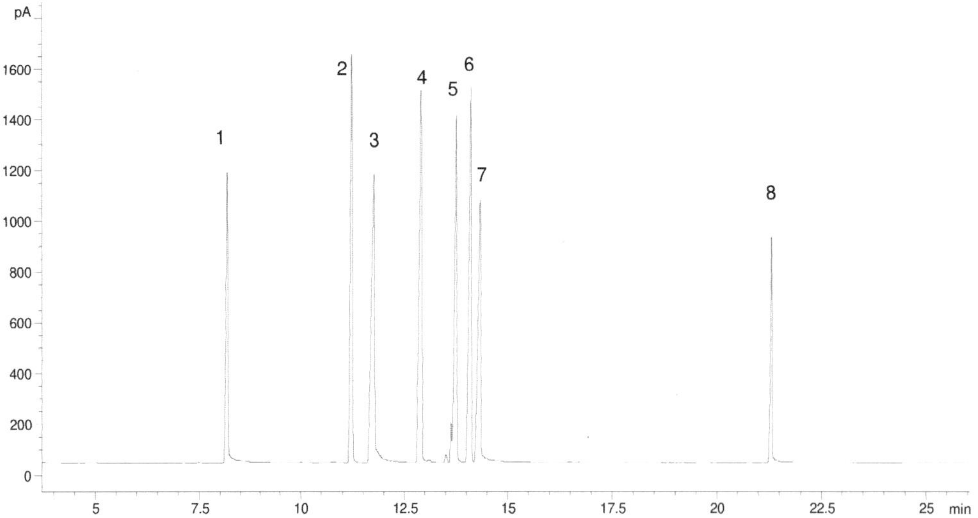


**Figure 1.** Mixture A separation chromatogram after extraction.

Absolute recoveries were determined by using external calibration with the standard methyl myristate. The results of the recovery studies are presented in table 1. The mean values are resulted from four (LLE) and five (SPE) extraction procedures and two injections of each extract. The relative standard deviation values were bellow 6% for LLE and 3% for SPE extraction procedure. The methods were applied for qualitative and quantitative determination of the organic compounds from the bitter studied.

**Table 1.** Comparative recovery mean values between two extraction procedures.

| Compound | $t_R$ (min) | LLE (%) | SPE (%) |
|---|---|---|---|
| 1. 3-hepten-2-one | 8.33 | 76.03 | 85.19 |
| 2. fenchone | 11.40 | 79.12 | 87.55 |
| 3. phenyl ethyl alcohol | 12.00 | 76.13 | 92.85 |
| 4. terpinen-4-ol | 12.80 | 80.97 | 90.76 |
| 5. citronellol | 13.60 | 82.62 | 89.84 |
| 6. carvone | 14.00 | 83.96 | 94.01 |
| 7. anisaldehyde | 14.40 | 87.50 | 102.45 |
| 8. methyl myristate (ES) | 21.00 | - | - |
| Mean values | - | 80.91 | 91.81 |

$t_R$ = retention time

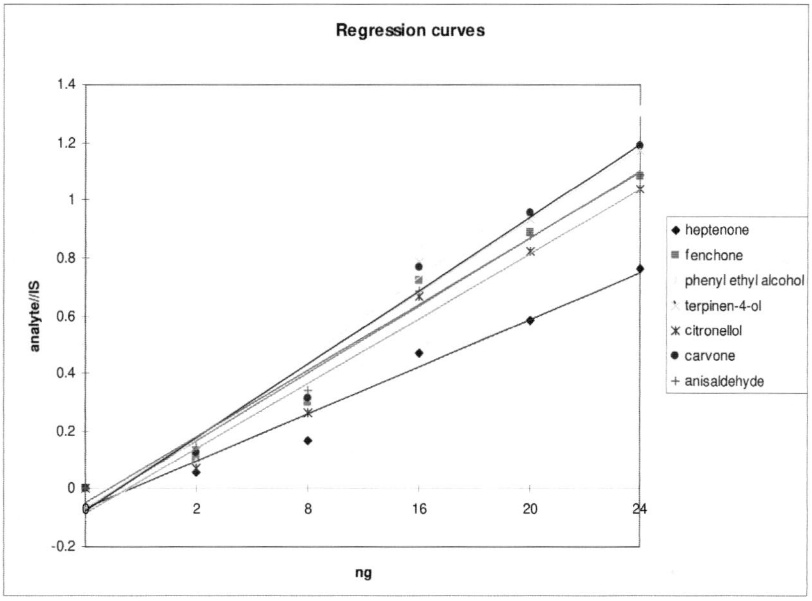


**Figure 2.** The regression curves of the studied standards.

Table 2 presents the compounds identified in the LLE extract from the bitter. Minor differences were observed between the two extraction methods in the analyzed standard mixture. The two extraction methods were applied for qualitative and quantitative determination of the major organic compounds in "Floratonic" bitter. Bitter shows sedative, antiseptic, anti-inflammatory, carminative, antirheumatic, antidepresive and tissue regenerating effects.

The high levels of lignan flavonoids lariciresinol and matairesinol (figure 3) in "Floratonic" bitter may contribute to the protective effect on coronary heart diseases and arteriosclerosis. The plant lignans, as well as their mammalian metabolites enterolactone and enterodiol, have antioxidative properties [5, 8].

**Table 2.** Compounds identified in "Floratonic" bitter by LLE.

| Compound | $t_R$ | ng/ml |
|---|---|---|
| beta-myrcene | 7.20 | 8.11 |
| ethyl caproate | 7.40 | 9.70 |
| 1,8 cineole | 8.09 | 106.53 |
| benzyl alcohol | 8.28 | 2.32 |
| phenyl acetaldehyde | 8.35 | 2.32 |
| fenchone | 9.22 | 8.35 |
| linalool | 9.40 | 1.24 |
| phenyl ethyl alcohol | 9.80 | 1.59 |
| isopinocarveol | 10.20 | 9.57 |
| camphor | 10.30 | 9.26 |
| pinocarvone | 10.60 | 1.59 |
| p-ethyl phenol | 10.77 | 1.59 |
| 4-terpineol | 10.93 | 20.04 |
| p-cymen-8-ol | 11.07 | 0.83 |
| alpha-terpineol | 11.15 | 36.96 |
| estragol | 11.28 | 0.83 |
| verbenone | 11.53 | 0.83 |
| 2,3dihydrobenzofuran | 11.74 | 22.57 |
| Z-citral | 12.10 | 0.83 |
| carvone | 12.20 | 0.99 |
| p-anisaldehyde | 12.34 | 0.74 |
| ethyl salicylate | 12.57 | 0.74 |
| dimethylhexynediol | 12.60 | 0.74 |
| p-ethyl guaiacol | 12.70 | 0.74 |
| trans-anethole | 12.80 | 16.45 |
| 4-vinyl 2-methoxyphenol | 13.33 | 0.74 |
| alpha-terpinyl acetate | 13.88 | 4.30 |
| eugenol | 14.05 | 0.74 |
| M=164 | 14.50 | 0.74 |
| methyl eugenol | 14.78 | 0.74 |
| vanillin | 14.80 | 0.74 |
| M=152 | 15.80 | 0.74 |
| M=152 | 16.35 | 2.71 |
| ethyl p-hydroxybenzoate M=166 | 16.77 | 5.98 |
| p-hydroxybenzoic acid +propyl p-hydroxybenzoate | 16.80 | 9.43 |
| M=180 | 17.20 | 3.41 |
| ethyl vanillate M=196 | 17.60 | 5.58 |
| alpha-cadinol M=222 | 18.60 | 0.74 |
| methyl myristate | 19.50 | 24.75 |
| ethyl m-hydroxycinnamate(Z) M=192 | 19.53 | 8.94 |
| erythrocentaurine M=176 | 19.55 | 9.73 |
| 7acetyl-2hydroxy-2methyl-5-isopropylbicyclo[4.3.0]nonane M=238 | 19.80 | 9.83 |
| furocoumarin M=186 | 20.30 | 8.35 |
| M=198 | 20.40 | 9.53 |
| ethyl m-hydroxycinnamate(E) M=192 | 21.00 | 78.50 |
| M=220 | 21.40 | 5.98 |
| M=222 | 22.00 | 3.41 |
| palmitic acid M=256 | 22.50 | 3.01 |
| dehydrocostuslactone M=230 | 23.08 | 18.62 |
| vanillosmin M=230 | 23.23 | 14.08 |
| (12R,13R)-8,12-epoxy-14-labden M=306 | 24.38 | 4.99 |
| osthol M=244 | 24.68 | 9.83 |
| M=278 | 24.80 | 18.52 |
| M=272 | 25.20 | 7.06 |

| | | |
|---|---|---|
| 2-(methoxyphenyl)-5-methylindazolin-3-one M=254 | 25.35 | 68.03 |
| M=246 | 25.50 | 4.79 |
| M=230 | 25.58 | 5.68 |
| m/z257 M=306 | 26.11 | 6.47 |
| 13-epitorulosol M=306 (m/z257) | 27.11 | 7.85 |
| M=286 | 27.17 | 0.74 |
| M=344 m/z137,138 | 30.19 | 11.41 |
| Matairesinol M=358 | 32.06 | 297.24 |
| Lariciresinol M=360 | 32.40 | 16.75 |

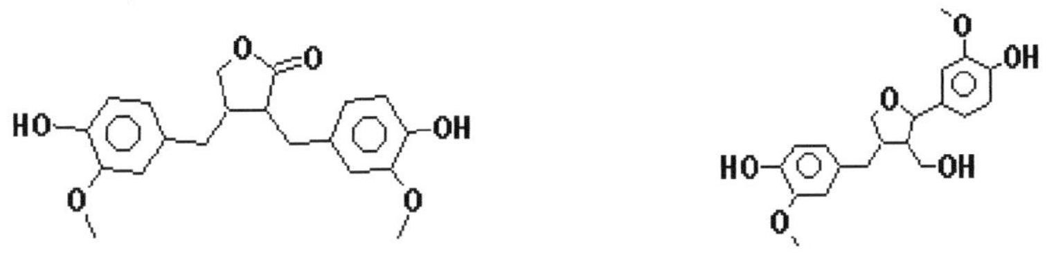

Matairesinol Lariciresinol

**Figure 3**. Lignans identified in bitter.

## 4. Conclusions

The methods presented are suitable for the determination of trace amounts of organic compounds in hydroalcoholic plant extract. Good recovery mean values were obtained by LLE (81%) using an appropriate mixture of solvents comparable with the recoveries of the other extraction method, SPE (92%). Precision gave RSD lower than 3%. Each of the extraction methods is simple, rapid, accurate and inexpensive with reduced solvent and time consuming. The methods presented are suitable for medicinal herb organic compounds determination. No major differences were observed between LLE and SPE in the plant extracts studied. Terpenic compounds, polyunsaturated fatty acids and flavonoids are some of the compounds responsible for anti-inflammatory, antioxidant, anticarcinogenic activity of plants used in traditional medicine.

## References

[1] Mauri P, Migliazza B and Pietta P 1999 *J. Mass Spectrom.* **34** 1361-7
[2] Karcher A and El Rassi Z 1999 *Electrophoresis* **20** 3181-9
[3] Kiss T, Gyemant G, Danos B and Nanasi P 1998 *Acta Pharm. Hung.* **68** 263-8
[4] Calixto J B, Santos A R, Cechinel Filho V and Yunes R A 1998 *Med. Res. Rev.* **18** 225-59
[5] Mazur W M, Wahala K, Rasku S, Salakka A, Hase T and Adlercreutz H 1998 *British J. Nutr.* **79** 37-45
[6] Kalogeropoulos N, Chiou A, Mylona A, Ioannou M S and Andrikopoulos N K 2007 *Food Chemistry* **100** 509-17
[7] Kostadinova E, Alipieva K, Stefova M, Antonova D, Evstatieva L, Stefkov G, Tsvetkova I, Naydenski H and Bankova V 2008 *Phcog Mag.* **4** 14 102-6
[8] Milder I E, Arts I C, van de Putte B, Venema D P and Hollman P C 2005 *Br. J. Nutr.* **93** 393-402.

# Tracing quantum trajectories of electrons in interaction with arbitrary-shape laser pulses

**Katalin Kovács and Valer Toşa**
National Institute for Research and Development of Isotopic and Molecular Technologies, 65-103 Donath , 400293 Cluj-Napoca, Romania
E-mail: valer.tosa@itim-cj.ro

**Abstract.** In this paper we present a theoretical and numerical tool for tracing quantum trajectories of electrons in interaction with arbitrary-shape laser pulses. The method calculates quantum trajectories using saddle-point analysis within the strong field approximation. The method is very general, thus it is applicable to arbitrarily shaped finite laser pulses. Example results and comparison with the classical trajectories are presented.

## 1. Introduction

High-order harmonic generation (HHG) is the phenomenon produced as a result of the interaction of atomic (molecular) systems with strong laser pulses. A suitable model that is simple enough but very successful in explaining the main processes involved in HHG is the well-known three step model [1]. Under the influence of an intense ($> 10^{13} W/cm^2$) laser field an electron first tunnels through the Coulomb barrier, becomes a free electron travelling in the laser field, then recombines with the parent system and emits a photon of high energy. This energy equals the kinetic energy gained by the electron during its excursion guided by the laser field plus the atomic/molecular ionization potential ($I_p$). Due to the interference between coherently emitted photons, only odd order harmonics of the fundamental radiation are obtained. Since the first and third steps in the three step model, namely the tunnel ionization and recombination are typical quantum phenomena, but the excursion of the electron in electromagnetic field is fairly a classical process, a reasonable approximation in the theoretical treatment of the HHG process is the strong-field approximation (SFA) [2]. This model relies on the basic assumption that the incoming laser field is much stronger than the atom's Coulomb potential, thus during its travel the electron is driven solely by the electromagnetic field.

A great deal of information and phenomenological insight regarding the mechanisms leading to HHG can be obtained by examining the trajectories of electrons contributing to harmonic photon emission. Remarkable papers were published on this topic [3, 4, 5, 6] using linearly and elliptically polarized laser radiation, or even bi-circular fields. In all these cases the authors report results obtained from analytical calculations and numerical simulations assuming that the radiation propagates as a plain wave and its electric field component has an expression that can be handled analytically. Moreover, the envelope of the pulse is not included, constant amplitude is assumed for all optical cycles.

In experiments and practical applications, however, atomic systems (dilute gases) are irradiated with very short (several tens of femtoseconds), intense, low frequency (for example a

$Ti:Sa$ laser operates at ≈800 nm central wavelength) laser pulses which are usually tailored to a complex shape and spatial-spectral profile in order to meet several requirements. Analytical calculations completely fail when one wants to trace electron trajectories realized in such realistic fields, and finding the suitable numerical method is not a trivial task. The purpose of the present paper is to present such a numerical tool and first results obtained for quantum trajectories of electrons in arbitrary laser fields, and to discuss and interpret them in comparison with the classically obtained trajectories.

## 2. Model and method

### *2.1. Strong-field approximation and saddle-point equations*

The harmonic spectrum is obtained in the framework of the SFA as the Fourier transform of the dipole moment:

$$\vec{x}(t) = i\int_0^t dt' \int d^3\vec{p}\quad \vec{d}^*(\vec{p}-\vec{A}(t))\cdot exp(-iS(\vec{p},t,t'))\cdot \vec{E}(t')\cdot \vec{d}(\vec{p}-\vec{A}(t)), \tag{1}$$

where in the complex phase term $(exp(-iS(\vec{p},t,t')))$ the quantity $S(\vec{p},t,t')$ is the quasiclassical action:

$$S(\vec{p},t,t') = \int_{t'}^{t} dt'' \left(\frac{[\vec{p}-\vec{A}(t'')]^2}{2} + I_p\right), \tag{2}$$

with $\vec{p} = \vec{v}(t) + \vec{A}(t)$ the canonical momentum, $\vec{A}(t)$ is the vector potential of the field $-\partial\vec{A}(t)/\partial t = \vec{E}(t)$, where $\vec{E}(t)$ is the electric component of the electromagnetic field.

A valid method to describe quantum trajectories within the SFA and dipole approximation is to perform saddle-point analysis of the quantum-mechanical transition amplitude [7]. The most probable trajectory of an electron that ionizes at time $t$, travels in the laser field, then recombines at time $t'$ occurs under the condition of stationary phase. The stationary phase condition leads to the following saddle-point equations:

$$\nabla_{\vec{p}} S(\vec{p},t,t') = 0 \quad\Rightarrow\quad \frac{\vec{\alpha}(t)-\vec{\alpha}(t')}{t'-t} = \vec{p}_s \tag{3}$$

$$\frac{\partial}{\partial t} S(\vec{p},t,t') = 0 \quad\Rightarrow\quad [\vec{p}_s - \vec{A}(t)]^2 = 2Ip \tag{4}$$

$$\frac{\partial}{\partial t'} S(\vec{p},t,t') = 0 \quad\Rightarrow\quad [\vec{p}_s - \vec{A}(t')]^2 = 2(n\omega - Ip), \tag{5}$$

with $\vec{\alpha}(t) = \int_t \vec{A}(t)dt$ and $\vec{p}_s$ the stationary canonical momentum which is a constant of motion.

The first saddle-point equation (3) expresses the return condition for the electron, contains the assumption that only those electrons contribute to the harmonic spectrum which return and recombine to the nucleus. The second (4) and third (5) equations represent the energy conservation at the moment of ionization and recombination, respectively. We need to solve the above nonlinear system of equations for variables $\vec{p}_s$, $t$ and $t'$. Since $I_p < 0$ it results immediately that all the quantities $\vec{p}_s, t, t'$ will be complex. Rearranging the equations and writing them for the real and imaginary parts, the system of equation becomes a set of four equations with four unknowns: $t_{re}, t_{im}, t'_{re}, t'_{im}$.

It can be assigned a physical meaning to the complex quantities $\vec{p}_s, t, t'$. The fact that they become complex is a natural consequence of including in the model the pure quantum phenomenon of tunneling. The imaginary part of $t$ "birth" time can be interpreted as tunneling time, while the imaginary part of $t'$ recombination time is related to the probability of the event. One expects that the real parts of the complex electron trajectories obtained after solving the saddle-point equations to be close to those trajectories obtained by pure classical calculations (see 2.2).

*2.2. Numerical procedure*

The starting point in the proposed numerical method for obtaining the electron's complex trajectories is the calculation of classical electron trajectories. This is a fair starting point, since it roughly indicates the range where the more realistic quantum trajectories have to be looked for. Classical electron trajectories are obtained by solving Newton's equation of motion with the return condition for the electron included. The resultant expression for the classical trajectory is [8]:

$$\vec{r}(t') = \vec{r_0} + \vec{\alpha}(t') - \vec{\alpha}(t) - \vec{A}(t)(t' - t) \tag{6}$$

In the first step we obtained, starting from the electric field given in numerical form, the quantities involved in the calculations, namely $\vec{A}(t)$ and $\vec{\alpha}(t)$. The next step is to solve the system of nonlinear equations for $t_{re}, t_{im}, t'_{re}, t'_{im}$, and for a given harmonic order, the initial guess being the solution of the classical equation of motion. For this purpose we used appropriate subroutines from the NAG numerical package [9] for solving nonlinear systems of equations.

Our purpose was to work out a general trajectory-finding method for arbitrary laser fields, that is, a field which can be handled numerically. A basic requirement is to be able to define and calculate the quantities involved ($\vec{E}(t)$, $\vec{A}(t)$ and $\vec{\alpha}(t)$) at a complex time. This was performed through a power series expansion starting from $\vec{E}(t)$ as a function of the real variable $t$.

## 3. Results and discussion

Intense research is going on both on theoretical/modeling and on experimental side to obtain single attosecond bursts [10, 11]. For this purpose laser pulses of different complex shapes have to be produced. One particular choice within the so-called polarization gating technique was recently proposed [12]. The results for quantum electron trajectories presented here are obtained by using the laser fields and parameters defined in [12], which induce a polarization gating in the leading edge of the pulse.

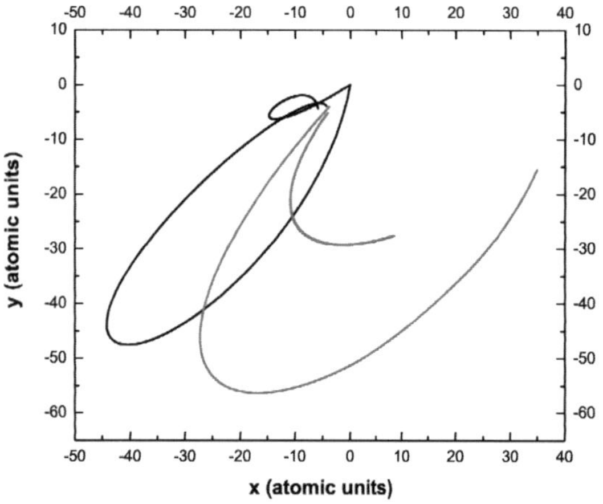


**Figure 1.** Short and long electron trajectories which generate the 35th harmonic. Black lines stand for quantum trajectories, red lines for classical trajectories. Travel times in units of optical cycles are 0.46 and 0.93 for the quantum trajectories; 0.48 and 0.91 for the classical short and long trajectories respectively.

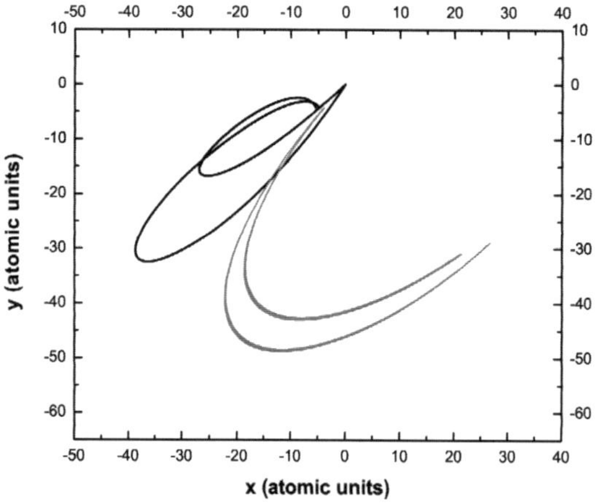


**Figure 2.** Electron trajectories near the cut-off which generate the 55th harmonic. Black lines stand for quantum trajectories, red lines for classical trajectories. Travel times in units of optical cycles are 0.62 and 0.79 for the quantum trajectories; 0.67 and 0.75 for the classical trajectories.

Figure 1. shows typical electron trajectories that generate plateau harmonics, while Figure 2. represents trajectories generating harmonics close to the cutoff. As general observations, valid for both cases we can affirm the followings: *(i)* Both graphs represent trajectories produced right after (almost) perfect linear polarization is lost. This is the explanation for the loop-like quantum trajectories instead of a straight line which would have occurred in case of linearly polarized resultant field. *(ii)* Altough the travel times of classical and quantum trajectory pairs are not very different, their actual itineraries differ substantially. The explanation for this lies in the fundamental difference between the two models. When we include quantum tunneling effect in the model, the ionized electron appears in continuum with non-zero initial velocity, contrary to the zero initial velocity assumption of the classical calculations. *(iii)* Due to the tunneling effect the electron appears in the continuum at a non-zero $\vec{r}_0$ position, but it returns almost exactly to zero (the assumed position of the nucleus). This realistic feature is not reproduced by the classical electron trajectories, they seem to recombine much farther from the nucleus then their birth position.

In the case of plateau harmonics (Figure 1.) the difference between short and long quantum trajectories is qualitative. Electrons appear in the continuum at different positions and with different initial velocities which predefine their whole excursion. As we approach the cutoff, differences between shorter and longer trajectories decrease both as birth time and initial velocity (Figure 2.). Exactly at the cutoff we would obtain one single family of trajectories with travel time $\approx 0.67$ optical cycles in the interval when the linear polarization gate is open. Since in classical calculations it is assumed that electrons appear in the continuum near the origin (nucleus) with zero initial velocity, classical trajectories do not show any qualitative change in their shape, except the duration of the travel times.

## 4. Conclusions

We have developed a novel general method for tracing quantum trajectories of electrons in interaction with strong low-frequency and *arbitrary-shape* laser fields. With this method we can obtain supplementary information concerning the conditions in which a pulse of the harmonic order of interest is generated. In particular, from the phase of the trajectory one can extract in which conditions phase matching is obtained.

As it was shown on the results, by including the quantum effect of tunnelling in the model, the obtained electron trajectories became significantly more realistic and accurate. The first results obtained with the help of this new instrument were presented here. More elaborated results reproducing different experimental conditions are left for further publications.

## References

[1] Corkum P B 1993 *Phys. Rev. Lett.* **71** 1994
[2] Lewenstein M *et. al.* 1994 *Phys. Rev.* A **49** 2117
[3] Kopold R, Becker W and Kleber M 2000 *Optics Communications* **179** 39
[4] Milošević D B 2000 *J. Phys. B: At. Mol. Opt. Phys.* **33** 2479
[5] Milošević D B, Becker W and Kopold R 2000 *Phys. Rev.* A **61** 063403
[6] Milošević D B, Paulus G G and Becker W 2005 *Phys. Rev.* A **71** 061404(R)
[7] Salière P *et. al.* 2001 *Science* **292** 902
[8] Milošević D B, Paulus G G and Becker W 2003 *Opt. Express* **11** 1418
[9] See the library documentation on *www.nag.co.uk*
[10] Tosa V, Yakovlev V S and Krausz F 2008 *New Journal of Physics* **10** 025016
[11] Kienberger R *et. al.* 2004 *Nature* **427** 817
[12] Altucci C, Esposito R, Tosa V and Velotta R 2008 *Optics Letters* **33** 2943.

# General aspects of solid on liquid growth mechanisms

**E Laux, J Charmet, H Haquette, O Banakh, L Jeandupeux, B Graf and H Keppner**

University of Applied Science Arc, Eplatures-Grise17, CH-2300 La Chaux-de-Fonds, Switzerland

E-mail: Herbert.Keppner@he-arc.ch

**Abstract:** Liquids, in general, tend to have a lower density as solids and therefore it is not straightforward to deposit solid over liquids in a way that the liquid becomes hermetically sealed under the solid layer. The authors review that several phenomena that can easily be observed in nature are only due to particular anomalies and solid on liquid is rather an exception as the rule. Natural solid on liquid systems are lacking of thermal, mechanical or chemical stability. It is not surprising, that one is not at all used thinking about to *e.g.* replace the gate oxide in a thin film transistor by a thin film of oil, or, to find in other microsystems functional liquids between a stack of thin solid films. However, once this becomes a serious option, a large variety of new Microsystems with new functionalities can be easily designed. In another paper (this conference and [1]) the authors pioneered that the polymer Parylene (poly(p-xylylene)) can be deposited on liquids coming already quite close to the above mentioned vision. In this paper the authors ask if one can synthesize other solid on liquid systems and surprisingly conclude, based on experimental evidence, that solid on liquid deposition seems to rather be the rule and not the exception.

## 1. Introduction

Daily observation seem to exclude coverage of solid layers on liquids due to the principle of Archimedes, that is based on density considerations. Exceptions are due to special designs of mechanically pre-formed cans or ships allowing to bottle liquids into solid containers and gravitation is not anymore the selection criterion for what will be above or below and what is substrate or coating. The most striking solid on liquid system that can be observed in nature is certainly based on the anomaly of water-density that is highest at 4°C. By that ice coverage of lakes as soon as phase transition from liquid to solid occurs. As another case stressing anomalies is the fact that mercury as a liquid has a particular high density and the successful experiment shown in figure 1 seems to be straightforward. In this experiment paint spry was applied over mercury droplets.

The analysis of the growth mechanism can be given as follow: the paint covers the surface of mercury and will, due to density consideration not sink to the bottom of the droplet. It could happen that the mercury first prevents polymerization at the interface, however the high rate application of paint let dominate its concentration and after an interfacial layer the pure paint will polymerize resulting to a first synthetic solid on liquid system.

Looking at low-density liquids it is no longer possible to count on a Mercury effect. Let us assume that precursors from a vapor or liquid phase arrive on the surface of the liquid. If a solid layer should be synthesized, two basic mechanisms could predict successful solid on liquid deposition: First, the

Journal of Physics: Conference Series **182** (2009) 012029 doi:10.1088/1742-6596/182/1/012029

precursors remain on the surface and cross-link in order to form a layer. In this case the liquid substrate behaves like a solid substrate.

**Figure 1.** a, b Paint spray on mercury after polymerization.

In this case linear growth of the solid layer is predictable. Second, the polymerization is induced by a precursor and the interaction between the molecules of the liquid substrate. In this case the growth is in the beginning reaction limited, later diffusion limited. Hence parabolic thickness-time relationship can be predicted. For the second mechanism the oxidation of liquid aluminum can be mentioned, hereby a real solid on liquid system is existing as long as the temperature remains above 650°C.

## 2. Experimental

The vapor phase generation for precursor-feed can be taken from standard deposition processes for solid on solid deposition. In order to have a big choice, chemical vapor deposition (CVD) or plasma enhanced CVD (PECVD) processes are preferable. Using vacuum systems for Low-Pressure CVD or glow-discharge plasma processes, the choice of liquid substrates is strongly limited and the vapor pressure of the liquid must be well below the chamber pressure. Suitable liquids are e.g. glycerol, bis(3,5,5-trimethylhexyl) phthalate, and bis(2-ethylhexyl) adipate. In order to have more choice of liquids, CVD and PECVD have to be carried out at atmospheric pressure. In the case of CVD the creation of precursors need high temperatures that let, again, evaporate the liquid substrat. In the case of atmospheric PECVD, a cold corona was applied in a so-called silent discharge configuration as proposed by [2].

Using liquid precursors to create solid on liquid systems is similar from the two layer-forming mechanisms described above. The following experimental conditions must be at hand:

1. Two non-miscible liquids.
2. Casting a lower density top liquid onto a higher-density bottom liquid.
3. The top liquid undergoes a phase transition either by itself or by a chemical reaction induced from the bottom liquid.

## 3. Results and discussion

Putting a glycerol droplet on the bottom electrode of a RF (13.56 MHz) driven parallel plate reactor using the radical of plasma-induced dissociation of ethylene gas at 50 Pa pressure, led immediately appear a solid amorphous carbon layer that covers the droplet. As shown in figure 2a, strong internal tensions let appear pronounced wrinkles: the liquid in its role as substrate is unable to compensate the stress induced from nucleation and subsequent polymerization of the solid layer.

In a second experiment of the same category, just by replacing ethylene gas by HMDSO (hexamethyldisiloxane) vapour, again at 50 Pa pressure, a similar phenomenon can be observed; in this case a layer similar to quartz is grown on the liquid. Again strong wrinkles indicate the internal tensions of the layer. The results are shown in figure 2b.

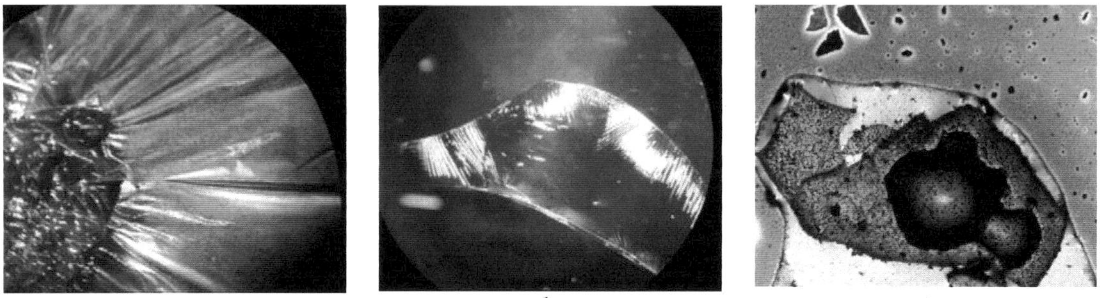


**Figure 2.** a) PECVD of a polymer carbon layer from an ethylene glow-discharge on a glycerol droplet; b) PECVD of $Si_xO_y$ from a HMDS-O glow-discharge on glycerol; c) atmospheric pressure PECVD of a fluor-polymer layer on glycerol from a Ar/$CF_4$ mixture corona discharge.

Going to atmospheric pressure PECVD, $CF_4$ is used as gas to be dissociated by the corona discharge. Again, a thin layer is grown and identified as fluorine-polymer layer. Is contains apparently less internal stress as the better replication of the solid layer on the liquid shows in figure 2c.

In order to find representative systems that rely on the second growth mechanism, liquid PDSM (polydimethylsiloxane) without curing agent is used as liquid substrate. Running an Ar / $H_2$ plasma at 50 Pa pressure, let atomic hydrogen induce polymerization of PDMS occurs in the vicinity of the surface because lack of curing agent in the liquid. Here a solid PDMS layer is grown as shown in figure 3. We show by deposition-time thickness measurements that the reaction in the beginning proceeds faster (reaction limited) and becomes diffusion-limited because the growing layer reduces the atomic hydrogen supply to the reaction zone.

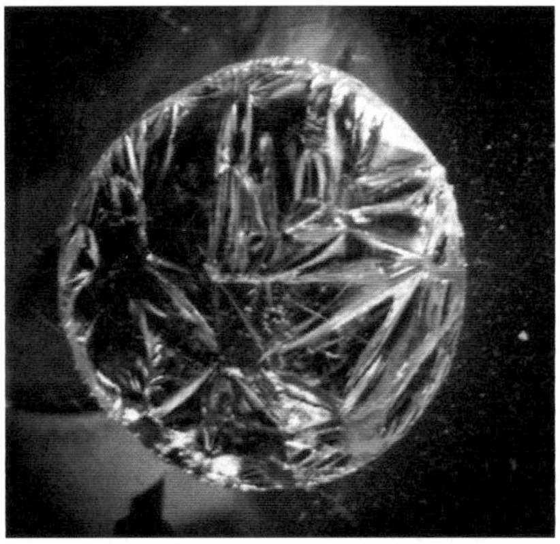

**Figure 3.** Surface polymerization of liquid PDMS due to atomic hydrogen exposure from a Ar/$H_2$ glow discharge plasma at 50 Pa and 13.56 MHz plasma excitation frequency. The wrinkles at the surface indicate the stress that is exerted due to polymerization.

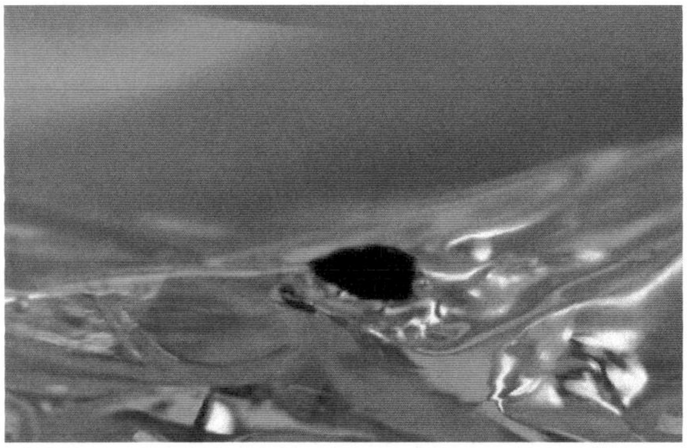

**Figure 4.** After evaporation of the solvent a stable THV 220 layer is formed over liquid of fluorescin.

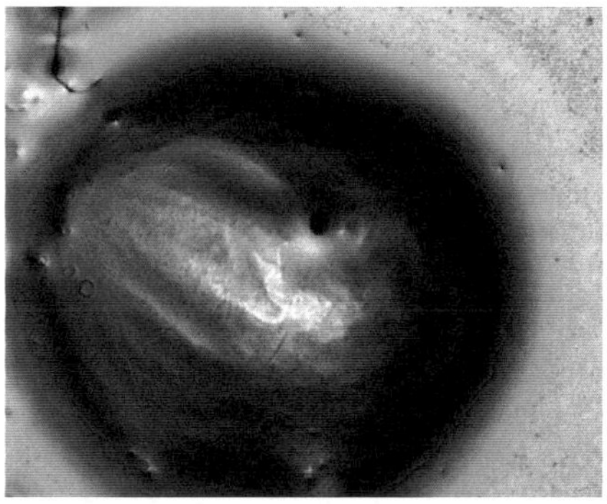

**Figure 5.** Solid on liquid system based on sol-gel solidification of the top liquid.

In the next experiments were carried out using liquid precursors. As a first approach the fluorinated polymer called THV 220 [3] is completely dissolved in *e.g.* ketone. THV 220 dissolved in MEK (methyl ethyl ketone) is casted on higher density liquids. Evaporation of the solvent let cross-link the macromolecules and replicate the macroscopic shape where the liquid was cast on, shown in figure 4. Here, a kind of bridge-building effect can be observed, starting from the outer border.

A further approach using liquid precursors is demonstrated using TEOS on an ionic liquid. The sol-gel reaction in the top liquid forms $SiO_2$ creating by that a solid layer that covers the bottom liquid as shown in figure 5.

## 4. Conclusions and outlook

It is foreseen that stable solid on liquid deposition (SOLID) systems have the potential to further expand the technological possibilities looking at architecture and design of next generation Microsystems. In such systems the liquid can just be passive as Volume- placeholder being subsequently removed as sacrificial liquid for caverns and channels creation, as needed in the field of fluidic systems. For other applications the liquid will be active, containing molecules or particles that remain functional in the liquid or that functionalize the overgrowing solid by chemical reaction. Based on simple models, the authors predicted growth mechanisms that have been experimentally brought to evidence in this paper. It can be assumed that e.g. the precursors coming from the gas phase will not necessarily disappear from the surface by sinking below the liquid surface since inter-molecular forces at the surface are much more important than gravitational effects. As soon as precursors accumulate at the surface and cross-link, linear polymer growth is observed. The authors could bring to evidence such a behavior for all tested CVD and PECVD examples that have been successfully tested without any exception. If the precursors react with the liquid, transition from reaction-limited to diffusion limited growth is observed.

In summary, our preliminary study let us conclude that solid on liquid deposition is rather the rule than the exception.

**Acknowledgements**
The project is funded by 6 EU project MULTIPOL NMP3-CT-200-033201.

**References**

[1] Keppner H and Benkhaira M 2006 Method for producing a plastic membrane device and the thus obtained device; patent WO/2006/063955
[2] Eliasson B and Kogelschatz U 1991 *Nonequilibrium Volume Plasma Chemical Processing, IEEE Transactions on Plasma Science* **19** 1063-77
[3] www.dyneon.com/fluorothermoplastics/THV.

# Synthesis and spectral characterisation of a new metal complex with a bidentate ligand

**Violeta Niculescu[1], Radu Tamaian[1], Nadia Păun[1], Adrian Pîrnău[2] and Laszlo Szabo[3]**

[1] National Research and Development Institute for Cryogenics and Isotopic Technologies, 4 Uzinei, POBox Raureni 7, 240050 Râmnicu Vâlcea, Romania

[2] National Institute for Research and Development of Isotopic and Molecular Technologies, 65-103 Donath, 400293 Cluj-Napoca, Romania

[3] Babes-Bolyai University, Faculty of Physics, 1 Mihail Kogalniceanu, 400084 Cluj-Napoca, Romania

E-mail: violeta@icsi.ro

**Abstract.** Literature mentions the important biologic-active, antimalaric, antiviral, antitumoural, tuberculostatic properties of 3-alkyl-and 3-alkenyl-substituted derivatives of 2-hydroxy-1,4-naphthoquinone and arylamino-1,4-naphthoquinones; the same properties are shown by the complexes that these ligands form with metal ions, which act in the biological structures as essential microelement. This paper reports the study of the structure of some new electron-transfer complexes with bidentate ligands of the naphthoquinone series by using IR, visible and UV spectra. A quantum-mechanical interpretation of the electronic transitions for the free and coordinated ligand has been performed in order to get information on the coordination of the heteroatoms to the M(II). The complexes of transitional metals with naphthoquinonic ligand have been prepared by following the procedure described by Jensen and Nielsen. The structural formula of the free and coordinated organic ligand has been modelled on the computer. The electronic spectra of the complexes studied lead to the conclusion that they are square-planar.

## 1. Introduction

The literature has offered new data related to the complex compounds formed by transition metals and ligands having conjugated double bonds, *i.e.* a system of delocalized electrons in their molecule. The special interest about these complexes is due, among others to the fact that the complex compounds readily participate in reversible electron-transfer reactions. Literature mentions the important biologic-active, antimalaric, antiviral, antitumoural, tuberculostatic properties of this kind of ligands; the same properties are shown by the complexes that these ligands form with metal ions, which act in the biological structures as essential microelements [1]. The importance of such compounds can be exemplified by the interesting biological activities associated with many 3-alkyl- or 3-alkylamino-substituted derivatives of 2-hydroxy-1,4-naphthoquinone or 2-chloro-1,4-naphthoquinones (figure 1) [2].

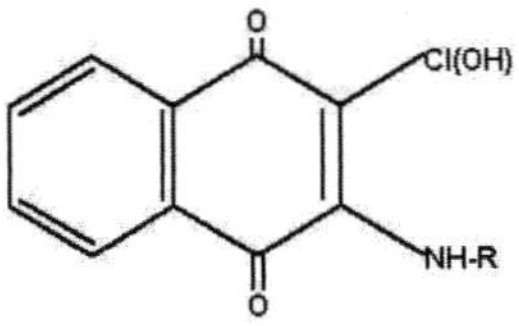


**Figure 1**. Structure of 2-chloro (or 2-hydroxi-)-3 amino-1,4-naphtoquinone.

Further need for a study of compounds of this type is illustrated by the observation that the ortho-amino quinoid unit is present in many antitumour antibiotics such as actinomycins, mitomycin C, porfiromycin and streptonigrin [3]. Because there are few literature [4, 5, 6] regarding the complexes of the general [M-N2S2] type, we hope that the data provided by this paper will contribute to enrich the knowledge in this insufficiently studied field.

## 2. Experimental

*2.1. Preparative methods*

In this work we have used the new naphthoquinonic ligand containing X=S, Y=N as donor atoms: 2-mercapto-3- nicotinamido-1,4-naphthoquinone (MNNQ) (figure 2). This ligand was synthesized according the scheme from (figure 3).

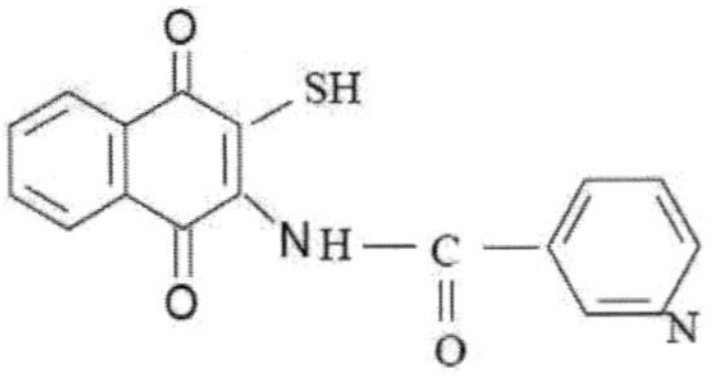


**Figure 2.** Structure of 2-mercapto-3-nicotinamido-1,4-naphthoquinone (MNNQ).

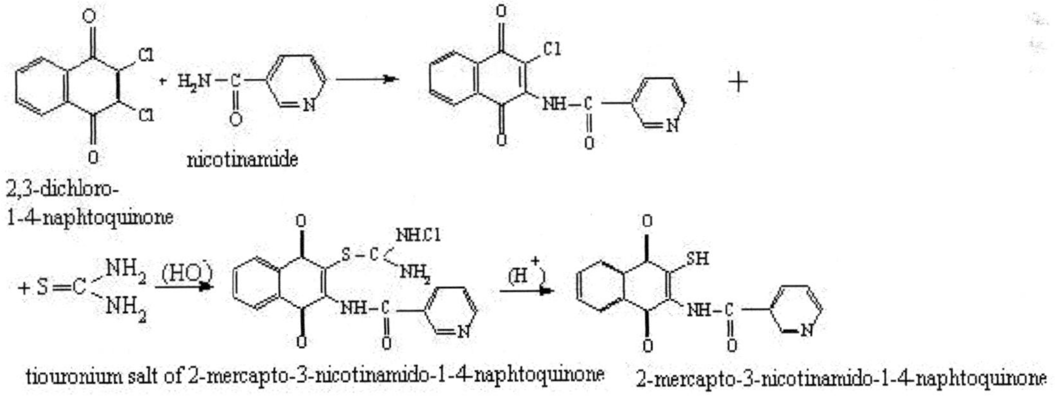


**Figure 3.** Synthesis of the MNNQ ligand.

The complex of Ni (II) with naphtoquinonic ligand (MNNQ) has been prepared by the procedure described by Jensen and Nielsen [7]. The complex is a microcrystalline powder, whose melting point is higher than that of the pure ligand (table 1). It is air-stable, insoluble in ordinary organic solvents, sparingly soluble in dichlorethane and dioxane and soluble in dimethylformamide (DMF) and dimethylsulphoxide (DMSO). The molar electrical conductivity of the complex was measured on a Model OK-102/1 Radelkis conductivity-meter, in $10^{-4}$ M DMF solutions, at 22°C.

**Table 1.** Preparation details, elemental analyses and molar conductance of complex.

| Compound | Mol wt calcd | mp (°C) | Molar conductance ($\Omega^{-1}cm^2mol^{-1}$) | Appearance | Analyses (%) Element | Calcd | Found |
|---|---|---|---|---|---|---|---|
| $[Ni(MNNQ)_2]$ | 676.71 | 122 | 2.98 | Deep-red microcrystalline | Ni | 8.68 | 8.97 |
| | | | | | C | 56.75 | 56.89 |
| | | | | | S | 9.46 | 9.22 |
| | | | | | N | 8.28 | 8.07 |
| | | | | | H | 2.66 | 2.84 |

*2.2. Spectral determinations*

The results of the elemental analyses show that these complex compound is of the type [ML2] where M = Ni(II) and L = MNNQ (figure 4). In order to explain this composition, we must admit that MNNQ act as bidentate ligand.

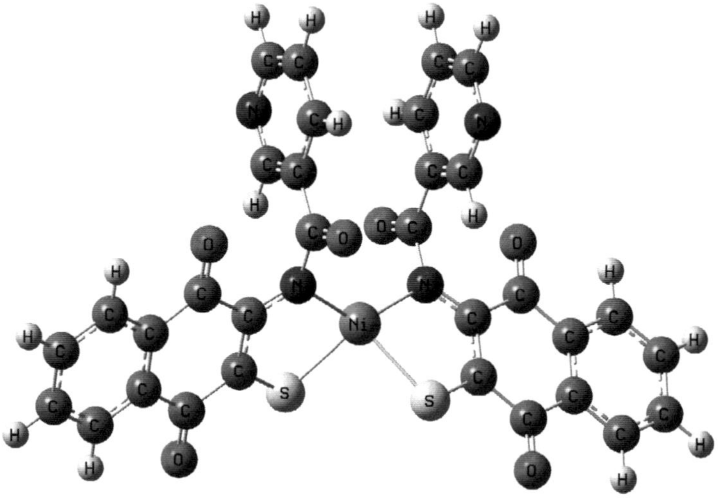

**Figure 4.** Molecular structure of Ni(II) complex.

The assumption has been confirmed by the physical-chemical analyses and quantum-mechanical studies described bellow. In order to obtain data regarding the nature of the chemical bonds and the type of the atoms involved in the coordination were performed infrared spectra for the free ligand and complex compound. These results are showed in (table 2).

**Table 2.** Infrared absorption frequencies ($cm^{-1}$) of ligand and of complex.

| Compound | $\nu_{N-H}$ | $\nu_{C=O}$ | $\nu_{CN}$ coupling C=C | $\nu_{CN}$ coupling $NH_2$ | $\nu_{SH}$ | $\nu_{C=C}$ | $\nu_{C-H}$ | $\nu_{NH}$ | $\nu_{C=S}$ |
|---|---|---|---|---|---|---|---|---|---|
| L | 3482 i<br>3370 i | 1670 i | 1580 i | 1356 i | 680 w | 1150 m | 1005 w | 755 vi | 670 m |
| $NiL_2$ | 3410 w<br>3310 w | 1670 i | 1585 m | 1355 m | 650 w | 1145 m | 1005 w | 758 vi | 678 w |

vi=very intense; i=intense, m=medium, w=weak

The absorption bands due to vibrations groups not involved in the coordination appear in the infrared spectra of the free ligand and of the complex in the same spectral regions, with unchanged or at most slightly modified intensities, because of the electromeric effects due to the coordination. The characteristic bands of the N-H and S-H groups, which appear in the IR spectra of the ligand as intense or medium intense bands in the complex spectra the bands are modified in intensity and appear at slightly modified frequencies, proving the involvement of these groups in the coordination. The metal complex is also characterized by the appearance of some new bands of medium and low intensity at 570 $cm^{-1}$ and 440 $cm^{-1}$, which can be assigned to ν(M-S) and ν(M-N) frequencies. So, IR spectra indicate the participation of both the sulphur and nitrogen atoms in coordination to the metal. The IR spectra were recorded on a Perkin Elmer FT-1600 Hewlett Packard instrument in anhydrous KBr pellets.

In order to establish the coordination geometry of the new complex, a spectral analyses in the visible and UV range was performed. The bands observed in the electronic absorption spectra of the studied complex were assigned according to Gray and Ballhausen [8] and Vanquickenborne [9]. An Unicam UV-visible 2-300 spectrophotometer was used to perform the electronic spectra, which were obtained with $10^{-3}$ M acetone solutions. The structure of the complex was explained using the molecular orbital approach. The metal orbitals involved in σ-bonding in square planar complexes are: $nd_{z^2}$, $nd_{x^2-y^2}$, $(n+1)s$, $(n+1)p_x$ and $(n+1)p_y$. Judging from the values of the overlap integrals, $nd_{x^2-y^2}$, $(n+1)s$, $(n+1)p_x$ and $(n+1)p_y$ account for most of the σ-bonds, and $nd_{z^2}$ makes only a minor contributions. The most important π-molecular orbital is formed by the $(n+1)p_z$ metal orbital and a combination of π-orbitals. In table 3 there are the assignments for the transitions occurring in the electronic spectra.

**Table 3.** Electronic spectra in UV-Vis range for the complex.

| Compound | $\nu_{max}$ ($cm^{-1}$) | $\lambda_{max}$ | Assignment |
|---|---|---|---|
| [Ni(MNNQ)₂] | 20517 | 185 | ${}^1A_{1g} \rightarrow {}^1A_{2g}[b_{2g}(\pi^*) \rightarrow b_{1g}(\sigma^*)]$ |
| | 29150 | 293 | ${}^1A_{1g} \rightarrow {}^1B_{1g}[b_{2g}(\pi^*) \rightarrow a_{1g}(\sigma^*)]$ |
| | 35714 | 880 | ${}^1A_{1g} \rightarrow {}^1B_{1u}[b_{2g}(\pi^*) \rightarrow a_{2u}(\sigma^*)]$ |
| | 37233 | 11200 | ${}^1A_{1g} \rightarrow {}^1E_{u}[e_{g}(\pi^*) \rightarrow a_{2u}(\pi^*)]$ |

*2.3. Computer modeling*

The structural formula of the free and coordinated organic ligand has been modelled on the computer. The molecular geometry has been optimized using the Molecular Mechanics approach (MM+) [10].

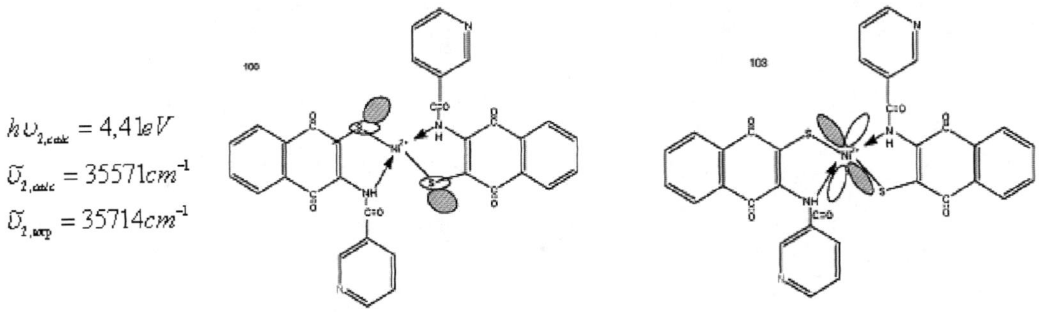


**Figure 5.** Transition 100-103.

In the complexes of Ni(II) the first absorption band is assigned to an electron transfer from an electron lone pair localized on the sulphur atom and the second one to an electron transfer from an electron lone pair localized on the nitrogen atom. These transitions are shifted toward higher wave numbers for the coordinated ligand just as a consequence of the coordination. These shifts are caused by the mixing of the antibonding states with d-metal orbitals as well as to the lowering the electron lone pairs energy that occurs after coordination. By comparing the spectra of the complex with the one of the free ligand, we observed that the first two bands are shifted, while the last two ones are unshifted, meaning that the oxygen atoms are not involved in the coordination, this one being made by the sulphur and nitrogen atoms of the ligand. The absorption maximum from 35714 $cm^{-1}$ is attributed to 100→103 transition, the absorbed photon energy being 4.41 eV, that corresponds to $\tilde{\upsilon} = 35571\ cm^{-1}$. This is a ligand-metal transition, the 100 orbital being practically an orbital with unparticipating electrons from sulphur and 103 being a molecular orbital from a d orbital of Ni(II) ion (figure 5). The absorption maximum from 20517 $cm^{-1}$ is attributed to a d-d transition between 102→103 orbitals, both being molecular orbitals from *d* orbitals of the central metallic ion (figure 6).

$$h\upsilon_{1,calc} = 2{,}52\,eV$$
$$\tilde{\upsilon}_{1,calc} = 20326\,cm^{-1}$$
$$\tilde{\upsilon}_{1,exp} = 20517\,cm^{-1}$$

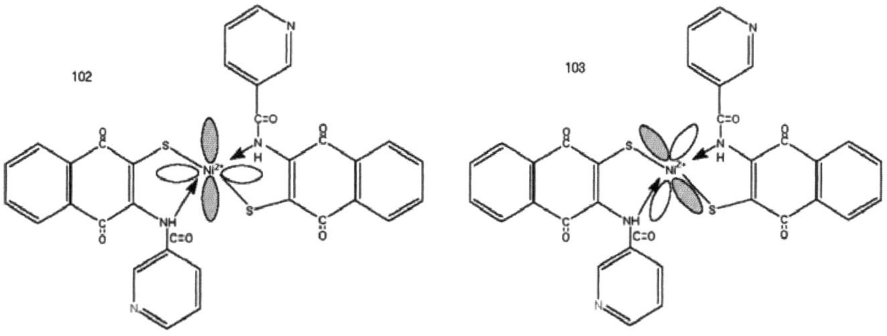


**Figure 6.** Transition 102-103.

## 3. Conclusions

Correlation between elementary analysis and physical-chemical determinations suggests that Ni complex is [ML2] type. Electronic spectra proved that it is a square-planar structure complex. So, on one hand, using computer analysis of animated vibrations, we proved that the molecule is square-planar and appears as trans isomer, and, on the other hand, using computer analysis for molecular orbitals, we proved that the coordination is assured by the two sulphur atoms from the "mercapto" group and the two nitrogen atoms from amidic group of the nicotinamide, situated in ortho- position toward the sulphur atoms.

## References

[1] Balch A L, Rohrscheid F and Holm R H 1965 *J. Am. Chem. Soc.* **87** 2301-8
[2] Truitt P, Mahon F, Hall R L and Eris T E 1960 *J. Org. Chem.* **25** 962-9
[3] Block E, Ofori-Okai G, Chen Q and Zubieta J 1991 *Inorg. Chem. Acta* **189** 137
[4] Holm R H, Balch A L, Davison A, Maki A H and Berry T E 1967 *J. Am. Chem. Soc.* **89** 2866-74
[5] Zheng Y Y, Saluja S, Yap G P, Blumenstein M, Rheingold A L and Francesconi L C 1996 *Inorg. Chem.* **35** 6656-66
[6] Burlov A S, Borodkina I G, Kuznetsova L I, Urae A I, Makarova N I, Vasilchenko I S, Borodkin G S, Altun O, Feiizoglu Abdullaev A and Garnovskii A D 2004 *Russ. J. Gen. Chem.* **74** 772-5
[7] Jensen K A and Nielsen P H 1966 *Acta Chem. Scand.* **20** 597-602
[8] Gray H B and Ballhausen C J 1963 *J. Am. Chem. Soc.* **85** 260-8
[9] Vanquickenborne L G 1981 *Inorg. Chem.* **20** 796-802
[10] Calzaferri G and Brande M 1992 *QPCE Bulletin* **12** 73-80.

# A study about illumination and colors vision

**Florin Toadere**

National Institute for Research and Development of Isotopic and Molecular Technologies, 65-103 Donath, 400293 Cluj-Napoca, Romania

E-mail: florin.toadere@itim-cj.ro

**Abstract.** In this paper we propose a software algorithm which can simulate the human eyes colors view. We present a spectral image processing algorithm, in which we use the color checker spectral image, the human eyes cones spectral sensitivity and the spectral power distribution for light sources. We use three spectrums of real lamps, made by EG&G PerkinElmer: sunlight 5500K, xenon flash and xenon CW, and three CIE light standard sources: D65, CIE A and F7. In order to validate the functionality of the algorithm we simulate the color checker picture hues difference, under different illuminations generate with the light spectrums presented in this paper.

## 1. Introduction

Human vision is sensitive to visible light, that part of the electromagnetic spectrum with wavelengths from about 400 to 700 nm. The illumination determines the amount of light that covers a surface. Color helps in the perception of the beauty of the digital image. The perceived color of the surface is determined not only by the color of the surface but also by the color of the light. Therefore, the type of lamp in the light fixture can significantly impact the perception of the color of the object. This effect should be taken into account when we see images on the computer screen. The perception of the object's shape differs with the light distribution on its surface and with the configuration of the resulting shadows. The direction of the light beam can easily affect the perception of the object's shape [1-4].

Color vision is the capacity of an organism or machine to distinguish objects based on the wavelengths of the light they reflect or emit. Color derives from the spectrum of light interacting in the eye with the spectral sensitivities of the light receptors. The nervous system derives color by comparing the responses to light from the three types of cone photoreceptors in the eye L, M, S (long, medium and short) equivalent to R, G, B (red, green and blue) colors [8]. Reflected color can be measured using a reflectometer, which takes measurements in the visible region of a given color sample. If the custom of taking readings at 3.7 nanometer increments is followed, the visible light range of 400-700nm will yield 81 readings [9]. These readings are typically used to draw the sample's spectral reflectance curve.

In order to see the image colors under different illuminations, we use a spectral image processing algorithm, which allow us to combine the spectral image with the cones sensibilities and the light sources spectral power distributions. We illuminate the color checker image with three EG&G PerkinElmer lamps and three CIE standards. Also we need to make compatibility between the monitor radiance and the human eyes perception. Using our algorithm we can chose any type of light sources.

## 2. The spectral image processing algorithm

In this paper we use a spectral image under the Spectral Binary File Format (.spb). This format has the following characteristics: file identifier is a 3 letter string SPB (Spectral Binary file) located at the beginning of the file. Image dimensions and wavelength values are stored in file header. Dimensions ($x$, $y$ and n) are stored in uint32 format and wavelength values in foat32 format. Spectral image values are reflectance values stored as float32. Spectral image values are scaled between 0 and 1, where 1 describes maximum reflectance. Image data is written to the file in column order and values are stored in little endian form [9].

If we perceive light that is reflected from a surface, instead of light that is directly emitted from a light source, our eyes receive result of the scalar product of reflectance and radiance spectrum [1-6]. In continuous case human eye response is:

$$c_i = \int_{\lambda \min}^{\lambda \max} S_i(\lambda) r(\lambda) l(\lambda) d\lambda \, , \; i = S, \, L, \, M \tag{1}$$

were:

$S_i(\lambda)$ is the function of sensitivity of the $i$-th type of cones

$r(\lambda)$ is the fraction of the reflected illuminant energy

$l(\lambda)$ is the spectral distribution of light

$L$, $M$, and $S$ are the responses of the long, medium, and short cones of the eye [1-4].

The image obtained using equation 1 is not enough from the monitor colors possibilities of representation. In order to remediate this deficiency we have to make compatibility between monitor possibility of colors generation and how the human eyes cones perceive the colors radiance. We need to specify how the displayed image affects the cone photoreceptors [4-8]. To make this estimate we need to know: the effect that each display primary has on your cones and the relationship between the frame-buffer values and the intensity of the display primaries (gamma correction).

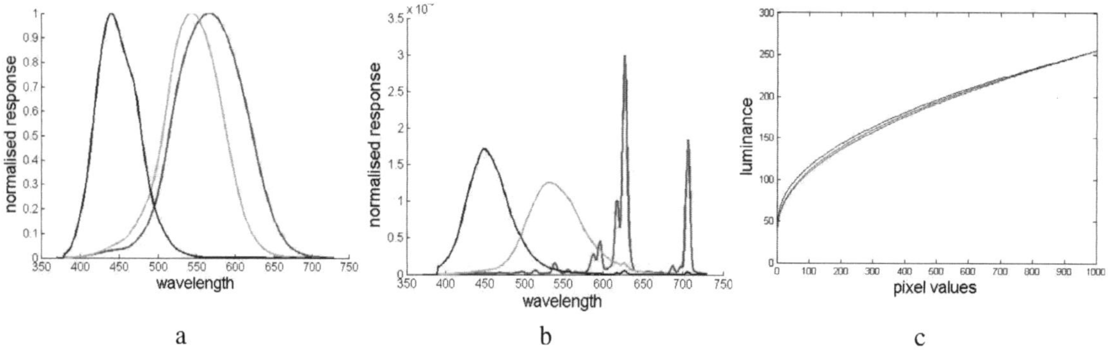


**Figure 1.** a) L, M, S cones spectral response, b) a CRT monitor spectral response, c) inverse gamma function.

To compute the effect of the display primaries on the cones, we need to know the spectral power distribution of the display; a CRT (cathode ray tube) monitor (figure 1b), and the relative absorptions of the human cones (figure 1a). Having this data, we can compute the 3x3 transformation [4] that maps the linear intensity of the display R, G, B signals into the cone absorptions L, M, S (equation 2).

$$\begin{bmatrix} 14.0253 & -13.5154 & 0.7385 \\ -4.1468 & 10.1490 & -1.3618 \\ -0.1753 & -0.5663 & 7.3776 \end{bmatrix} \tag{2}$$

In addition, the characteristics of the display device (screen display) where the digital image is viewed also affect the intensity distribution and interrelationship of contrast between light and dark regions in the specimen. The effects are characterized by a variable known as gamma [8].

$$V_{out} = V_{in}{}^{\gamma} \tag{3}$$

Phosphors of monitors do not react linearly with the intensity of the electron beam. For CRT display monitors and televisions, the luminance produced at the face of the display is a power function, which is proportional to the voltage applied to the faceplate grid raised to an exponential power. The numerical value of the exponent of this power function is known as gamma. By definition gamma is a nonlinear operation used to code and decode luminance or tristimulus values in video or image systems.

In conformity with the equations (1)-(3) we have the next algorithm steps:

1. Load the data into Matlab (spectral image, lamps spectrums and cones response)
2. For each lamp spectrum illumination, we compute the human eye color response using equation (1).
3. We make compatibility between monitor possibility of colors generation and how the human eyes cones perceive the colors radiance using equation (2).
4. We make gamma correction, to correct the monitor luminance using equation (3).

## 3. Simulation results

### *3.1. Illumination simulation*

In this paper we use as spectral image the color checker picture and three light sources from EG&G PerkinElmer: sunlight 5500 K, xenon flash lamp and xenon CW. These lamps spectrums spread in UV an IR but in our simulation we focus our attention on visible light between 380 and 780 nm, because the spectral image is defined only on visible range. The lamps spectral powers distributions are represented in figure 2b and further information can be found in [8, 10]. Also we simulate the illumination using the CIE standards: CIE A, D65 and F7 (figure 2a)) [2, 7].

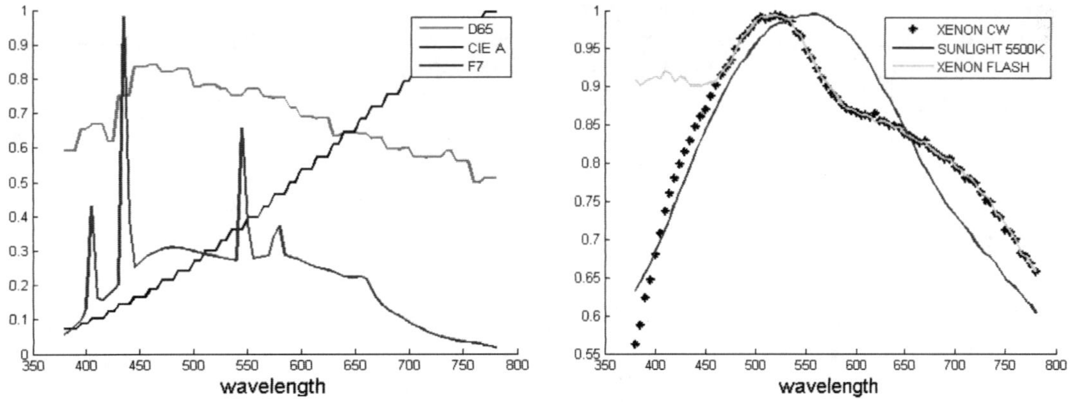


**Figure 2.** a) CIE spectrums, b) EG&G PerkinElmer lamps spectrums.

In order to present the functionality of the spectral image processing algorithm, we simulate the original color checker image illuminated with EG&G PerkinElmer lamps spectrums. In figure 3a we have the color checker original image. Because the lamps spectrums cover the entire visible spectrum and have little differences in shape (figure 2b), we expect that the illuminations simulations to have little differences. In figures 3b-d, we present the simulations make under the sunlight 5500 K illumination, xenon flash illumination and xenon CW illumination. On the other hand we have the light sources spectrums standardized by CIE. We use three CIE standards: CIE A, D65 and F7. The CIE A standard represents a classical incandescent bulb, the D65 standard represents the midday sky without clouds and F7 represents a fluorescent source. Those light sources spectrums (figure 2a) are very different and we expect that the illumination effect (colors hues) to be very different (figure 4).

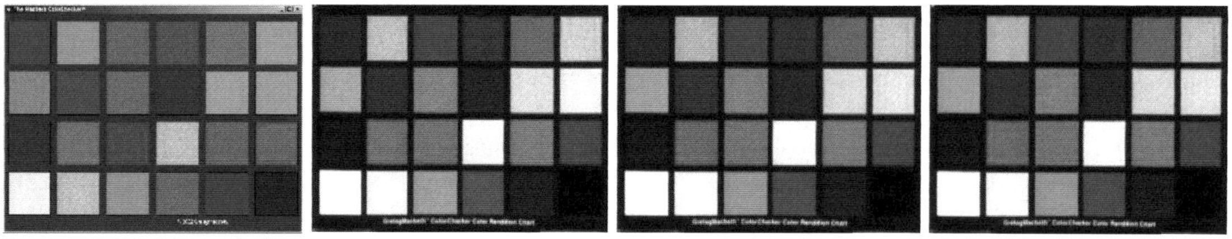

a b c d

**Figure 3.** the original color checker image, b) sunlight 5500 K illumination c) xenon flash illumination, d) xenon CW illumination.

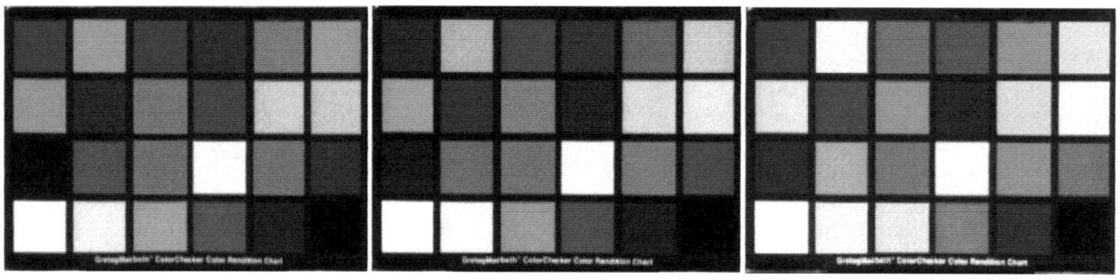

a b c

**Figure 4.** a) CIE A illumination, b) D65 illumination, c) F7 illumination.

## 4. Conclusion

As we can see from day by day experience the light sources and lamps have different colors hues, in function of the spectrums they cover. In this paper we image simulated the color checker characteristics (colors hues) under different light sources illumination. We simulate the illumination with three EG&G PerkinElmer lamps and we see that those lamps try to copy the sun light at 5500K. From figure 3 we see that the illuminated images have little colors hues differences. In figure 4 we simulated the color checker picture illuminated with CIE standards and we see that those standards effects are very different. In general our algorithm supports as input data any visible spectrums and works in real time.

## References

[1] Fairchild M D 2004 *Color Appearance Models* 2nd Edition (Wiley)
[2] Malacara D 2004 *Color Vision and Colorimetry Theory and Applications* (SPIE Press)
[3] Shevell S K 2003 *The science of color* (Elseiver)
[4] Wandell B 1995 *Foundations of Vision* (Sinauer Associates)
[5] Toadere F 2008 *ATOM-N 2008 4th Int. Conf. Advanced Topics in Optoelectronics, Microelectronics and Nanotechnologies* (Constanta)
[6] Toadere F and Szopos E 2008 *2008 IEEE- TTTC Int. Conf. on Automation, Quality and Testing, Robotics AQTR 2008 (THETA 16)* (Cluj Napoca)
[7] http://en.wikipedia.org/wiki/Category:Color_space
[8] http://optoelectronics.perkinelmer.com/content/whitepapers/XenontheFullSpectrum.pdf
[9] http://spectral.joensuu.fi/index.php?page=spi
[10] http://home.earthlink.net/~pubspectra/#Spectral_Data_Links.

Journal of Physics: Conference Series **182** (2009) 012032 doi:10.1088/1742-6596/182/1/012032

# Raman and EPR studies of calcium-phosphate glasses doped with manganese ions

**Dana Toloman[1], A R Biris[1], Adriana Popa[1], Oana Raita[1], L M Giurgiu[1] and I Ardelean[2]**

[1] National Institute for Research and Development of Isotopic and Molecular Technologies, 65-103 Donath, 400293 Cluj-Napoca, Romania

[2] Babes-Bolyai University, Faculty of Physics, 1 Mihail Kogalniceanu, 400084 Cluj-Napoca, Romania

E-mail: dana.toloman@itim-cj.ro

**Abstract**. The structure of $xMnO\cdot(1-x)[2.5P_2O_5\cdot CaO]$ glass system with $1 \leq x \leq 10$ mol % was prepared and investigated by Raman and EPR spectroscopy. The characteristic Raman bands of these glasses due to the stretching and bending vibrations were identified and analyzed by increasing of MnO content. The intensity and frequency variations for the characteristic phosphate group vibrations have been correlated with the changes of the structural units present in these glasses. All the investigated samples exhibit EPR signals which are characteristic to the $Mn^{2+}$ ions. The shapes of spectra are also changed with the increasing of manganese ions content.

## 1. Introduction

Phosphate glasses have been of large interest for a variety of technological applications due to several unique properties such as high thermal expansion coefficient, low viscosity, UV transmission or electrical conduction [1]. Important biological applications for calcium phosphate glasses exist. Also, it was demonstrated that they are biocompatible as bones and dental implants [2, 3].

The properties that make phosphate glasses candidates for so many applications are related to their molecular – level structure [4]. The basic building blocks of amorphous phosphates are the P-tetrahedra [4]. They result from the formation of $sp^3$ hybrid orbital by the P outer electrons ($3s3p^3$). The fifth electron is promoted to a $3d$ orbital where strong $\pi$ – bonding molecular orbital are formed with oxygen 2p electrons. These tetrahedra link through covalent bridging oxygens to form various phosphate anions. The tetrahedra are classified using the $Q^n$ terminology, where "n" represents the number of bridging oxygen per tetrahedron [5]. Specific properties of phosphate glasses could be understood only if the behavior over the entire composition range is known, which is also of interest with regard to changes of certain macroscopic properties.

The understanding of the glasses structure on different length scales remains difficult mainly due to the lack of periodicity inherent to glasses. Only the suitable correlation of experimental data obtained from several investigations provided by different methods allows the finding of helpful structural information.

In order to evidence the structural changes induced by MnO in the phosphate glass network, a xMnO·(1-x)[2.5P2O5·CaO glass system was prepared and investigated by means of Raman and EPR spectroscopy.

## 2. Experimental

xMnO·(1-x)[$2.5P_2O_5$·CaO] glass system with $1 \leq x \leq 10$ mol % have been prepared from corresponding quantities of $MnCO_3$, $(NH_4)_2HPO_4$ and $CaCO_3$ of p.a. purity, by melting at 1200 C for 5 min. The molten liquid was poured onto a cooper plate and pressed with another one very quickly to preserve the local disorder characteristic to the liquid state.

The Raman spectra were collected at room temperature using a JASCO NRS-3300 micro-Raman Spectrometer with an air cooled CCD detector in a backscattering geometry and using a 600/mm grating. The microscope objective used for the studies was 100X. As excitation, it was used a 785 nm laser line with the power at the sample surface of 85 mW.

The EPR experiments were performed at room temperature in the X-band (9.46 GHz) on a Bruker ELEXSYS E500 spectrometer.

## 3. Results and discussion

The Raman spectra of the investigated samples are displayed in figure 1. It can be observed that their shapes are changed with the increase of MnO content. On the other hand the majority of the bands are large and asymmetric and they present some shoulders.

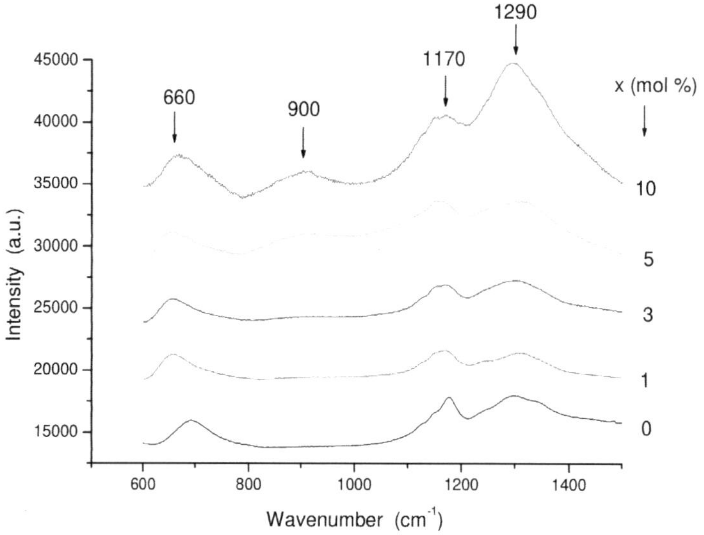


**Figure 1.** Raman spectra of xMnO·(1-x)[$2.5P_2O_5$·CaO] glasses.

The most important bands observed in our glasses are: ~ 660 $cm^{-1}$ assigned to the P-O-P symmetric stretching vibrations; ~ 1170 $cm^{-1}$ due to the symmetric stretching motions of two non-bridging oxygen (NBO) atoms bonded to phosphorous atoms ($PO_2$) in the $Q^2$ phosphate tetrahedron; ~ 1300 $cm^{-1}$ due to the P=O symmetric stretching vibrations[6-11]. For high content of MnO ($x \geq 5$ mol %) a new band appear at ~ 900 $cm^{-1}$ due to the ($PO_4$) symmetric stretching vibrations of the $Q^0$ species [9]. The increase of $Q^0$ groups is due to the breaking of phosphate chains [8].

The asymmetric 660 $cm^{-1}$ band assigned to P–O–P symmetric stretching vibrations of the long-chain phosphate glasses is shifted to 702 $cm^{-1}$ due to the decreasing of phosphate chains lengths [10-14]. The shift of this band may be also attributed to a change in the in-chain P–O–P bond angle as an effect of the MnO network modifier on the glass structure [14, 15]. The higher wavenumber of the P–

O–P symmetric band is a result of the smaller P–O–P bond angle, characteristic for shorter phosphate chain length [6, 7].

Recorded EPR spectra show resonance line due to the $Mn^{2+}$ ($3d^5$, $^6S_{5/2}$) paramagnetic ions for all range of concentration. The structure of the spectra strongly depends on the manganese content of the samples. The EPR spectra are presented in figure 2. The EPR spectra observed in the calcium phosphate glass matrix consist of a well resolved sextet centered at g = 2.0 with a broad background signal underneath the sextet and very small absorption centered at g = 4.3, which is characteristic for isolated $Mn^{2+}$ ions in rhombic distorted octahedral environment, as shown in figure 2.

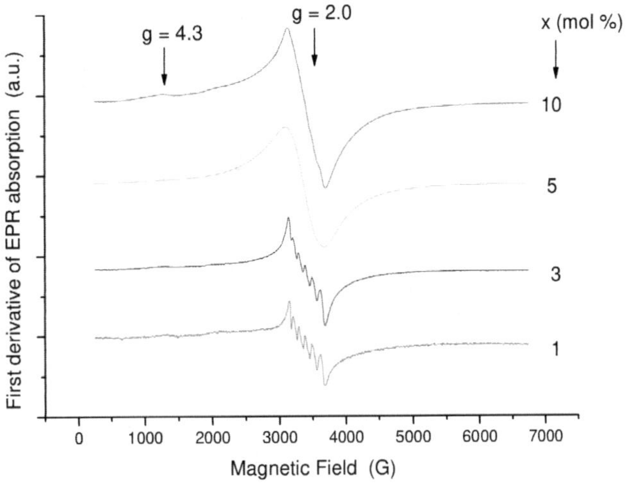


**Figure 2.** EPR spectra for $xMnO \cdot (1-x)[2.5P_2O_5 \cdot CaO]$ glasses .

The g = 2.0 resonance line is known to arise from the central $M_s = |-1/2\rangle \rightarrow |+1/2\rangle$ transition where $M_s$ is the effective spin component. The six-line multiplet results from the hyperfine interaction of the electron spin with the $^{55}Mn$ nucleus (I = 5/2). This isotropic signal at g = 2.0 is due to $Mn^{2+}$ ions in an environment close to octahedral symmetry

The hyperfine structure (hfs) superimposes on a large absorption line. It could be considered as the envelope of all contributions at the resonance absorption having g = 2.0 caused by a random distribution of $Mn^{2+}$ in the $2.5P_2O_5 \cdot CaO$ glass matrix. When the MnO content increases (x = 5 mol %) the hfs is smeared out due to both, the distribution of crystal field parameters and the dipole-dipole interaction. The g = 2.0 linewidth increases with the MnO content until x = 5 mol % due to the increase of the dipolar interaction between the $Mn^{2+}$ ions. The reduction in the increase tendency of the g = 2.0 linewidth for the $5 \leq x \leq 10$ mol % compositional range, indicates the possible presence of superexchange interactions between the $Mn^{2+}$ ions (table 1) [17].

**Table 1.** The EPR parameters of the resonance line corresponding to g = 2.0.

| X (% mol) | ΔH (G) | $I_{EPR}$ x $10^{10}$ (a.u.) |
|---|---|---|
| 1 | 527 | 3.3742 |
| 3 | 540 | 7.9008 |
| 5 | 578 | 43.8666 |
| 10 | 565 | 22.4684 |

In table 1 is also presented the absorption line intensity, $I_{EPR}$, (obtained after the integration of the area under the corresponding signal) for the investigated samples. In the concentration range $1 \leq x \leq 5$ mol %, the increase of MnO content in the samples determines an increase of $I_{EPR}$. Since the signal intensity is proportional to the number of EPR active species involved in the resonance absorption, the observed increase of $I_{EPR}$ reflects an increase of the concentration of $Mn^{2+}$ ions involved in other structural vicinities. For the concentrations higher than x = 5 mol %, the increase of the MnO content determines a corresponding decrease of the line intensity. It could be due to the decrease of the number of $Mn^{2+}$ ions and probably, to the appearance in the investigated glasses of the $Mn^{3+}$ ions which do not manifest in EPR spectra at room temperature [18].

## 4. Conclusions

The glass network structure of the investigated sample is mainly formed by P-O-P symmetric stretching vibrations; $(PO_2)$ in the $Q^2$ phosphate tetrahedron and P=O symmetric stretching vibrations. For higher content of MnO, due to the $(PO_4)$ symmetric stretching vibrations of the $Q^0$ species a new band appears. The vibrational spectral features indicate a depolymerization process of the phosphate network with increasing of the manganese content. This process implies an increase of the number of non-bridging oxygen.

The EPR investigations performed on these glasses have indicated the presence of two signals: an isotropic one at g = 2.0 which is due to $Mn^{2+}$ in an environment close to the octahedral symmetry and a very small absorption centered at g = 4.3 which is characteristic for isolated $Mn^{2+}$ ions in cubic symmetric sites slightly tetragonally or rhombically distorted. In the samples with up to 5 % mol MnO, the isolated $Mn^{2+}$ ions in well structured vicinities and the dipolar coupled ones were identified. For higher concentrations, the superexchange interaction between the $Mn^{2+}$ ions could be present and probably, $Mn^{3+}$ ions appear in the investigated glasses.

## References

[1] Sales B C 1987 *Mater. Res. Soc. Bull.* **12** 32
[2] Hench L L, Splinter R J, Allen W C and Greenlee T K 1972 *J. Biomed. Mater. Res. Symp.* **2** 117
[3] Ogino M and Hench L L 1980 *J. Non-Cryst. Solids* **38-39** 673
[4] Brow R K 2000 *J. Non-Cryst. Solids* **263-264** 1
[5] Liebau F 1981 *Structure and Bonding in Crystals II* (New York: Academic Press) p 197
[6] Koo J, Bae B and Na H K 1997 *J. Non-Cryst. Solids* **212** 173
[7] Scagliotti M, Villa M and Chiodelli G 1987 *J. Non-Cryst. Solids* **93** 350
[8] Cozar O, Magdas D A, Nasdala L, Ardelean I and Damian G 2006 *J. Non-Cryst. Solids* **352** 3121
[9] Karakassides M A, Saranti A and Koutselas I. 2004 *J. Non-Cryst. Solids* **347** 69
[10] Pemberton J E, Latifzadeh L, Fletcher J P and Risbud S H 1991 *Chem. Mater.* **3** 1995
[11] Hudgens J J, Brow R K, Tallant D R and Martin S W 1998 *J. Non-Cryst. Solids* **223** 21
[12] Reis S T, Faria D L A, Martinelli J L, Pontuschka W M, Day D E and Partiti C S M 2002 *J. Non-Cryst. Solids* **304** 188
[13] Ilieva D, Jivov B, Bogachev G, Petkov C, Penkov I and Dimitriev Y, 2001 *J. Non-Cryst. Solids* **283** 195
[14] Bertoluzza A, Battaglia M A, Simoni R and Long D A 1983 *J. Raman Spectrosc.* **14** 178
[15] Rouse Jr G B, Miller P J and Risen W M 1978 *J. Non-Cryst. Solids* **28** 193
[16] Griscom D L 1980 *J. Non-Cryst. Solids* **40**, 211.

# New solid-state methods for investigating molecular structure: the CHCC approach

**Carmen Tripon, Xenia Filip, Vasile V Morariu and Claudiu Filip**

National Institute for Research and Development of Isotopic and Molecular Technologies, 65-103 Donath, 400293 Cluj-Napoca, Romania

E-mail: carmen.tripon@itim-cj.ro

**Abstract.** A new experimental technique based on cross polarization under magic angle spinning (CP/MAS) is introduced. This method, called CHCC, is used to quantify $^{1}H$–$^{13}C$ polarization transfer efficiency in high resolution 2D ($^{13}C$–$^{13}C$) correlation spectra of uniformly $^{13}C$-labeled compounds. Comparing the intensities of the cross-peaks obtained by techniques that are usually employed to determine inter-nuclear distances in natural abundance compounds, and the CHCC experiment, it is shown that the latter has the advantage of measuring larger proton-carbon distances. This is experimentally demonstrated on $^{13}C$ uniformly labelled L-Tyrosine hydrochloride.

## 1. Introduction

Structural refinement of supramolecular aggregates is one of the most important achievement of solid-sate nuclear magnetic resonace (SSNMR). Recent work on this field reports on structure elucidation of different categories of materials, such as non-crystalline compounds [1-3], nanocomposites [4,5], paramagnetic systems [6], surface bond peptides [7], and proteins [8-10].

Several experimental approaches based on polarization transfer (CP) between $^{1}H$–$^{13}C$($^{15}N$) spin species, combined with fast magic angle sample spinning (MAS), are used to measure inter-atomic distances, and we mention here the CHHC[11,12], CPPI [13] and LG-CP [14-17] experiments. For instance, the LG-CP MAS experiment was successfully employed to measure the carbon signal enhancement, and implicitly the $^{13}C$–$^{1}H$ heteronuclear interactions in naturally abundant compounds. The situation becomes more complex in uniformly $^{13}C$-labeled samples, where this procedure can be applied only for the case of strongly coupled or spatially isolated CH spin pairs. Moreover, in 2D LG-CP HETCOR correlation spectra individual $^{13}C$ spin contributions are difficult to separate, due to the poor $^{1}H$ spectral resolution.

Therefore, in the present paper we will introduce a new experimental approach, called CHCC, which is based on the polarization transfer between $^{1}H$ and $^{13}C$ under MAS conditions. This new technique has the following advantages: (a) it encodes $^{1}H$–$^{13}C$ polarization transfer efficiency in 2D ($^{13}C$–$^{13}C$) correlation spectra with increased selectivity compared to 2D ($^{1}H$–$^{13}C$) spectra, due to the superior $^{13}C$ resolution; (b) it gives more accurate structural details, being able to measure larger $^{1}H$–$^{13}C$ distances than LG-CP MAS HETCOR. The above mentioned advantages will be verified experimentally on L-Tyrosine·HCl.

## 2. Experimental

Solid state NMR spectra of uniformly $^{13}C$ labeled L-Tyrosine·HCl were recorded using the CHCC pulse sequence depicted in figure 1. All the experiments were carried out at room temperature, on a Bruker AVANCE-400 spectrometer (100 MHz $^{13}C$ Larmor frequency). Standard CP/MAS was calibrated at a spinning frequency $\upsilon_R$ = 10.5 kHz, using a $^1H$ $90^0$ pulse length of 3.75 μs. The 2D CHCC spectra were acquired under two-pulse phase-modulated (TPPM) $^1H$ decoupling at 60 kHz by averaging 16 scans. The CP transfer was optimized for the first Hartmann – Hahn matching condition ($\omega_{1C} = \omega_{1H} - \omega_R$), where the $^1H$ and $^{13}C$ *rf* fields have been calibrated to 50 and 40 kHz. A contact pulse of 1 ms was used for the first CP, and a shorter pulse-width has been employed for the second CP step (65 μs). Two values for the mixing time which corresponds to the third CP were considered, namely 500 μs and 1 ms, respectively. Suitable phase-cycling was applied during the experiment for selecting the polarization transfer path illustrated in figure 1 with dotted lines.

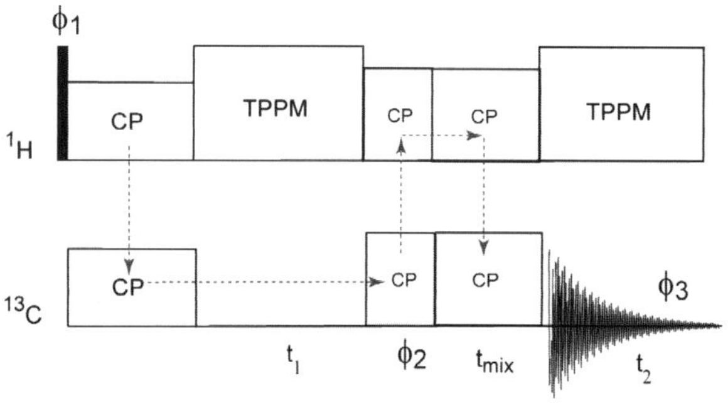


**Figure 1.** The CHCC pulse sequence employed in the present work. Dotted line represents the selected $^1H$-$^{13}C$ inter-nuclear polarization transfer path.

## 3. Results and discussion

CHCC cross-peaks are generated following a three step procedure: (*i*) standard $^1H$ – $^{13}C$ CP is followed by $t_1$-evolution under $^{13}C$ chemical shifts and simultaneous $^1H$ TPPM decoupling, which allows chemical site resolution on the carbon dimension; (ii) subsequently, a short CP pulse is applied to transfer the polarization from the $^{13}C$ nuclei mostly to their directly bonded $^1H$; (*iii*) finally, by incrementing the last CP time ($t_{mix}$) this $^1H$ polarization is transferred back not only to the strongly bonded carbon with the polarized proton, but also, to other neighboring $^{13}C$ nuclei. The individual CP transfer efficiencies are reflected in the corresponding cross-peaks intensities. Also, the evolution with the mixing time, $t_{mix}$, of all cross-peak buildup curves are expected to depend on the distances between the polarized $^1H$ nucleus and individual $^{13}C$ sites.

The CHCC technique is used to selectively measure the efficiency of proton-carbon polarization exchange in uniformly $^{13}C$-labeled compounds, where the initial polarization comes primarily from the protonated carbons. To test its accuracy in providing structural information reflected by ($^{13}C$, $^{13}C$) correlations, numerical simulations using the SPINEVOLUTION [18] program were performed on representative spin systems. For this purpose, the simulated cross polarization buildup curves are obtained using the LG-CP HETCOR pulse sequence, and CHCC sequence, respectively. Comparing the intensities of the CP signal for different C-H distances, in the LG-CP HETCOR case the value of the buildup curve intensity becomes almost zero for a $^{13}C$–$^1H$ distance of ~3 Å, while in the CHCC case the buildup curve, which corresponds to a 4.5 Å distance, is still observable. These preliminary results clearly show the increased sensitivity of the CHCC experiment to structural details in molecular systems. To analyze the precision of this newly introduced procedure, the CHCC 2D ($^{13}C$-$^{13}C$) spectra of 100% U[$^{13}C$]L-Tyrosine·HCl are discussed next.

The experimental 2D ($^{13}C$–$^{13}C$) CHCC correlation spectra of uniformly $^{13}C$-labeled L-Tyrosine·HCl, for two values of the mixing time, $t_{mix}$, *i.e.* 0.5 ms and 1 ms, respectively are presented in figures 2a, 2b. The intra-molecular $^1H$–$^{13}C$ distances span a relatively broad range, from 2.13 Å (H7–C8), to 5.65 Å (H3–C8). The discussion of the experimental results is confined to inter-molecular $^1H$–$^{13}C$ distances less than 6 Å.

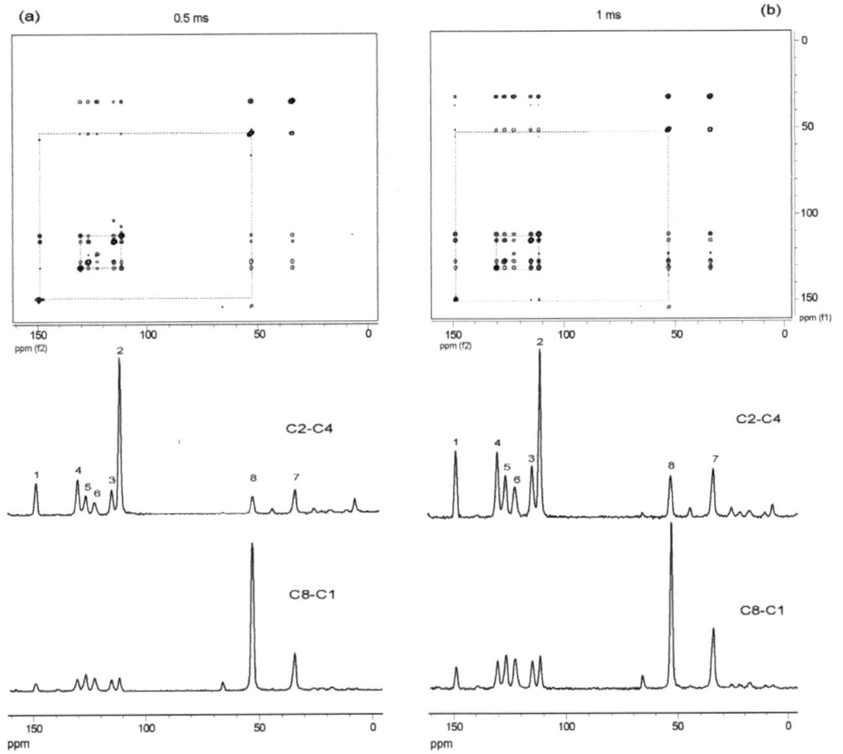


**Figure 2.** 2D CHCC spectra recorded at 0.5 (a) and 1 ms (b) mixing times on uniformly-$^{13}$C labeled L-Tyrosine·HCl. 2D ($^{13}$C–$^{13}$C) correlation CHCC spectra were acquired under TPPM $^{1}$H decoupling at 60 kHz by averaging 16 scans. Horizontal slices for the C2 and C8 diagonal peaks are shown in the lower side of the figure, which enables one to illustrate the C2–C4, and C8–C1 cross-peaks discussed in the text.

At a close examination of the CHCC spectra, one can notice that the most intense cross-peaks (C7–C8, C3–C5, C2–C4) are those which correspond to shortest $^{1}$H–$^{13}$C distances, of about 2.15 Å, indicating that the selected mixing time encodes information about short intra-molecular contacts. This is in agreement with the theoretical findings that such cross-peaks will dominate the spectrum at all mixing time values (see figure 2b). However, cross-peaks which are due to the correlations between more distant spins are also well shaped, such as C5 (C4)–C7 (2.7 Å), C3–C4 (3.9 Å), and C2–C7 (4.68 Å).

The comparison between theoretical and experimental results will be confined in the remainder of this work to two relevant CHCC cross-peaks, namely C2–C4 and C8–C1, which imply short- or medium-size distances (C2–C4), as well as large distances (C8–C1). The intensities of these cross-peaks are due to both contributions, the intra-molecular $^{1}$H–$^{13}$C contacts (2.1 Å for C2–C4; 5.1 Å for C8–C1), and inter-molecular contacts (4 Å for C2–C4; 5.6 Å for C8–C1). The most significant outcome achieved using this approach is the occurrence of C8–C1 cross-peak, proving the ability of the CHCC experiment to generate the correlation between two carbon sites which imply large intra- and inter-molecular distances, up to 6 Å, which are commonly found in complex solid-sate compounds. In contrast with the 2D LG-CP HETCOR experiment, where the information is limited by the $^{1}$H poor spectral resolution, this approach being sensitive to proton-carbon distances up to 3 Å, in the CHCC case the cross-peaks corresponding to long range carbon-carbon correlations are well defined. The present approach hence, allows for the study of structural environment in supra-molecular aggregates with relevance in biophysics and chemistry.

## 4. Conclusions

In summary, we have introduced here the so called CHCC mixing sequence, designed to probe the $^{1}H$–$^{13}C$ polarization transfer efficiency in high resolution SS-NMR 2D ($^{13}C$, $^{13}C$) correlation spectra of uniformly-labeled compounds. This method was implemented using numerical simulations on representative spin systems, and verified experimentally on U[$^{13}C$] L-Tyrosine·HCl. For the investigated values of the mixing time, the experimental results indicate that the CHCC technique can be used to qualitatively describe large inter-molecular distances in solid-state systems within a broad range, up to 6 Å. Therefore, the increased accuracy of the CHCC experiment in investigating intermolecular conformation in complex systems makes this approach suitable for structural refinement in a large variety of supra-molecular compounds.

## Acknowledgments

Financial support from the ANCS is gratefully acknowledged.

## References

[1] Cadars S, Lesage A and Emsley L 2005 *J. Am. Chem. Soc.* **127** 4466
[2] Alam T M and Lang D P 2001 *Chemical Physics Letters* **336** 385
[3] Zeyer M, Montagne L and Jaeger C 2003 *Solid State Nucl. Magn. Reson.* **23** 136
[4] Schmidt-Rohr K, Rawal A and Fang X W 2007 *The Journal of Chemical Physics* **126** 054701
[5] Karp E S, Newstadt J P, Chu S G and Lorigan A 2007 *J. Magn. Reson.* **187** 112
[6] Wickramasinghe N P and Ishii Y 2006 *J. Magn. Reson.* **181** 233
[7] Lange A, Seidel K, Verdier L, Luca S and Baldus M 2003 *J. Am. Chem. Soc.* **125** 12640
[8] Balayssac S, Bertini I, Lelli M, Luchinat C and Maletta M 2007 *J. Am. Chem. Soc.* **129** 2218
[9] Goldbourt A, Gross B J, Day L A and McDermott A E 2007 *J. Am. Chem. Soc.* **129** 2338
[10] Ganapathy S, van Gammeren A J, Hulsbergen F B and de Groot H J M 2007 *J. Am. Chem. Soc.* **129** 1504
[11] Lange A, Luca S and Baldus M 2002 *J. Am. Chem. Soc.* **124** 9704
[12] Tycko R and Ishii Y 2003 *J. Am. Chem. Soc.* **125** 6606
[13] Ladizhansky V and Vega S 2000 *J. Chem. Phys.* **112** 7158
[14] Palmas P, Tekely P and Canet D 1993 *J. Magn. Reson.* **A 104** 26
[15] Brus J and Jakes J 2005 *Solid State Nucl. Magn. Reson.* **27** 180
[16] Dvinskikh S V, Zimmermann H, Maliniak A and Sandstrom D 2005 *The Journal of Chemical Physics* **122** 044512
[17] Hong M, Yao X, Jakes K and Huster D 2002 *J. Phys. Chem.* **B 106** 7355
[18] Veshtort M and Griffin R G 2006 *J. Magn. Reson.* **178** 248.

Journal of Physics: Conference Series **182** (2009) 012034 doi:10.1088/1742-6596/182/1/012034

# Simplified portable 4 MHz RF plasma demonstration unit

**Cristian D Tudoran**

National Institute for Research and Development of Isotopic and Molecular Technologies, 65-103 Donath, 400293 Cluj-Napoca, Romania

E-mail: cristian.tudoran@itim-cj.ro

**Abstract.** This paper documents the design of an RF plasma demonstration unit. The paper shows that there are significant challenges associated with the generation of high frequency plasma under atmospheric pressure, but it yields some interesting results and a simple and elegant design. The generator makes use of a standard power MOSFET in a modern switching amplifier design to produce the required RF power and drive the resonator to produce the high frequency discharge.

## 1. Introduction

Over the last decade, the interest in plasmas generated at atmospheric pressure has increased. Atmospheric plasmas offer high excitation selectivity and energy efficiency in plasma chemical reactions. They are sources of UV, Vis and IR radiation, free radicals such as O, OH and ozone that can play important roles in various techniques. This simplified RF plasma unit was built for several reasons : to prove that it could be done, despite many technical difficulties, to see how the appearance and behaviour of the electrical discharge differ at MHz frequencies, compared to "traditional" DC or low frequency corona discharge, to make the system compact and portable, easy to set up and use for demonstrations. The operating frequency of 4 MHz was chosen because it is sufficiently far into the RF spectrum so that the generated discharge will have a different behaviour than the low frequency corona.

## 2. The RF plasma source

Conventional power amplifiers for use at several MHz usually employ specially designed RF power semiconductors and operate in class A or class C. These devices typically use special fabrication techniques that reduce the stray capacitances down to acceptable levels to permit efficient operation at such high frequencies. The plasma unit presented in this paper was built using only general purpose components that are easily available. However, there are real challenges associated with getting standard switch-mode power MOSFETs to switch efficiently at 4 MHz. At this frequency it is difficult to drive the gate with fast edges due to its large capacitance. Large die power MOSFETs typically have gate capacitances of several nanofarads. It takes a considerable power to quickly charge and discharge this capacitance 4 million times every second. The "Miller effect" adds to this difficulty due to capacitance from drain to gate if the device is operated from a high voltage supply, such as in this particular application. Another problem are the high switching losses when the switching time represents a significant portion of the total period. A typical switching time of 50 ns is acceptable at 200kHz where the switching time only represents 1% of the total period. However the same 50 ns switching time represents 20% of the period at 4 MHz. This means that the device would spend a considerable portion of the total period passing through the linear region dissipating a lot of power. Also, there are switching losses due to repeated discharging of the MOSFETs output capacitance. Every time a MOSFET is switched on the energy stored in its output capacitance is discharged into the MOSFET channel. This is particularly bad for devices being operated at high voltage and high frequencies, when the stored charge can be considerable.

### *2.1. The circuit explained*

The system diagram can be broken down into the following functional blocks: power supply, 4 MHz crystal oscillator, high current MOSFET driver, MOSFET power stage and the Tesla resonator, tuned at 4 MHz. The diagram is shown in figure 1.

The power supply has tree outputs: +5 V for the crystal oscillator, +12 V for the driver stage and +300 V for the MOSFET power stage. The two low voltage outputs are filtered and stabilized. The high voltage output is filtered by an LC cell. The crystal oscillator module is based on the SN7405 TTL integrated circuit (hex-inverter with open collector outputs) and provides a highly-stable square wave at 4 MHz. It was decided to drive the circuit from a fixed frequency oscillator because such a small Tesla resonator will not produce enough corona even at full power to greatly detune itself. Also, the circuit should be less sensitive to the surrounding environment detuning the resonator. The system can always be tuned by adding or removing a turn from the Tesla resonator, or fine adjustments made by raising or lowering the breakout point.

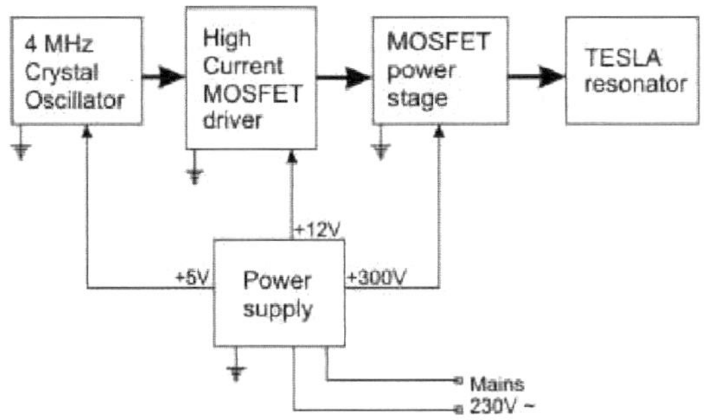


**Figure 1.** The system diagram.

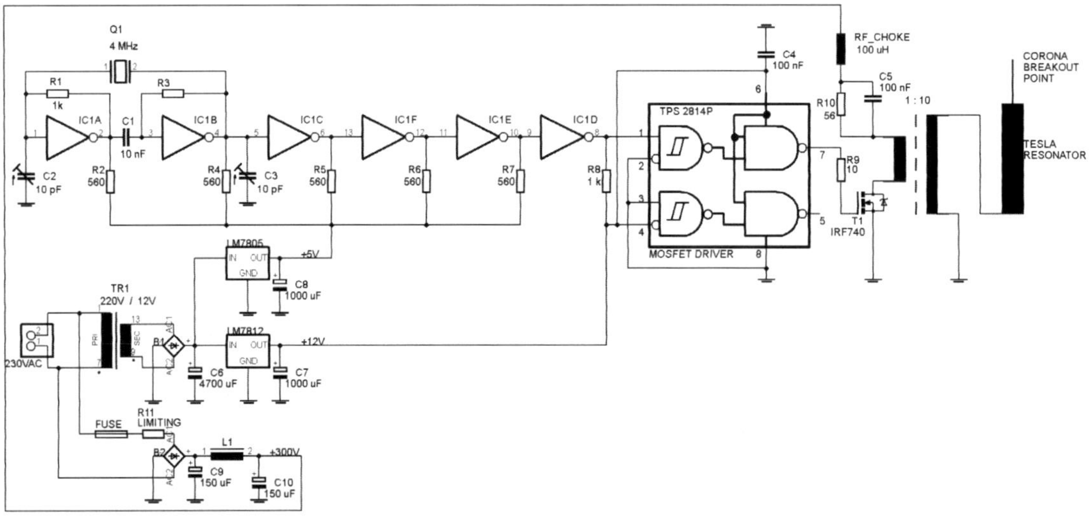


**Figure 2.** Generator schematics.

The high current MOSFET driver module is based on the TPS2814P integrated circuit and it is capable of delivering peak currents of 2 A into highly capacitive loads. The power stage is built around the IRF740 power MOSFET transistor. Figure 2 shows the full schematics.

### *2.2. Tesla resonator and the corona discharge*

The Tesla resonator is a high frequency air cored resonant transformer consisting of two windings, named primary and secondary. The primary winding is where the RF power developed by the power stage is coupled into the Tesla resonator. The dimensions, number of turns, and coupling coefficient to the resonator are critical to correct operation. Both the coupling and the inductance of this primary winding determine the power throughput of the system. The value of the coupling coefficient also effects the tuning of the resonator. The reason for this may not be immediately apparent, but increasing the coupling coefficient ties up more of the lower turns of the resonator in transformer action by magnetic coupling to the primary. As the coupling coefficient is increased there is less of the resonator's total inductance free to resonate, and so the apparent resonant frequency increases. The secondary winding – the resonator, has to be build so that it self-resonates at the desired frequency. This means that its inductance and stray-capacitance forms a series-resonant LC circuit, “tuned” at exactly 4 MHz. For this application the support for the coil was chosen a 50 mm diameter PVC tubing pipe. The coil was wound with 0.4 mm copper-enameled wire. At 4 MHz the skin depth in copper is only 0.033 mm. The 0.2 mm radius of the wire represents 6 skin depths so there would be little benefit in using a thicker wire, as the largest portion of the current is already flowing in the outermost skin of the wire. The parameters of the resonator coil were calculated using the following equations:

$$L = \frac{(N \cdot R)^2}{9R + 10H} \quad \text{(air cored solenoid inductance)} \tag{1}$$

where: L is the inductance in μH, N is the number of turns, R is the radius of the coil in inch, H is the length of the coil in inch.

$$C = (0.29 \cdot H) + (0.41 \cdot R) + 1.94\sqrt{\frac{R^3}{H}} \quad \text{(Medhurst self capacitance)} \tag{2}$$

where: C is the self capacitance in pF, R is the radius of the coil in inches, H is the length of the coil in inches. The resonance frequency is given by Thompson’s equation:

$$f = \frac{1}{2\pi\sqrt{LC}} \tag{3}$$

where: f is the frequency in Hz, L is the inductance in H, and C is the capacitance in F.

The correct number of turns was chosen by a series of approximations. Table 1 presents the final specifications of the Tesla resonator. Figure 3b presents the Tesla resonator and the corona appearance. The corona is a silent flickering orangey-lillac plasma flame, with a length of about 20mm, at an input power of 100 W.

**Table 1.** Tesla resonator specifications.

| | |
|---|---|
| Diameter | 50 mm |
| Length | 60 mm |
| No. of turns | 150 |
| Wire gauge | 0.4 mm |
| Wire length | 23.2 m |
| Inductance | 690.18 μH |
| Selfcapacitance | 2.35 pF |
| Natural res. freq. | 3.95 MHz |
| DC resistance | 2.84 Ω |

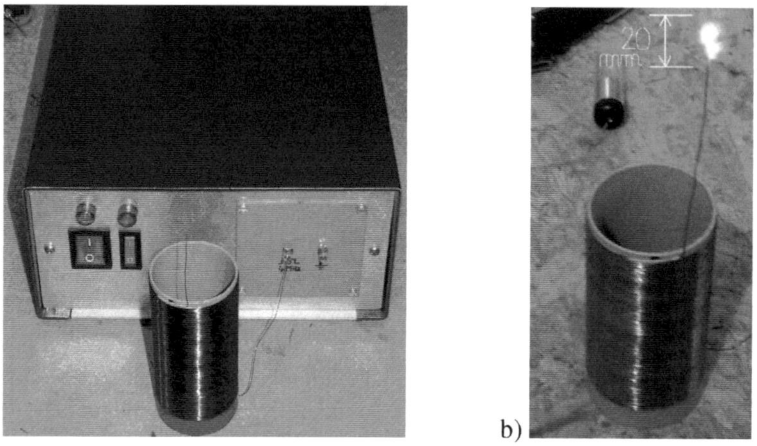

a) b)

**Figure 3.** a) Generator coupled to Tesla resonator. b) generated corona discharge at 100W input power.

Figure 3a presents the generator-unit being coupled to the Tesla resonator. Weight no more than 3 kg, the unit is compact, portable and easy to set up for demonstrations. The setting up procedure is very easy, consisting of supplying the unit with 230 VAC and connecting the Tesla resonator to the output. The unit has two control buttons on the front panel. The red button is the main power switch and the green button controls the driver module, turning on or off the corona discharge.

## 3. Future development – amplitude modulated RF corona and applications

The nature of the corona discharge at these high frequencies makes it the ideal candidate for low frequency modulation. The modulation of any high frequency plasma is possible by varying the RF output power proportionally with an incoming signal. This varying power causes the air around the discharge to vary in temperature, expand and contract producing waves. Modulation of the system is best implemented using high level modulation of the +300 V supply to the final power stage. An interesting effect which will be studied is the modulation of a microwave signal by a low frequency modulated RF plasma. When microwaves propagate through a plasma in which electron density and electron collision frequency periodically vary, the propagated wave is modulated in amplitude and phase. An interesting application of the modulated RF corona discharge in experimental physics is a method for making superhydrophobic substrates on polymeric or non-polymeric articles.

## 4. Conclusions

This 4 MHz RF plasma unit was built to study the behaviour of the corona discharge at high frequencies compared to traditional DC or low frequency corona discharge. It was built using only general purpose components that are easily available. The generator is driven by a stable crystal oscillator working at 4 MHz. The overall characteristics are: simplicity, elegant design, high portability and easy set up for demonstrations.

## References

[1] Stoffels E, Flikweert A J, Stoffels W W and Kroesen G M W 2002 *Plasma Sources Sci. Technol.* **11** 383
[2] Leveille V and Coulombe S 2005 *Plasma Sources Sci.Tehnol.* **11** 467
[3] Anghel S D and Simon A 2007 *Plasma Sources Sci. Tehnol.* (submitted)
[4] Denicolai M 2002 *Rev. Sci. Instrum.* **73** 3332

# Quasi-ballistic light scattering - analytical models *versus* Monte Carlo simulations

**Ioan Turcu[1] and Mikhail Kirillin[2]**

[1] National Institute for Research and Development of Isotopic and Molecular Technologies, 65-103 Donath, 400293 Cluj-Napoca, Romania

[2] Institute of Applied Physics of the Russian Academy of Sciences, 603950 Nizhny Novgorod, Russia

Email: ioan.turcu@itim-cj.ro

**Abstract**. Approximate analytical solutions for the light scattering in a plan parallel geometry, where each scattering behaves according to a Henyey-Greenstein (HG) phase function, are presented and compared with Monte Carlo simulations. Analyzing each *n*th order scattered flux, the obtained angular spreading is very well described also by a HG phase function. However, the total scattered flux deviates from the HG type dependence revealing the limits of the approximations.

## 1. Introduction

The migration of photons in different particulate media is generally described by the radiative transfer equation (RTE) [1] that expresses the energetic balance between the incident, outgoing, absorbed, and scattered photons. Since in realistic cases there are very few available analytical solutions for the RTE equation [2], several numerical techniques have been developed. Among the most widely used numerical methods the statistical Monte Carlo (MC) techniques have become ubiquitous representing a stochastic realization of the RTE being able to approximate the exact solution for a given problem with any desired accuracy. Their main drawback is the need to simulate a large number of photon paths in order to reach a high accuracy level. There are many MC codes (capable to perform direct or inverse MC simulations) that have been developed in various fields of science and technology that deal with radiation transfer in turbid media [3-7]. Recently the Photon Transport Simulator MC software has been used to obtain the directions of the successive orders, multiple scattered photons using one-term and two-term Henyey-Greenstein (HG) as single particle phase function [8].

In a far field experiment the scattering volume, as a whole, can be regarded as a "macroscopic particle" characterized by an effective scattering phase function which can be described analytically [9-11]. The effective phase function (EPF) describing multiple scattering processes can be obtained by adding together the contributions for all scattering orders. The procedure has been applied using the HG single particle phase function. The accuracy of the approximate analytic solutions for the EPF obtained in the quasi-ballistic light scattering (QBLS) regime has been systematically assessed [12], the theoretical results being tested experimentally [12] on a well known model: polystyrene spheres in suspension.

The main goal of the present paper is to evaluate the correctness of the EFP analytic solutions by comparing them with simulation data given by MC codes.

## 2. The quasi-ballistic scattering

### 2.1. The effective phase function

If the absorption is neglected, the QBLS can be understood as a steady state redistribution process attained by balancing the populations of photons indexed by the scattering order [10, 11]:

$$\phi(\tau,\mu)=\phi_0(\tau,\mu)+\phi_s(\tau,\mu) \quad ; \quad \phi_s(\tau,\mu)=\sum_{n=1}^{\infty}\phi_n(\tau,\mu) \tag{1}$$

In equation (1) $\phi_0$ stands for the reduced incident flux, $\phi_s$ for the scattered flux and $\phi_1;\phi_2;...;\phi_n;...$ for the first, second and $n$th order scattered fluxes, respectively. The two variables are the optical thickness $\tau$ and the cosine of the polar angle $\mu=\cos(\theta)$.

In axially symmetric geometries the directional distribution of scattered photons is frequently described by the $\mu$-dependent single particle HG phase function:

$$f_{HG}(\mu)=\frac{1}{4\pi}\frac{1-g^2}{\left(1-2g\mu+g^2\right)^{3/2}}=\frac{1}{2\pi}\sum_{l=0}^{\infty}(l+\tfrac{1}{2})g^l P_l(\mu) \tag{2}$$

given by its expansion in terms of Legendre polynomials, where $g=\langle\mu\rangle$ is the anisotropy parameter. The $n$th order scattered fluxes are obtained as products of two functions:

$$\phi_n(\tau,\mu)=\frac{\tau^n}{n!}e^{-\tau}\cdot f_n^{(HG)}(\mu) \tag{3}$$

where the functions $f_n^{(HG)}(\mu)$ are the $n$th order phase functions describing the angular spreading of the corresponding scattered flux:

$$f_n^{(HG)}(\mu)=\sum_{l=0}^{\infty}(l+\tfrac{1}{2})g^{nl}P_l(\mu)=\frac{1}{2}\frac{1-g^{2n}}{\left(1-2g^n\mu+g^{2n}\right)^{3/2}}. \tag{4}$$

The angular flux of scattered photons can be rewritten in a more compact form:

$$\phi_s(\tau,\mu)=(1-e^{-\tau})f_{eff}(\tau,\mu) \tag{5}$$

using the EPF $f_{eff}(\tau,\mu)$ which can be given explicitly by [9, 11]:

$$f_{eff}^{(HG)}(\tau,\mu,g)=\sum_{n=1}^{\infty}\frac{\tau^n}{n!}\frac{f_n^{(HG)}(\mu)}{e^{\tau}-1}=\sum_{l=0}^{\infty}(l+\tfrac{1}{2})\frac{(e^{\tau g^l}-1)}{e^{\tau}-1}P_l(\mu)\;, \quad g_{eff}^{(HG)}(\tau,g)=\frac{e^{\tau g}-1}{e^{\tau}-1} \tag{6}$$

together with the $\tau$-dependence of the effective anisotropy parameter.

Unfortunately, although each $f_n(\mu,g)$ may be expressed as HG type phase function, the coefficients in equation (6) don't have the usual power type dependence in the polynomial order $l$ which is characteristic for a HG phase function. To express the EPF as HG type phase function, the coefficients in equation (9) have been approximated by power type dependencies [10]:

$$\frac{e^{\tau g^l}-1}{e^{\tau}-1}\approx g^{lG(\tau)} \quad ; \quad G(\tau)=\frac{(\tau-1)e^{\tau}+1}{(e^{\tau}-\tau-1)} \tag{7}$$

As a direct consequence, the EPF can be also approximated by a HG phase function:

$$f_{eff}^{(HG)}(\tau,\mu,g) \approx f_a(\tau,\mu,g) = \frac{1}{2}\frac{1-g^{2G(\tau)}}{\left[1-2\mu g^{G(\tau)}+g^{2G(\tau)}\right]^{3/2}},\quad g_a(\tau)=g^{G(\tau)} \qquad (8)$$

and by a fairly accurate expression of the anisotropy parameter.

### *2.2. Monte Carlo simulation*

The Monte Carlo method applied to light propagation in scattering media is based on the calculation of trajectories of a large number of photons ($5\cdot10^7$ in our simulations) randomly migrating in a particulate medium (a suspension containing red blood cells in our case). The main hypothesis was that the single scattering event is suitably described by a HG phase function. The simulations were performed using a previously developed Monte Carlo algorithm [7, 14, 15], a 1 mm thick plan-parallel cuvette, and a haematocrit $H$ varying between $10^{-4}$ ($\tau$ = 0.076) and $10^{-2}$ ($\tau$ = 7.6). The upper limit of the $H$ values has been fixed low enough to maintain the relatively strong anisotropy of the scattering process. The red blood cells and the surrounding medium are supposed to be non-absorbing.The basic input parameters for MC simulations are the optical properties of the medium: the scattering coefficients of $\mu_s = 760\cdot H$ [mm$^{-1}$] (for photons with $\lambda$ = 633 nm), the single scattering anisotropy factor $g=0.789$ and the used HG type phase function. The output parameters are the angular polar distributions (with a resolution of 0.05 degrees) of the photons scattered $n$ times considering the scattering process as axially symmetric.

## 3. Results

Monte Carlo simulation of scattered photons using a HG single scattering phase function showed good correlation with the theoretical predictions. Representative results are shown in figure 1 where the two

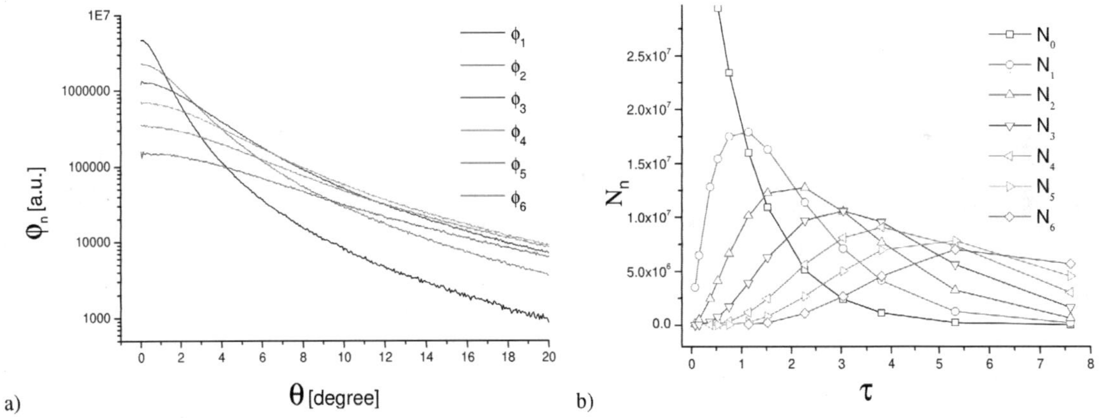


**Figure 1.** Typical raw data for the MC simulation of the quasi-ballistic light scattering process. a) Angular spreading of the photons scattered quasi-ballistically on a 1mm thick red blood cell suspension with a relatively small haematocrit $H = 5\cdot10^{-3}$. b) Typical $\tau$ -dependences of the scattered photons. $N_0$ stand for unscattered photons and $N_1$-$N_6$ denote the total number of photons corresponding to the first several scattering orders.

types of MC simulations are presented: the angular spreading of the scattered photons (left side) and the $\tau$ -dependence of the number of photons scattered $n$ times (right side). The angular spreading of each scattering order is very well described by HG phase functions and the obtained anisotropies are very close to the predicted values, as can be observed in figure 3.a. Deviations from the theoretical predictions appear in the case of samples with higher optical thickness and/or for higher scattering orders. The corresponding errors are shown in figure 2. (the theoretical fluxes are given by equation(3))

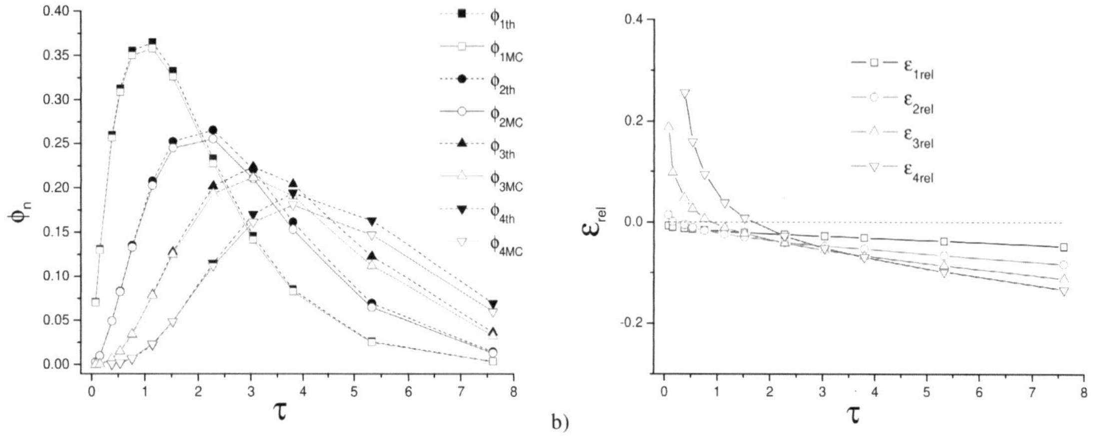


**Figure 2.** The accuracy of the analytical model assessed by MC simulations. a) The $\tau$ -dependence of the normalized fluxes of scattered photons. We show the analytic results obtained according to equation (3) together with the corresponding MC simulated values for the first 4 scattering orders. b) The flux relative error $\varepsilon_{rel} = (\phi_{th} - \phi_{MC}) / \phi_{MC}$ describing quantitatively the discrepancies between the theoretical predictions and the data obtained by MC simulations.

Figure 3b presents the decrease of the anisotropy parameter with the increase of the sample optical thickness. One can see from the beginning that there are differences between the two analytic functions $g_a(\tau)$, $g_{HG}(\tau)$ and also that the values obtained by MC simulations are systematically higher as compared to the predicted ones. The differences are reasonably small for optical thicknesses smaller than 1 but after that become more pronounced increasing with the increase of $\tau$.

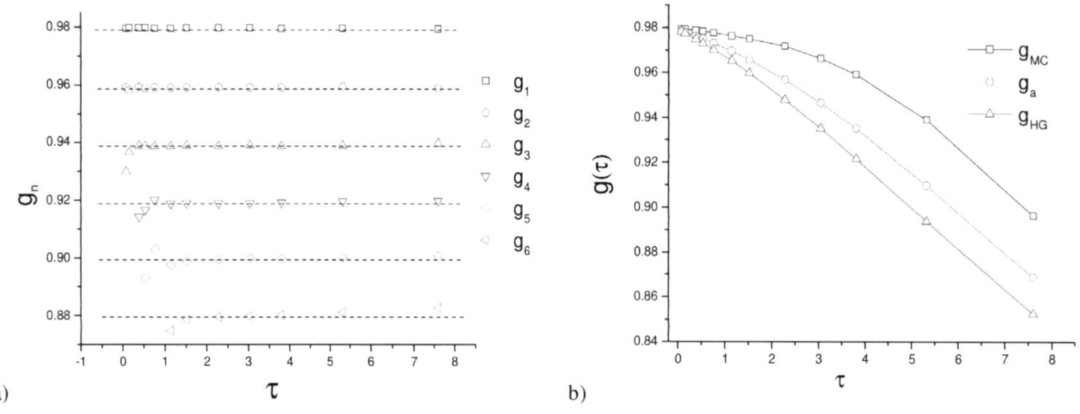


**Figure 3.** The assessment of the accuracy of the theoretical predictions concerning the anisotropy of the scattering process. a) The $n^{th}$ order anisotropy (for $1 < n < 6$) obtained by fitting the angular spreading of MC scattered photons is almost independent on the optical thickness of the sample. The obtained values are also very close to the theoretical predictions (dashed lines). b) The $\tau$ -dependence of the predicted decrease of the scattering anisotropy. The MC simulations ($g_{MC}$) are compared with the two theoretically obtained formula $g_a$ and $g_{HG}$.

## 4. Conclusions

The angular spreading of photons is very well described by HG type phase functions even for scattering orders larger than 10. The scattering anisotropy is also given by values that are very close to theoretical predictions and even more are almost unaffected by the increase of the optical depth (see figure 3a).

On the other hand the MC simulated values of the photon fluxes corresponding to different scattering orders are smaller compared to theoretical predictions. The relative errors become significant by the increase of the optical depth and/or by the increase of the scattering order. The overall effect refers to a super-estimation of the contribution of the multiple scattered photons introduced artificially in the theoretical approach. The errors introduced by the used approximations are transferred in the corresponding unreasonable fast decrease of the anisotropy functions $g_a(\tau)$, $g_{HG}(\tau)$ compared to the corresponding values obtained by the MC simulations.

## References

[1] Chandrasekhar S 1960 *Radiative Transfer* (London: Oxford University Press)
[2] Kokhanovsky A 2002 *Meas. Sci. Technol.* **13** 233-40
[3] Boas D, Culver J, Stott J and Dunn A 2002 *Opt. Express* **10** 159-70
[4] Ramella-Roman J, Prahl S and Jacques S 2005 *Opt. Express* **13** 4420-38
[5] Yaroslavsky I V, Yaroslavsky A N, Goldbach T and Schwarzmaier H-J 1996 *Applied Optics* **35** 6797-809
[6] Wang L and Wu H 2007 *Biomedical Optics: Principles and Imaging* (Wiley)
[7] Sergeeva E A, Kirillin M Yu and Priezzhev A V 2006 *Quantum Electron.* **36** 1023-31
[8] Pfeiffer N and Chapman G H 2008 *Opt. Express* **16** 13637-42
[9] Turcu I 2004 *J. Opt.* A*: Pure Appl. Opt.* **6** 537-43
[10] Turcu I 2006 *Appl. Opt.* **45** 639-47
[11] Turcu I, Pop C V L and Neamtu Silvia 2006 *Appl. Opt.* **45** 1964-71
[12] Turcu I and Bratfalean R 2008 *J. Opt.* A*: Pure Appl. Opt.* **10** 015002
[13] Turcu I, Bratfalean R and Neamtu Silvia 2008 *J. Opt.* A*: Pure Appl. Opt.* **10** 075007
[14] Meglinski I, Kirillin M, Kuzmin V and Myllylä R 2008 *Optics Letters,* **33** 1581-3
[15] Kirillin M Yu, Priezzhev A V and Myllylä R 2008 *Quantum Electron.* **38** 486-90.

Journal of Physics: Conference Series **182** (2009) 012036 doi:10.1088/1742-6596/182/1/012036

# Method validation for determination of heavy metals in wine and slightly alcoholic beverages by ICP-MS

**Cezara Voica, Adriana Dehelean and A Pamula**

National Institute for Research and Development of Isotopic and Molecular Technologies, 65-103 Donath, 400293 Cluj-Napoca, Romania

E-mail: cezara.voica@itim-cj.ro

**Abstract.** The *Organisation International de la Vigne et du Vin* (OIV) fixed an uppermost level for some heavy metals in wine. Consequently, the need to determine very low concentration of elements that may be present in wine in trace and ultra trace levels occurred. Inductively coupled plasma mass spectrometry ICP-MS is considered an excellent tool for detailed characterization of the elementary composition of many samples, including samples of drinks. In this study a method of quantitative analysis for the determination of toxic metals (Cr, As, Cd, Ni, Hg, Pb) in wines and slightly alcoholic beverages by ICP-MS was validated. Several parameters have been taken into account and evaluated for the validation of method, namely: linearity, the minimum detection limit, the limit of quantification, accuracy and uncertainty.

## 1. Introduction

Wine is a complex matrix, which besides water, sugar and alcohol, contains a great variety of components, inorganic as well as organic. The composition of wine is influenced by many factors related to the specific production area: grape varieties, soil and climate, culture, winemaking, transport and storage [1, 2]. Moderate wine consumption contributes to the daily intake of many essential metals, including Ca, Co, Cr, Cu, Fe, K, Mg, Mn, Mo, Ni, Zn. Wine contains macro-elements $c > 10$ mg/l (Na, K, Mg, Ca), micro-elements 10 mg/l $> c >$ 10 μg/l (Fe, Cu, Zn, Mn, Pb) and ultra-microelements $c < 10$ μg/l (Cr, As, Cd, Ni) [3, 4]. It should be noted that the levels of certain contaminants elements, such as Cu, Zn, As, Cd, Pb at different stages of the winemaking process are of great concern because of legal requirements [5]. The validation of the analytical method has become a basic prerequisite for those laboratories that work in an official food and beverage control [6]. The aim of this study is the validation of an analytical method for quantitative determination of toxic metals (Cr, As, Cd, Ni, Hg, Pb) in wines and slightly alcoholic beverages using the technique of inductively coupled plasma mass spectrometry ICP-MS. Thirteen wine assortments were investigated: five red wines and eight white wines.

## 2. Experimental

Mass spectrometry with inductively coupled plasma (ICP-MS) is a multi-element technique for analyzing liquid samples, characterized by high selectivity, sensitivity and detection limits much lower than other multi-element techniques [7, 8].

Journal of Physics: Conference Series **182** (2009) 012036 doi:10.1088/1742-6596/182/1/012036

### *2.1. Methods*

All the determinations were carried out by inductively coupled plasma quadrupole mass spectrometry (ICP-Q-MS). A Perkin Elmer ELAN DRC (e) instrument was used with a Meinhart nebulizer and silica cyclonic spray chamber and continuous nebulization. The operating conditions are: Nebulizer Gas flow rates: 0.86 l/min; Auxiliary Gas Flow: 1.2 l/min; Plasma Gas Flow: 15 l/min; Lens Voltage:7.25 V; ICP RF Power: 1100 W; CeO/Ce = 0.031; $Ba^{++}/Ba^{+}$ = 0.016.

### *2.2. Materials*

Ultra-pure de-ionized water (18 $M\Omega cm^{-1}$) from a Milli-Q analytical reagent-grade water purification system (Millipore) and ultra-pure $HNO_3$ 60% (Lot No B0157318 Merck) were used. All plastic lab ware used for the sampling were new or cleaned by soaking 24 h first in 10% $HNO_3$ then in ultra-pure water. In order to validate the method for determining the concentration of heavy metals in wines high toxicity and alcoholic beverages weak, a synthetic slightly alcoholic sample (1% alcohol, using ethanol - reagent for analysis, minimum 95%) of 2 μg/l concentration As, Be, Cd, Cr, Hg, Ni, Pb, Se, Tl was prepared. To the ICP-MS analysis of wines, samples are diluted by 10 ÷ 20x volume of ultrapure water. Since wines are generally 10-12% alcohol, a synthetic sample of 1% alcohol reproduce with sufficient fidelity real matrix samples. A multielement Merck 100 mg/l As, Be, Cd, Cr, Hg, Ni, Pb, Se, Tl standard (Multi-element Standard Solution IX CertiPUR; Matrix 6.4% $HNO_3$) was used which to prepare firstly a solution of 10 μg/l. Using this solution, six alcoholic solutions of different concentrations (1, 2, 2.5, 5, 7, 9) μg/l were then prepared. The solutions of 1, 2.5, 5, 7, 9 μg/ l were used as calibration solutions and the solution of 2 μg /l was employed as a synthetic sample.

## 3. Results and discussions

Several parameters have been taken into account and evaluated for the validation of the analytical methods for quantitative determination of toxic metals in wines and slightly alcoholic beverages (Cr, As, Cd, Ni, Hg, Pb), namely: applicability range and linearity, the minimum detection limit, the limit of quantification, repeatability, accuracy, uncertainty.

*Linearity domain*: using calibration solutions, calibration curves y = ax + b were determined (y is the signal intensity, x is the know concentration of the given analyte in the calibration solution). The linearity of the calibration curve was considered acceptable when the correlation factor R > 0.999.

*The minimum detection limit*: is the lowest concentration or quantity of analyte which can be measured with reasonable statistical certainty. To determine the limit of detection 3SD, a method developed by Perkin Elmer was used. Ultra pure water of 18.2 $M\Omega cm^{-1}$ was aspired and signal intensities for blank were recorded. A solution of 10 μg/l As, Ni, Cd, Hg, Pb, Cr was aspired and the signal intensities for these analytes were recorded (table 1).

**Table 1.** The limit of detection for As, Ni, Cd, Cr, Pb, Hg.

| Elements | $I_{blank}$ | $RSD_{blank}$ | $SD_{blank}$ | $I_{net}$ | LOD (μg/L) |
|---|---|---|---|---|---|
| **As** | 15.73 | 10.95 | 1.72 | 57626.89 | 0.001 |
| **Ni** | 70.06 | 8.10 | 5.67 | 29042.42 | 0.006 |
| **Cd** | 55.13 | 8.06 | 4.45 | 23136.62 | 0.006 |
| **Hg** | 1110.16 | 4.74 | 52.66 | 63118.47 | 0.025 |
| **Pb** | 421.14 | 9.94 | 41.86 | 196287.20 | 0.006 |
| **Cr** | 15.78 | 8.73 | 1.37 | 22584.22 | 0.002 |

The limit of detection was calculated by the equation (1), where: $SD_{blank}$ is the standard deviation for the signal recorded on the blank for the element studied, $conc_{sample}$ is the concentration [μg/l] of the analyte in the sample, $I_{sample}$, $I_{blank}$ are the signal intensities recorded for the sample and blank respectively.

$$LOD = 3 \cdot SD_{blank} \cdot conc_{sample} / [I_{sample} - I_{blank}] \quad (1)$$

*The limit of quantification* is the lowest concentration that can be quantitatively determined with an acceptable level of repeatability accuracy. The quantification limit is generally considered to be approximately ten times the minimum detection limit. In order to verify that, three standard solution of 0.01, 0.05, 0.1 µg/l As, Cr, Ni, Hg, Pb, Cd were prepared and aspired in the ICP-MS inlet system.

*The maximum measurement limit* is conditioned by the dynamics of the spectrometer detectors and limited by the requirement that the total amount of the dissolved solid must not exceed 0.2% in the sample solution (unless clogging of the nebulizer nozzels would lead to instabilities and loss of sensitivity). To test the maximum measurement limit, a slightly alcoholic sample of 2 mg/l was prepared, of which there were determinations for the elements taken into account, using calibration solutions of 0.5, 1, 2.5, 5 mg/l.

*Repeatability* was obtained measuring the same sample, with the same method, in the same laboratory, with the same equipment, by the same operator, in short intervals of time. To determine the repeatability, the concentration of six metals in the synthetic sample (2 µg/l) described above was determined. The obtained standard deviation was in the range 0.09-0.5 µg/l. Intermediate repeatability was obtained measuring the same sample, with the same sample, in the same laboratory, but by different operators and in different days. Standard deviation was found to lie between 0.05-0.12 µg/l.

*Accuracy* expresses the correlation between the arithmetic mean of the measured values and the accepted reference value. Considering the "true" concentration of 2 µg/l, the bias (%) was calculated for the elements taken into consideration (0.8-17 %).

*Uncertainty estimation.* All the possible sources of uncertainty have to be carefully identified and taken into account. While measuring the concentrations by ICP-MS with external standard, fluctuations in the measurement of ionic currents occurring as a result of the electrical noise in the detector, instabilities in plasma discharge, instabilities of the electrical parameters of the analyzer, lead to uncertainties in the determination of the parameters of the calibration line. Possible errors in the preparation of the calibration solutions increase these uncertainties. Let us assume that the calibration line has the form: y = ax + b, where y is the signal intensity and x is the concentration of the calibration solution. Using the equations (2) and (3), the standard deviation of the parameters a, b and the standard deviation unknown concentration were determined (n is the number of points that determine the calibration curve).

$$\sigma_a = \sqrt{\frac{n}{n\sum x^2 - \left(\sum x\right)^2}} \cdot \sigma_y$$
$$\sigma_b = \sqrt{\frac{\sum x^2}{n\sum x^2 - \left(\sum x\right)^2}} \cdot \sigma_y \tag{2}$$
$$\sigma_y = \sqrt{\frac{\sum (y - b - ax)^2}{n}}$$

$$\sigma_x = x\sqrt{\left(\frac{\sigma_{y-b}}{y-b}\right)^2 + \left(\frac{\sigma_a}{a}\right)^2} = x\sqrt{\frac{\sigma_y^2 + \sigma_b^2}{(y-b)^2} + \left(\frac{\sigma_a}{a}\right)^2} \tag{3}$$

**Table 2.** Uncertainty for As, Cd, Hg, Pb Ni,Cr measurements by ICP-MS.

| Metal | a | b | $\sigma_a$ | $\sigma_b$ | Y | $\sigma_y$ | X(µg/L) | σX(µg/L) | 2·σX*(µg/L) |
|---|---|---|---|---|---|---|---|---|---|
| **As** | 6255.48 | 26.85 | 86.07 | 387.95 | 12368.91 | 121.69 | 1.973 | 0.070 | 0.141 |
| **Cd** | 2706.56 | -160.46 | 50.06 | 225.63 | 4987.42 | 82.85 | 1.902 | 0.095 | 0.191 |
| **Hg** | 13950.70 | 946.40 | 274.47 | 1652.52 | 33925.86 | 213.74 | 2.364 | 0.128 | 0.256 |
| **Pb** | 24703.50 | 595.31 | -16336.6 | 2683.05 | 54523.05 | 1275.57 | 2.183 | 1.449 | 2.897 |
| **Ni** | 3476.13 | -290.32 | 61.00 | 273.07 | 6325.46 | 88.78 | 1.903 | 0.089 | 0.178 |
| **Cr** | 8038.78 | 0.96 | 39.00 | 10.06 | 7890.32 | 81.43 | 1.998 | 0.023 | 0.046 |

* Expanded uncertainty (k = 2) represents interval that there is a probability of 95% to be the true value.

It is important to observe that the uncertainty estimation calculated in this manner should be associated with the concentration domain taken into account and with the analyst skill. Table 2 presents the uncertainty for As, Cd, Hg, Pb, Ni and Cr in the 2 µg/l concentration range by ICP-MS. The extended uncertainty is between 0.04-0.25 µg/l (except for lead).

## 4. Application of the method to wine samples

In an attempt to characterize Romanian wines from the point of view of their heavy metal content, 13 samples representing different wine assortments were studied. Thirty metals were identified with concentrations ranging from 0.02 µg/l to 6 mg/l. Light metals: Na, Mg, K, Ca were the largest share, showing concentrations of the order of mg/l (ppm). Some metals considered to be micronutrients (Cu, Zn) were found in concentration of the order of several µg/l. Heavy metals with high toxicity like Pb, Cd, As showed concentrations below the limits imposed by OIV (table 3).

**Table 3.** Metals content of some Romanian wines.

| Vineyard | Assortment of wine | Type of wine | Concentration [µg/L] | | | | | |
|---|---|---|---|---|---|---|---|---|
| | | | As | Cd | Hg | Pb | Ni | Cr |
| Cotnari, Moldova | Feteasca Alba 2005 | semi-sweet white | 3.26 | 1.42 | 81.14 | 21.74 | 31.32 | 73.68 |
| | Selected wine 2005 | semi-sweet white | 7.66 | 1.22 | 200 | 1.98 | 7.66 | 400.50 |
| | Tamâioasa romaneasca 2008 | sweet white | 9.44 | 0.44 | <LOD | 28.30 | 17.92 | 203.04 |
| | Tamâioasa romaneasca 2005 | sweet white | 0.29 | 0.40 | <LOD | 16.8 | 16.80 | 99.74 |
| | Busuioaca de Moldova 2003 | sweet white | 2.08 | 0.38 | <LOD | 13.44 | 34.46 | 70.90 |
| | Muscat Ottonel 2004 | semi-sweet white | 12.2 | 0.84 | 53.94 | 24.16 | 78.02 | 144.42 |
| Recas | Burgund Mare, Frankovka | semi-sweet red | 5.26 | 0.28 | 10.4 | 9.82 | 93.94 | 46.38 |
| | Rosu Dulce | sweet red | 6.30 | 0.28 | <LOD | 14.76 | 14.76 | 66.02 |
| Prahova | Vinul Casei Riesling 2007 | dry white | 4.80 | 0.28 | <LOD | 36.58 | 35.36 | 244.64 |
| | Cabernet Sauvignon 2007 | dry red | 0.96 | 0.08 | <LOD | 10.48 | 37.32 | 81.64 |
| | Cabernet Sauvignon 2005 | dry red | 10.64 | 0.38 | <LOD | 22.74 | 69.94 | 210.64 |
| Jidvei | Dry Muscat 2006 | semi-dry white | 5.90 | 0.64 | <LOD | 47.38 | 68.58 | 102.04 |
| Murfatlar | Pinot Noir 2007 | semi-sweet red | 3.42 | 0.72 | 0.643 | 15.82 | 15.82 | 128.88 |

## 5. Conclusion

In this study a method of quantitative analysis for the determination of toxic metals (Cr, As, Cd, Ni, Hg, Pb) in wines and slightly alcoholic beverages by ICP-MS was validated and applied. Several statistical parameters have been taken into account and evaluated for the validation of method: the limit of detection ranged between 0.002-0.03 for the six metals studied ensures the limit of quantification required for quantitative determinations of the concentrations of these elements in slightly alcoholic beverages; good linearity (correlation factor $0.9998 \geq R \geq 0.9994$) for most elements recommend the method described for determination at trace and ultra-trace level; maximum limit of measurement is limited by the requirement that the total amount of solid dissolved in the sample solution must not exceed 0.2%. The uncertainty estimation takes into account the uncertainty influence in determining the parameters of the calibration curve; values determined lies between 0.04-0.25 mg/l. The results obtained for some Romanian wines are in agreement with the values reported in literature by other authors [9, 10]. Wide variability in the concentration of metals at trace level suggests the possibility to correlate it with the origin of wine.

## References

[1] Sperkova J, Suchanek M 2005 *Food Chem.* **93** 659-63
[2] Catarino S and Curvelo-Garcia A S, Bruno de Sousa R 2006 *Talanta* **70** 1073-80
[3] Cozub G and Rusu E 1996 (Chişinău: Litera)
[4] Sîrghi C D, Găină B S, Bălănuţă A P, Carpov S S, Căldare Gh A, Musteaţă Gr A, Palamarciuc L F, Pănăşescu I S, Stratulat O A, Sverdlic V N and Prida I A 1992 *Cartea vinificatorului* (Chişinău)
[5] *Compendium of International Methods of Analysis* – OIV 2007 **2** MA-E-C1-01-LIMMAX
[6] D'Ilio S, Petrucci F, D'Amato M, Di Gregorio M, Senofonte O, Violante N 2008 *Anal. Chim. Acta* **624** 59-67

[7] Becker J S and Dietze H J 2003 *Int. J. Mass Spectrom.* **228** 127-50
[8] Ammann A A 2007 *J. Mass Spectrom.* **42** 419-27
[9] Muranyi Z and Kovacs Z 2000 *Microchem. J.* **67** 91-6
[10] Kment P, Mihaljevic M, Ettler V, Šebek O, Strnad L, Rohlova L 2005 *Food Chem.* **91** 157-65.

# Optical soundings of cars' exhaust and urban atmosphere with laser-plasma light and open-path spectrophotometry

**Antonio Borghese** and **Luciano Speranza**[1]

Instituto Motori – Consiglio Nazionale delle Ricerche, Via Marconi 8, 80125 Napoli, Italy

E-mail: a.borghese@im.cnr.it

**Abstract.** Optical soundings of i.c. engines' exhaust and urban atmosphere have been carried out, using an Open-Path UV Spectrophotometer, based on a Laser-Produced Plasma as the light source. The measured optical depths are numerically inverted in terms of known species, by means of the Non-negative Least-Squares (NLS) algorithm. Concentrations of major UV-absorbing species are evaluated at the tailpipe of in-use vehicles (NO, $NO_2$, EC) and in urban atmosphere over a 200m-long optical path ($O_2$, $O_3$. $NO_2$, EC, nitrates). Also, we report on the detection in both environments of an additional compound, the nanometric organic carbon (NOC), known to form in fuel-rich flames, which is not considered yet in ordinary emissions inventories. Also, the problem of dilution/dispersion of primary species is addressed, by tracking the combustion-formed $CO_2$ at the tailpipe and in urban atmosphere.

## 1. Introduction

Emission of combustion by-products (exhaust) from internal combustion (i.c.) engines into the atmosphere results in a) the dilution/dispersion of the burnt gases, b) the physical and chemical evolution of some primary species and c) the formation of new (secondary) compounds, their identities and amounts depending on complex reaction pathways. On one side, relevant primary species consist mainly of carbon, nitrogen and sulphur oxides, along with hydrocarbons (H/C) and elemental carbon (EC), the latter believed to derive mostly from diesel engines. On the other side, air quality standards consider primary $NO_2$ and $SO_2$, secondary $O_3$, and the overall $PM_x$ aerosols; the latter are known to be composed of internal/external mixtures of both inorganic (sulphates/nitrates) and carbonaceous matter. Albeit cars' emissions and air pollution are clearly linked by a cause-to-effect relationship, yet the processes a-c) have only partially been understood and still deserve further investigations, particularly on the sources and the amounts of atmospheric OC [1].

Reported on here are optical soundings of engines' exhaust and urban atmosphere, using an Open-Path UV Spectrophotometer, based on a Laser-Produced Plasma as the light source. We measure concentrations of exhaust emissions (NO, $NO_2$, EC and OC), at the tailpipe of in-use vehicles on a chassis-dynamometer, and of some relevant species ($O_2$, $O_3$, $NO_2$, OC, EC, nitrates) over a 200 m long optical path in urban atmosphere. Also, the problem of the dilution/dispersion of primary species from the emission sources to the urban atmosphere is addressed, by considering the combustion-formed

[1] Now with ARPA Lombardia, Brescia, Italy.

$CO_2$ for internal calibration of the relative abundances of species at the tailpipe and in urban atmosphere.

## 2. Open-path Spectrophotometry

Most species of environmental concern, whether gases or particles, absorb and/or scatter light in the UV wavelength band (200-400 nm). The resulting attenuation – or extinction – of the travelling light is measured by the *optical depth* $\tau = \log(I_0 / I_T)$, where $I_0$ and $I_T$ are the light intensities incident and transmitted, respectively, through the medium. The optical depth $\tau$ is a function of the light wavelength $\lambda$ and describes the medium in terms of the optical-path length $L$, the number concentrations $N_s$ and the extinction cross-sections $\sigma_{ext,s}(\lambda)$ of the S species, according to the composition model :

$$\tau(\lambda) = L \cdot \sum_S N_s \cdot \sigma_{EXT,S}(\lambda) \pm \varepsilon \tag{1}$$

In open atmosphere, the low values of the number concentrations of some relevant species and of their extinction cross-sections require optical paths $L$ of the order of several hundreds meters, in order to provide detectable attenuation. On the other hand, the strong absorption of atmospheric oxygen and ozone at $\lambda < 300\ nm$ sets a natural barrier for long-path spectroscopic measurements. The emission intensity of commercially available light sources (*e.g.* Xenon or $D_2$ lamps) either vanish below 250 nm or are too weak at all to work as required. Instead, Laser-Produced Plasmas (LPP's) constitute a breakthrough, since they have an extremely high and broadband emission intensity [2], even extending into the Vacuum Ultra-Violet region [3].

An LPP-based UV Spectrophotometer has been developed for applications of Open-path Extinction Spectroscopy to track gases and aerosols from cars' exhaust to the urban atmosphere. A typical lay-out is shown in Figure 1. The instrument comprises a Transmitter, where the UV emission from an LPP, fired by a Nd-YAG Q-switched laser, is collimated by a parabolic mirror, and a Receiver, which includes a collection optics, a spectrograph and an MCP-intensified CCD detector, all embedded in a compact case. Two retroreflectors $RR_1$ and $RR_2$, set at the distances $L_1$ and $L_2$ along the optical axis, reflect two fractions of the transmitted light back to the receiver.

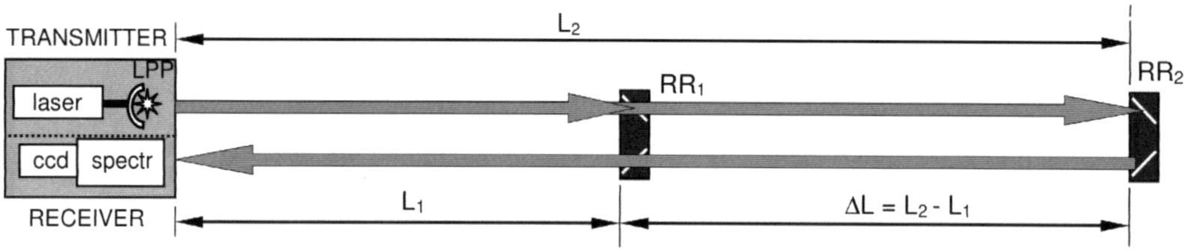


**Figure 1.** Optical lay-out of the Open-path Spectrophotometer.

Using two retroreflectors adds key-features to the optical configuration: *i*) the transmitter and the receiver can be housed in the same compact case; ii) the incident light $I_0$ and the instrumental response $\eta$, which are of difficult evaluation, are canceled out; *iii*) media at remote distances $L_1$ can be optically probed.

## 3. Numerical inversion of optical data

For each light pulse, two spectra $I_1(\lambda)$ and $I_2(\lambda)$, corresponding respectively to the light returns from $RR_1$ and $RR_2$, are acquired and processed to give the optical depth $\tau$ due to G gaseous species and P size-distributions $n_p(d)$ of particles:

$$\tau(\lambda) = \log(I_1/I_2) = 2 \cdot \Delta L \cdot \sum_G N_g \cdot \sigma_{EXT,g}(\lambda) + 2 \cdot \Delta L \cdot \sum_P \int n_p(d) \cdot \sigma_{EXT,p}(\lambda, d) \cdot \partial d \pm \varepsilon \tag{2}$$

The number concentrations $N_g$ and $n_p(d)$ of S = G + P species are retrieved by numerically inverting the optical data, using the NLS (Non-negative Least Squares) routine [4]; the algorithm is based on: 1) a composition model of S species with known cross-sections; 2) the non-negativity constraint of the solutions $N_g, n_p(d) \geq 0$; 3) recursive minimization of the error $\varepsilon$.

## 4. Results at the cars' exhaust and in the urban atmosphere

Figures 2A-B show two exemplary results of Open-path UV Spectrophotometry, obtained over a 5 m long round-trip optical path on the exhaust of vehicles – production year 2004 – equipped with spark-ignited (A) and diesel (B) engines. The vehicles are run at the chassis dynamometer over the New European Driving Cycle (NEDC), which accounts for both urban and highway driving. Air/exhaust dilution of 10:1 avoids condensation of the combustion-formed water vapor. Besides the known spectra of NO, $NO_2$ and EC, the composition model requires for best fit a fourth component, shown as NOC in figures 2A-B, which will be discussed later on.

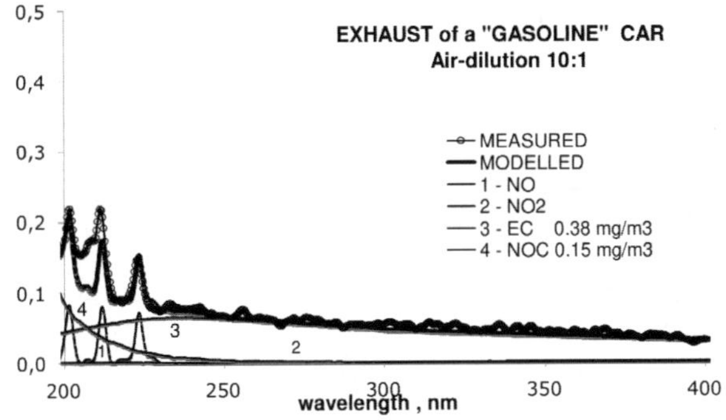


**Figure 2A.** Measured (dots) and modelled (thick line) optical depth in the full UV spectral band (200-400 nm) of air-diluted exhaust emitted from a car equipped with a spark-ignited engine. Also shown (thin lines) are the four contributions to the total extinction from NO, $NO_2$, EC and NOC. Numerical inversion of the optical data has been carried out with the NLS algorithm (see text).

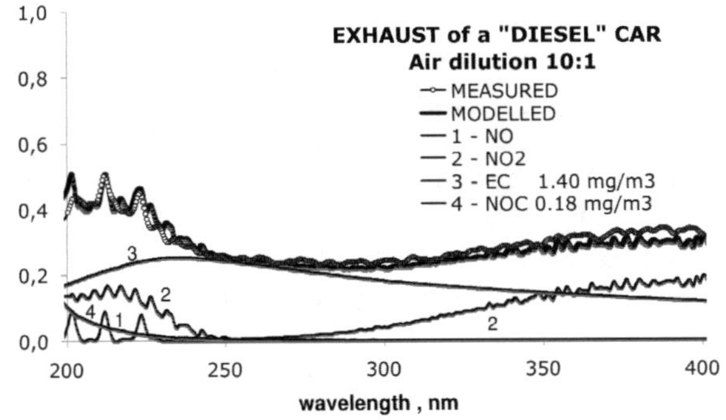


**Figure 2B.** Measured (dots) and modelled (thick line) optical depth in the full UV spectral band (200-400 nm) of air-diluted exhaust emitted from a car equipped with a diesel engine. The four contributions to the total extinction from NO, $NO_2$, EC and NOC, shown with thin lines, have been obtained as those in figure 2A.

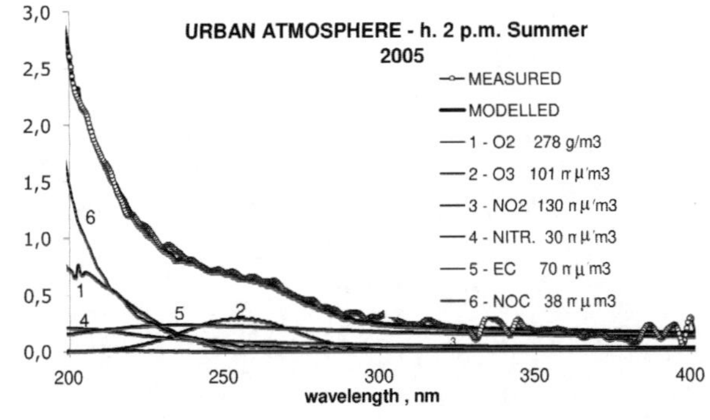


**Figure 3.** Measured (dots) and modelled (thick line) optical depth in the full UV spectral band (200-400 nm) of urban atmosphere over 2x100m-long optical path. The contribution to the total extinction from $O_2$, $O_3$, $NO_2$, EC, NOC and nitrates, (thin lines) have been obtained using the NLS algorithm.

Figure 3 shows another exemplary result of Open-path UV Spectrophotometry, now obtained over a 100m-long round-trip optical path in urban atmosphere, at 2.00 p.m. on a summer day in 2005. In addition to the primary species shown in figures 2A-B, other (secondary) compounds: $O_2$, $O_3$ and nitrates, need to be included in the composition model. The resulting best fit predicts nitrate particles with mean size $d \approx 100\ nm$ and standard deviation $\sigma_d \approx 1.5$. Here again, the 'excess' optical extinction observed below $\lambda \approx 250nm$ can only be fitted by considering the NOC contribution as with data of figures 2A-B.

## 5. Discussion and conclusions

Data in figures 2A-B show the amounts of known species (NO, $NO_2$, EC) emitted from i.c. engines, which are consistent with those obtained with standard analytical instrumentation. As expected, spark-ignited engines emit amounts of EC far smaller, yet measurable, than diesel engines. Data in figure 3 show a more complex composition, comprising both primary ($NO_2$, EC) and secondary ($O_3$, nitrates) species, reflecting the intervention of atmospheric photochemistry.

In all the cases examined, the experimental data could be fitted only by providing the composition models with the spectral contribution denoted as NOC in figures 2A-B and figure 3. NOC is an acronym for Nanometric Organic Carbon, a precursor of EC particles as well as a stable by-product of slightly fuel-rich combustion, which has been thoroughly described in previous studies on laboratory flames [5,6]. Compared to EC, NOC has much smaller particles size ($d < 5nm$), as well as quite different chemical composition (organic) and optical properties (transparent below $\lambda \approx 250nm$). Due to such properties and since substantial amounts are emitted from i.c. engines and practical combustors, NOC is currently considered by nano-toxicologists as a pollutant with potential health risk [7].

A further point refers to the quantitative relationship between the species concentrations measured at the exhaust and those resulting from dilution/dispersion in urban atmosphere. A viable approach to the problem considers primary and chemically stable species, *e.g.* EC, NOC and $CO_2$, which are expected to conserve their relative abundances from source to sink. At the air-diluted exhaust in figures 2A-B, the combustion-formed $CO_2$ concentration is nearly 1.3% and the evaluated abundances of EC and NOC, relative to $CO_2$, are $[EC]/[CO_2] \approx 1/20{,}000$ and $[NOC]/[CO_2] \approx 1/60{,}000$. On the other hand, in conditions of urban atmosphere (figure 3), the measured $CO_2$ concentration exceeds the background level ($\approx$ 390ppmv) by 80 ppmv ($\approx$150mg/m$^3$), an 'excess' $CO_2$ mostly due to traffic emissions. As a result, the relative abundances are estimated at $[EC]/[CO_2] \approx 1/2{,}000$ and $[NOC]/[CO_2] \approx 1/5{,}000$. Such values are at least ten times higher than those evaluated at the exhaust of new cars (figures 2A-B); however, the discrepancy is only apparent, since *i*) the circulating fleet is much older and more emissive than the new vehicles tested in this work, and *ii*) the drive style of most cars is more irregular, hence again more emissive, than the well-controlled conditions of a standard driving cycle at the chassis dynamometer.

## References

[1] Bond T, Streets D G, Yarben K F, Nelson S M, Woo J-H and Klimont Z 2005 *J. Geophys. Res.* **109** D14203
[2] Borghese A and Merola S S 1998 *Appl. Opt.* 1998 **37** 1-7
[3] Borghese A and Di Palma T M 2007 *Appl. Opt.* **46** 4948-53
[4] Paatero P 1997 *Chem. Int. Lab. Syst.* **37** 23-35
[5] Sgro L A, Basile G, Barone A C, D'Anna A, Minutolo P, Borghese A and D'Alessio A 2003 *Chemosphere* **51** 1079-90
[6] Sgro L A, Borghese A, Speranza L, Barone A C, Minutolo P, Bruno A, D'Anna A and D'Alessio A 2008 *Environ. Sci. & Technol.* **42** 859-63
[7] Oberdörster G, Oberdörster E, Oberdörster J 2005 *Environ. Health. Perspect.* **113** 823-39.

Journal of Physics: Conference Series **182** (2009) 012038 doi:10.1088/1742-6596/182/1/012038

# Stable oxygen and hydrogen isotopes measurement by CF-IRMS with applications in hydrology studies

**Diana Costinel[1], Voicu V Grecu[2], Raluca Vremera[1] and Stela Cuna[3]**

[1] National Research and Development Institute for Cryogenics and Isotopic Technologies, 4 Uzinei, POBox Raureni 7, 240050 Râmnicu Vâlcea, Romania

[2] University Bucharest, Faculty of Physics, Department of Atomic and Nuclear Physics, 405 Atomistilor, CP MG 11, 077125 Bucharest-Magurele, Romania

[3] National Institute for Research and Development of Isotopic and Molecular Technologies, 65-103 Donath, 400293 Cluj-Napoca, Romania

E-mail: diana@icsi.ro

**Abstract.** The major changes in isotopic composition of natural waters occur in the atmospheric part of the water cycle and in surface waters which are exposed to the atmosphere. This study demonstrated the utility of the Continuous Flow – Isotope Ratio Mass Spectrometry method for measuring natural variation of the occurring isotopes of hydrogen ($^2H$) and oxygen ($^{18}O$) in meteoric waters. The variation of $\delta^{18}O$ and $\delta^2D$ values from precipitation fallen in Raureni-Valcea area between May-December 2007 and September 2008-March 2009 were measured together with the $\delta^{18}O$ and $\delta^2D$ values from the Bistrita River. The Local Meteoric Water Line was reported for this area. Also, the variation of $\delta^{18}O$ and $\delta^2D$ values was correlated with the temperature and humidity in the same period.

## 1. Introduction

Stable and radioactive isotopes can be used as tracers, marking a water body or a certain quantity of water. A good example is the phenomenon that the rain water during a heavy storm is often depleted in the heavy isotope (deuterium or oxygen 18) with respect to the most abundant isotope ($^1H$ and $^{16}O$, respectively). This offers the possibility to follow the rain water in the surface runoff and even quantitatively analysing the runoff hydrograph. Craig's global meteoric water line defines the relationship between $^{18}O$ and $^2H$ in worldwide fresh surface waters:

$$\delta^2H=8\delta^{18}O+10\ (‰)\ [vs\ \text{SMOW}]$$

The meteoric relationship of $^{18}O$ and $^2H$ arises from fractionation during condensation from the vapour mass. The Rayleigh distillation during rainout is responsible for the partitioning of $^{18}O$ and $^2H$ between warm and cold regions. Craig's line is only global in application, and is actually an average of many local or regional meteoric water lines, which differ from the global line due to varying climatic and geographic parameters. This partitioning was soon recognized as a tool for characterizing groundwater provenance studies.

Journal of Physics: Conference Series **182** (2009) 012038 doi:10.1088/1742-6596/182/1/012038

## 2. Sampling and sample preparation

The samples were collected monthly from precipitation fallen in Raureni-Valcea area (latitude 45.6°N, longitude 24.22°E) between May-December 2007 and September 2008-March 2009 and from Bistrita River in the same period. The Bistrita River is the source water from population who lives in Raureni-Valcea. The main concern during sampling, transport and storage was to avoid isotope fractionation through evaporation or diffusive loss of water vapour, and/or isotope exchange with the surroundings as well as with the bottle material. Only for transportation, we used food-quality PET bottles, available pre-sterilised and pre-treated with sodium thiosulphate for the neutralisation of chlorine in samples. Each bottle was clearly identified with batch code, daily date, hour, and temperature.

## 3. Materials and methods

The conversion of any type of organic or inorganic compound into simple gases is the basic principle for IRMS. We analysed the water samples using the Gas Bench II device together with Continuous Flow Isotope Ratio Mass Spectrometer (CF-IRMS) Delta V Plus. The Gas Bench II is a flexible continuous flow preparation device and inlet system with Repetitive loop injection of gases. It is controlled by a robotic using septum-sealed sample vials.

The most common sample preparation method for $^{18}O/^{16}O$ in water is the $CO_2$-$H_2O$ equilibration, in which about a mill mole amount of $CO_2$ (10 to 20 ml) is brought into isotopic equilibrium with a few micro litres of sample water and reference material (generally at 25.0 ± 0.2°C):

$$H_2{}^{18}O + C^{16}O_2 \Leftrightarrow H_2{}^{16}O + C^{18}O^{16}O$$

Similar to the equilibration method for $^{18}O/^{16}O$ analysis, the $^2H/^1H$ ratio of a water sample can be transferred to $H_2$ by isotopic exchange (Coplen *et al* 1991):

$$^2H^1HO + {}^1H_2 \Leftrightarrow {}^1H_2O + {}^2H^1H$$

The results are reported as deviations versus international standards in the δ, notation in ‰:

$$\delta * X/X(‰) = \left( \frac{(^*X/X)_{sample}}{(^*X/X)_{standard}} - 1 \right) x1000$$

The $^2H/^1H$ and $^{18}O/^{16}O$ ratios are measured and reported as $\delta^2H$ and $\delta^{18}O$ relative to Standard Mean Ocean Water (SMOW) with 1SD of ± 0.1 $^{13}C/^{12}C$ ratios is measured and reported as $\delta^{13}C$ relative to Pee Dee Belemnite (PDB) from South Carolina, USA.

The experimental conditions for analysing the D and $^{18}O$ using Gas Bench II were:

| a) Deuterium analysis | | b) $^{18}O$ Analysis | |
|---|---|---|---|
| Sample size | 200 µl of water extract from sample | Sample size | 500 µl of water extract from sample |
| $H_2$/He | 2% | $CO_2$/He | 0.36% |
| Equilibration | 60 min at 25°C | Equilibration | 20h at 25°C |
| Sampling loop | 10 µl | Sampling loop | 10 µl |
| Replicates (loop switches) | 10 | Replicates (loop switches) | 10 |
| GC column | Pora Plot Q 25 m | GC column | Pora Plot Q 25 m |
| Primary standard | VSMOW, GISP, SLAP | Primary standard | VSMOW, GISP, SLAP |

## 4. Results and discussion

The experimental results obtained for precipitation together with mean temperature and humidity are shown in table 1.

The isotopic composition of oxygen and hydrogen in rainwater varies with time of the year. According with Craig's relation, we obtained the Local Meteoric Water Line, as is shown in figure 1.

**Table 1.** The results of measurements the D/H and $^{18}O/^{16}O$ isotope ratios from precipitation in period September 2008-March 2009 in Raureni-Valcea area, the values for local mean temperature and humidity (latitude 45.6°N, longitude 24.22°E), the extend incertitude for $^{18}O$ measurements $u_c$ = 1% (k = 2) and for deuterium measurements $u_c$ = 0.65% (k = 2).

| Samples | $\delta^{18}O/^{16}O$ (‰)$_{vs\ SMOW}$ | $\delta^{2}H/^{1}H$ (‰)$_{vs\ SMOW}$ | Mean Temp. (°C) | Humidity (%) |
|---|---|---|---|---|
| 15.09.08 | -8.47 | -57.62 | 12.3 | 91.5 |
| 16.09.08 | -8.96 | -61.55 | 18.4 | 80.5 |
| 17.09.08 | -9.63 | -66.97 | 13.7 | 90.0 |
| 05.10.08 | -10.07 | -70.56 | 12.6 | 87.5 |
| 26.11.08 | -6.91 | -45.07 | 5.3 | 92.5 |
| 03.12.08 | -4.46 | -25.25 | 7.9 | 96.5 |
| 14.12.08 | -8.05 | -51.51 | 3.9 | 92.0 |
| 15.12.08 | -7.22 | -50.64 | 4.8 | 90.0 |
| 22.01.09 | -5.58 | -37.85 | 11.5 | 85.0 |
| 28.01.09 | -5.15 | -35.26 | 8.8 | 88.0 |
| 29.01.09 | -9.05 | -59.41 | 5.9 | 86.5 |
| 02.02.09 | -16.99 | -125.92 | 1.5 | 84.0 |
| 03.02.09 | -11.85 | -87.86 | 3.0 | 89.5 |
| 04.02.09 | -5.03 | -18.43 | 8.6 | 85.0 |
| 18.02.09 | -16.19 | -117.82 | 3.0 | 82.5 |
| 19.02.09 | -15.12 | -110.99 | 2.7 | 84.5 |
| 20.02.09 | -14.61 | -106.85 | 5.3 | 64.0 |
| 06.03.09 | -9.14 | -62.61 | 13.3 | 75.0 |
| 07.03.09 | -11.31 | -80.21 | 12.0 | 78.0 |
| 09.03.09 | -11.26 | -79.77 | 13.1 | 69.5 |
| 10.03.09 | -14.69 | -107.49 | 5.1 | 84.0 |
| 11.03.09 | -14.23 | -103.78 | 9.0 | 90.0 |
| VSMOW | 0.00 ± 0.03 | 0.0 ± 0.3 | - | - |
| GISP | -24.76 ± 0.09 | -189.5 ± 1.2 | - | - |
| SLAP | -55.50 ± 0.03 | -428.0 ± 0.3 | - | - |
| IA-R052 | -19.64 ± 0.11 | - | - | - |
| IA-R053 | -10.18 ± 0.20 | - | - | - |
| IA-R054 | +0.56 ± 0.23 | - | - | - |

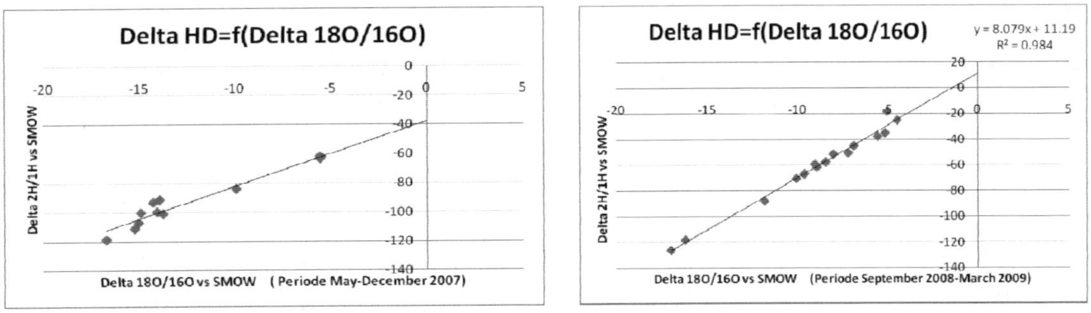


**Figure 1.** The results of measurements the D/H and $^{18}O/^{16}O$ isotope ratios from precipitation in period May-December 2007 (left) in Raureni-Valcea area (latitude 45.6°N, longitude 24.22°E), compared with the results of measurements the D/H and $^{18}O/^{16}O$ isotope ratios from precipitation in period September 2008-March 2009 in the same area (right).

This equation for "meteoric water line" which traced the isotopic compositions of natural waters originating from atmospheric precipitation in Raureni-Valcea area in period September 2008-March 2009 (latitude 45.6°N, longitude 24.22°E) is:

$$\delta^2 H = 8.079\, x\, \delta^{18} O + 11.9 \quad (R^2 = 0.984)$$

But, the dependencies of isotope compositions on temperature and amount of precipitation are now considered to be more complex (Aggarwal *et al* 2004, Alley and Cuffey 2001, Hendricks *et al* 2000) than were suggested in earlier reviews of GNIP data. These complexities arise from the much greater influence of moisture source and transport patterns on isotope composition compared to the influences of temperature and/or humidity alone. We have such of dependence, as are shown in figure 2.

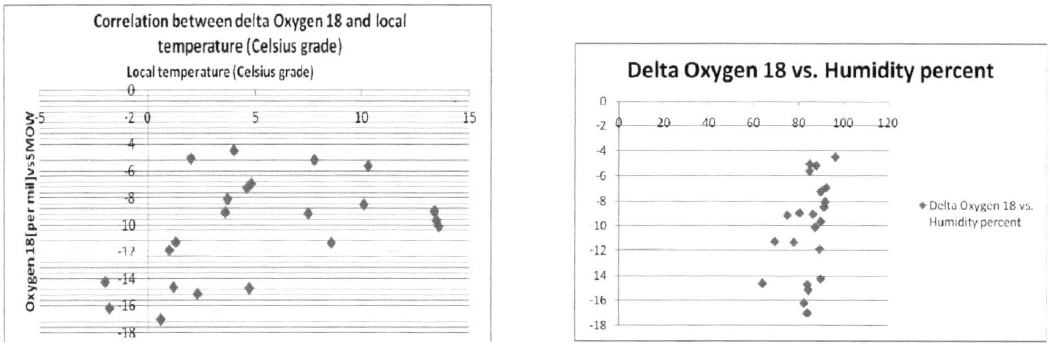


**Figure 2.** Correlation between $\delta^{18}O$ in precipitation, local air temperature (left) and humidity percent (right) in period September 2008-March 2009 in Raureni-Valcea area (latitude 45.6°N, longitude 24.22°E).

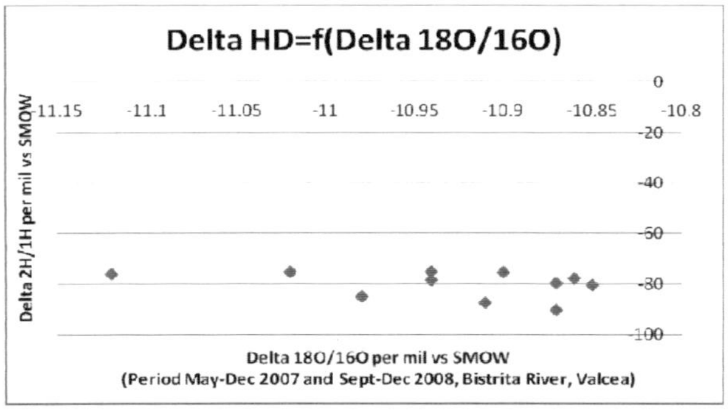


**Figure 3.** The results of measurements the D/H and $^{18}O/^{16}O$ isotope ratios from Bistrita River (Valcea-Romania) in period May-December 2007 and September-December 2008.

In Figure 3 we can see a little seasonal variation in isotopic composition from water content in Bistrita River. The average for these values represents the isotope fingerprint for water content in Bistrita River: $\delta^2H/^1H$ $(‰)_{vs\ SMOW} = -80.02 \pm 0.52$ (k = 2) and $\delta^{18}O/^{16}O$ $(‰)_{vs\ SMOW} = -10.93 \pm 0.11$ (k = 2).

## 5. Conclusions

The Isotope Ratio Mass Spectrometer (Delta V Plus, Thermo-Bremen), coupled with the continuous flow GasBench II unit, was used for the measurements of the D/H and $^{18}O/^{16}O$ isotope ratios in water collected from precipitation and Bistrita River during 2 years, 2007-2009. The Local Meteoric Water Line was reported for this area. Also, the variation of $\delta^{18}O$ and $\delta^2D$ values was correlated with the temperature and humidity in the same period.

## References

[1] Aggarwal P K, Froehlich K, Kulkarni K M and Gourcy L L 2004 *Geophys Res. Lett.* **31** L08203
[2] Coplen T B, Wildman J and Chen J 1991 *Anal. Chem.* **63** 910-2
[3] Clark I and Fritz P 1997 *Environmental Isotopes in Hydrogeology* (New York: Lewis Publishers) p 36
[4] Candelone J P, Hong S and Boutron C F 1994 *Anal. Chim. Acta* **299** 9
[5] Nelson S T 2000 *Rapid Commun. Mass Spectrom.* **14** 1044-6
[6] Schwikowski M, Barbante C and Doring T 2004 *Environ. Sci. Technol.* **38** 957.

# Photodegradation of the indoor organic pollutants by UV irradiation using $TiO_2$ catalysts

**Ioana C Glajar and Z Moldovan**

National Institute for Research and Development of Isotopic and Molecular Technologies, 65-103 Donath, 400293 Cluj-Napoca, Romania

E-mail: ioana.glajar@itim-cj.ro

**Abstract.** Volatile organic compounds (VOCs) are a major environmental concern, because of their carcinogenic and toxic effects on human health. The most frequent types of VOCs found in indoor air are, according to literature, Choloroform, p-dichlorbenzene, tetrachloroethylene, formaldehyde, NOx. Another VOCs found very often mentioned in the literature are ethanol and acetone or BTEX compounds. The investigated compounds used in this work for studding the photodegradation effect are toluene and cholorobenzene. In the present work were calculated the photodegradation rates of the compounds mentioned above using UV radiation and $TiO_2$, as catalyst. The obtained results are discussed based on comparative values of removal quantities for few time intervals for different types of catalysts based on $TiO_2$ aerogel.

## 1. Introduction

Photocatalytic oxidation process has been intense studied and is a promising method for removing and destroying various VOCs [1, 2, 3] in the polluted air. This technique using $TiO_2$ is cheap and doesn't require special conditions, the experiment is made usually at room temperature and pressure and the catalyst is stable. According to previous literature, using $TiO_2$ for organic pollutants degradation proved to be very efficient [4, 5, 6].

In this paper we studied the photodegradation of toluene and chlorobenzene compounds which are often found in the indoor air. Source materials for toluene vapors are paints, adhesives, gasoline and combustion sources, but for chlorobenzene are pesticides, plasticizers, preservatives, dyes and disinfectants manufactures [7, 8]. All experiments were made in a home made glass reactor.

## 2. Experimental

The used reactor volume was of 44 l, it had an air-blower, which maintained the homogenization of the compounds during the experiment. At the bottom of the reactor are two evacuation valves used for elimination of sample after every experiment. At the top there is a septum system for introduction of sample and collection of sample at determined time intervals. The reactor is provided with a UV source, which emits light in the range of 350-380 nm at a power of 60 W.

Usually in previous studies of photocatalytic degradation processes, several types of catalyst have been reported, such as: $TiO_2$, ZnO, $Fe_2O_3$ or ZnS [10]. In our experiments three different types of catalysts based on $TiO_2$ was used: $TiO_2$ Degussa P25, $TiO_2$-N doped and $TiO_2$-N doped thermic treated at 450°C ($TiO_2$-N doped TT). Catalyst support was glass plates having the 2 mm thickness, and 10x15 cm surface. The support plates were covered with three layers: first one of $TiO_2$ (for a good adhesion), the next two $TiO_2$ aerogels (doped or undoped). The experiment was performed on pairs of plates with catalysts, total area of 300 $cm^2$ at room temperature (22°C) and humidity (around of 35% RH).

The initial concentration of compounds was 200 ppm. The measurements were made in a time interval of 300 min, the samples being collected from reactor at every 30 min. For quantitative determinations 1 ml of gas sample was injected into a Perkin Elmer 990 model gas-chromatograph equipped with a polar column of 6 m length and 3 mm inside diameter, made from stainless steel. The column temperature was programmed from 75°C to 200°C. The concentrations left in the photoreactor were obtained from chromatogram peak area having FID as detector.

## 3. Results and discussions

In the early reported studies of the photocatalytic processes was shown that based on oxidation and reduction reaction as final compounds result $CO_2$ and $H_2O$ respectively [10]. The major physical parameters that have influenced the photocatalytic kinetics are: wavelength, concentrations of reactants, temperatures and the radiant flux [8].

According to the literature [11] the intermediate products detected from the UV photodegradation of gaseous chlorobenzene included hydrochloric acid, acetic acid, formic acid, phenol and chlorophenol. In the toluene photodegradation process as intermediate compounds were shown benzaldehyde and *p*-methylphenol [12].

We made four experiments: the first one was realized with no catalyst and we considered this as a blank experiment. In the next three experiments we tested the nanocatalysts that we have mentioned before. All the results are graphical represented in figure 1. Work conditions were as in section 2.

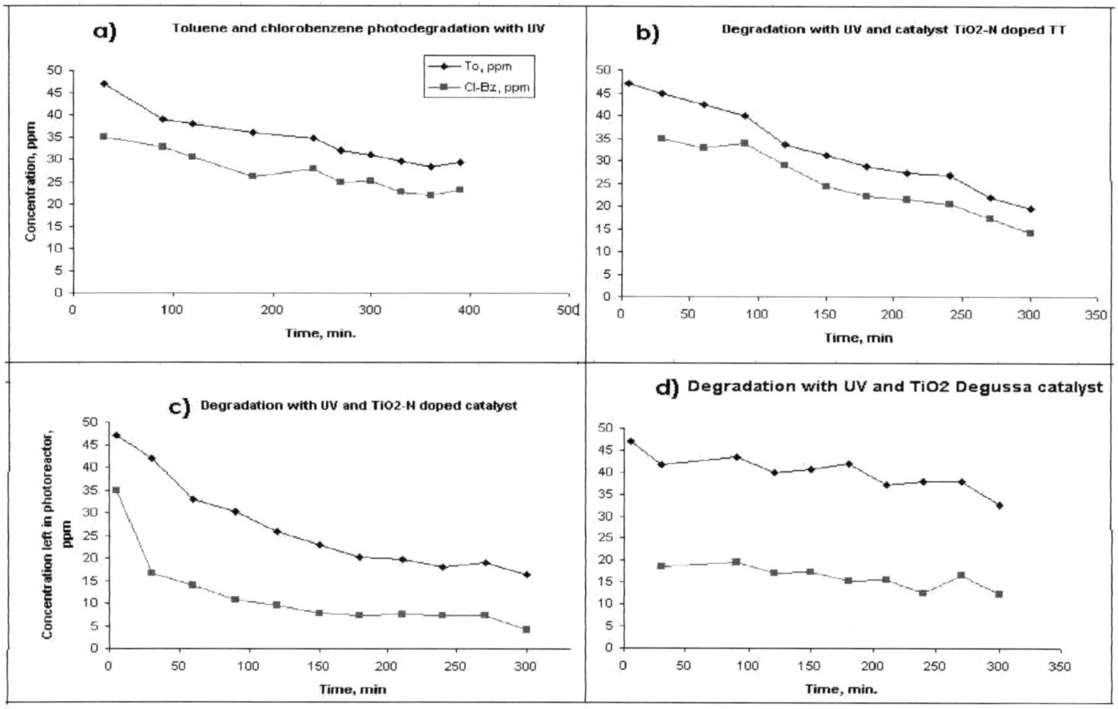


**Figure 1.** Toluene and Chlorbenzene photodegradation under UV light and $TiO_2$ catalysts. a) without catalyst; b) N-doped $TiO_2$ thermic treated; c) N-doped $TiO_2$; d) Degussa P25 catalyst.

Analyzing the above graphics as well as those obtained on blank samples (no UV or catalyst) we concluded that removal of compounds from photoreactor is due to two processes: adsorption on enclosure walls and the proper photodegradation.

The photodegradation process can be mathematical described by an exponential equation [13]: which can be liniarised using logarithm function:

$$\ln C = \ln C_0 - k \cdot t \tag{1}$$

where $C$ is removed concentration by photodegradation process, $C_0$ is the starting photodegradation concentration, $k$ is the constant rate and $t$ is time. Drawing the logarithm of concentrations obtained from experiment as a time function the photodegradation process (linear segment of drawing) can be separated of adsorption process. Table 1 lists some specific value, obtained for hourly time intervals.

**Table 1.** Average photodegradation rate (% of $C_0$) for Toluene (T) and Chlorobenzene (CB) under UV light and different types of $TiO_2$ catalysts.

| Time | $TiO_2$-N doped TT | | $TiO_2$-N doped | | $TiO_2$ Degussa P25 | |
|---|---|---|---|---|---|---|
| (min) | T | CB | T | CB | T | CB |
| 60 | 9.7 | 13.9 | 13.4 | 13.2 | 3.5 | 6.02 |
| 120 | 18.5 | 26.4 | 25.0 | 26.2 | 6.9 | 13.3 |
| 180 | 26.3 | 37.0 | 35.0 | 37.2 | 10.2 | 19.3 |

As can be seen in the table 1 the removal efficiency of compounds is improved by using N doped $TiO_2$ catalyst relative to $TiO_2$ Degussa P25. In the situation of N doped $TiO_2$ thermal treated a discriminant effect of Chlorbenzene removal efficiency relative to Toluene must be mentioned. The effect can be explained based on stereo configuration of involved compounds.

## 4. Conclusions

Using $TiO_2$ catalysts under UV light, a significant degradation of VOCs compounds was observed. Quantity of compound photodegradated depends of catalyst type and organic pollutant structure. Doping $TiO_2$ is increasing the photocatalytic activity (in our case $TiO_2$-N doped is the most efficient catalyst). Usually a stronger photodegradation was observed for chlorbenzene comparing with toluene. When $TiO_2$-N doped is used the photodegradation rate values for both compounds are much closed. All experiments were performed at normal conditions of temperature and humidity. At decreasing concentrations of compound two processes compete: adsorption and photodegradation.

## References

[1] Boulamanti A K, Korologos C A and Philippopoulos C J 2008 *Atmos. Environ.* **42** 7844
[2] Jo W-K and Kim J-T 2009 *J.Hazard. Mater.* **164** 360
[3] Ao C H, Lee S C, Mak C L and Chan L Y 2003 *Appl. Catal. B Environ.* **42** 119
[4] Jeong J *et al* 2005 *J. Photochem. Photobiol A Chem.* **169** 279
[5] Kim S B *et al* 2002 *Chemosphere* **48** 437
[6] Shiraishi F and Ishimatsu T 2009 *Chem. Eng. Sci.*, doi: 10.1016/j.ces.2009.02.0248.
[7] Holcomb L C and Seabrood B S 1995 *Indoor Environ.* **4** 7
[8] Wang S, Ang H M and Tade M O 2007 *Environ. Intern.* **33** 694
[9] Ibrahim H and Lasa H 1999 *Ind. Eng. Chem. Res.* **38** 3211
[10] Zhao J and Yang X 2003 *Build. Environ.* **38** 645
[11] Wang C, Xi J and Hu H 2008 *Chemosphere* **73** 1167
[12] Addamo M, Augugliaro V Coluccia S, Di Paola A, Garcia-Lopez E, Loddo V, Marci G, Martra G and Palmisano L 2006 *Int. J. Photoenerg.* ID 39182, 1
[13] Doll T E and Frimmel F H 2003 *Chemosphere* **52** 1757.

# The improvement of removal effects on organic pollutants in Wastewater Treatment Plants (WWTP)

**O Marincas[1], P Petrov[2], T Ternes[3], V Avram[1] and Z Moldovan[1]**

[1] National Institute for Research and Development of Isotopic and Molecular Technologies, 65-103 Donath, 400293 Cluj-Napoca, Romania

[2] Water Treatment Station Siluet B, 21 Pencho Slaveikov Street, Varna 9000, Bulgaria

[3] Federal Institute of Hydrology, Am Mainzer Tor 1, 56068 Koblenz, Germany

E-mail: olivian.marincas@itim-cj.ro

**Abstract.** Purpose of this study is to improve the efficiency of removal in wastewater treatment plants of some organic pollutants like pharmaceuticals, antioxidants, pesticides (triazines, phenylurea herbicides), personal care products (PCPs) musk fragrances (galaxolide and tonalide) and estrogens using zeolites with excellent absorption capacity. The zeolite selected for all experiments was Szedimentin-MW. The experiment took place in three stages: no zeolite addition, zeolite added at the end of the bioreactor and zeolite added at the start of the bioreactor. The water samples were pre-concentrated with solid phase extraction (SPE) procedure and analyzed with analytical system Gas Chromatography/Mass Spectrometry (GC/MS).

## 1. Introduction

Wastewater treatment systems have been designed to minimize the environmental impacts of discharging untreated wastewater. The choose of solutions for wastewater treatment lead to different performance in pollutants removal and therefore to different impacts on the environment [1].

In the recent decades, the increase in human population brought about increased quantity of wastewaters and increased demands on the quality of wastewater treatment products, the effluent and sewage sludge [2].

Persistent pollutants like pharmaceuticals, antioxidants, pesticides (triazines, phenylurea herbicides), musk fragrances (galaxolide and tonalide) and estrogens are very frequent detected in industrial and domestic wastewater and show great damage to enviroment. In recent years, interest has been focused on the removal of this organic pollutants from aqueous/sludge systems.

The widespread presence of pharmaceuticals and personale care products in the aquatic environment can be explained by their extensively us in medical practices and incomplete removal in Wastewater Treatment Plant (WWTP). Their annual consummation ranges between a few kilograms and several hundred tons per individual compounds.

It is assumed that PCPs could act as persistent compounds, because of their continual discharge into aquatic media via WWTP effluents, which sustain a multigenerational exposure for the resident organisms.

Also, the polycyclic synthetic musk as Galaxolide (HHCB) and Tonalide (AHTN) has large applications as fragrances in soaps, perfumes, air fresheners, detergents, fabric softener and other cleaning products. There are indications that the polycyclic musks are not only to be found in various environmental compartments but also in the aquatic food chain as well as in fatty tissue and mothers' milk [3].

Zeolite is one of the excellent substances with a great absorption capacity, which is produced at mines naturally, and is also synthesized by artificial process. There are various types of zeolite by its crystal structure. Zeolite itself is often utilized as water purification material for its incomparable ion-exchange capacity [4].

They have been used as adsorbents, molecular sieves, membranes, ion exchanger and catalysts in municipal and industrial pollution control, as well as in horticulture, agriculture, and environmental soil remediation but their primary use has been in water and wastewater treatment [5].

Zeolite refers to a group of minerals that are basically hydrated calcium potassium sodium aluminosilicates in which the water is held in cavities in the lattice. The lattices are negatively charge and they loosely hold cations. Their ability to exchange one cation for another is known as their "cation-exchange capacity" or "CEC". Total cation-exchange capacities in natural zeolites vary from 25 to 300 meq/100 g [6].

The present paper concern is the study of removal efficiency of Szedimentin-MW zeolite on few class of compounds early described, in WWTP serving an equivalent population of 300.000 (Varna, Bulgaria). The water samples were pre-concentrated with solid phase extraction (SPE) procedure and analyzed with analytical system Gas Chromatography/Mass Spectrometry (GC/MS).

## 2. Experimental

*The full-scale experiments* have dual purpose: optimisation of the dose of zeolite (experiments at various concentrations) and the zeolite dosing points.

The simplified scheme of experimental WWTP is shown in figure 1:

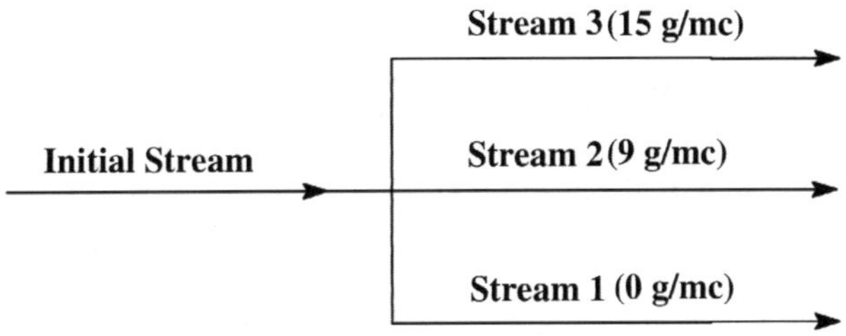


**Figure 1.** Scheme of WWTP. Stream 3: Dose to maintain 7.5 % concentration of zeolite in the activated sludge. Stream 2: Dose to maintain 5.0 % concentration of zeolite in the activated sludge. Stream 1: Reference stream. 0 % concentration of zeolite in the activated sludge. Zeolite dosing points: Stream 1 – No zeolite addition; Stream 2 – Zeolite added at the end of the bioreactor; Stream 3 – Zeolite added at the start of the bioreactor.

The parameters of the experiments were: Sludge concentration in the bioreactor: 2.0 to 2.8 g/L and in the recycled sludge: 3.0 to 4.2 g/L. The sludge retention time was 8 days on average and the hydraulic retention time 10 hours on average, 3-4 hours at peak loads (no inlet buffer tank). The measurements of selected compounds were done in three consecutive days.

### *2.1. Reagents and materials*

All solvents were of HPLC grade. Dichloromethane was purchased from JT Baker (Deventer, Holland). Acetonitrile, Methanol, HPLC Water and Isooctane from Merck (Darmstadt, Germany). Glass microfiber filters (GF 6, Ø 47 mm, Lot EU0940-1) were obtained from Whatman (Dassel,

Germany). The OASIS HLB 3cc cartridge containing the 30 mg macroporous copolymer [poly(divinylbenzene-co-N-vinylpyrrolidone)], was purchased from Waters Corporation (Milford, Mass USA).

*2.2. Instrumentation*
A Thermo Finnigan Trace GC interfaced with a Polaris Q ion trap mass spectrometer was used for all samples. The gas chromatograph was equipped with a (5%-Phenyl)-methylpolysiloxane GC column (HP-5MS column, 30 m x 0.25 mm *i.d.*, 0.25 µm film thickness) purchased from Agilent Technologies (Santa Clara, U.S.A.). The conditions of the analysis were as follows: injection volume (2.0 µl) with splitless injection, injector temperature: 280°C and transfer line temperature: 300°C. The carrier gas was helium with flow rate of 2 ml/min at constant flow with vacuum compensation. The temperature programme: the initial temperature was 80°C (1 min), than increased to 150°C at 15°C/min, to 192°C at 3°C/min, to 193°C at 0.2°C/min and to 300°C at the rate of 10°C/min, where it was held for 10 min.

*2.3. Solid phase extraction method*
The most common sample preparation for trace analysis of organic pollutants in water involves solid-phase-extraction (SPE). The extraction efficiency of SPE is found to be higher than the traditional liquid–liquid extraction [7].

In our research, we used for extraction, the OASIS HLB cartridge described in section 2.1. The samples were acidified with 2 N HCl to pH 2 and filtrated on a glass fibre filter.

*2.3.1. SPE procedure*
The divinylbenzene/N-vinylpyrrolidone copolymer SPE cartridge (Oasis HLB) was prewashed with methanol and organic-free water with the flow rate of 2 ml/min. The water sample was then percolated through the cartridge with a flow rate of 3-5 ml/min under vacuum. Finally, after drying, the cartridge was eluted with methanol [8].

After elution the extract was concentrated by rotavapor system and transferred into isooctane. The final extract was transferred to vials and the excess of solvent was evaporated under a stream of dry nitrogen. The samples were injected (2 µl of sample in iso-octane) in the GC/MS system without derivatization.

## 3. Results and discussion
GC conditions allowed the complete separation of the target compounds.

**Table 1.** Concentration of antioxidants (1,2), triazines (3-6), phenyl urea herbicides (7,8), fragrances (9), antimicrobials (10) and antiepileptic drugs (11) in ng/l. The symbol (*) is for concentration under LOQ, (**) the symbol BHA and BHT denoted the compounds Buthylhydroxyanisole and Buthylhydroxytoluene respectively.

| Nr. crt. | Compound | LOQ (ng/l) | Day 1 | | | | Day 2 | | | | Day 3 | | | |
|---|---|---|---|---|---|---|---|---|---|---|---|---|---|---|
| | | | P0 | P1 | P2 | P3 | P0 | P1 | P2 | P3 | P0 | P1 | P2 | P3 |
| 1 | BHA** | 30 | 824 | 717 | 825 | 413 | 1007 | 866 | 945 | 475 | 805 | 749 | 709 | 714 |
| 2 | BHT** | 30 | 560 | 343 | 220 | 88 | 306 | 269 | 215 | 137 | 264 | 237 | 196 | 206 |
| 3 | Prometon | 7 | * | * | * | * | * | * | * | * | * | * | * | * |
| 4 | Atrazin | 40 | * | * | * | * | * | * | * | * | * | * | * | * |
| 5 | Propazin | 13 | * | * | * | * | * | * | * | * | * | * | * | * |
| 6 | Ametrin | 30 | * | * | * | * | * | * | * | * | * | * | * | * |
| 7 | Diuron | 20 | * | * | * | * | * | * | * | * | * | * | * | * |
| 8 | Isoproturon | 7 | * | * | * | * | * | * | * | * | * | * | * | * |
| 9 | Tonalide | 15 | 251 | 349 | 150 | 159 | 166 | 251 | 199 | 154 | 213 | 199 | 182 | 144 |
| 10 | Triclosan | 7 | 762 | 264 | 476 | 215 | 490 | 459 | 505 | 288 | 496 | 472 | 531 | 254 |
| 11 | Carbamazepine | 10 | 72 | 77 | 77 | 112 | 105 | 114 | 105 | 113 | 60 | 37 | 44 | 58 |

Quantities were calculated using the chromatographic area obtained in the full scan mode. It has been used as internal surrogate standard Diphenylamine in concentration of 1000 ng/L. The results obtained are shown in table 1.

We observe that:

- the concentrations of compounds in P1 was close to P0 (except for triclosan in day 1);
- the concentrations of compounds in P2 was around of 72% in day 1, 87% in day 2 and 85% in day 3 (relative to P0);
- the concentrations of compounds in P3 was around of 62% in day 1, 69% in day 2 and 76% in day 3 (relative to P0);
- the minimum of concentration was observed for BHT day 1, point 3 (16%) and triclosan, day 1, point 3 (28 %);
- in all situation carbamazepine was observed in concentration close to P0;

the results shown that more best retention was in the point 3, day 1.

## 4. Conclusions

For stream 2 (9 g/mc) we observe an average removal quantity of 19%. For stream 3 (15 g/mc) we observe an average removal quantity of 31%. To increase the efficiency of removal we observe that is good to use a big amount of zeolite (> 15g/mc), added at the start of the bioreactor. In all situation carbamazepine removal was 0%. Removal quantity was related to stream 1 (no zeolite added). The studied waste water originate main from domestic activity. This is the explanation of the fact that in all samples pesticides classes was not detected.

## Reference

[1] Antoniadis A, Takavakoglou V, Zalidis G and Poulios I 2007 *Catal. Today* **124** 260-5
[2] Sasakova N, Ondrasovicova O, Ondrasovic M, Vargova M, Kottferova J, Culenova K and Papajova I 2005 *ISAH* vol 2
[3] Moldovan Z 2006 *Chemosphere* **64** 1808-17
[4] Kimochi Y, Masada T, Mikami Y, Tsuneda S and Sudo R 2008 *Water Science & Technology – WST* **58.4** 847-51
[5] Metes A, Kovacevic D, Vujevic D and Papic S 2004 *Water Res.* **38** 3373-81
[6] Osmanlioglu A E 2006 *J. Hazard. Mater.* B **137** 332-5
[7] Ma W T, Fu K K, Cai Z and Jiang G B 2003 *Chemosphere* **52** 1627-32
[8] Mezcuza M, Agüera A, Lliberia J L, Cortés M A, Bagó B and Fernández-Alba A R 2006 *J. Chromatogr.* A **1109** 222-7.

# Structural elucidation of organic contaminants by chemical ionisation mass spectrometry

**Zaharie Moldovan**

National Institute for Research and Development of Isotopic and Molecular Technologies, 65-103 Donath, 400293 Cluj-Napoca, Romania

E-mail: zaharie.moldovan@itim-cj.ro

**Abstract.** The PI-CI mass spectra formation for a new family of aromatic amines, with general formula: $R_1$-Ph-NH-Ph-$R_2$ is discussed in correlation with the $R_1$ and $R_2$ structure. The compounds where isolated from some environmental samples by GC/MS technique. The characteristic ions are produced by rearrangement processes involving olefin and alkane neutral molecule elimination from $[M+H]^+$ and sole olefin molecule elimination from $[M+C_2H_5]^+$.

## 1. Introduction

In the chemical ionization mass spectrometry (CI-MS), the ionization of a substance takes place by ion-molecule reactions rather than electron impact (EI) or field ionization (FI). The CI spectrum is different from the spectra produced by other ionization techniques and the analytical and scientific utility of the method stems from the unique character of the chemical ionization spectra. By comparison, the molecule-ions initially formed by EI processes are odd-electron ions, whose decomposition is influenced by the presence of the odd electron. Evidently, the ions initially formed by even-electron CI processes may be expected to decompose by rather different networks of pathways. Another reason for difference is the fact that whereas in EI ionization, the energy transfer distribution can include values with 10 eV above ionization potential, the energy transfer in CI processes exceeds only 5 eV [1]. As a consequence, complementary structural information can be obtained by CI processes.

The ion decomposition depends only on its internal energy (described by probability function P(E)) and of the rate constant k(E) of the particular reaction. Decompositions of molecular ions in the mass spectrometer are generally produced by two ways: *i*) simple cleavage of a bond, and *ii*) by prior rearrangement of atoms and elimination of neutral molecule. In the rearrangement reaction two new bonds are formed offsetting the energy required to cleave other two. The critical energy for a reaction involving only a simple bond cleavage is higher. Based on this reason it results that by CI the reactions involving rearrangement of atoms are predominant relative to the simple fission reactions [2].

## 2. Results and discussion

The paper discusses the positive ionization CI (PI-CI) mass spectra (methane as reactive gas) for five members from a new family of aromatic amines shown in table 1. The compounds where isolated from some environmental samples by GC/MS technique[3]. The EI mass spectra early reported[3] are dominated of ions [M-15]+ and [M-71]+, produced by simple cleavage (benzylic bond), corresponding to elimination of radical $-CH_3$ and $-C_5H_{11}$ respectively from R.

**Table 1.** The studied compounds structure $R_1$-Ph-NH-Ph-$R_2$.

| Compound | $R_1$ | $R_2$ | Symbol |
|---|---|---|---|
| 1 | H | $C(CH_3)_3$ | (0,4) |
| 2 | $C(CH_3)_3$ | $C(CH_3)_3$ | (4,4) |
| 3 | H | $C(CH_3)_2C_5H_{11}$ | (0,8) |
| 4 | $C(CH_3)_3$ | $C(CH_3)_2C_5H_{11}$ | (4,8) |
| 5 | $C(CH_3)_2C_5H_{11}$ | $C(CH_3)_2C_5H_{11}$ | (8,8) |

*2.1. Mechanism of production of ions visible in PI-CI mass spectra*

*2.1.1. Pseudo molecular ion*

The principal reagent ions formed from methane at 1 torr are $CH_5^+$ and $C_2H_5^+$ [1]. The two types of pseudo molecular ions are produced by interaction of the neutral compounds with different types of ions resulting from reactive gas:

$R_1$-Ph-NH-Ph-$R_2$ + $CH_5^+ \rightarrow R_1$-Ph-$N^+H_2$-Ph-$R_2$ $[M+1]^+$ (proton transfer)

$R_1$-Ph-NH-Ph-$R_2$ + $C_2H_5^+ \rightarrow R_1$-Ph-HN($C_2H_5$)-$R_2$ $[M+29]^+$ (electrophilic addition) [4]

*2.1.2. Fragment ions*

Having $[M+1]^+$ ion as precursor, two reactions of decomposition are visible:

a. Rearrangement reaction involving an H transfer in a *four-membered transition state* and the elimination of *olefin* molecules of type [R-H] ($C_4H_8$, $C_8H_{16}$) depending of $R_1$ and $R_2$ structures. The difference in mass units relative to $[M+1]^+$, $\Delta_1$ is 56 an 112 respectively.
b. Rearrangement with H transfer also in a *four-membered transition state* and elimination of *alkane* molecule of type $CH_4$ or $C_5H_{12}$ depending of atom subgroups present in $R_1$ and $R_2$. Mass difference $\Delta_1$ being 16 and 72 respectively.
c. From $[M+29]^+$ only the olefinic molecules of type [R-H] ($C_4H_8$, and respectively $C_8H_{16}$) are eliminated after H transfer. The mass difference relative to $[M+29]^+$, $\Delta_{29}$ is also 56 and 112 corresponding to neutral elimination.

The mechanism of fragmentation of the ions $[M+1]^+$ and $[M+29]^+$ is shown in scheme I.

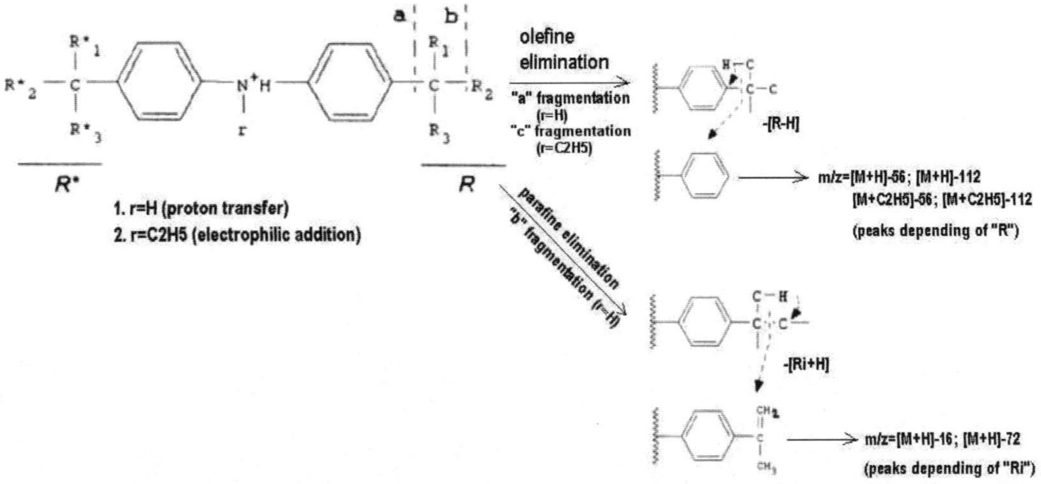


**Scheme I**. Fragmentation of the ions $[M+1]^+$, r = H (*a* and *b* fragmentation) and $[M+29]^+$, r = $C_2H_5$ (only *c* fragmentation).

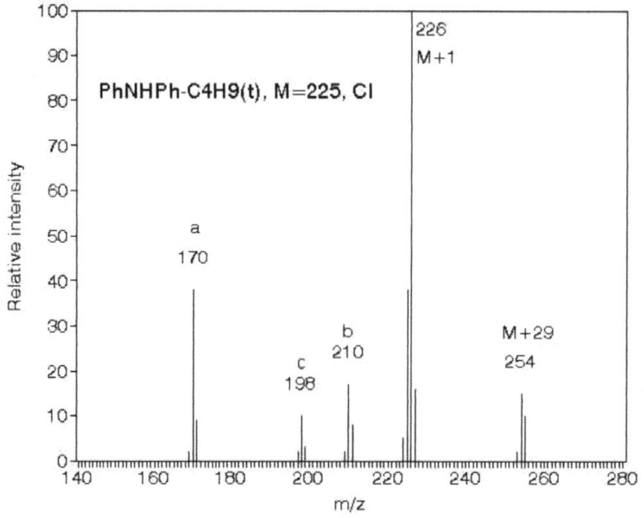


**Figure 1.** PI-CI mass spectrum of Compound 1(0,4). The ions *a* and *c* correspond to olefin $C_4H_8$ elimination but the ion *b* to alkane $CH_4$ elimination from R.

The mass spectrum of compound 1 is shown in figure 1. The mass spectrum presents tree fragment ions: *ion a, m/z 170* (elimination of $C_4H_8$ from $[M+1]^+$) *ion b, m/z 210* (elimination of $CH_4$ from $[M+1]^+$) and *ion c, m/z 198* (elimination of $C_4H_8$ from $[M+29]^+$. The ions *a* and *b* are diagnostic ions for *R*. The difference between $[M+1]^+$ and *a* as well as between $[M+29]^+$ and *c* is the same ($\Delta_1 = \Delta_{29} = 56$) corresponding to neutral $C_4H_8$. The ion *b* is a diagnostic ion for detailed structure of *R*. The difference $\Delta_1$ between $[M+1]^+$ and *b* is 16 corresponding to atom group $CH_4$ as subgroup from R.

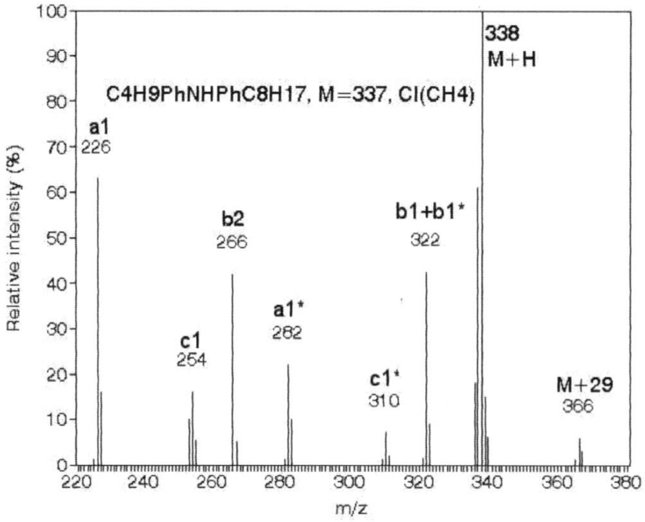


**Figure 2.** PI-CI Mass Spectrum of compound 4 (4*,8). The type of ions are indicated by *a, b* and *c* and *a**, *b** and *c** for contribution of R and R* respectively.

**Table 2.** Number of fragment ions related to molecular structure.

| Compound | $N_a$ | $N_b$ | $N_c$ | $N_t$ |
|---|---|---|---|---|
| 1: (0,4) | 1, $C(CH_3)_3$ | 1, $CH_3$ | 1, $C(CH_3)_3$ | **3** |
| 2: (4,4) | 1, $C(CH_3)_3$ | 1, $CH_3$ | 1, $C(CH_3)_3$ | **3** |
| 3: (0,8) | 1, $C(CH_3)_2C_5H_{11}$ | 2, $CH_3$ + $C_5H_{11}$ | 1, $C(CH_3)_2C_5H_{11}$ | **4** |
| 4: (4,8) | 2, $C(CH_3)_3$+$C(CH_3)_2C_5H_{11}$ | 2, $CH_3$ + $C_5H_{11}$ | 2, $C(CH_3)_3$ + $C(CH_3)_2C_5H_{11}$ | **6** |
| 5: (8:8) | 1, $C(CH_3)_2C_5H_{11}$ | 2, $CH_3$ | 1, $C(CH_3)_2C_5H_{11}$ | **4** |

The mass spectrum of compound 4 (4*,8) is shown in figure 2. The type of ions is denoted by a, b, c and a*, b*, c*, originating from R and R* respectively. The number of fragment ions observed in mass spectrum are six: *ion a, m/z 226,* $\Delta_1=112$*; c, m/z 254,* $\Delta_{29}=112$*;* $b_1$*, m/z 266,* $\Delta_1=72$*; a*, m/z 282,* $\Delta_1=56$*; c*, m/z 310,* $\Delta_{29}=56$ *and* $b_2$*, m/z 322,* $\Delta_1=16$. The ions *a, b,* and *c* are diagnostic ions for R but *a*, b** and *c** are diagnostic ions for R*. The values $\Delta_1$ and $\Delta_{29}$ corresponding to detected ions are of 112 (elimination of $C_8H_{12}$ from R), 56 (elimination of $C_4H_9$ from R*) 72 (elimination of $C_5H_{12}$ as subgroup from R) and 16 (elimination of $CH_4$ as subgroup from R and R*).

## 3. Conclusions

The ions present in PI-CI mass spectra of compounds 1-5 are produced by rearrangement processes involving H atom transfer by a four membered transition state. Two type of ions were observed: the first one formed by elimination of olefin group with structure [R-H] and a second one produced by elimination of a paraffin group with structure [$R_i$+H]. $R_i$ being a subgroup of atoms from R. Between the number of peaks detected in PI-CI mass spectra and the structure of R groups the following correlations exist: $N_a$ (peaks *a*) = 1 x $N_R$ (R structures); $N_b$ (peaks *b*) = 1 x $N_{Ri}$ ($R_i$ structures); $N_c$ (peaks *c*) = 1 x $N_R$(R structures). $N_t$ (total) = $N_a$ (peaks *a*) + $N_b$ (peaks *b*) + $N_c$ (peaks *c*). For compounds 1-5, a practical relation between total numbers of peaks observed in mass spectrum and R structures was established (table 2).

## References

[1] Chapman J R 1985 *Practical Organic Mass Spectrometry* (Chichester: Wiley)
[2] McLafferty F W and Turecek F 1993 *Interpretation of Mass Spectra* (Mill Valley: University Science Books)
[3] Moldovan Z and Bayona J M 2000 *Rapid Commun. Mass Spectrom.* **14** 379
[4] Moldovan Z Maldonado C and Bayona J M 1997 *Rapid Commun. Mass Spectrom.* **11** 1077.

# Numerical model to solve impurities' migration in water pipes

**Valer Toşa and Katalin Kovács**

National Institute for Research and Development of Isotopic and Molecular Technologies, 65-103 Donath, 400293 Cluj-Napoca, Romania

E-mail: valer.tosa@itim-cj.ro

**Abstract.** We present a novel numerical method based on finite differences to solve the problem of impurities' migration from a hollow core multilayer cylinder filled with a known liquid medium. We expose the numerical method and the developed application. The validation is performed by calculating the migration in a configuration which allows an analytic solution. Applications are foreseen in modeling impurities' migration from modern multilayer water pipes into drinking water.

## 1. Introduction

The use of polyethylene (PE) pipes in drinking water distribution networks and household installations causes a contamination of the consumers' tap water which is a major concern of water pipes producers. Two types of PE pipes are used in the drinking water network; high density PE (HDPE) pipes in the distribution mains and service pipes, and/or cross linked PE (PEX) in the household installations as hot and cold water pipes. More than 100 different organic compounds are found to migrate into water from polyethylene materials like HDPE and PEX (see for example [1, 2]).

The model presented here is similar to that used to calculate migration of additives from food packaging polymers into the food [3] developed in one dimension and taking into account the European Commission's recommendations [4]. In these cases the main chemical parameters of importance are the migrant's diffusion coefficient in the polymer and its partition coefficient at the interface. The concentration of the migrant in this model is assumed to have no boundary layer resistance between packaging and food [5]. The basic difficulty, however, lies in the composition of the media from which the migration takes place, i.e. food packaging or water pipes, as these media are nowadays constituted of multilayer materials. This structure introduces new parameters for the modeling like the diffusion coefficients of the migrant in different layers and the partition coefficient between two adjacent layers as defined by the ratio between migrant solubility in the two layers.

One should stress that the case of diffusion in a multilayer system of cylindrical geometry is present not only in water pipes but also in food packaging. Indeed, in many cases, liquid food like beverages, oil, fruit juices, etc. are packed in PET bottles and the calculation of migrated substance from the bottles into the food can be influenced to a certain extent by this geometry.

The paper is organized as follows: in Section 2 we present the actual problem to be solved numerically, then in Section 3 we validate the used numerical procedure and present the obtained results. Finally, we conclude and refer to further developments and applications of the model.

## 2. Exposition of the Problem

The equation of diffusion in cylindrical geometry has the form:

$$\frac{\partial c}{\partial t} = \frac{1}{r}\frac{\partial}{\partial r}\left(rD\frac{\partial c}{\partial r}\right). \tag{1}$$

A numerical solution of the diffusion equation is obtained by performing the following steps: (i) define a numerical grid; (ii) use a numerical method (for example Lagrange polynomials) to approximate finite differences at mesh points; (iii) discretize the diffusion equation using the method of finite differences; (iv) solve the resultant system of algebraic equations. For our case the algebraic equations have the form:

$$c_i^{n+1} - c_i^n = D\delta t \cdot F\left[\theta\left(Ac_{i-1}^{n+1} - Bc_i^{n+1} + Ec_{i+1}^{n+1}\right) + (1-\theta)\left(Ac_{i-1}^n - Bc_i^n + Ec_{i+1}^n\right)\right], \tag{2}$$

where $c_i^n$ denotes the migrant concentration at mesh point $i$ and time $n\delta t$, while $A, B, E, F$ are constants resulting from the geometry of the system and the implementation of the Lagrange polynomials for the derivatives.

Discretization on a numerical grid of the diffusion equation for the case of multilayer systems is not trivial at the interfaces between adjacent layers. Due to the non-unitary value of $K$ partition coefficient (in general), the concentration profile of the migrant is not continuous at the interface between layers. A suitable method to treat this numerical difficulty is to implement the fictitious point method. Both the general finite difference and fictitious point methods are presented in Ref. [3] for planar geometry.

The method presented above was further developed in a computer program application which was tested by modeling a case in which analytical solution can be derived. The results of the validation of the method will be presented in the following section.

## 3. Method Testing and Results

Our goal is to test and validate the numerical method sketched in the previous section. For this purpose we need to find a particular case which can be treated analytically. In his book Crank [6] discusses the problem of migration in a cylinder and gives the solution under different conditions. The closest case to our practical situation is the migration of impurities from a well-stirred solution surrounding concentrically a solid cylinder. In other words the solution is playing the role of the hollow core pipe while the in inner cylinder plays the role of water. The initial concentration of migrant in the solution is $C_0$ and the water is initially free from migrant. This problem can be treated analytically and would correspond to the two-layer case in the numerical simulation. The amount $M_t$ of migrant in the cylinder after time $t$ is expressed as a fraction of the corresponding amount $M_\infty$ after infinite time by the relation:

$$\begin{aligned}\frac{M_t}{M_\infty} = 1 \quad &- \quad \frac{\gamma_3}{\gamma_3+\gamma_4} exp\left[4\gamma_3^2\frac{Dt}{r^2\alpha^2}\right] erfc\left[\frac{2\gamma_3}{\alpha}\left(\frac{Dt}{r^2}\right)^{\frac{1}{2}}\right] \\ &- \quad \frac{\gamma_4}{\gamma_3+\gamma_4} exp\left[4\gamma_4^2\frac{Dt}{r^2\alpha^2}\right] erfc\left[-\frac{2\gamma_4}{\alpha}\left(\frac{Dt}{r^2}\right)^{\frac{1}{2}}\right],\end{aligned} \tag{3}$$

where $\alpha = V_{pipe}/(K \cdot V_{water})$ with $K$ partition coefficient, $\gamma_3 = \frac{1}{2}\left[(1+\alpha)^{1/2}+1\right]$ and $\gamma_4 = \gamma_3 - 1$. The inner and outer radii of the pipe are $r$ and $R$, respectively. As shown in Ref. [6] this solution is valid only for small values of $Dt/r^2$ and $\alpha$ moderate. This limitation will become visible also on our results.

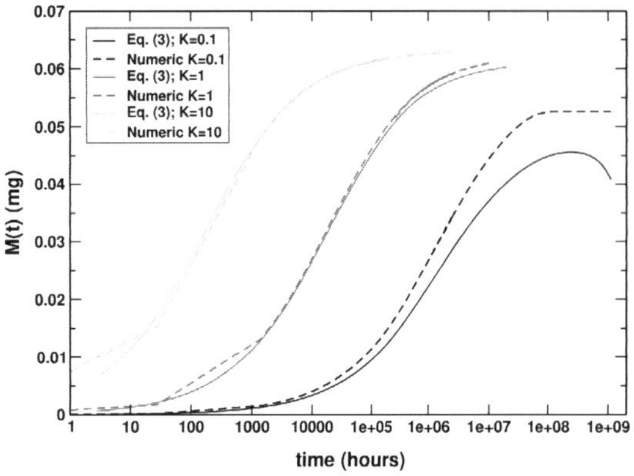


**Figure 1.** Temporal evolution of the total amount of migrant in the hollow core of the cylinder: comparison of analytical results Eq. (3). (solid lines) and numerical results (dashed lines) for different values of $K$.

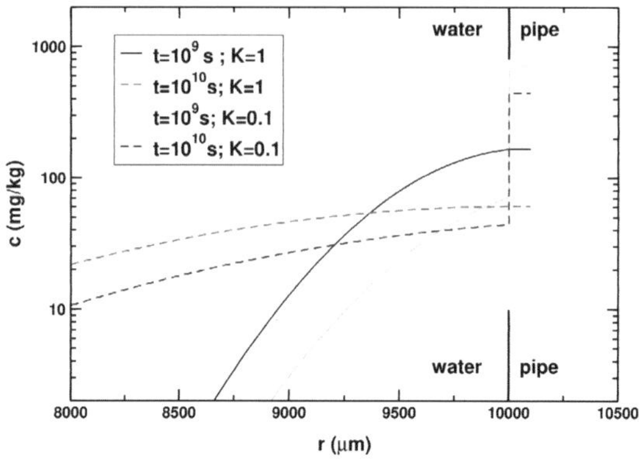


**Figure 2.** Numerical results: spatial profile of the migrant mass concentration in the vicinity of the layer boundary for different values of $K$ after different times of migration.

In Fig. 1. we represent the temporal variation of the total amount of migrant in the water for different values of the partition coefficient at the pipe/water boundary. A value $K$ here means that the concentration of the migrant in the water just at the interface with the pipe is $K$ times that in the pipe at the interface with the water. Other parameters of the model are: $r = 1\ cm$, $R = 1.01\ cm$, $C_0 = 1000\ mg/kg$, $D = 10^{-12}\ cm^2/s$.

One can observe a reasonable agreement of the numerical results (dashed lines) with the curves corresponding to Eq. (3) for partition coefficient $K = 1$ and 10. However, we remark that this agreement is limited to times less then 10000 hours in the case of partition coefficient $K = 0.1$. We can read from the graph that for very large times ($t > 10^8$ hours) the analytical formula completely fails. Indeed, according to Crank [6] Eq. (3) works only for small $Dt/r^2$, which is $3.6 \cdot 10^{-4}$ for $t = 10^5$ hours, but for $t = 10^9$ hours is as high as 3.6. On the other hand, the analytical condition to keep $\alpha$ "moderate" is not so well-defined. In our case $\alpha = 0.00201$ for $K = 10$, $\alpha = 0.0201$ for $K = 1$ and $\alpha = 0.201$ for $K = 0.1$. Based on Fig. 1. this latter value proves to be no more "moderate".

Another very important constraint to obtain reasonable agreement between analytical results using Eq. (3) and the numerical results is to fulfill the condition of well-stirred solution of migrant in the pipe. We implement this constraint numerically by setting the diffusion coefficient of migrant inside the pipe to a value many orders of magnitude larger than its diffusion coefficient in water. Results presented on both figures are obtained with $D_{water} = 10^{-12} cm^2/s$ and $D_{pipe} = 10^{-7} cm^2/s$. Obviously, these values are completely unphysical. In reality diffusion coefficients governing migration of impurities in water are much larger then the corresponding values for diffusion in the PE pipes. The only reason we have yet chosen these unrealistic values was to fit in the constraints for the validity of the analytical formula, such that direct comparison can be performed and the numerical results can be tested.

On Fig. 2. we represent numerical results for the spatial profile of the migrant concentration for different $K$ at different times. One can remark the continuity of the concentration curve at the water/pipe interface when $K = 1$, as expected. In the case when $K = 0.1$ we have a finite jump in the migrant concentration at the interface. Regardless of the amount of time elapsed from the beginning of the migration the concentration-jump at the interface is conserved at the fixed value $c_{water}/c_{pipe} = K = 0.1$ validating the correctness of the numerical model.

The generalization of the numerical method to simulate migration in multilayer systems is straightforward, while we have seen the serious limitations of the analytical treatment.

## 4. Conclusions

In this paper we exposed the problem of migration of impurities from hollow pipes into their cylindrical inner core. The most important practical occurrence of this configuration is the case of PE water pipes and any kind of foodstuff in cylindrically shaped plastic packaging. Experimental measurements involving real systems are very expensive and time consuming. The availability of a suitable model, however, can overcome both of these difficulties. Here we presented such a model, tested and validated it by implementing a very simple case which can be treated analytically and we were able to directly compare the analytical and numerical results.

As a conclusion we can state that the presented numerical model is suitable to simulate migration of impurities through multilayer systems with cylindrical geometry. In comparison with the limited possibilities of analytical treatment this numerical method has several major advantages: (i) its validity range is not limited in time, nor in what concerns any other parameter of the real system (diffusion and partition coefficients, layer thickness, etc.) and (ii) it can be very easily generalized to arbitrary number of cylindrical layers which is the real situation in case of PE water pipes and polymer food packaging materials as well.

The next step of this research will be to simulate multilayer cases with realistic parameters, to fit them with experimental data, thus we hope to obtain good estimates of quantities which are very difficult to be measured experimentally.

## Acknowledgements

The present work was financially supported by research project PN II IDEI no. 372.

## References

[1] Brocca D, Arvin E and Mosbæk H 2002 *Water Research* **36** 3675
[2] Skjevrak I, Due A, Gjerstad K O and Herikstad H 2003 *Water Research* **37** 1912
[3] Tosa V, Kovacs K, Mercea P and Piringer O 2008 *AIP Conference Proceedings* **1048** 802
[4] European Commission 2003 *Food contact materials – Practical Guide* available at `http://ec.europa.eu/food/food/chemicalsafety/foodcontact/practical_guide_en.pdf`
[5] Denberg M, Arvin E and Hassager O 2007 *Journal of Water Supply: Research and Technology* **56** 235
[6] Crank J 1975 *The Mathematics of Diffusion* (New York: Oxford University Press).

# Identification and measurement of pesticide contaminants in food products by electron impact GC/MS

**Florina Tusa[1], Zaharie Moldovan[1] and Mircea Vlassa[2]**

[1] National Institute for Research and Development of Isotopic and Molecular Technologies, 65-103 Donath, 400293 Cluj-Napoca, Romania

[2] Faculty of Chemistry and Chemical Engineering, Babes Bolyai University, 11 Arany Janos, 400028 Cluj-Napoca, Romania

E-mail: zaharie.moldovan@itim-cj.ro

**Abstract.** The paper concern is determination of eight pesticides in food products samples. The target compounds are: Lindane, Heptachlor, Aldrin, *o,p*-DDE, Dieldrin, Endrin, *p,p'*-DDT, and Methoxychlor. The compounds quantities were performed from chromatographic area obtained in full scan GC/MS mode after baseline separation and by comparation with surrogate internal standard area (Diphenylamine). The samples were concentrated by extraction with organic solvents (acetone) by Solid-Liquid Extraction (SLE) procedures the recovery factors being better than 80% except for Heptachlors. The coefficient of correlation of detector response function was better than 0.913 and LOQ under 0.015 μg/g. The method enables to determine pesticides at low μg/g in food supplements.

## 1. Introduction

Pesticides comprise a large number of substances that belong to different chemical groups – the only common characteristic is that they are effective used against pests [1]. This class of compounds are widely used to combat diseases and pests, and may have main effect in the production of vegetable and animal foodstuff. By their large use residues of those compounds can reach the environmental compartments or human food [2, 3].

Most of the chemical compounds released into the environment via human activity are among the most dangerous chemicals. They are highly toxic, causing a number of adverse effects, notably death, diseases and birth defects between human and animals. Specific effect can include cancer, allergies and hypersensitivity, damage to the central and peripheral nervous systems, reproductive disorders, and disruption of the immune system [4, 5].

The determination of pesticide residues in agricultural products, plant and environmental samples has been major subject for many years because of their potential risk on human health, persistence and tendency to bio-accumulate [6, 7].

The analysis of this class of compounds is carried out by means of several steps as: extraction from matrix, concentration of target compounds and finally instrumental separation and quantitatively determination. The quality of every process usually contributes to quality of the analytical results. During the extraction and enrichment step, the trace solutes are isolated from the matrix, and their concentration is increased to enable their identification or quantification. In environmental and food safety fields a large variety of extraction and enrichment techniques were reported [8, 9, 10].

The present method concern is determination of target compounds: Lindane, Heptachlor, Aldrin, *o,p*-DDE, Dieldrin, Endrin, *p,p'*-DDT, and Methoxychlor in some food solid sample of commercial interest using Solid-Liquid Extraction procedures prior GC/MS analysis [11, 12, 13].

## 2. Materials and methods

### *2.1. Instruments*

Analysis was performed on a Thermo Finnigan Trace GC interfaced with a Polaris Q ion trap mass spectrometer, operated in electron impact ionization (EI) mode. Electron energy was set at 70 eV and ion source at 230ºC. Separations were carried out using a DB-5MS fused silica capillary column (30 m x 0.25 mm *i.d.*, 0.5 μm film thickness). Helium was employed as the carrier gas with flow rate of 1.5 °ml/min at constant flow with vacuum compensation. One micro litres of a standard solution or sample extract were injected in the splitless mode at 250°C using the following program: 150°C, then 5°C/min ramp to 300°C. Total run time was 30 min.

### *2.2. Extraction*

A quantity of 1 g of every sample of food supplement was weighed in a 10 ml centrifuge tube and than 5 ml of acetone was added. The tube was vigorously shaken for 3 min using a Vortex mixer and afterwards, centrifuged at 9.000 rpm for 5 min. The extracted samples were concentrated to dryness and than dissolved in isooctane. The final extract was adjusted to 0.5 ml in a dry flow of argon and than injected (2 μl of extract) in a GC/MS system. Diphenylamine was used as surrogate standard.

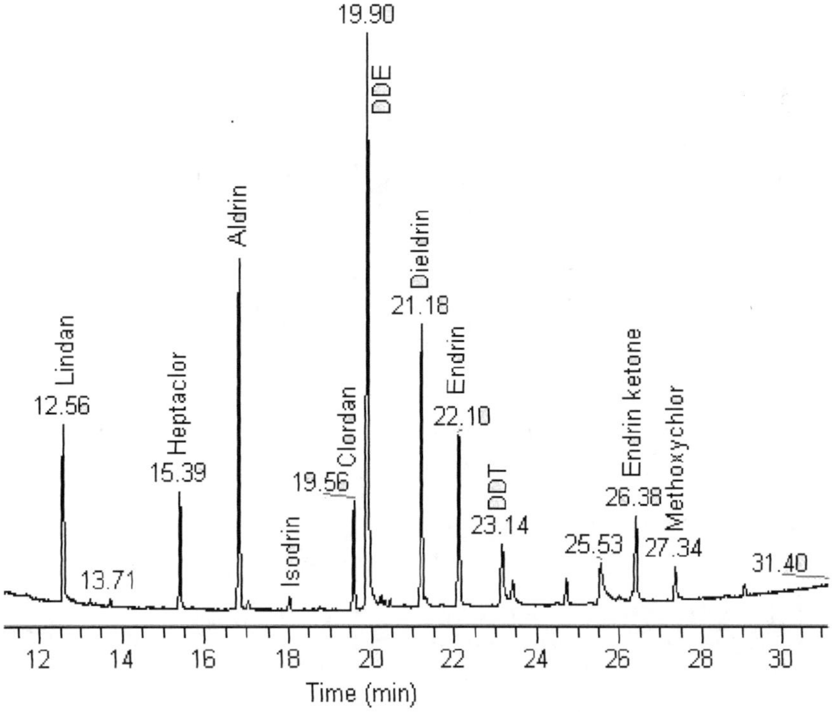


**Figure 1.** Separation of pesticides using GC/MS system

## 3. Results and discussion

The main objective of this study was to develop a rapid method based on the solid-liquid extraction and GC/MS system for analysis of common pesticides in food samples at low mg/kg. The lindane, heptachlor, *o,p*-DDE, *p,p'*-DDT, aldrin, dieldrin, endrin, metoxychlor were chosen because their appear frequently in the food products [14, 15].

Figure 1 shows the separation of the selected groups of pesticides by GC/MS system.

The recoveries were determined by using a food sample spiked with different amounts of analytes and the calculated quantities compared with the spiked concentrations. Every point was obtained as a mean of three different determination of the same concentration. Recovery factors values are grater than 86%, except for heptachlor for which the recovery factor is of 75%.

The relation between compound concentration and chromatographic peak area in all shows a linear equation having correlation coefficient ($R^2$) better than 0.913 in the range from 0.04 μg/g to 4 μg/g.

The limit of quantification (LOQ) was obtained for every compounds corresponding at an amount of a signal three times greater than noise. The obtained values are under 0.040 μg/g.

The experimental values obtained for retention time ($t_R$), correlation coefficient ($R^2$), recovery factors ($R_f$) and limit of quantification (LOQ) are presented in table 1.

**Table 1.** Experimental values for Retention time ($t_R$), Correlation coefficient ($R^2$), recovery factors ($R_f$), and limit of quantification (LOQ).

| Compound | $t_R$ (min.) | $R^2$ | $R_f$ (%) | LOQ μg/g) |
|---|---|---|---|---|
| Lindane | 12.56 | 0.9311 | 90.44 | 0.008 |
| Heptachlor | 15.39 | 0.9735 | 75.00 | 0.041 |
| Aldrin | 16.80 | 0.9134 | 86.38 | 0.015 |
| *o,p*-DDE | 19.90 | 0.9199 | 80.27 | 0.002 |
| Dieldrin | 21.18 | 0.9451 | 98.63 | 0.012 |
| Endrin | 22.10 | 0.9284 | 88.88 | 0.040 |
| *p,p'*-DDT | 23.14 | 0.9291 | 86.44 | 0.016 |
| Metoxychlor | 27.20 | 0.9516 | 84.33 | 0.015 |

The results obtained for few samples of food supplement are shown in table 2.

Three target compounds were determined above of LOQ in the samples of plant extract $S_1$ (raspberry), $S_2$ (maize), $S_3$ (cranberry), $S_4$ (rose) and $S_5$ (horsetail). The detected quantity was under 0.23 μg/g. The detected compounds are: Lindane, DDT and Metoxychlor.

**Table 2.** The average values concentration of target compounds determined in the samples $S_1$-$S_5$ μg/g.

| Compound | $S_1$ | $S_2$ | $S_3$ | $S_4$ | $S_5$ |
|---|---|---|---|---|---|
| Lindane | - | 0.230±0.025 | - | - | 0.076±0.009 |
| Heptachlor | - | - | - | - | - |
| Aldrin | - | - | - | - | - |
| *o,p*-DDE | - | - | - | - | - |
| Dieldrin | - | - | - | - | - |
| Endrin | - | - | - | - | - |
| *p,p'*-DDT | - | 0.142±0.018 | - | 0.110±0.021 | - |
| Metoxychlor | 0.031±0.002 | 0.052±0.007 | 0.080±0.013 | 0.011±0.003 | - |

## 4. Conclusion

Using suitable solvent in a solid-liquid extraction procedure of pesticides, the food samples can be investigated in a reasonable time (around of few hours) by GC/MS. In every investigated food sample was detected traces of chlorinated pesticides. The maximum concentration was for Lindane (0.23 μg/g in sample $S_2$).

## 5. References

[1] Tomlin C 2000 *The Pesticide Manual* (Surrey, UK: British Crop Protection Council) p 133
[2] Stenersen J 2004 *Chemical Pesticides Mode of Action and Toxicology* (Boca Raton, Fl: CRC Press) p 10
[3] Tadeo J L 2008 *Analysis of Pesticides in Food and Environmental Samples* (Boca Raton, Fl: CRC Press) p 2
[4] Cairns T and Sherma J 1992 *Emerging strategies for pesticide analysis* (Boca Raton, Fl: CRC Press) p 20
[5] Konstantinou I K, Hela D G and Albanis T A 2006 *Environ. Pollut.* **141** 555
[6] Ahmed F E 2001 *Trends Anal. Chem.* **20** 649
[7] Zambonin C G, Quinto, M, De Vietro N and Palmisano F 2004 *Food Chemistry* **86** 269
[8] Henríquez M, Becerra J, Barra R and Rojas J 2006 *J. Chil. Chem. Soc.* **51** 938
[9] Hernandez-Borges J, Cabrera J C, Rodriguez-Delgado M A, Hernandez-Suarez E M and Sauco V G 2009 *Food Chemistry* **113** 313
[10] Beyer A and Biziuk M 2008 *Food Chemistry* **108** 669
[11] Schreck E, Geret F, Gontier L and Treilhou M 2008 *Talanta* **77** 298
[12] Torres C M, Pico Y and Manes J 1996 *J. Chromatogr. A* **754** 301
[13] Gabo A B, Kurz M H S, Pizzutti I R, Adaime M B and Zanella R 2004 *J. Braz. Chem. Soc.* **15** 945
[14] Aysal P, Ambrus A, Lehotay S J and Cannavan A 2007 *J. Environm. Sci. and Health Part. B* **42** 481
[15] Lambropoulou D A and Albanis T A, 2007 *Anal. Bioanal. Chem.* **389** 1663.

# Non-destructive analysis in a study of the religious art objects

**Nicoleta Vornicu[1], Maria Geba[2] and Cristina Bibire[1]**

[1] Metropolitan Center of Research TABOR, The Metropolitanate of Moldavia and Bukovina, Closca 9, 700066 Iaşi, Romania

[2] Cultural Heritage Restoration – Conservation Centre "Moldova", National Museum Complex, Iasi, Romania

E-mail: cmctaboriasi@yahoo.com

**Abstract**. The icon *Descending of the Saint Spirit* from Bucium Church, dating in the year 1814 and was done in tempera on wood technology. The characterization of cultural heritage materials is essential for the comprehension of their degradation mechanisms. The present study aims at identifying the pigments in the various layers, establishing the possible existence of an organic binder and scientifically evaluating the state of preservation. To this end, were used non-destructive methods, as: microscopic (SEM), XRF and spectroscopic (FTIR).

## 1. Introduction

The characterization of cultural heritage materials is essential for the comprehension of their degradation mechanisms. X-Ray Fluorescence spectrometry (XRF) allows a rapid and simple determination of the elemental composition of a material. As a non destructive tool, it has been extensively used for analysis in art and archeology since the early 1970. Whereas it is commonly used for qualitative analysis, recent efforts have been made to develop quantitative analysis even with portable systems.

**Figure 1.** Icon *Descending of the Saint Spirit.*

However the interpretation of the results obtained with this technique can turn out to be problematic in the case of layered structures such as easel paintings. The use of differential X-ray attenuation allows modeling the various layers: indeed the absorption of X-rays through different layers will result in modification of intensity ratio between the different characteristic lines.

To characterize and identify the pigments present in religious painting (icons of wood support) of the XVIII$^{th}$ century the collection of Bucium Church, Iasi.

## 2. Experimental

The goal of this research was to know which pigments were applied in this artwork. For this study non-destructive techniques testing were used: microscopic (SEM), X-Ray Fluorescence and FT-Infrared Spectroscopy (FTIR) and chromatographic (HPLC). These analyses have been made with the help of a portable X-ray fluorescence spectrometer Innov X Alpha Series. Source excitement: X-ray tube, the anode Ag W, 10-40 kV, 10-50 μA, up to 5 filters. Detector: Si PIN diode, < 230 eV FWHM at 5.95 keV Mn Kα line and FTIR spectrometer Vertex 70 Bruker, 30-25 000 cm$^{-1}$.

The retable was analysed in 15 points.

The pigments applied were recognized on the bases of characteristic chemical elements from the XRF spectra. The comparison of the counts per second of different elements with regard to the background offers the possibility to obtain semi-quantitative results.

**Table 1.** Summary of the results – XRF system portable.

| Element | Line | Colour | Energy | Can be found in spectrum no. |
|---|---|---|---|---|
| Fe | $K_\alpha$ $K_\beta$ | Blue | 6.40; 7.06 | 5, 9, 12 |
| Hg | $K_\alpha$ $K_\beta$ | Cinnabar | 8.64; 9.57 | 14, 15 |
| Pb | $L_\alpha$ $L_\beta$ | White lead | 10.5; 12.6 | all |
| Ti | $K_\alpha$ $K_\beta$ | White titanium | 4.51;4.93 | 1-9 |
| Au | $L_\alpha$ $L_\beta$ | Gold | 9.71; 11.4 | 1, 2, 3 |
| Ag | $K_\alpha$ $K_\beta$ | Silver | 23; 25 | 6, 14 |
| Fe<br>Cr | $K_\alpha$ $K_\beta$<br>$K_\alpha$ $K_\beta$ | Yellow ochre | 6.40; 7.06 | 1, 2, 6, 7, 8 |

XRF spectrum of the blue pigment obtained by the same instrument from the same position indicates the existence of Fe, Ni, Zn (figure2).

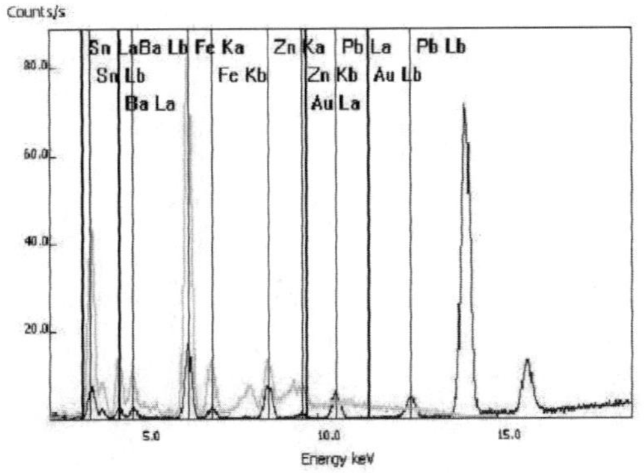


**Figure 2.** XRF spectra pigment blue.

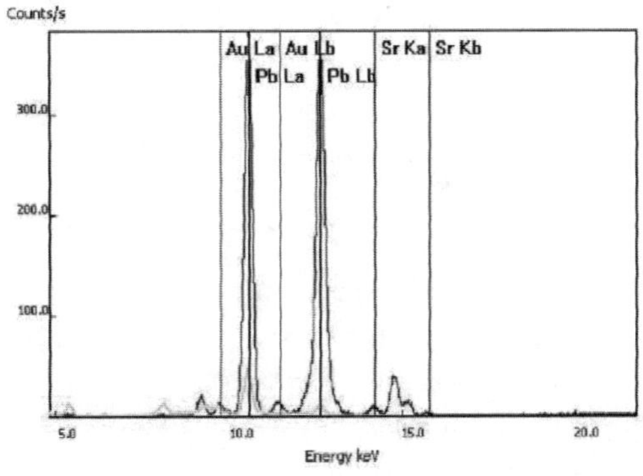


**Figure 3.** XRF gold spectra.

Journal of Physics: Conference Series **182** (2009) 012044 doi:10.1088/1742-6596/182/1/012044

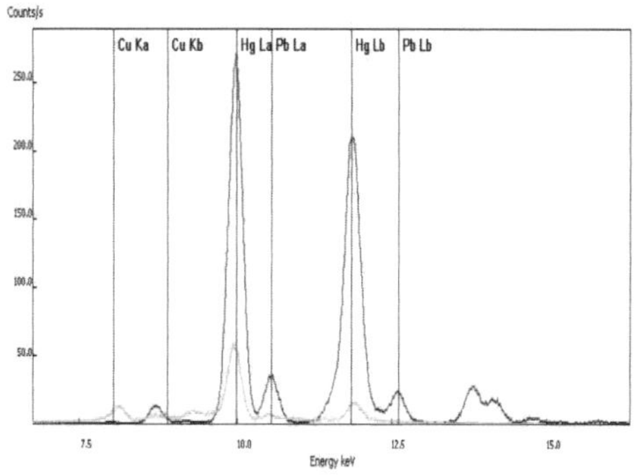


**Figure 4.** XRF spectra pigment cinnabar.

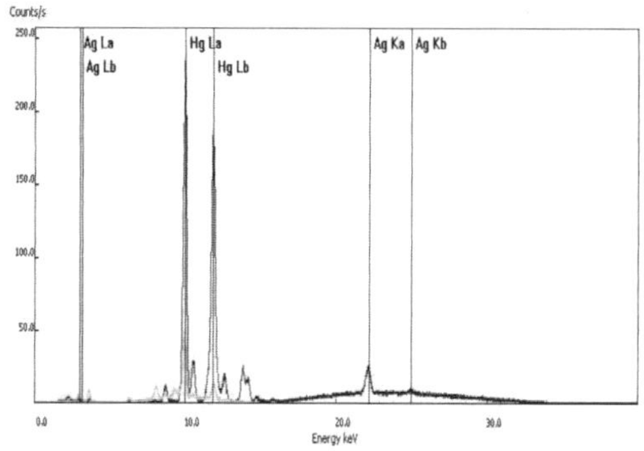


**Figure 5.** XRF spectra silver.

Pigments were subjected to spectral analysis in the field (400-4000 $cm^{-1}$), using the FTIR, 620 JASCO, Japan. Technique was applied in direct transmission process using KBr pellet technique.

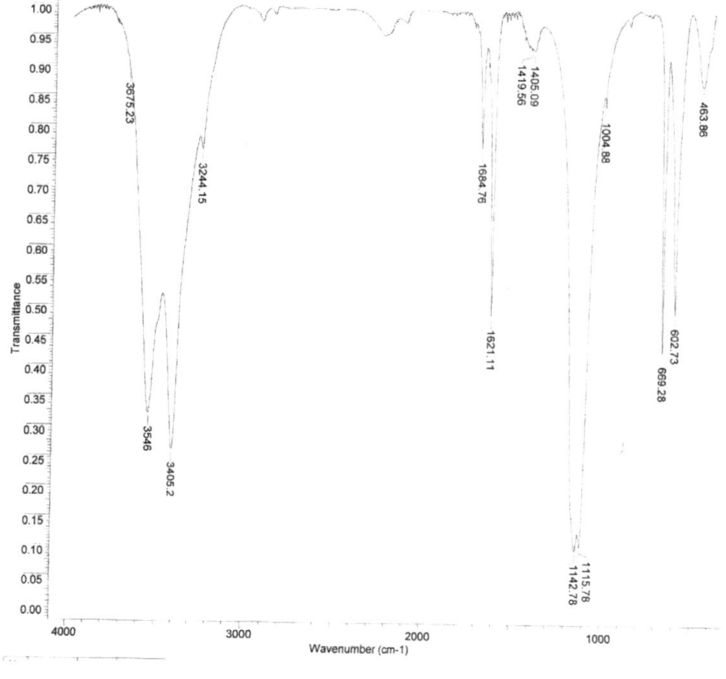


**Figure 6.** FT-IR spectra to identification of the pigment.

This comparison allowed the identification of protein peaks in all the icons study (N–H asymmetrical stretching between 3400-3200 $cm^{-1}$ and C–N stretching 1419-1004 $cm^{-1}$ range.

Using analysis SEM-EDX could highlight the state of degradation of pigments analyzed; for example in figure 7 is the SEM image of red pigment, cinnabar.

Performances of Tesla BS300 electron microscopy scan used were: 60 A resolution, depth of 1.2 mm, continuous increase in the 20-30000x.

The analysis of pigments by microscopy SEM and FTIR spectroscopy reveals morphological changes induced by endogenous and exogenous factors.

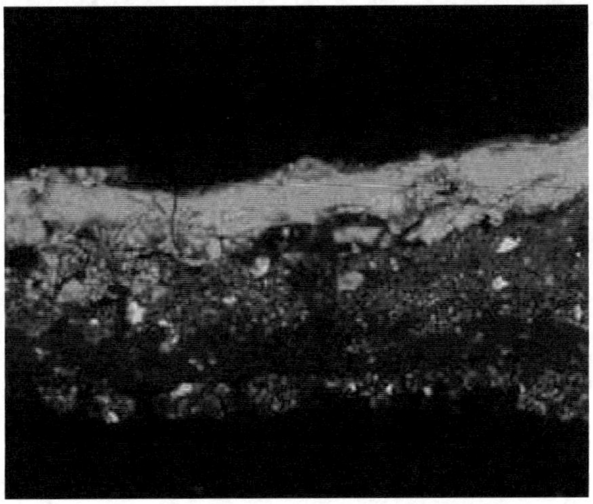

**Figure 7.** SEM image a cross-section of a micro-sample, pigment red – cinnabar.

## 3. Conclusions

The pigments used originally are all commonly applied in the XVIII$^{th}$ century: cinnabar, lead white, yellow ochre, gold, silver blue Prussian and black organic. The presence Pb in all spectra shows that lead white was used also to lighten other pigments, as well as applied in the painting preparation and sometimes as a dryer. Non-destructive analytical techniques have enabled the identification of natural pigments to use icon and allowed restorers use treatment the conservation-restoration adequate. Results have contributed to the completion of a given covering pigment, binders and primers used in XVII$^{th}$-XIX$^{th}$ centuries.

## References

[1] Derrick M 1999 *Infrared spectroscopy in conservation science* (The Getty Conservation Institut)
[2] Comelli D, D'Andrea C, Valentini G, Cubeddu R, Colombo C and Toniolo L 2004 *Appl. Opt.* **43** 2175
[3] Comelli D, Valentini G, Nevin A, Farina A, Toniolo L and Cubeddu R 2008 *Rev Sci. Instr.* **79** 086112
[4] Alberti R, Florini C, Guazzoni C, Klatka T and Longoni A 2007 *Nucl. Instrum. Meth.* **580** 1004
[5] Thoury M, Elias M, Frigerio JM and Barthou C 2007 *Appl. Spectr.* **61** 12.

Journal of Physics: Conference Series **182** (2009) 012045 doi:10.1088/1742-6596/182/1/012045

# Hydrogen storage in PrNiIn and related alloys

**O Ardelean[1], G Blanita[1], G Filoti[2] and P Palade[2]**

[1] National Institute for Research and Development of Isotopic and Molecular Technologies, 65-103 Donath, 400293 Cluj-Napoca, Romania

[2] National Institute for R&D of Materials Physics, 105 bis Atomistilor, PO Box MG-7, 077125 Bucuresti, Romania

E-mail: ovidiu.ardelean@itim-cj.ro

**Abstract.** The effect of partial substitution of Fe for Ni, Sn for In and total substitution of Nd for Pr on hydrogen absorption by PrNiIn type alloys are reported. In these hydrides pairs of H atoms are found at distances of 1.5-1.6 Å, violating the so called "2-Å rule" for metal hydrides. The storage capacity does not increase in the substituted alloys, suggesting that the pair formation is a local phenomenon, which can not lead to high storage capacity.

## 1. Introduction

For mobile applications to fuel cell powered vehicles, on-board storage materials with hydrogen absorption/desorption capacities of at least 6.5% H are needed. To reach the high volumetric and gravimetric density suitable for mobile applications, most of the research is focused on: high-pressure gas cylinders, liquid hydrogen, physisorbed on a solid surface (*e.g.* porous metal-organic frameworks, activated carbons), metal hydrides of the metals or intermetallics, complex hydrides of light elements (alanates, boranates) alone or mixed with amides, storage via chemical reactions and some other methods. Despite the research and development of thousands of storage alloys produced now at commercial scale, they are not able to store reversibly more than ~3% H. However, they are still interesting for niche applications (stationary or mobile).

Recently, the NdNiIn type compounds [1-5] attracted interest due to the very short H-H distances of 1.5-1.6 Å observed in their hydrides, violating the so called "2-Å rule" for metal hydrides; the model calculations showed that the interactions is strongly repulsive for such H-H pairs [2]. It might not be inferred straightforwardly that this short local H-H distance might really lead to a high increase in the overall storage capacity but these compounds still deserve attention as storage media. In this work, the characteristics of some $RENi_{1-x}M_xIn_{1-y}M_y$ (RE = Rare Earth) are reported and discussed comparatively with literature data [3].

## 2. Experimental

Compounds: $PrNi_{0.9}Fe_{0.1}In$, $PrNiIn_{0.8}Sn_{0.2}$, $NdNiIn_{0.8}Sn_{0.2}$ were prepared by arc melting under pure argon. p-c-T isotherms were measured by the volumetric Sieverts' method. The lattice parameters were calculated with the program Mc Maille and the Mössbauer spectra were obtained using the $^{57}Co$ (in transmission geometry).

## 3. Results and discussion

$PrNi_{0.9}Fe_{0.1}In$, $PrNiIn_{0.8}Sn_{0.2}$, $NdNiIn_{0.8}Sn_{0.2}$ have been synthesized to determine the effect of substitution on hydride characteristics and especially for the hydrogen atoms pair H–H forming, at unusually short distances.

### $PrNi_{0.9}Fe_{0.1}In$

The Fe substitution for Ni results in the same hexagonal structure of $PrNi_{0.9}Fe_{0.1}In$ (P-62m – No 189) with a=7.563 Å and c=4.012 Å, a little higher than for the PrNiIn.

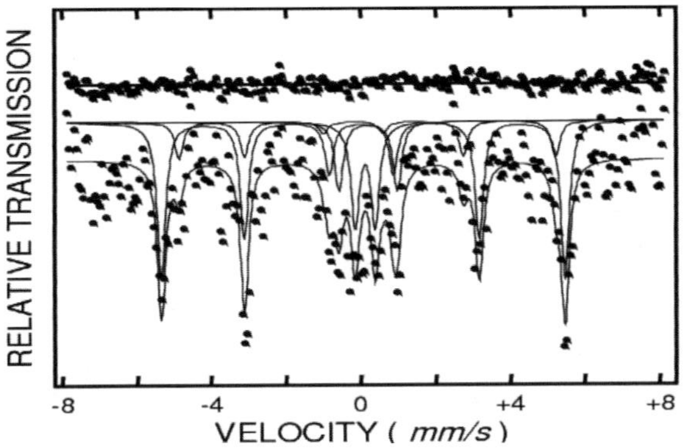


**Figure 1.** Mossbauer spectra of hydrided $PrNi_{0.9}Fe_{0.1}$ and its components at ambient temperature.

Mössbauer spectra, see figure 1, revealed two different sites for the Fe atoms, showing that part of the iron segregates as bcc α-Fe after hydriding. Two hydrides belonging to the same structure were identified: $PrNi_{0.9}Fe_{0.1}InH_{1.2}$ and $PrNi_{0.9}Fe_{0.1}InH_{0.6}$ with a = 7.3004 Å, c = 4.5627 Å and respectively a = 7.594 Å, c = 4.074 Å, in agreement with the highly anisotropic expansion characteristic to these compounds [2]. Desorption isotherms for PrNiIn and $PrNi_{0.9}Fe_{0.1}In$ reported in figure 2, exhibit three regions.

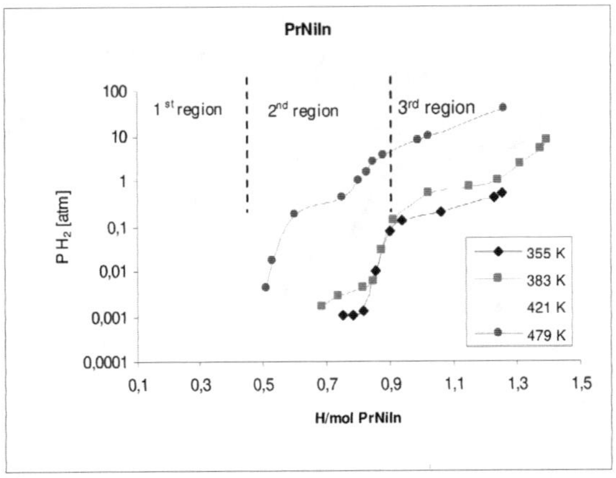


(a)

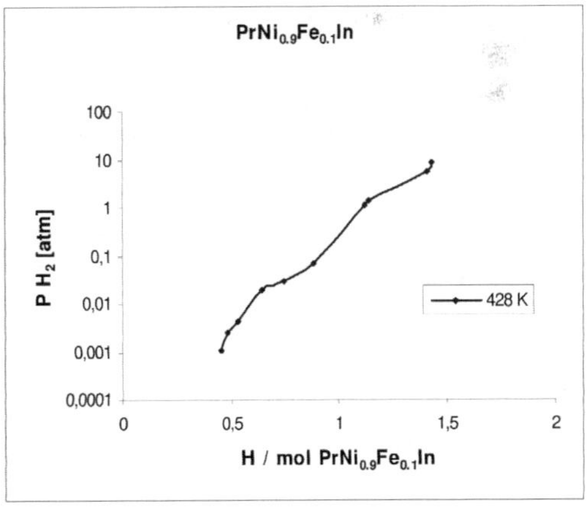


(b)

**Figure 2.** Desorption isotherms for a) PrNiIn – $H_2$ and b) $PrNi_{0.9}Fe_{0.1}In$ – $H_2$ systems.

Literature [4] shows that, at room temperature, 1.33 H/mol was achieved at ~1 bar $H_2$ but at higher pressures the composition can reach 1.63 H/mol [5]. The desoption isotherms is composed of 3 regions: 3rd region, hydrogen is desorbed, at high pressure, from octahedral sites $Pr_3NiIn_2$, which corespunds to a less stable hydride (high pressure: low ΔH); 2nd region, hydrogen is desorbed from bipiramidal sites $Nd_3Ni_2$, which corresponds to H-H pair decomposition; 1st region, at high temperature and low pressure, the rest of the hydrogen from the half-occupied bipiramids is desorbed.

For PrNiIn the second region gives ΔH= -34.2 kJ/molH and ΔS= -65.8 cal/K.molH for the **β↔γ** transformation region corresponding to the decomposition of the H-H pairs, in good agreement with the literature data for LaNiIn (-34.6 kJ/molH and -70.7 cal/K.molH). From the van't Hoff plots for $PrNi_{0.9}Fe_{0.1}In$, ΔH = -34.0 kJ/molH and ΔS = -64.4 cal/K.molH were obtained, showing that at this Fe concentration, the substitution does not influence significantly the H-H interaction.

**$PrNiIn_{0.8}Sn_{0.2}$**
After activation of the material, the total absorption capacity at room temperature corresponds to $PrNiIn_{0.8}Sn_{0.2}H_{1,33}$.

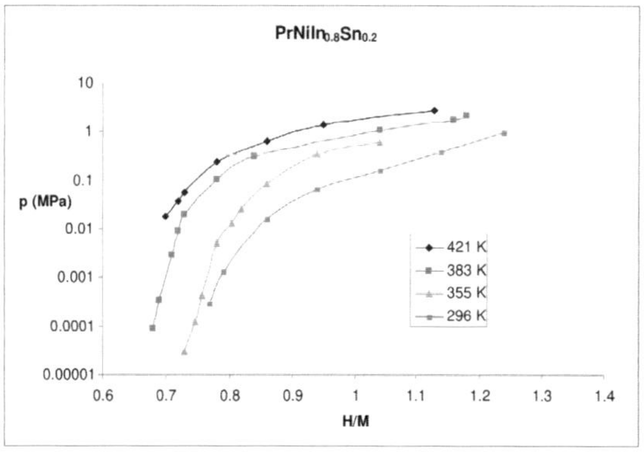


Figure 3. Desorption isotherms for $PrNiIn_{0.8}Sn_{0.2}$.

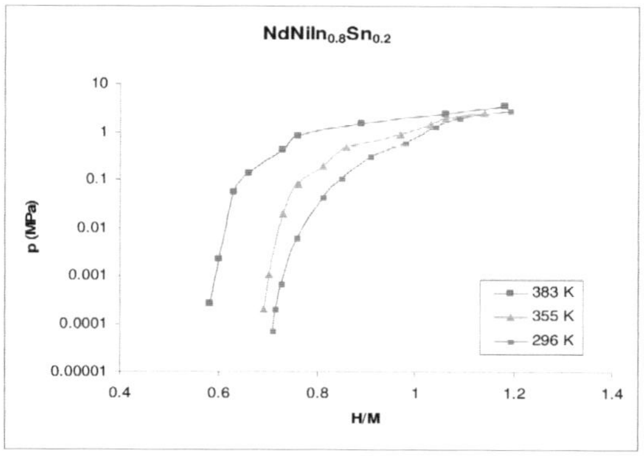


**Figure 4.** Desorption isotherms for $NdNiIn_{0.8}Sn_{0.2}$.

$PrNiIn_{0.8}Sn_{0.2}$ exhibits a single region in the range 0.67-1.2 H/mole. An evaluation of the desorption enthalpy, from the van't Hoff equation, give ~ 5 kcal/mol $H_2$, a less stable hydride, but due to the point scattering, this value is not very accurate.

The lattice parameters for $PrNi_{0.9}Fe_{0.1}InH_{1.2}$ show anisotropic expansion similar to those reported for $PrNiInH_{1.33}$ [2], characteristic to H-H pairs. However, the Sn substitution for In results in much lower ***Δc/c(%)*** than 12-16% probably due, either to an instability of the pairs as in $NdNiIn_{0,75}Al_{0,25}$ hydride (for Al/In>1/9, with H located in other sites) [3] or to formation of other type of hydrides like $LaNiSnD_2$ [1].

**Table 1.** Effects of hydrogen on lattice parameters.

| | a | c | c/a | *Δa/a (%)** | *Δc/c (%)** | *ΔV/V (%)** |
|---|---|---|---|---|---|---|
| PrNiIn | 7.5410 | 3.9500 | 0.5238 | - | - | - |
| $PrNiInH_{1.33}$ | 7.2600 | 4.5600 | 0.6281 | −3.73 | 15.40 | 7.01 |
| $PrNi_{0.9}Fe_{0.1}In$ | 7.5630 | 4.0120 | 0.5300 | - | - | - |
| $PrNi_{0.9}Fe_{0.1}InH_{1.2}$ | 7.3004 | 4.5627 | 0.6250 | -3.47 | 13.70 | 5.96 |
| $NdNiIn_{0,8}Sn_{0,2}$ | 7.4990 | 3.9480 | 0.5265 | - | - | - |
| $NdNiIn_{0,8}Sn_{0,2}H_x$ | 7.4320 | 4.2340 | 0.5700 | -0.90 | 7.20 | 5.34 |

* Variation of the parameter for hydrogenated alloy compared to the non-hydrogenated alloy

**$NdNiIn_{0.8}Sn_{0.2}$** was chosen to observe the influence of the lanthanide on the hydride stability. At 3 MPa, it has a $NdNiIn_{0.8}Sn_{0.2}H_{1.26}$ composition. The desorption isotherms, figure 4, has the same shape as $PrNiIn_{0.8}Sn_{0.2}$ suggesting that no H-H pair was formed.

Briefly, it can be said that with Sn ($r_0$ = 1.862 Å) substitution for In ($r_0$ = 1.841 Å), higher atomic radius should lead to a more stable hydride, but the plateau pressure of $PrNiIn_{0.8}Sn_{0.2}$ shows the opposite trend, suggesting different types of hydrogen sites accommodating H atoms.

Nd ($r_0$ = 1.821 Å) substitution for Pr ($r_0$ = 1.828 Å) results in a less stable hydride due to the compression of the lattice, as shown by the higher plateau pressure of $NdNiIn_{0.8}Sn_{0.2}$ compared to $PrNiIn_{0.8}Sn_{0.2}$.

## 4. Conclusions

The influence of the substitutions in compounds of NdNiIn type hydrides are in agreement with the lattice parameters variations. The enthalpies corresponding to the decomposition of the hydrides prove a repulsive interaction between H-H pairs, in agreement with literature [2-4] and experimental data for other compounds of this type [3]. The substitutions of Fe for Ni or Sn for In do not result in an increase of the H storage capacity: the formation of H-H pairs in this type of hydrides is a characteristic at atomic scale and does not result in materials with high hydrogen storage capacity, suitable for applications.

## Acknowledgements

The authors acknowledge the National Authority for Scientific Research for financial support for this work under program CEEX, project MENER 707/2006.

## References

[1] Yartys V A, Denys R V, Hauback B C, Fjellvåg H, Bulyk I I, Riabov A B and Kalychak Ya M 2002 *J. Alloys Compd.* **330-332** 132
[2] Vajeeston P, Ravindran P, Vidya R, Kjekshus A, Fjellvag H and Yartys V A 2003 *Phys. Rev.* B **67** 014101
[3] Riabov A B, Denys R V, Sato M, Delaplane R G and Yartys V A 2005 *J. Alloys Compd.* **404-406** 107
[4] Sato M, Denys R V, Riabov A B and Yartys V A 2005 *J. Alloys Comp.* **400** 184
[5] Denys R V, Riabov A B, Yartys V A, Hauback B C and Brinks H W 2003 *J Alloys Compd.* **356-357** 65.

# A photoelectrochemical cell for the study of the photosensitive materials used in solar-hydrogen energy

**P Ardelean[1], E Indrea[1], T D Silipas[1], C Ardelean[2], Gh Mihailescu[1], Ramona-Crina Suciu[1], Simina Virginia Dreve[1], Z Moldovan[1], Marcela Corina Rosu[1]**

[1] National Institute for Research and Development of Isotopic and Molecular Technologies, 65-103 Donath, 400293 Cluj-Napoca, Romania

[2] Technical University of Cluj-Napoca, Department of Automatization and Computers, 26-28 Gheorghe Baritiu, 400027 Cluj-Napoca, Romania

E-mail: petru.ardelean@itim-cj.ro

**Abstract.** It was built a versatile photoelectrochemical cell devoted to the comparative study of the photosensitive materials used as photoelectrodes in solar-hydrogen production. The experimental arrangement makes possible a relative evaluation of the electrodes properties by the measurement of the electric parameters, giving directly I = f (U) for the cell electric circuit with and without an external electrical bias. It also gives a direct measurement of the volume of the evolved gases, and an on-line analysis of the gases by the coupled gas chromatograph, or of-line, by a mass spectrometer.

## 1. Introduction

Considerations concerning the emission of greenhouse gases and their influence on the global warming, goes to a conclusion that it is necessary to develop technologies for the generation of new clean sources of energy [1, 2, 3] to supplement and ultimately to replace the fossil fuels. It must be taken into account also the fact that the fossil fuels have not a promising future: 40 years for the oil and about 62 years for the natural gases.

As it seems, hydrogen will replace fossil fuels as energy carrier [4, 5]. The production of hydrogen from the splitting of water, using solar energy, so called solar-hydrogen, is very convenient from both environmental and economic considerations [1, 6]. The solar-hydrogen technologies exhibits some important advantages: the process requires supplies of only solar energy and water, the associated infrastructure is simple and does not have any moving parts, maintenance being minimal, the necessary infrastructure is simple and the process does not cause pollution [1]. The difficulty consists in the development of a special class of new photosensitive materials for an efficient and clean conversion of solar energy.

$TiO_2$ (titania) is a promising candidate for the development of photoelectrodes for solar-hydrogen production due to its advantages [7, 8, 9]. It is less expensive than other photosensitive materials and titania reserves are abundant. Also $TiO_2$, unlike others materials, has outstanding resistance to corrosion and photocorrosion in aqueous solutions [1] and this could ensure that its functional properties remain unchanged over a very large period of time.

The main disadvantage of this material consists in the necessity of increasing its low value of energy conversion efficiency from ~ 0.5% obtained initially [7] to an acceptable value (around 10%). This low value is due to its low photosensitivity given by the relatively large band gap ~3eV [9] and consequently its poor absorption of visible light [1, 10]. So, its functional properties must be optimized in order to modify the band gap at a convenient value between 1.8÷2.2 eV [1].

In this direction intensive research, in many laboratories, is dedicated to increase its visible light sensitivity [9, 10].That is because it is considered that the performance electrodes are expected to be obtained by the modification of $TiO_2$ rather than on the development of new valence semiconductors [1]. Encouraging results [11, 12] were obtained, but also some skepticism [1] was reported. However, the incorporation of aliovalent ions [11, 13] results in a reduction of band gap. Also the properties of $TiO_2$ can be modified by varying the defect chemistry (and the related electronic structure) [1, 14].

## 2. Description

In our laboratory there is an interest for the development of the materials with desired properties in order to be used as higher performance photosensitive materials. Our work on titania is mainly aimed at the optimizations of its functional properties. In this purpose, it was built a photoelectrochemical cell provided with some facilities and included in an experimental arrangement in order to allow a comparative study of $TiO_2$ and $TiO_2$ based materials properties.

The experimental arrangement figure 1, include: the fotoelectrochemical cell F, a gas chromatograph GC, a variable DC stabilized power supply P, a digital voltmeter V and a digital miliampermeter M, and a light source S.

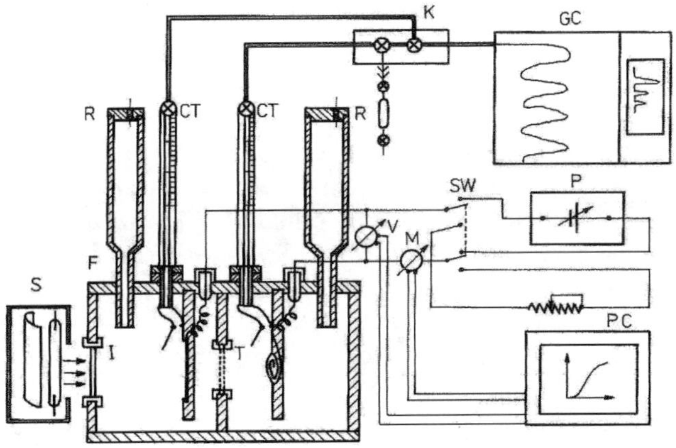


**Figure 1**. The experimental arrangement of the photoelectrochemical cell. F the photoelecrochemical cell; GC the gas chromatograph; P the power supply; K the three way valve system; S the light source; I the quartz window; T the Nafion proton transfer membrane; R electrolyte reservoirs; CT calibrated tubes.

**Figure 2**. The photoelectrochemical cell.

The parts of the photoelectrochemical cell (figures 1 and 2) are given in the following part. The frame, is a two-compartment Plexiglas reactor provided with a quartz window I, and having a Nafion proton transfer membrane T on the separation wall. The separation of the cell in two compartments allows to use of two different electrolytes in the performed studies and this make possible to provide a small chemical bias assisting the studied processes. The cover of the frame is provided with two calibrated tubes with stopcocks CT on the lines of the collected gases, which allows the measurement of the volume of the produced gases. The cover is also provided with two electrolyte reservoirs R to supply the content of the two compartments cell as the electrolyte is consumed in the cleavage process (and to maintain the electrolyte levels in the calibrated tubes), and with two electric feedthroughs connected to the photoanode and respectively to the photocathode which allows to connect the electrodes (the anode and the cathode) to the external electrical circuit (figure 1). The photoanode and the photocathode could be easily fixed and taken down from the supports, and the supports positions could be adjusted by screwdrivers. The measurement of the electrical parameters of the circuit could give important data concerning the electrode material properties. As it can be observed from figure 1, there are two ways to perform the measurements of the electrical parameters. It is known that, without a bias supply, the efficiency of the cell is very low. So we can use in performing the measurements an external variable DC electrical supply (position „up" of the double electric switcher SW) and we can use only one electrolyte. We can use also two different electrolytes in the two compartments (NaOH at anode and $H_2SO_4$ at cathode) and we have an „internal" chemical bias instead an external one (position „down"). The measurement of the electrical parameters of the circuit could be visualized on PC as curves I = f(U), (I being the measured current intensity in mA and U the value of the voltage in V), or as tables, and they are useful for appreciating the parameters of electrodes materials. The exit of the two calibrated tubes for gas collection is connected by a three way valve system K to the gas chromatograph or, to a sampling ampoule. In this way the gas could be analyzed directly by the chromatograph or could be sampled and analyzed off-line by a mass spectrometer.

In order to homogenize the electrolyte each compartment of the cell is provided with a magnetic core glass stirrer. The radiation is generated by a homemade light source S which is provided with a mercury-arc lamp and it can give a beam intensity which can be regulated (by distance modification) between 75 mW/cm$^2$ and 1 mW/cm$^2$.

## 3. Experimental results and discussions

In figure 3 it is given the current-voltage characteristics obtained with an external electrical supply. The anode is an ITO ($In_2O_3$:$SnO_2$) substrate with deposited nanocristaline $TiO_2$ based electrode, synthesized via wet-chemical techniques, by screen-printing, and the cathode was a Pt plane spiral wire. In both compartments electrolyte with pH = 13.5 was used. The curves *a* and *b* are obtained with and respectively without light beam. As it can be observed, the influence of the light beam becomes more important as U increases. So, the difference between the I values with and respectively without light beam, starts from 5% at 4 V and it increases at 20% at 8 V remaining afterwards at a value over 10%. After the system reaches the equilibrium (~ 40 minutes) the correspondence between the intensity of the current and the production of hydrogen is in close relationship. For example at U = 8.5 V for an current intensity of 38.9mA, which correspond to 2.43 x 10$^{17}$ electrons/s, will be expected a number of 1.216 x 10$^{17}$ $H_2$ molecules/s (0.272 cm$^3$/min). As the measured hydrogen production rate was 0.24 cm$^3$/minute, it results that about 88% of hydrogen molecules are detected. The small discrepancy is due probably to some gas leakage. A chromatogram of the gas obtained from the cathode compartment is given in figure 4. Argon was used as carrier gas. It is observed that $O_2$ and $N_2$ are also present, from residual air, and theirs concentrations go lower in time. In the mass spectrum of a sample taken from the evolved gas in the cathode compartment, an isotope effect was observed. As it is known, in the water splitting, an isotope effect takes place. So, the $H_2O$ molecules are easier split than HDO. In this case the value of HD/$H_2$ observed in mass spectra is ~70 ppm, HDO/$H_2O$ being as it knows ~150 ppm in the natural water, the isotope effect being more pronounced that it was expected.

In the case of working with internal chemical bias (position down of the electric switcher S), the observed open circuit voltage was 0.7 V and the short circuit current was 11 mA/cm$^2$. The active area of the electrode was 6.25 $cm^2$. The concentration of the electrolytes was 1 mol/l NaOH in the anode compartment and 1 mol/l $H_2SO_4$ in the cathode compartment. The measurements were performed after the system reaches equilibrium (~ 40 min). The work is in progress.

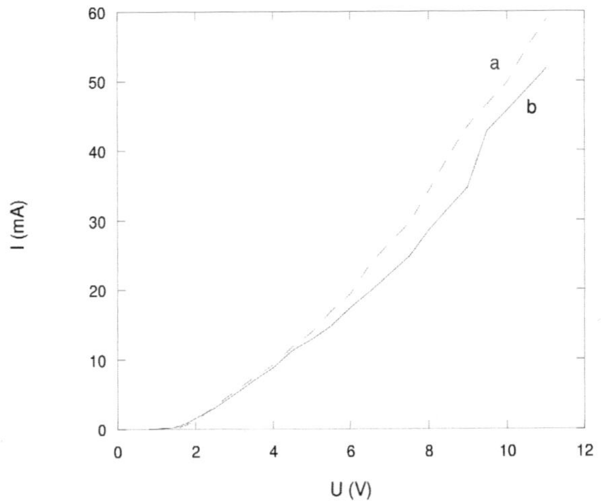


**Figure 3.** The current-voltage characteristics with external electrical supply: a with the light beam, b without the light beam.

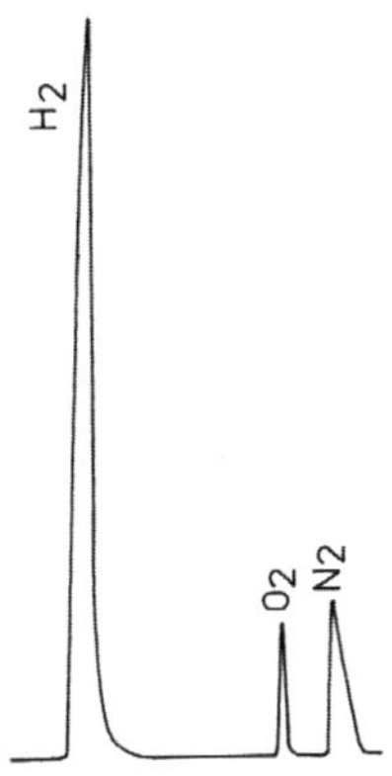


**Figure 4**. The chromatogram of the gas obtained from the cathode compartment

## 4. Conclusions

A two-compartment photoelectrochemical cell was constructed and tested with a nanocristaline titania as anode and a plane spiral Pt wire cathode, devoted to the comparative study of the titania based materials necessary as photoelectrodes. The electrodes supports, permits a facile replacement of the electrodes and position adjustment. The experimental arrangement allows a comparative study of the electric properties of titania based materials under chemical bias or with an external electrical bias as well as a measurement of the volume of the evolved gas and its analysis.

## References

[1] Nowotny J, Bak T, Novotny M K and Sheppard L R 2007 *Int. J. Hydrogen Energy* **32** 2609-29
[2] Friedli H, Lotscher H, Oescher H and Siegenthaler U 1996 *Nature* **324** 237-9
[3] Russ M 1994 *Series of Unwelttechnik* (Aachen: Shaker) pp 1-180
[4] Nowotny J and Sheppard L R 2007 *Int. J. Hydrogen Energy* **32** 2607-8
[5] Veziroglu T N 1998 *Int. J. Hydrogen Energy* **23** 1077-8
[6] Veziroglu T N 2008 *Int. J. Hydrogen Energy* **33** 1-2
[7] Fujishima A, Kohayakawa K and Honda K 1975 *J. Electrochem. Soc.* **122** 1487-9
[8] Nozic A J 1975 *Nature* **257** 383-6
[9] Nowotny J, Sorrell C C, Sheppard L R and Bak T 2005 *Int. J. Hydrogen Energy* **30** 521-44
[10] Bak T, Nowotny J, Rekas M and Sorrel C C 2002 *J. Hydrogen Energy* **27** 991-022
[11] Khan S U M, Al-Shahry M and Ingler W B 2002 *Science* **297** 2243-4
[12] Nazeeruddin M K and Gratzel M 2001 *J. Photochem. Photobiol.* A: *Chem.* **145** 79-86
[13] Nakamura R, Tanaka T and Nakato Y 2004 *J. Chem. Phys.* B **108** 10617-20
[14] Bak T, Nowotny J and Sorrell C C 2003 *J. Phys. Chem. Solids* **64** 1069-87.

# The effect of solution/free volume ratio on the MOF-5 characteristics

**G Blanita[1], D Lupu[1], M Lazar[1], A R Biris[1], V Pascalau[2], O Ardelean[1], I Coldea[1], I Misan[1], G Popeneciu[1] and M Vlassa[2]**

[1] National Institute for Research and Development of Isotopic and Molecular Technologies, 65-103 Donath, 400293 Cluj-Napoca, Romania

[2] Babes-Bolyai University, Faculty of Chemistry and Chemical Engineering, 11 Arany Janos, 400028 Cluj-Napoca, Romania

E-mail: gabriela.blanita@itim-cj.ro

**Abstract**. The influence of solution/free volume ratio on the specific surface area and pore volume of MOF-5 was investigated. MOF-5 was synthesized, by solvothermal reaction, using an identical starting mixture, at different solution/free volume ratio of 0.245, 0.807 and 3.429. The specific surface area decreased from 3018 $m^2/g$ to 2680 $m^2/g$ and 1099 $m^2/g$ respectively. The nitrogen adsorption capacity decrease in the same manner from 817.35 $cm^3/g$ to 740.75 $cm^3/g$ and 294.56 $cm^3/g$ respectively.

## 1. Introduction

Metal-organic frameworks are a well known class of porous materials due to their extremely high surface area and pore volume. Among them, MOF-5, $Zn_4O$(1,4-benzenedicarboxylate)$_3$, originally described in 1999, is the most studied metal-organic framework and was intensively investigated due to its high surface area as well as the cheap and raw materials [1]. In 2003, the first investigation of hydrogen storage in MOFs was reported for MOF-5 [2]. Since then, different synthetic and activation procedures were developed with the aim of obtaining maximum hydrogen storage capacity [3-6]. The specific surface area and pore volume are two factors that govern the hydrogen storage in MOF materials.

This work presents the influence of solution/free volume ratio on the specific surface area and pore volume of MOF-5 synthesized by solvothermal method.

## 2. Experimental

### *2.1. General*

The studied materials were treated under argon atmosphere, using a vacuum line. The $Zn(NO_3)_2 \cdot 6H_2O$ was purchased from Acros, the terephtalic acid (98%) was purchased from Aldrich. CH2Cl2 and dimethylformamide (DMF) were purchased from Merck.

The prepared MOF-5 samples were characterized by X-ray powder diffraction (XRD), TG analysis, IR spectroscopy and N2 adsorption/desorption. The XRD patterns of the dried samples were recorded on Bruker D8 Advance powder X-ray diffraction apparatus using Cu Kα radiation.

Nitrogen adsorption/desorption isotherms were obtained on a Sorptomatic sorptometer at liquid nitrogen temperature. The samples were degassed at 493°C for 3 h prior to adsorption. The specific surface areas of the samples were calculated by the Brunauer-Emmett-Teller method in the $p/p_0$ range of 0.02-0.1. IR spectra were recorded as KBr pellets on a Jasco FT/IR-610 spectrophotometer. TG analysis was carried out in argon with a heating rate of 10°C/min using a thermogravimetric analyzer SDT Q600.

### *2.2. Synthesis*

MOF-5 materials were synthesized by modified method given in literature, at different solution/free volume ratio of 0.245, 0.807 and 3.429 [6]. In a typical synthesis, 1.8 g $Zn(NO_3)_2{\cdot}6H_2O$ and 0.34 g 1,4-benzenedicarboxylic acid were dissolved in 196 ml DMF and 4 ml $H_2O$. Then the reaction mixture was loaded into a 400 ml autoclave, sealed and placed in an oven. The autoclave was heated at 100°C for 7 h. After the completion of the reaction, the mixture was allowed to cool down to room temperature and transferred in an argon-filled flask. All subsequent manipulations were preformed under argon atmosphere. The solvent was siphoned off and the remaining solid washed six times with anhydrous DMF, each time letting the solid soak in DMF for 12 h. Then the DMF was siphoned off and the solid washed with anhydrous $CH_2Cl_2$ using the same procedure as with DMF. After the final $CH_2Cl_2$ wash, the solvent was siphoned off and the included $CH_2Cl_2$ was removed under vacuum ($10^{-2}$ mbar, 6 h) to give $Zn_4O$(1,4-benzenedicarboxylate)$_3$.

## 3. Results and discussion

$Zn_4O$(1,4-benzenedicarboxylate)$_3$ materials were synthesized, by solvothermal reaction, at 100°C for 7 h, using an identical starting mixture, at different solution/free volume ratio of 0.245 (MOF-5b), 0.807 (MOF-5a) and 3.429 (MOF-5c).

**Table 1.** Specific surface area and pore specific volume of MOF-5a, MOF-5b and MOF-5c based on nitrogen adsorption/desorption isotherms.

| Compound | $S_{BET}$ ($m^2$/g) | Pore specific volume ($cm^3$/g) |
|---|---|---|
| MOF-5a | 2680 | 1.14 |
| MOF-5b | 3018 | 1.36 |
| MOF-5c | 1099 | 0.42 |

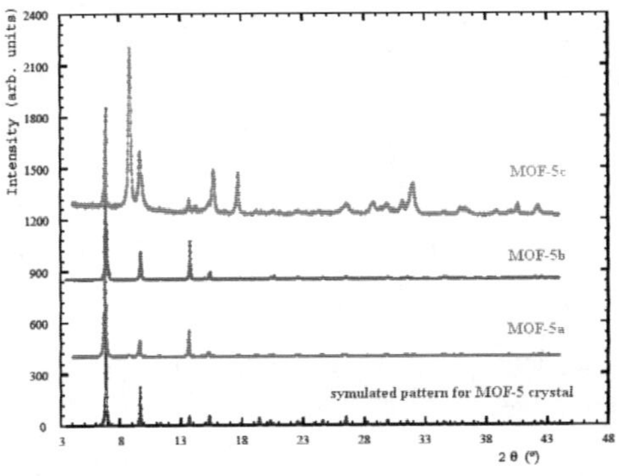


**Figure 1.** XRD patterns of MOF-5a, MOF-5b, MOF-5c and simulated XRD pattern for MOF-5 crystal.

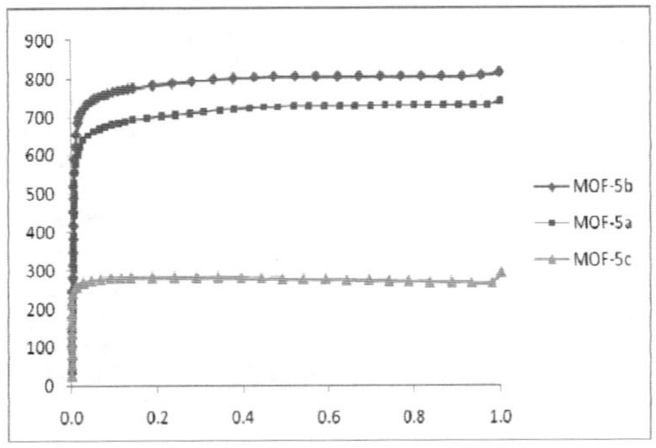


**Figure 2.** Nitrogen gas sorption isotherm at 77K for MOF-5a, MOF-5b and MOF-5c.

XRD patterns of MOF-5a, MOF-5b and MOF-5c are shown in figure 1. Simulated XRD pattern for MOF-5 are also included [7]. Clearly MOF-5a and MOF-5b have the same XRD patterns as MOF-5. The powder XRD pattern of MOF-5c deviates significantly from the simulated pattern for MOF-5 crystal.

MOF-5a and MOF-5b are stable until 400°C, as TGA curves revealed, which are consistent with the reported value for MOF-5 [1]. The thermal decomposition of MOF-5c starts at 300°C.

The IR spectra of MOF-5a, MOF-5b and MOF-5c show the expected strong characteristic bands [8] for the $\nu_{sym}$(CO) (1391 cm$^{-1}$, 1389 cm$^{-1}$ ,1389 cm$^{-1}$, respectively) and $\upsilon_{as}$(CO) (1573 cm$^{-1}$, 1579 cm$^{-1}$, 1579 cm$^{-1}$, respectively), adsorbed water (3400-3200 cm$^{-1}$) and $\nu_{sym}$(Zn4O) (533 cm$^{-1}$, 524 cm$^{-1}$ and 523 cm$^{-1}$, respectively) [9].

Nitrogen adsorption/desorption isotherms were measured by the standard volumetric method at 77K. The corresponding data for surface area and specific pore volume are shown in Table 1. MOF-5b synthesized at solution/free volume ratio of 0.245, has the highest specific surface area and pore volume. The surface area decrease from 3018 m$^2$/ g (MOF-5b) to 1099 m$^2$/ g (MOF-5c) with the increasing solution/free volume ratio from 0.245 (MOF-5b) to 3.42 (MOF-5c).

It is thus clear that the solution/free volume ratio influences gas adsorptive capacity of MOF-5. As shown in figure 2, MOF-5b adsorbs 817.35 cm$^3$ $N_2$/g, followed by MOF-5a with 740.75 cm$^3$ $N_2$/g and by MOF-5c with 294.56 cm$^3$ $N_2$/g.

All the data obtained from XRD pattern, TGA curve, IR spectra and $N_2$ adsorption isotherm reveal that MOF-5c is a "MOF-5 like" material.

## Conclusions

This study demonstrates that solution/free volume ratio in solvothermal synthesis affects the structure, surface area, pore specific volume and gas adsorption capacity of MOF-5.

## Acknowledgements

The authors thank the National Authority for Scientific Research (ANCS) for granting the project PN 09-440104.

## References

[1] Li H, Eddaoudi M, O'Keeffe M and Yaghi O M 1999 *Nature* **402** 276
[2] Rosi N L, Eckert J, Eddaoudi M, Vodak D T, Kim J, O'Keeffe M and Yaghi O M 2003 *Science* **300** 1127
[3] Panella B, Hirscher M, Putter H and Muller U 2006 *Adv. Funct. Mater.* **16** 520
[4] Rowsell J L C, Millward A R, Park K S and Yaghi O M 2004 *J. Am. Chem. Soc.* **126** 5666
[5] Wong-Foy A G, Matzger A J and Yaghi O M 2006 *J. Am. Chem. Soc.* **128** 3494
[6] Kaye S S, Dailly A, Yaghi O M and Long J R 2007 *J. Am. Chem. Soc.* **129** 14176
[7] Eddaudi M, Kim J, Rosi N, Vodak D, Wachter J, O'Keeffe M and Yaghi O M 2002 *Science* **295** 469
[8] Reineke T M, Eddaoudi M, fehr M, Kelly D and Yaghi O M 1999 *J. Am. Chem. Soc.* **121** 1651
[9] Berkesi O, Andor J A, Jayasooriya U A and Cannon R D 1992 *Spectrochim. Acta* A **48** 147.

# Gold nanoparticles bind to porphyrins. A potential dye-sensitized solar cell

**Sorina Garabagiu and Gheorghe Mihailescu**

National Institute for Research and Development of Isotopic and Molecular Technologies, 65-103 Donath, 400293 Cluj-Napoca, Romania

E-mail: sorina.garabagiu@itim-cj.ro

**Abstract**. Porphyrins could be used for designing dye sensitized solar cells. The correlation between their band gap HOMO-LUMO and the conduction band of the semiconductor ($TiO_2$) allows the electron transfer from the dye to the electrode, through the semiconductor. Our work reported here has revealed that a strong and large absorption band is obtained when a complex between gold nanoparticles (GNP) and porphyrin dye is formed.

## 1. Introduction

As a renewable alternative energy source, solar energy is a potential choice for solving the shortage of fossil energy. In this regard, solar cells could convert solar energy into electric power. Currently, the most popular methods are silicon based solar cells.

The production of electrical power from sunlight in organic solar cells involves the following: firstly, sunlight photons are absorbed within a photoactive layer leading to the formation of excitons. The excitons migrate to the interface of a donor-acceptor semiconductor hetero-junction and dissociate to form free charges, consisting of electrons and holes, and these charges are moved to the electrodes to eventually yield a photocurrent in an external circuit.

Dye-sensitized solar cells have many advantages over the conventional silicon solar cells [1]. These cells work on principles similar to the processes in natural photosynthesis. They use an organic dye to absorb the incoming light and produce excited electrons which will generate electrical current. The absorption band of the dye has to be as large as possible, from 350 nm to 900 nm, in the visible and near-infrared region of the solar spectrum. $TiO_2$-Ru dye sensitized based on solar cells have limited availability and undesirable environmental impact, due to Ru atoms. These facts led to search for cheaper and safer organic-based dyes. Given their efficacy in photosynthesis, porphyrin dyes have great potential in this regard [2-4].

In this work, we investigate the coupling between gold nanoparticles and porphyrins, and these new materials will be helpful to the design of dye sensitized solar cells (DSSE).

## 2. Experimental

### *2.1. Materials and methods*

Hydrogen tetrachloroaurate ($HAuCl_4{\cdot}3H_2O$) was purchased from the Institute of Chemistry in Cluj-Napoca, whereas sodium citrate, dichloromethane and tetrabutylammonium perchlorate from Sigma Chemie GmbH Germany. The porphyrin dye ZnP:Zn(II)-tetrapyridylporphyrin was synthesized at the Institute of Chemistry, in Timisoara. OEP: 2,3,7,8,12,13,17,18-octaethyl-porphyrin was purchased from Aldrich. Double distilled water was used in all experiments.

UV-Vis and fluorescence spectroscopic studies were carried out using a Jasco 550 and Jasco FP-6500. Gold nanoparticles were prepared using Turkevich method [5]. TEM images (using a JEOL JEM 1010 Transmission Microscope) demonstrated that we have obtained 30 nm gold nanoparticles.

*2.2. Discussion and results*

After dissolving the porphyrin molecules in dichloromethane, we acquired the absorption spectra of these dyes. The porphyrin dyes have an intense and narrow absorption band at 420 nm (Soret band) and a large, weak band between 500-700 nm (Q band), as seen in figure 1.

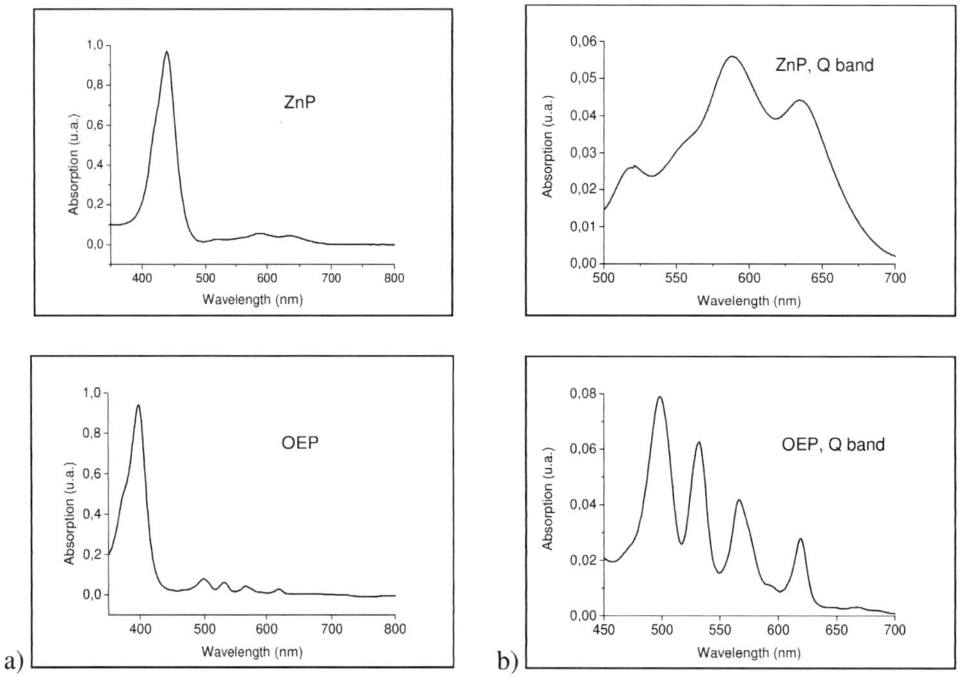


**Figure 1.** UV-Vis spectra of porphyrins: a) the entire spectra, b) Q band of porphyrins.

Analising the Soret band, with the λ-onset method, we determinate the energy gap between HOMO and LUMO ($\Delta E_g$). Using cyclic voltammetry we obtained the electrochemical spectra of the porphyrins (figure 2), solved in dichloromethane.

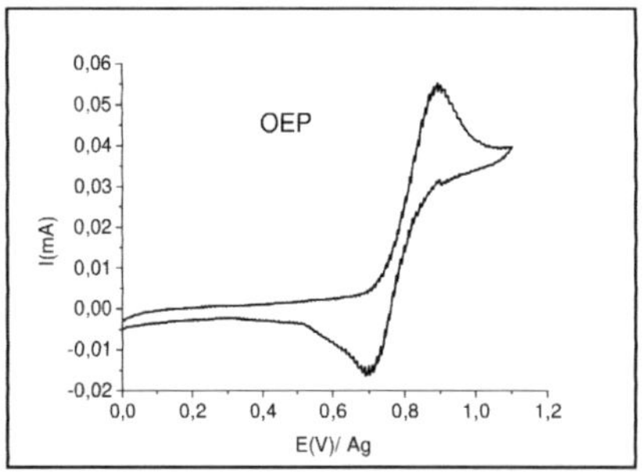


**Figure 2.** Cyclic voltammogram of OEP.

An Ag wire was used as reference electrode. From the cyclic voltammogram, we calculated the redox potential $E_0$. A conversion was made to NHE. With $\Delta E_g$ and $E_0$ known, it is possible to determine the energy levels HOMO and LUMO, and to compare them with the levels of the semiconductor ($TiO_2$). The results are shown in table 1.

**Table 1.** $E_0$, HOMO, LUMO for porphyrins, $TiO_2$ and electrolyte, with NHE reference.

| Porphyrin | λ onset (nm) | $\Delta E_g$ (eV) | $E_0$ (eV) | HOMO (eV) | LUMO (eV) (conduction band) |
|---|---|---|---|---|---|
| ZnP | 469 | 2.64 | - | - | - |
| OEP | 423 | 2.93 | 0.972 | 0.972 | -1.958 |
| $TiO_2$ | - | 3.20 | - | 3.000 | -0.200 |
| Electrolyte | - | - | 0.300 | - | - |

To design dye-sensitized solar cells, it is necessary that the dye absorb the visible light and then transfer the photoelectron from his LUMO state to the semiconductor. This orbital must have a higher energy than the conduction band of the semiconductor, so the electron could pass to the semiconductor and then from the electrode, in the external circuit. Also, HOMO energy level of the dye must be lower than the redox potential of the electrolyte.

As can be seen in table 1 (with NHE reference), LUMO energy state of OEP has a higher value than the conduction band of $TiO_2$, which means that the electronic transfer between dye and semiconductor could happen. Also, the value of redox potential of the electrolyte ($E_0$) is higher than the corresponding values of the dyes and this facilitate the electron transfer to the oxidized dye.

One way to encompass a broader absorption band is the attachment of functional groups on the basis of porphyrins. However this action involves organic synthesis. Gold colloids bearing surface negative charge readily bind the functional groups. This negative charge must be compensated by adding an acidic solution. These conditions facilitate functionalization and linking of GNP to nitrogen atoms in the porphyrinic structure.

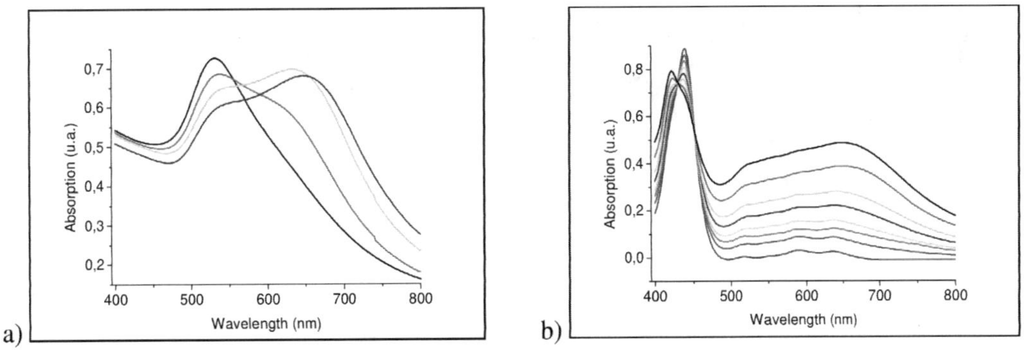


**Figure 3.** UV-Vis absorption spectra of: a) $GNP+ZnP_{drops}$; b) $ZnP+GNP_{drops}$.

The influence of pH on the reaction between GNP and porphyrins was investigated. There is no shift in the maximum peak position in the UV-Vis absorption spectra of gold nanoparticles until the pH value is 2.5. Our experiments were made at pH = 2.52 and the porphyrin we used was ZnP. In figure 3 is presented the UV-Vis spectra for GNP and the sequential addition of ZnP, with modified pH. One can see that the plasmon surface peak of GNP in figure 3a is attenuated and a signal for the newly formed complexes appear between 500-700 nm. When we mixed a fixed quantity of porphyrin solution with GNP, by adding small drops of the GNP solution, the spectra in figure 3b demonstrate that GNP bind to the porphyrinic ring, as a shift appears in the maximum peak position.

This could be due to the existence in the porphyrinic ring of GNP. Metal atoms in the porphyrinic ring are decisive for the lifetime of the electronic excited states, which influence the electron transfer between the dye and the conduction band of $TiO_2$. Zn porphyrins are more efficient than Cu porphyrins for solar cells because they have a longer lifetime of the electronic excited states [2].

## 3. Conclusion

In this work we demonstrated that porphyrins could be used for $TiO_2$ sensitization. An electronic transfer from the porphyrinic dye to $TiO_2$ is possible due to their HOMO-LUMO energy states. We also proved that some porphyrins capture in the porphyrinic ring gold nanoparticles, and a strong and large absorption band is obtained when a complex between gold nanoparticles and porphyrin dye is formed. This suggests that these structures (porphyrin + gold nanoparticles + $TiO_2$) could be used as new materials for photovoltaic cells.

## References

[1] Gratzel M 2001 *Nature* **414** 338
[2] Campbell W M, Burel A K, Officer D and Jolley K W 2004 *Coord.Chem. Rev.* **248** 1363
[3] Gervaldo M, Fungo F, Durantini E N, Silber J J, Sereno L and Otero L 2005 *J. Phys. Chem. B* **109** 20953
[4] Fish J R, Kubaszewski E, Peat A, Malinski T, Kaczor J, Kus P and Czuchajowski L 1992 *Chem. Mater.* **4** 795
[5] Turkevich J, Stevenson P C and Hillier J 1951 *Faraday Discuss.* II **55**.

# Preparation and characterization of nickel based multicomponent catalysts

**Mihaela Lazar, Maria Mihet, Monica Dan, Valer Almasan and Petru Marginean**

National Institute for Research and Development of Isotopic and Molecular Technologies, 65-103 Donath, 400293 Cluj-Napoca, Romania

E-mail: diana.lazar@itim-cj.ro

**Abstract** Two series of modified alumina supported nickel catalysts were prepared and characterized. First series consist in bimetallic Ni-Au and Ni-Ag catalysts supported on alumina; the second series consist in Ni catalysts supported on bicomponent oxides. The catalysts were prepared by impregnation: $Ni/Al_2O_3$, $Ni/CeO_2$-$Al_2O_3$ and $Ni/La_2O_3$-$Al_2O_3$ (10 wt % Ni and 6 wt % additional oxide) and co-impregnation: $Ni$-$Au/Al_2O_3$ and $Ni$-$Ag/Al_2O_3$ (10 wt % Ni and 1 wt % Au or Ag). The catalytic surface was characterized by $N_2$ adsorption-desorption isotherms. The catalytic sites were characterized by hydrogen Thermo Programmed Desorption (TPD) method.

## 1. Introduction

The usual catalyst for methane steam reforming is based on Ni as the active element [1]. The main problem with this type of catalysts is that Ni also favors the formation of graphite, which deactivates the catalysts and may eventually lead to their destruction. One solution is to selectively change the surface structure through the formation of a superficial alloy [2]. Metals such as Ag [3] and Au [4] on Ni substrate have been tested and shown promising for coke minimization in the steam reforming (of) hydrocarbons. In order to obtain catalysts with high activity and stability, it is fundamental to maintain the dispersion of the active phase and the stability of the support. Both properties can be induced by addition of a second oxide to the support. $La_2O_3$ is able to stabilize alumina and it can also avoid Pt agglomeration [5]. The addition of ceria to Pt supported catalysts improves the catalytic properties due to the formation of a larger Pt-$CeO_2$ interface [6]. $Pd/Al_2O_3$ promoted with ceria and lanthana showed higher activity for steam reforming of methane [7]. In this paper we present the results obtained in preparation and characterization of $Ni/Al_2O_3$ catalysts modified by addition of Ag and Au to Ni and $CeO_2$ and $La_2O_3$ to alumina support.

## 2. Experimental

For preparation of multicomponent nickel catalysts the wet impregnation method was used. Alumina supported Ni catalysts (10 wt % Ni) modified with Ag (1 wt %) and Au (1 wt %) were prepared by co-impregnation of $\gamma$-$Al_2O_3$ (St = 135 $m^2/g$) with aqueous solutions of $Ni(NO_3)_2 \cdot 6H_2O$ and $AgNO_3$ ($HAuCl_4 \cdot 3H_2O$ respectively). The impregnated samples were dried at room temperature overnight, calcined under nitrogen at 500°C for 4 h and reduced in H2 at 400°C for 4 h. The mixed supports containing $CeO_2$ (6 wt %) and $La_2O_3$ (6 wt %) loadings were prepared by impregnation of $\gamma$-alumina with aqueous solution of $Ce(NO_3)_2$ or $La(NO_3)_3$ followed by calcinations in nitrogen at 500°C for 4 h.

Nickel (10 wt %) was than added by impregnation of mixed supports with aqueous solutions of $Ni(NO_3)_2$, as described below.

Total surface area ($S_t$), pore volume ($V_p$) and pore radius ($r_p$) of the samples were obtained from $N_2$ adsorption-desorption isotherms. The isotherms were measured using a Sorptomatic 1990 apparatus. The BET method was used for calculation of $S_t$, and the Dollimore-Heal method for determination of porosity parameters.

The TPD experiments were performed in an installation composed by: gas introduction unit, tubular reactor and thermal conductivity detector (TCD). The gases used in these experiments were 99.999% purity. An extra-purification was made by passing the gases through a $Ni/Cr_2O_3$ catalyst followed by a molecular sieve, just before entering in the reactor. The catalysts sample was activated in hydrogen at 350°C and then cooled down. The hydrogen was adsorbed on catalyst's surface at 0°C for about 1 h and then Ar was passed through the reactor for about 30 min., at 0°C, in order to remove the un-chemosorbed hydrogen from the system. The thermodesorption was then started by heating the sample in argon with heating rate 5°C /min and argon flow 25 ml/min. The desorbed hydrogen was monitored by TCD.

## 3. Results and discussions

The surface area and pore volume of the $Ni\text{-}Me/Al_2O_3$ catalysts have similar values with the simple $Ni/Al_2O_3$. That's means that the deposed metal nanoparticles are very small and well dispersed on the support surface, including the pore's walls. For $Ni/M_xO_y\text{-}Al_2O_3$ these values are smaller (table 1) which lead as to conclusion that either the second oxide deposited on alumina fills some smaller pores, or the repeated thermal treatment modify the surface area.

**Table 1.** Total surface area and pore volume for multicomponent nickel catalysts.

| | Catalysts | | | | |
|---|---|---|---|---|---|
| | $Ni/Al_2O_3$ | $Ni\text{-}Ag/Al_2O_3$ | $Ni\text{-}Au/Al_2O_3$ | $Ni/CeO_2\text{-}Al_2O_3$ | $Ni/La_2O_3\text{-}Al_2O_3$ |
| $S_t$ ($m^2/g$) | 113 | 115 | 114 | 105 | 109 |
| $V_p$ ($cm^3/g$) | 0.151 | 0.150 | 0.153 | 0.145 | 0.140 |

The adsorption-desorption isotherms are of type IV which corresponds to mesoporous solids (figure 1). The pore size distribution is relatively uniform with a medium diameter of 50 Å (figure 2). The medium size of pores does not differ very much between the 5 catalysts. It is possible that the additional oxide and/or metal in $Ni/M_xO_y\text{-}Al_2O_3$ catalysts to be deposited on the pore's walls, but to don't fill them completely, explaining the slight decrease in total surface area.

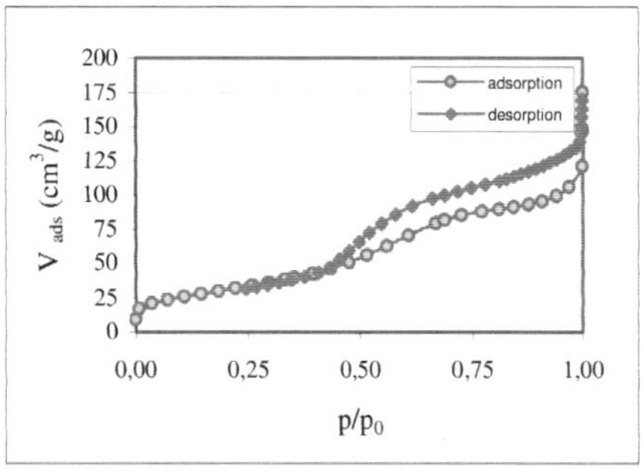


**Figure 1.** Adsorption-desorption isotherms of $N_2$ for $Ni\text{-}Au/Al_2O_3$.

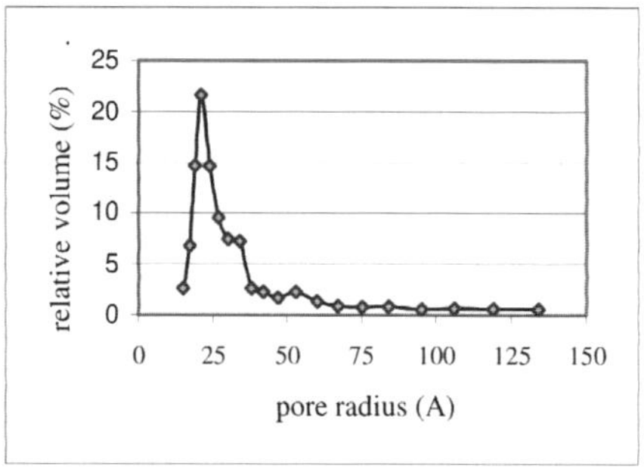


**Figure 2.** Pore distribution for $Ni/CeO_2\text{-}Al_2O_3$.

The heterogeneous supported metal catalyst surface includes a number of active sites showing different adsorptive and catalytic activities. Temperature-programmed desorption is a helpful technique for investigating these heterogeneous surfaces. The hydrogen atoms are desorbed from the surface at different temperatures depending on the strength of the hydrogen-metal bond. The addition of the second metal or the second oxide modifies the surface properties which are reflected in the modifying of the catalytic sites. Figure 3 presents the TPD plots for the five studied catalysts.

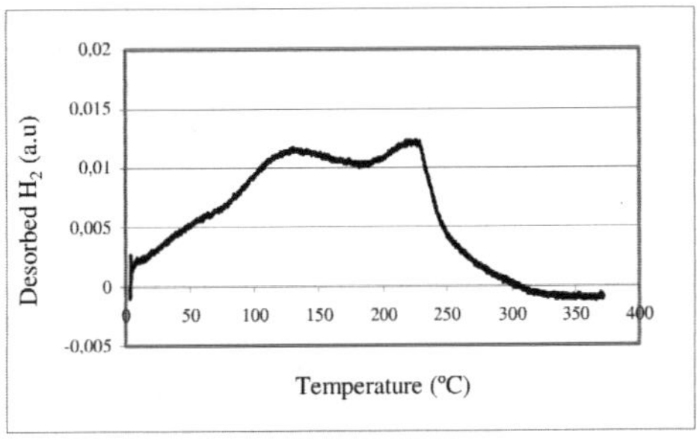


$Ni/Al_2O_3$

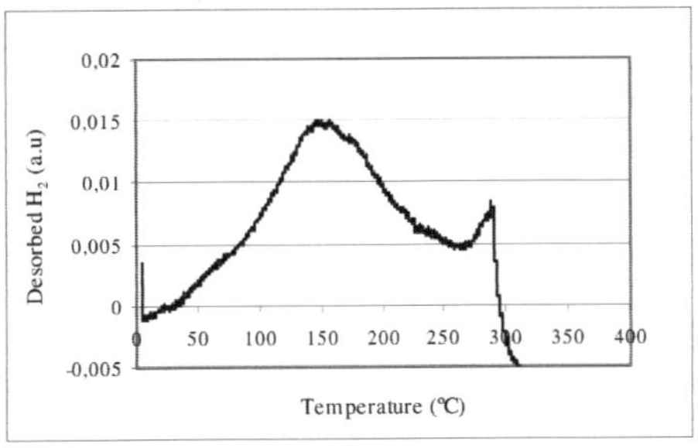


$Ni/CeO_2 - Al_2O_3$

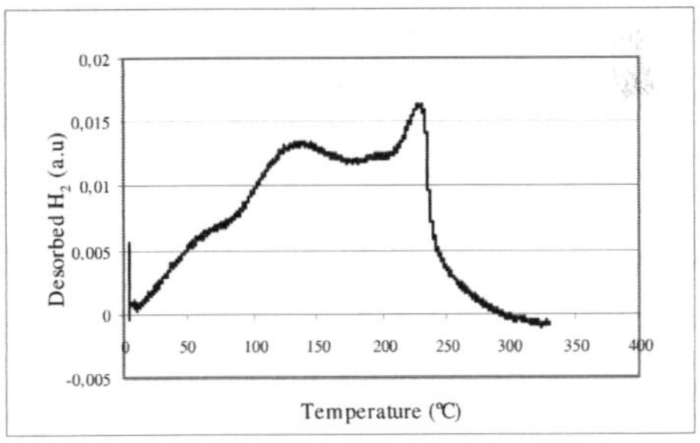


$Ni/La_2O_3 - Al_2O_3$

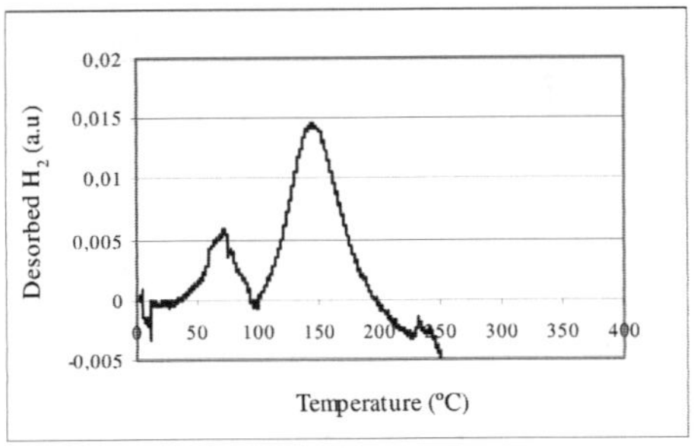


$Ni - Ag/Al_2O_3$

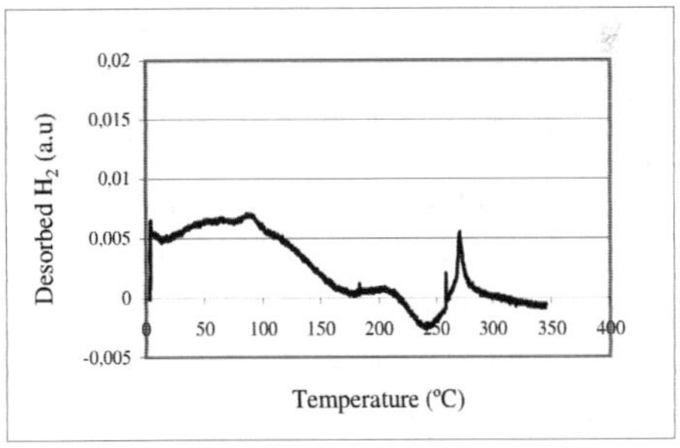


$Ni - Au/Al_2O_3$

**Figure 3.** Hydrogen thermodesorption spectra for multicomponent nickel catalysts.

The TPD plots show two desorption peaks indicating that hydrogen desorbs from the surface in two steps. This situation corresponds to the presence of at least two types of active sites on the catalysts surface. Additional components to $Ni/Al_2O_3$ catalyst change the shape of thermodesorption spectra. By adding the second oxide to the catalyst support, the first desorption peak shifts ~ 20°C to higher temperatures (from ~ 130°C to ~ 150°C). The nickel supported on mixed oxides adsorb hydrogen stronger that $Ni/Al_2O_3$. The second desorption peak decreases in intensity by adding ceria to alumina support and increases by adding lanthanum oxide. The silver and gold change dramatically the shape of TPD plots. By adding Ag, two distinct peaks are obtained: first one having smaller intensity desorbs at 74°C and the second one, more intense, desorbs at 145°C. Silver decreases very much the desorption temperature (about 74°C) indicated that a part of adsorbed hydrogen is much weaker bonded to the catalytic sites. Adding Au to the nickel catalyst, the desorption temperature and the total amount of desorbed hydrogen are very much decreased. Both these parameters indicate a weaker capacity of Ni-Au catalysts to adsorb hydrogen comparing to Ni catalysts.

## 4. Conclusions

Two series of modified alumina supported nickel catalysts were prepared and characterized by $N_2$ adsorption-desorption and hydrogen Thermo Programmed Desorption. The catalysts were prepared by impregnation: $Ni/Al_2O_3$, $Ni/CeO_2$-$Al_2O_3$ and $Ni/La_2O_3$-$Al_2O_3$ and co-impregnation: Ni-Au/$Al_2O_3$ and Ni-Ag/$Al_2O_3$. The total surface area and pores volume are not influenced by the addition of Ag and Au, but decrease at the addition of the transitional oxide. The TPD results are influenced by the addition of second metal and oxide. The Ni supported on mixed oxides have the desorption peaks at higher temperature than $Ni/Al_2O_3$ indicating stronger hydrogen adsorption sites. The addition of Ag and Au have the opposite effect decreasing much the desorption temperature and desorbed amount of hydrogen in the case of gold. These data will be correlated with catalytic activity results in steam reforming reaction in order to better describe the catalytic mechanism.

Future experiments using different concentration of noble metals and transitional oxides will allow us to design a catalyst with improved qualities for hydrogen production by methane steam reforming.

**Acknowledgements**

This research was supported by the Romanian ANCS Department: PN II Program under contract 21004/2007.

## References

[1] Rostrup-Nielsen J 1984 *Catalysis, Science and Technology* vol 5, eds J R Anderson and M Boudart (Berlin: Springer) p 1
[2] Besenbacher F, Chorkendorff I, Clausen B S, Hammer B, Molenbroek A M, Nørskov J K and Stensgaard I 1998 *Science* **279** 1913
[3] Parizotto N V, Rocha K O, Damyanova S, Passos F B, Zanchet D, Marques C M P and Bueno J M C 2007 *Appl. Catal.* A **330** 12
[4] Jarrah N A, van Ommen J G and Lefferts L 2006 *J. Catal.* **239** 460
[5] Zhang X, Walters A B and Vannice MA 1995 *J. Catal.* **155** 290
[6] Santos A C S F, Damyanova S, Teixeira G N R, Mattos L V, Noronha F B, Passos F B and Bueno J M C 2005 *Appl. Catal.* A **290** 123
[7] Cassinelli W H, Feio L S F, Araujo J C S, Hori C E, Noronha F B, Marques C M P and Bueno J M C 2008 *Catal. Lett.* **120** 86.

Journal of Physics: Conference Series **182** (2009) 012050 doi:10.1088/1742-6596/182/1/012050

# Hydrogen desorption from $NaAlH_4$ catalyzed by ball-milling with carbon nanofibers

**D Lupu, G Blanita, I Misan, O Ardelean, I Coldea, G Popeneciu and A R Biris**

National Institute for Research and Development of Isotopic and Molecular Technologies, 65-103 Donath, 400293 Cluj-Napoca, Romania

E-mail: dan.lupu@itim-cj.ro

**Abstract.** The effect of ball-milling $NaAlH_4$ with carbon nanofibers on the hydrogen desorption rates is reported. The rates are strongly dependent on the ball to powder ratio and milling time. The use of an Arrhenius equation corrected for the dependence of hydrogen pressure allows a comparison with other literature data, performed in different conditions. It shows that ball-milling at high energy is the most favorable route to increase the desorption rate and lower the activation energy for hydrogen desorption.

## 1. Introduction

A technical challenge for the development of fuel cell powered vehicles is a suitable material for on-board hydrogen storage with 6.5 wt%H storage capacity. Sodium alanate has been considered as a potential hydrogen storage material with its theoretical capacity of 5.6 wt%H resulting from the first two desorption steps. The discovery of the catalytic effect of Ti on hydrogen desorption kinetics from $NaAlH_4$ by Bogdanovic and Schwickardi [1] attracted the interest in this relatively cheap hydrogen storage material because the desorption can be lowered near the working temperature of the PEM fuel cells. A large number of studies recently reviewed [2] have been focused on a wide range of catalysts and procedures, aiming to lower the desorption temperature and to increase the desorption rates. Ball-milling with doping agents proved to result both in high capacity reversible storage with excellent kinetics and a „one-step direct synthesis method" from NaH and Al powder [3].

A loss of capacity appears due to the irreversible reaction between the hydride and some dopants [4,5] which led to investigations of various forms of carbon as catalyst [6-8]. The positive effect of carbon was assigned to the inhibition of grain aggregation [8]. This paper reports the effects of ball-milling (BM) $NaAlH_4$ with carbon nanofibers on hydrogen desorption.

## 2. Experimental

In a glove box under dry nitrogen, 2 g Aldrich tech. $NaAlH_4$ were mixed with 0.2 g herringbone type carbon nanofibers (CNF) of 100-500 nm diameter, synthesized as described elsewhere [9], in the vial of 25 $cm^3$ and 10 bar of pure hydrogen. The characteristics of the planetary mill are: revolution radius 150 mm, vial radius 12.5 mm, rotation to revolution angular velocity $W_v/W_p = -2.47$.

With 24 balls of 6 mm diameter the ball/powder ratio is 10.3 and the transferred power calculated according to [10] is $P^* = 8$ Wh/g which is in the range of medium energy ball-milling.

Alternatively, Pd (10%)/CNF were used, obtained by impregnation of CNF with Pd nitrate, calcined at 500°C and reduced under hydrogen flow at 300°C. The hydrogen desorption was performed in a Sievert type apparatus, with a 6000 $cm^3$ volume, to ensure a pressure increase not higher than ~0.1 bar at the end of reaction, allowing thus measurements at rather constant pressure. The desorption was measured at constant temperature, starting from 1.1 bar $H_2$ pressure. The rehydrogenation was performed at, 120 bar using 99.999% $H_2$, 20 h at 110°C; each desorption run was started after maintaining the reactor for at least 1 h at the reaction temperature at a hydrogen pressure usually 50 bar $H_2$ higher than the equilibrium desorption pressure calculated from [11].

## 3. Results and discussion

The literature reports two types of (overall) kinetic curves of hydrogen desorption corresponding to the first desorption step from $NaAlH_4$: sigmoid type, associated to a nucleation and growth mechanism or

$$NaAlH_4 \leftrightarrow 1/3\ Na_3AlH_6 + 2/3\ Al + H_2 \leftrightarrow NaH + Al + 3/2\ H_2 \quad (1)$$

curves with decreasing slopes, corresponding to reactions proceeding evenly on the entire surface of all reactant particles [12]. However, in the later case a rather constant slope is observed for $NaAlH_4$ in a wide concentration range. In some cases, the type of curve changes with cycling or the catalyst as can be seen from data reported by Wang *et al* [13]. This might indicate that the catalyst is involved sometimes in a reaction as suggested by the comments on the differences between Sc and Ti catalysts [13] where the positive effect of Sc was assigned to the inhibition of aggregation of Al particles. The segregation of Al particles was reported by Bogdanovic *et al* [11] and Brinks *et al* [14] proposed that the formation of stable Ti-Al compounds would prevent the reverse (hydrogenation) reaction, leading to a loss of storage capacity.

The scope of this paper has not been to elucidate the reaction mechanism but rather to find some correlations with the hydride processing, able to identify simple and convenient ways to improve the overall reaction rates. Some typical hydrogen desorption curves obtained are given in figure 1. They are of the second type noted above, similar to other reported isothermal desorption curves for carbon [6] or other dopants such as $TiCl_3$ [15], Ti(Zr) [16]. This suggests that carbon is not implied in a reaction with Al, as reasonably may be assumed, in agreement with the observation of Cento *et al* That the role of carbon "could be rather physical than chemical probably inhibiting the grain aggregation" [7].

Because for hydrides the absorption/desorption kinetics depends on the hydrogen pressure [12], a comparison of the activation energies and pre-exponential factors from different literature results (often performed in different conditions, for example desorption in [13] is against 3.4 bar), the reaction rate must take into account the pressure dependence. Therefore, the activation energies and pre-exponential factors from our data and other reported results with available details of the experiments, have been evaluated from the desorption rate $r^c$ containing the term $(P_{eq}-P)$ as thermodynamic driving force of the desorption, with $P_{eq}$ – the hydrogen equilibrium pressure for the first step, evaluated from [11] and P – the pressure at which the desorption is measured. The Arrhenius equation as in [17] has been used

$$r^c = A(P_{eq}-P)\exp(-E_a/RT) \quad (2)$$

where A is the pre-exponential factor, Ea – the activation energy and R – the universal gas constant. The results obtained with the equation (2), by dividing the literature desorption rates with (Peq-P) are given in table 1. The CNF show activation energy and pre-exponential factors rather close to the literature data [6] for 10% carbon. The most important finding is that the overall desorption rate increases.

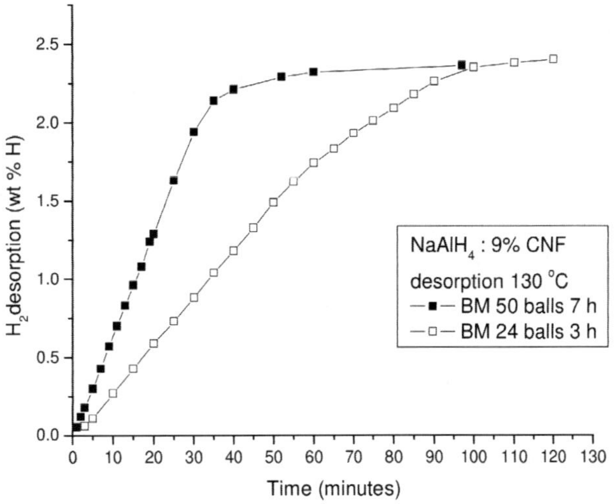


**Figure 1.** Hydrogen desorption at 130°C from ball-milled $NaAlH_4$ with 9% „herringbone" type carbon nanofibers obtained after 5 absorption/desorption cycles.

**Table 1.** Activation energy ($E_a$), pre-exponential factor (A) and balls/$NaAlH_4$ weight ratios RBP for hydrogen desorption from 2 g $NaAlH_4$ ball-milled with carbon nanofibers (CNF).

| Catalyst | RBP | $E_a$ (kJ/mol$H_2$) | A | r at 110°C (wt%H/h) | Conditions | Ref |
|---|---|---|---|---|---|---|
| 9%CNF | 10.3 | 51.5 | 12.2 | 0.36 | 24 balls 3h | This work |
| 9%Pd/CNF | 10.3 | 58.5 | 14.0 | 0.25 | 24 balls 3h | This work |
| 9%CNF | 21.5 | 53.1 | 13.7 | 1.0 | 50 balls 7h | This work |
| 10%Carbon | 15 | 57.0 | 13.7 | 0.32 | Vibratory BM | [6] |
| 2mol%$TiCl_3$ | - | 45.2 | 11.2 | 1.2 | - | [15] |
| 4% Sc | - | 38.7 | 10.2 | 2.6 | High-energy BM | [13] |
| 8%Ti+2%Zr | 30 | 24.9 | 5.9 | 3.0 | - | [16] |
| ~54% $ScCl_3$ | - | - | - | 5.6 | - | [3] |

Some literature data obtained from recalculated $r^c$ and measured desorption rates r at 110°C are also given for comparison significantly with the increasing RBP and milling time, parameters which increase the efficiency of BM due to a higher energy transfer from balls to powder [10]. The comparison with literature also show that high energy BM [13] or higher RBP [16] results in even higher desorption rates and a significant decrease of activation energy. This is consistent with the type of desorption curves suggesting a reaction which proceeds over the entire surface, when the initial slope is a product of the rate constant at the interface with the initial surface area [12] as the surface area can be increased by a more efficient BM. It should be noted that though Pd is efficient for hydrogen absorption/dissociation and transfer to the alanate, it does not result in any improvement of desorption characteristics.

It was noted [3] that $NaAlH_4$ seems to be closest to the demands for automotive applications in combination with advanced fuel cells working at 150°C. However, the heat management for absorption/desorption at these temperatures brings about additional difficulties and the calculations show [17] that the total weight of the storage tank increases by a factor of 1.6-2.

## 4. Conclusions

The carbon nanofibers are good catalysts for improving the hydrogen desorption kinetics of $NaAlH_4$ by ball-milling. The desorption rates can be increased by increasing the ball/powder ratio and milling time. An Arrhenius equation involving the dependence on the hydrogen pressure allows a unitary evaluation of the activation energy and pre-exponential factors for a comparison of literature data. High-energy ball-milling is a suitable way to achieve high desorption rates. Though rather close to the requirements for mobile applications, the heat management is a serious disadvantage for $NaAlH_4$. Materials with good storage capacity but low heat of hydrogen absorption seem much more suitable for a lower total weight of storage tank.

## Acknowledgements

The authors thank the National Authority for Scientific Research (ANCS) for granting the project MENER 707/24.07.2006.

## References

[1] Bogdanovic B and Schwickardi M J 1997 *J. Alloys Comp.* **253** 1
[2] Bogdanovic B, Eberle U, Felderhoff M and Schüth F 2007 *Scr. Materialia* **56** 813
[3] Bogdanovic B, Felderhoff M, Pommerin A, Schüth F, Spielkamp N and Stark A 2009 *J. Alloys Comp.* **471** 383
[4] Sandrock G, Gross K and Thomas G 2002 *J. Alloys Comp.* **339** 299
[5] Bellosta von Colbe J M, Bogdanovic B, Felderhoff M, Pommerin A and Schüth F 2004 *J. Alloys Comp.* **370** 104
[6] Zaluska A, Zaluski L and Ström-Olsen J O 2000 *J. Alloys Comp.* **298** 125
[7] Cento C, Gislon P, Bilgili M, Masci A, Zheng Q and Prosini P 2007 *J. Alloys Comp.* **437** 360
[8] Dehouche Z, Lafi L, Grimard N, Goyette J and Chahine R 2005 *Nanotechnology* **16** 402
[9] Olenic L, Mihailescu G, Pruneanu S, Lupu D, Biris A R, Margineanu P, Garabagiu S and Biris A S 2009 *J. Mater. Sci: Mater. Med* **20** 177
[10] Burgo N, Iasonna A, Magini M, Martelli S and Padella F 1991 *Il Nuovo Cimento* **13D** 459
[11] Bogdanovic B, Felderhoff M, Germann M, Härtel M, Pommerin A, Schüth F, Weidenthaler C and Zibrowius B 2003 *J. Alloys Comp.* **350** 246
[12] Gerard N and Ono S 1992 *Hydrogen Intermetallic Compounds II* (Topics in Applied Physics vol 67) ed L Schlapbach (Berlin: Springer) chapter 4 p 167
[13] Wang T, Wang J, Ebner A D and Ritter J A 2008 *J. Alloys Comp.* **450** 293
[14] Brinks H W, Hauback B C, Srinivasan S S and Jensen C M 2005 *J. Phys. Chem. B* **109** 15780
[15] Sandrock G, Gross K, Thomas G, Jensen C, Meeker D and Takara S 2002 *J. Alloys Comp.* **330-332** 696
[16] Xiao X, Chen L, Wang X, Li S, Chen C and Wang Q 2008 *Int. J. Hydrogen Energy* **33** 64
[17] Ahluwalia R K 2007 *Int. J. Hydrogen Energy* **32** 1251.

# Mobility of hydrogen species on Ni supported catalysts

**Maria Miheț, Mihaela Lazăr and Valer Almășan**

National Institute for Research and Development of Isotopic and Molecular Technologies, 65-103 Donath, 400293 Cluj-Napoca, Romania

E-mail: maria.mihet@itim-cj.ro

**Abstract**. A previously developed theoretical model has been used to evaluate metal to support transport rate coefficients of spilt over hydrogen species in case of Ni supported catalysts. Four Ni/oxide ($Ni/Cr_2O_3$, $Ni/ZrO_2$, $Ni/SiO_2$, and Ni/MgO) catalysts were investigated by isotopic transient kinetic method for H/D exchange reaction. A good agreement between experimental data and theoretical ones has been put into evidence, both cases showing an important influence of oxide nature on the quantity of spilt over hydrogen. Fitting the experimental with the calculated theoretical data permitted the evaluation of transport rate coefficients from the spillover initiator (metal) to the acceptor of migrating species (the oxide support).

## 1. Introduction

Mobility of species on the catalyst surface is a well-known phenomenon in heterogeneous catalysis. These mobile species can be reactants, intermediates as well as reaction products. This mobility is a result of chemisorbed species on a metallic surface that migrate on the support without any loss of activity. This phenomenon is termed spillover and has been observed for several species such as hydrogen, oxygen, nitrogen, carbon monoxide or other small organic species [1, 2].

Ni supported catalysts are known to be very active hydrogenation catalysts. Moreover, hydrogenation catalysts have been demonstrated to be hydrogen spillover initiators due to the presence of even small amounts of metal nanoclusters [3]. In general, hydrogen spillover leads to the formation of new OH groups on the support, as oxygen ions serve as acceptors of hydrogen species (atoms or ions). Thus, the effectiveness of spillover to an oxide support is determined by the hydrogen affinity of the acceptors on the support. However, hydrogen migration from the metal initiator to the support as the acceptor is very difficult to put into evidence as there is a continuous migration of species and oxide supports already contain hydroxyl groups. On the contrary, use of hydrogen isotopes (which present same chemical potentials, but differ in mass) permits 'marking' the migrating species. Thus, H/D isotopic exchange reaction is a very relevant solution for the investigation of hydrogen spillover on Ni supported catalysts. Moreover, use of isotopic exchange reaction has been previously reported for hydrogen (H/D) as well as for oxygen ($^{16}O/^{18}O$) as a tool for the study of surface dynamics [4, 5].

## 2. Experimental

For the purpose of this work, four nickel/oxide catalysts (80 wt % Ni), prepared by co precipitation, were investigated. The characteristics of these Ni supported catalysts are presented in table 1. These catalysts were tested in the H/D exchange reaction by Steady State Isotopic Transient Kinetic Analysis (SSITKA), which implies following up the evolution in time of the involved species until they reach steady state.

The experimental set-up consisted in a gas supply unit, a plug flow reactor and a quadrupole mass spectrometer (QMS) (figure 1). The gas supply unit was able to deliver pure Ar, $H_2$, and $D_2$ or any mixture of these gases. The reactor consisted in a cylindrical furnace, 11 cm of length, into which was placed 1 g of Ni catalyst. The catalyst was first activated in hydrogen, for 10 hours at 350°C, and then the temperature was reduced to the operating temperature (50°C, 100°C, 150°C, 200°C, 250°C and 300°C respectively). A sequence of 3 gas compositions was passed over the catalyst bed. In the first step the gas consisted in a binary mixture of 2 vol% $D_2$ in Ar, which was maintained until the concentration of $D_2$ in the effluent gas reached steady state. The second gas switch of pure Ar was maintained until no peak of $D_2$ or HD was detected. In the third sequence, a binary mixture of 2 vol% $H_2$ in Ar was passed over the catalyst. The evolution in time of the isotopic species was recorded by the online QMS after switching to the last gas mixture.

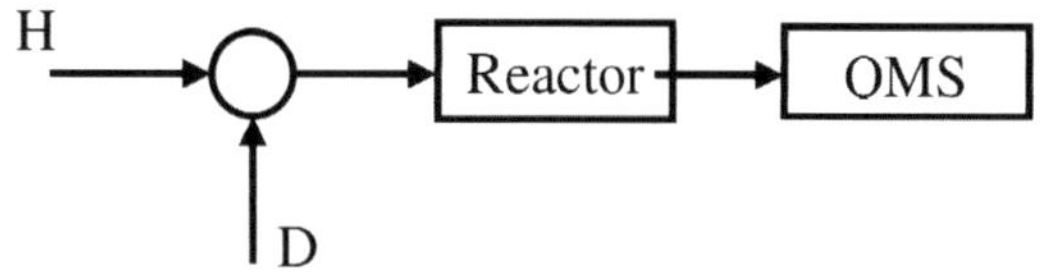


**Figure 1.** Scheme of the experimental set-up.

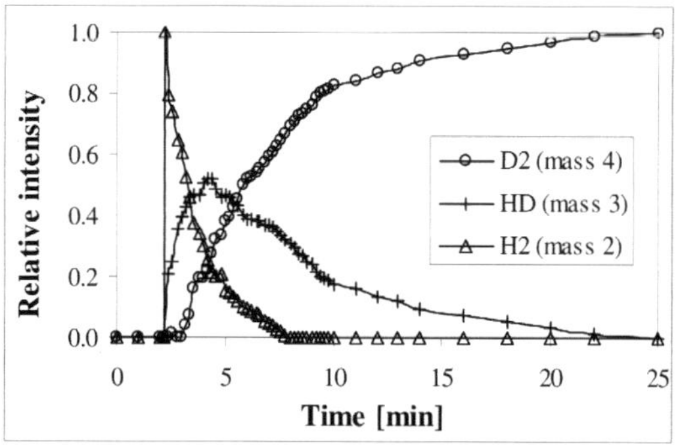


**Figure 2.** MS diagram for Ni/MgO at 250°C.

## 3. Results and discussions

An experimental MS diagram for Ni/MgO at 250°C is presented in figure 2. The last feed switch being $H_2$ in Ar, one would expect $H_2$ peak to be the first to appear in the MS diagram. The sequence of hydrogen species may be explained as follows. (a) During the first step ($D_2$ + Ar), all H chemisorbed on the metal and all H from the oxide surface as OH was replaced with D by H/D exchange. (b) During the second step, when pure Ar is passed over the catalyst, all H from the metal is washed away, while the OD adspecies on the support remain unchanged. (c) Finally, during the third sequence ($H_2$ + Ar) H is chemisorbed on the free metal surface, after which it spills over the oxide support. H/D exchange between the spilt over H and OD adspecies takes place on the support. D thus liberated from the support passes to the metal by reverse spillover and then desorbs into the gas phase. Because of the increasing probability for D species to get back on the metal, the first species in the effluent gas to be monitored are those containing D ($D_2$ and HD).

The MS diagrams were integrated in order to evaluate the quantity of deuterium desorbed from the support (table 1). For each of the four catalysts taken into account, an increase in operating temperature leads to a decrease of desorbed deuterium from the oxide support.

The experimental data were compared with theoretical curves calculated by means of an analytical model previously developed [6]. This theoretical model considers that in the catalyst bed hydrogen can be present in three phases: (*i*) in gas phase; (*ii*) chemisorbed on metal; and (*iii*) as OH/OD on the support. This analytical model based on mass balance equations for hydrogen species in the 3 phases was implemented and solved in MATLAB using the finite difference method. A good agreement between experimental and theoretical data was revealed, for each of the four Ni/oxide catalysts, and at each operating temperature. An example of such an agreement is presented in figure 3.

**Table 1.** Specific volume of desorbed deuterium.

| Temperature (°C) | $V_{des}$ ($cm^3$ D/$m^2$ support) | | | |
|---|---|---|---|---|
| | $Ni/Cr_2O_3$ [a]$S_{Ni}$ = 29.7 $m^2/g$ [b]$S_t$ = 160 $m^2/g$ | $Ni/ZrO_2$ $S_{Ni}$ = 19.4 $m^2/g$ $S_t$ = 85 $m^2/g$ | $Ni/SiO_2$ $S_{Ni}$ = 61.9 $m^2/g$ $S_t$ = 194 $m^2/g$ | $Ni/MgO$ $S_{Ni}$ = 41.4 $m^2/g$ $S_t$ = 98 $m^2/g$ |
| 50 | 0.191 | 0.386 | 0.209 | 0.426 |
| 100 | 0.186 | 0.329 | 0.161 | 0.408 |
| 150 | 0.170 | 0.310 | 0.136 | 0.389 |
| 200 | 0.166 | 0.295 | 0.109 | 0.259 |
| 250 | 0.152 | 0.196 | 0.094 | 0.293 |
| 300 | 0.130 | 0.162 | 0.110 | 0.294 |

[a] Ni surface area; [b] B.E.T. surface area.

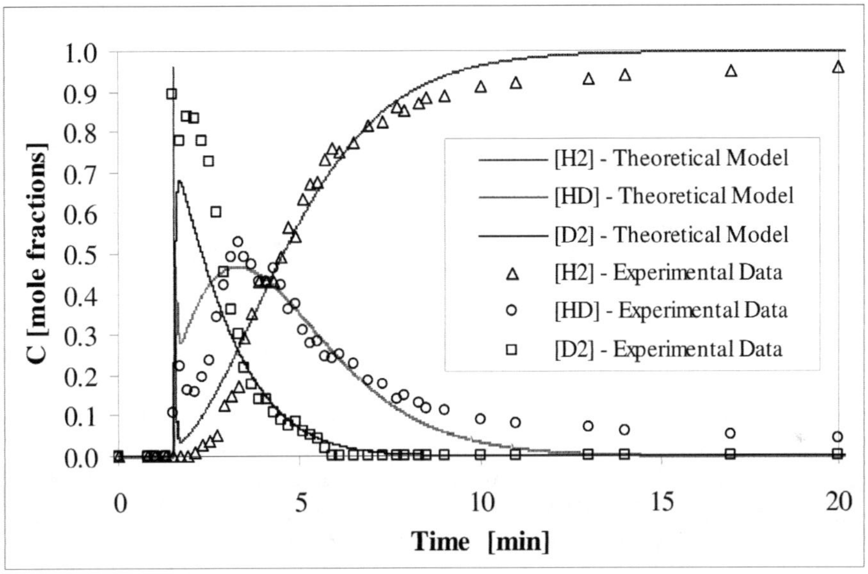


**Figure 3.** Comparison of experimental data with the analytical model for $Ni/ZrO_2$ at 250°C.

Fitting the experimental data with the calculated curves, for each catalyst and at each operating temperature, permitted the determination of rate coefficients for the transport of hydrogen species from metal to support and vice versa. This rate coefficient is actually a resistance opposed to transport of species from one phase to the other, meaning that a greater value of the rate coefficient leads to a lower quantity of hydrogen which migrates between the metal and the support in either way. Except for the $Ni/SiO_2$ catalyst, the calculated transport rate coefficients are in agreement with the calculated volumes of desorbed deuterium from the oxide support (figure 4 and figure 5).

The low rate coefficient in case of Ni/MgO could be accounted for the basic character of the oxide, that is $O^{2-}$ from its structure easily bonds to hydrogen species present in the catalyst space. This means that a greater quantity of hydrogen spills over onto MgO, the support being thus able to desorb a larger volume of deuterium afterwards (for the case of the experiment described before). $Cr_2O_3$ and $ZrO_2$ present an amphoteric character, which means that the electronic properties of the oxides do not easily permit the formation of OH or OD species on the surface. Larger transport rate coefficients in case of $Ni/Cr_2O_3$ and $Ni/ZrO_2$ catalysts, leading to smaller volumes of desorbed deuterium might be thus explained. The exception recorded in case of $Ni/SiO_2$ might be caused by the fact that Si is a semimetal, while $SiO_2$ presents an acid character. Moreover, virtual immobility of silanol groups (Si-O-H) was previously reported [4].

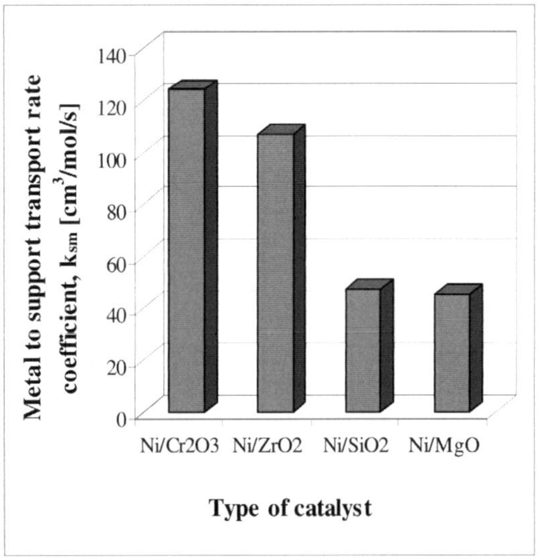


**Figure 4.** Metal to support transport rate coefficients at 250°C.

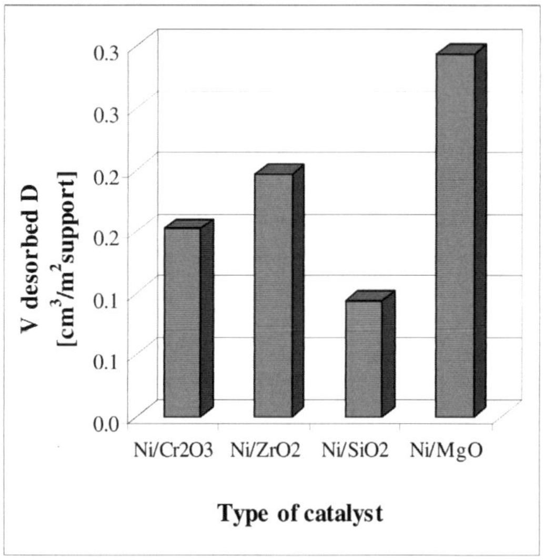


**Figure 5.** Specific volume of desorbed deuterium from the oxide support at 250°C.

## 4. Conclusions

Investigations on spillover of hydrogen species in Ni/oxide catalysts have shown that both temperature and nature of oxide support influence the mobility of species. Metal nanoclusters in the catalysts initiate hydrogen spillover, while the oxide support acts as the acceptor of these species. The quantity of spilt over hydrogen on the oxides differs significantly in case of the four catalysts taken into account. Experimental data as well as theoretically calculated ones have proven this fact as the magnitude of transport rate coefficients for hydrogen migrating from metal to support and vice versa are correlated with the volumes of desorbed deuterium from the oxide supports. Thus, the physico-chemical properties of the support are of great importance in the design of efficient heterogeneous catalysts. Further investigations on hydrogen spillover considering the electronic interactions between the activated species and the support, as well as those between the metal and support are needed.

## References

[1] Conner W C and Falconer J L 1995 *Chem. Rev.* **95** 759-88
[2] Roland U, Braunschweig T and Roessner F 1997 *J. Molec. Catal.* A: Chem **127** 61-84
[3] Rozanov V V and Krylov O V 1997 *Russ. Chem. Rev.* **66** 107-19
[4] Duprez D 2006 *Catal. Today* **112** 17-22
[5] Galdikas A, Descorme C and Duprez D 2004 *Solid State Ionics* **166** 147-55
[6] Almăşan V, Hodor I and Mărginean P 1997 *Appl. Surf. Sci.* **120** 335-9

# Temperature influence on the global and local structure of the chromia supported nickel catalysts

**Stelian Pintea[1,2], Vasile Rednic[1,2], Petru Mărginean[1], Xie Yaning[3] and Nicolae Aldea[1]**

[1] National Institute for Research and Development of Isotopic and Molecular Technologies, 65-103 Donath, 400293 Cluj-Napoca, Romania

[2] Babes-Bolyai University, Faculty of Physics, 1 Mihail Kogalniceanu, 400084 Cluj-Napoca, Romania

[3] Beijing Synchrotron Radiation Facilities of Beijing Electron Positron Collider National Laboratory, People's Republic of China

Email: stelian.pintea@itim-cj.ro

**Abstract**. The changes induced by the heat treatment on the global and the local structures of the supported nickel catalysts are investigated. The global structure parameters were obtained by processing the X-ray diffraction patterns of the samples using the Scherrer formula. The local structure parameters were determined by X-ray absorption spectra processing. Both types of measurements were carried out using synchrotron radiation. The nickel catalysts supported on chromia were prepared by coprecipitation method. The nickel/chromium atomic percent was 70/30. After preparation, the samples were thermally treated for three hours at 350, 650 and 950°C. A strong correlation between previous catalytic activity information and the global and local structure is evidenced. Also the correlation between the nickel crystallite size and the active metal-oxide support interaction is discussed.

## 1. Introduction

Supported catalysts are widely used in chemical processes. Reactions like hydrogenation processes [1, 2], isotope separation [3] and production of the deuterium labelled compounds [4] requires supported metal catalysts that chemisorbs hydrogen dissociatively.

The mechanism of the heterogeneous catalysis is very complex, but one of the conditions to have a high catalytic activity is to have also a high metallic surface. In order to obtain a high metallic surface, the average crystallite size of the active metal must be as small as possible [5,6]. The present contribution contains the results for the calculations of the metallic nickel crystallite sizes. The calculations were processed by the Scherrer formula on X-ray diffraction (XRD) measurements using synchrotron radiation.

The presence of active metal-oxide support interaction is also a particularity of supported metal catalysts, also present at supported nickel catalysts. This interaction induces a deformation of the local vicinity of the active metal, evidenced in extended X-ray absorption fine structure

(EXAFS) as a decreasing in the coordination number of the first coordination shell for the central nickel atoms [7].

## 2. Sample preparation

For the preparation of the chromia supported nickel catalysts coprecipitation method [7] was used. This method involves the coprecipitation of the mixture of nickel nitrate and chromium nitrate with a solution of sodium carbonate/hydroxide. The precipitate was thoroughly washed with double distilled water in order to eliminate the sodium ions. Then the precipitate was dried at 105°C in oven, calcined in nitrogen flow at 340°C and reduced in flowing hydrogen at 350°C. The catalyst samples obtained in this way are pyrophoric and they need passivation for handling them in air. The passivation was performed in nitrogen flow with low content of oxygen. By this process nickel ions with positive valence appear as a thin layer at the metal surface. The thermal treatments of the catalysts have been done in hydrogen flow at different temperatures for 3 h. In this way were prepared the nickel supported on chromium oxide catalysts samples thermally treated at 350, 650 and 950°C with 70% atomic percent of nickel and 30% atomic percent of chromium.

## 3. Experimental measurements

The XRD and EXAFS measurements were carried out on the 4W1B and 4W1C beamlines at the Beijing Synchrotron Radiation Facilities (BSRF).

The mentioned beamlines are operating at 30-50 mA and 2.2 GeV at room temperature [8]. The 4W1B beamline is an unfocussed monochromatic X-ray beam with 4 mrad of horizontal acceptance and has the following features: energy range of 3.5-22 keV; energy resolution of $\Delta E$ = 1-3 eV at E = 10 keV; the Bragg angle range of 5-70°.

The absorption coefficient of nickel K edge was determined using the Si(111) monochromator crystal on a transmission setup. The intensities of the incident and the transmitted beams were monitored by ionization chambers. As standard sample a nickel foil obtained by electrochemical method was used. Harmonics were rejected by detuning of the monochromator. For EXAFS analyses of the absorption coefficient different computer codes were used: EXAFS31 and EXAFS36 [9] from the library of the group and CDXAS [10].

XRD measurements were carried out using synchrotron radiation with 1.54 Å wavelength.

## 4. Results and discussion

### *4.1. XRD Results*

The XRD patterns for the nickel supported on chromia catalysts are presented in figure 1. Notations (1), (2) and (3) correspond to the patterns of the $Ni/Cr_2O_3$ catalysts thermally treated at 350, 650 and respectively 950°C. As standard sample a silicon crystal was used (see the XRD pattern of it in the inlet of the figure 1).

The most intense X-ray line profiles (XRLPs) present in these patterns corresponds to nickel in metallic state. One can see the (111), (200) and (220) XRPLs around the following values for the $2\theta$ angle of incidence: 44.97°, 51.85° and 76.38°.

As a second conclusion one can evidence that all the nickel content of the sample is in metallic state as long as there are no evidences for any other type of nickel present in the samples. Small XRLPs corresponding to chromium oxide can be observed at the 54.95, 63.58 and 65.23° values in the $2\theta$.

In table 1 the crystallite sizes calculated with the Scherrer formula for the nickel nanoclusters are given. As we can see, with the increasing of the thermal treatment temperature, the crystallite size is increasing leading to the diminution of the nickel metallic surface. These results are in good agreement with previous catalytic activity results [11]. It is also important to mention that the crystallite sizes are different for different crystalline directions. This conclusion can determine us to think about preferential orientations of the nickel clusters but the ratio between the intensities of the three peaks mentioned above are close to those for the pure nickel in the face centred cubic lattice so, preferential growth during the sample preparation or/and thermal treatment can be excluded. These different values for the crystallite sizes for different orientations can be useful information for further investigations regarding the shape of the crystallites.

**Table 1.** The crystallite sizes of the metallic nickel nanoclusters for different XRLPs calculated using the Scherrer formula.

| **Sample** | | **D [Å]** | | |
|---|---|---|---|---|
| | **XRLP** | **(111)** | **(200)** | **(220)** |
| $Ni/Cr_2O_3$ treated at 350°C | | 138 | 174 | 148 |
| $Ni/Cr_2O_3$ treated at 650°C | | 308 | 235 | 267 |
| $Ni/Cr_2O_3$ treated at 950°C | | 800 | 571 | 800 |

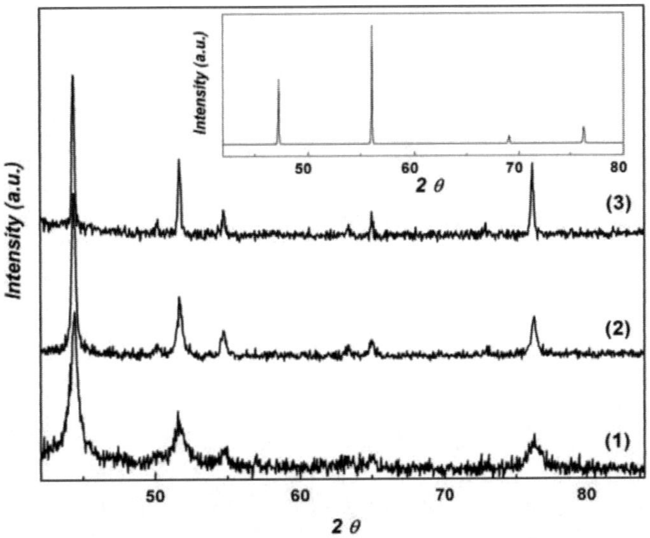


**Figure 1.** XRD patterns of the supported nickel catalysts thermally treated at: (1) 350°C, (2) 650°C and (3) 950°C. As inlet appears the XRD pattern of the Si standard sample.

*4.2. EXAFS Results*

The local structure of the nickel catalysts samples were investigated by X-ray absorption spectroscopy (XAS) and the resulted local parameters were extracted using a computer package program dedicated to EXAFS spectra processing developed in our working group.

In figure 2 the absorption coefficient of standard nickel sample obtained by electrochemical processes is represented.

Figure 3 includes the (XAS) spectra for the thermally treated samples. There are no evidences of the nickel oxides, as a proof that all the nickel oxide was reduced during the reduction step of the catalysts preparation process.

After data processing, the EXAFS functions of the first coordination shells corresponding to the catalysts samples were obtained. These first coordination shells are represented in figure 4.

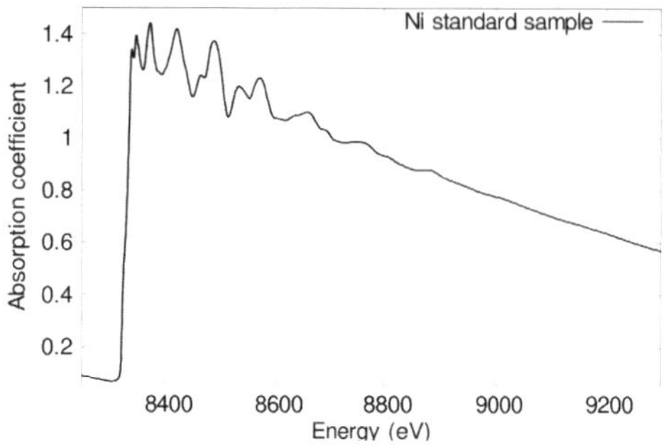


**Figure 2.** The X-ray absorption coefficient of the nickel standard sample.

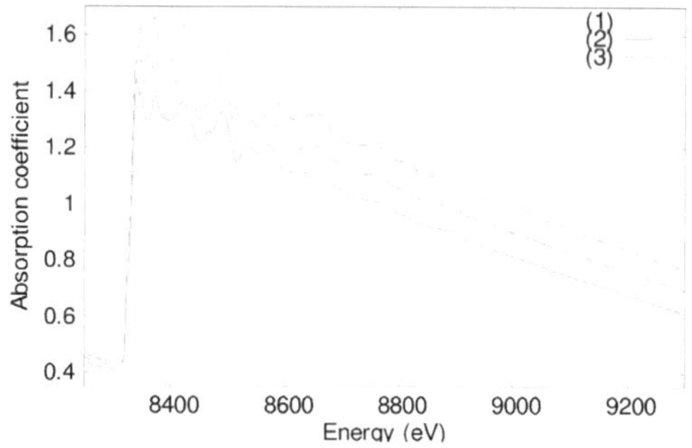


**Figure 3.** The X-ray absorption coefficient of the nickel catalysts thermally treated at: (1) 350°C, (2) 650°C and (3) 950°C.

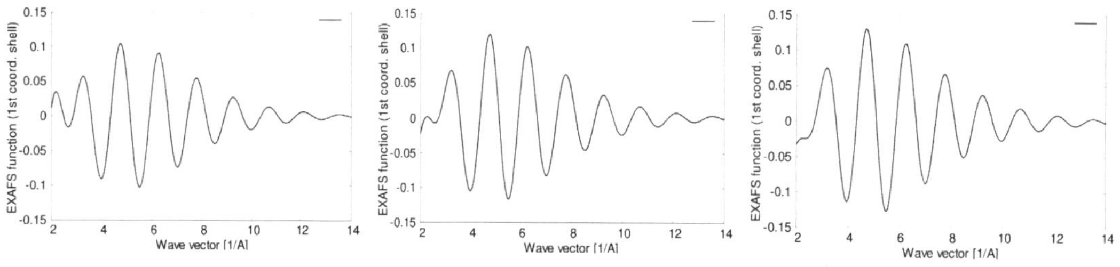


(1) (2) (3)

**Figure 4.** The EXAFS function of the first coordination shell for the samples thermally treated at: (1) 350°C, (2) 650°C and (3) 950°C.

The pure nickel has a face centered cubic lattice. In this configuration, the number of atoms in the first coordination shell is 12. In the catalysts a metal-support interaction could appear. This interaction leads to the diminution of the coordination number. The metal-support interaction is as big as the coordination number decreases. As one can see in table 2, the coordination number of the nickel atoms, $N_1$, increases when the thermal treatment temperature increases and this value tends to the value for the pure nickel. The coordination number increases as a sign of the recrystalization processes that occur during the heat treatment. In the same time, the coordination shell radius, $R_1$, does not change significantly. Only a small variation of this parameter is observed, as we can conclude looking in the table 2.

**Table 2.** The local structure parameters obtained by EXAFS spectra processing.

| Sample | $N_1 \pm \Delta N_1$ | $R_1 \pm \Delta R_1$ [Å] | $E_0 \pm \Delta E_0$ [eV] |
|---|---|---|---|
| Nickel standard | 12 | 2.49 | 8333.915 |
| $Ni/Cr_2O_3$ treated at 350°C | 8.89 ± 0.083 | 2.511 ± 0.003 | 8334.334±0.0211 |
| $Ni/Cr_2O_3$ treated at 650°C | 10.17 ± 0.068 | 2.497 ± 0.002 | 8333.986±0.0001 |
| $Ni/Cr_2O_3$ treated at 950°C | 11.09 ± 0.058 | 2.499 ± 0.002 | 8333.983±0.0003 |

The investigated samples from table 2 have about the same value of threshold energy as the K edge of pure nickel, but their positions are shifted to higher energy compared with the binding energy, $E_0$ = 8333 eV, of the standard nickel sample. This feature is due to the strong electronic interaction of the nickel nanocrystallites surrounded by oxide supports.

## 5. Conclusions

The local and the global structure are playing an important role in the catalytic activity. Variations of both of them with the thermal treatment were evidenced.

The strong dependence of the crystallite size on the treatment temperature was evidenced. The different values for the crystallite size corresponding to different orientations of the crystallites could be important information for further studies regarding the crystallites shape.

As long as in the XRD and XAS spectra of the supported nickel catalysts are no evidences of nickel oxides but only nickel in metallic state, we can consider the preparation method as an efficient one from the nickel consumption point of view.

The decrease in the number of atoms corresponding to the first coordination shell points out the existence of an electronic interaction between the nickel nanoclusters and the surrounding oxides. This interaction is evidenced also by the shift in the edge energies of the supported nickel catalysts investigated in the present contribution.

## References

[1] Pajonk G M and Teichner S J 1986 *Catalytic Hydrogenation* ed I Cerveny (Amsterdam: Elsevier) p 277
[2] Ponec V and Bond G C 1995 *Catalysis by Metals and Alloys* (Amsterdam: Elsevier)
[3] Shimizu M, Kioto S and Ninomiya R 1992 *Proc. Int. Symp. on Isotope Separation and Chemical Exchange Uranium Enrichment* eds Y Fujii *et al* (Tokyo) p 560
[4] Allen J and Voges R 1995 *Synthesis and Applications of Isotopically Labelled Compounds* (New York: John Wiley and Sons)
[5] Niemantsverdriet J W 2007 *Spectroscopy in Catalysis* (Weinheim: Wiley-VCH)
[6] Roduner E 2006 *Nanoscopic Materials. Size-dependent Phenomena* (Stuttgart: RSC Publishing)
[7] Aldea N, Gluhoi A, Marginean P, Cosma C and Yaning X 2000 *Spectrochim. Acta* B **55** 997
[8] BSRF *Activity Report 1991-1998* Beijing Electron Positron Collider National Laboratory, China
[9] Aldea N and Indrea E 1990 *Comput. Phys. Comm.* **60** 145
[10] San-Miguel A 1995 *Phys.* B **208-209** 177
[11] Gluhoi A, Marginean P and Stanescu U 2005 *Appl. Catal*, A: *General* **294** 208-14.

# Investigation on a three-stage hydrogen thermal compressor based on metal hydrides

**G Popeneciu, V Almasan, I Coldea, D Lupu, I Misan and O Ardelean**

National Institute for Research and Development of Isotopic and Molecular Technologies, 65-103 Donath, 400293 Cluj-Napoca, Romania

E-mail: gabriel.popeneciu@itim-cj.ro

**Abstract**. In this paper we report our recent investigation about a there-stage hydrogen thermal compressor based on metal hydrides (HTC) in order to reach an overall compression ratio 28:1. The research was focused to: (*i*) elaborate hydride alloys with good storage capacity and higher thermodynamic characteristics acquired by tailoring of their properties; (*ii*) develop new technical solutions based on advanced materials, and fast mass and heat transfer for a hydrogen storage-compression reactor; (*iii*) built up a prototype of the HTC. Cyclic performance of the hydrogen compressor is studied following up the operating parameters: supply pressure, storage volumes, cold and hot fluid temperatures, cycle duration. The experiments show that the HTC can attain a high overall compression ratio 28:1, it will raise the hydrogen pressure from 2 bars to 56 bars, using three hydride compression stages working between 20 and 80°C. Cycling the compressor at a short absorption-desorption cycle, about 2 minutes, a satisfactory hydrogen flow rate was obtain 10 l/cycle, which ensures a hydrogen flow rate about 300l/hour using a small quantity of hydride alloy, about 360 g. To improve the efficiency and economics of compression process, HTC prototype based on metal hydrides must operate in conjunction with advanced hydrogen production technologies from renewable resources.

## 1. Introduction

Hydrogen as an economically energy alternative, in the context of continuous decrease of fossil energy resources, environmental concerns related to release of gases responsible for the greenhouses effect and public health issues, must be considered a serious candidate over a time frame of decades. The vision of the hydrogen economy is based on two expectations: that hydrogen can be produced from domestic energy sources in a manner that is affordable and environmentally benign, and that applications using hydrogen (for example fuel cell vehicles) can gain market share in competition with the alternatives [1].

In Hydrogen Economy an important element of large scale hydrogen production, storage and utilization infrastructure is hydrogen compression. Considering that the future of hydrogen fuel cells become closer, energetic efficient compression technologies can contribute to hydrogen pipelines transport to the point of use and to distribute hydrogen by refuelling stations.

In order to use hydrogen in practice, besides production it must be efficiently stored and pressurized to increase the energy density. The most promising hydrogen storage methods are: absorption, compression, liquefaction and chemical binding [2]. Metal hydrides as absorbents for hydrogen absorption, offer some main advantages like high volumetric density and reversible absorption/desorption carried out at low pressure and low temperatures. Hydrogen compression based on the reversible hydrogen absorption/desorption ability of metal hydrides is investigated as a reliable process to compress hydrogen at high pressure, without contamination and at low energy costs [3-5].

In this paper we report the development of a novel thermal hydrogen compressor (HTC) that may offer advantages for compressing hydrogen produced from renewable resources using advanced production techniques. Our three-stage thermal compressor is an absorption based system that uses the properties of reversible metal hydride alloys in a cyclical manner: hydrogen is absorbed into an alloy bed at ambient temperature and is desorbed at elevated pressure when the bed is heated with hot water.

Our research was focused on three main directions in order to: (*i*) elaborate hydride alloys with high storage capacity, resistant to poisoning, and higher thermodynamic characteristics acquired by the modeling of their properties; (*ii*) elaborate new technical solutions based on advanced materials, and fast mass and heat transfer for a hydrogen storage-compression reactor; (*iii*) develop a functional model of a thermal hydrogen compressor.

## 2. Principle of operation

Usage of the metal hydrides to hydrogen thermal compression, is based on the equilibrium pressure $P_{eq}$ variation with temperature T, described by van't Hoff equation:

$$\ln P_{eq} = \frac{\Delta H}{RT} - \frac{\Delta S}{R} \tag{1}$$

where: $\Delta H$ is the enthalpy variation and $\Delta S$ is the entropy variation associated to absorption process (exothermal) or desorption (endothermic) of the metal hydride and R is the ideal gas constant. As shown in figure 1 the operation of a one-stage hydride compressor consists of four processes namely: DA – hydrogen absorption at low temperature ($T_L$) and low pressure ($P_L$); AB - sensible heating and compression from cold fluid temperature ($T_L$) to hot fluid temperature ($T_H$); BC - desorption of compressed hydrogen at hot fluid temperature ($T_H$) and high pressure ($P_H$); CD - sensible cooling from $T_H$ to $T_L$. Compressor starts to discharge hydrogen when the hydride equilibrium pressure exceeds the preset value of the delivery pressure. In this case, the storage pressure is constant at the preset value of the delivery pressure.

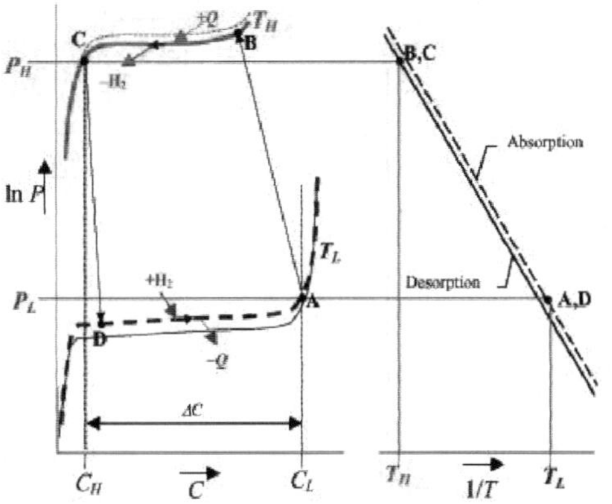


**Figure 1.** Principle of operation of a single stage hydrogen $MH_X$ compressor.

According to equation (1), the pressure increases exponentially with increasing temperature, and large pressure values can be obtained by moderate temperature changes. The metal hydride based hydrogen compressor can be tailored to cover a wide range of operating pressures and pressure ratios by selecting suitable alloys.

To have a high outlet pressure, more hydride units can be serial connected, each unit with a different alloy and successive higher operating pressure. The most important properties of an alloy suitable for hydrogen compression are good hydrogen absorption - desorption rate, smaller process enthalpy, fast reaction kinetics, great structural stability during the cycles. For compression, metal hydrides with large pressure to temperature gradients are desired, especially in the range of low temperatures. The efficiency of a single stage hydrogen metal hydride compressor is given by:

$$\eta = \frac{m(RT_H \ln P_H - RT_L P_L)}{mC_p(T_H - T_L) + \Delta H} \quad (2)$$

where: m is the mass of metal hydride, $C_p$ is the specific heat of metal hydride, $P_L$ is the equilibrium pressure at absorption temperature $T_L$, $P_H$ is the equilibrium pressure at desorption temperature $T_H$, $\Delta H$ is the enthalpy variation of metal hydride and R is the ideal gas constant.

## 3. Selection of alloys

To operate at low heat source temperatures and low hydrogen pressure inlet, the desorption set point of 80°C was chosen while the absorption temperature is 20°C. Metal hydride materials available for these conditions are Ti-Fe, $AB_2$ and $AB_5$ alloys. The $AB_5$ alloys were selected because they combine good thermodynamic properties and high reaction rates over a large range of plateau pressures, easy activation and long term stability [6].

To reach the desired compression ratio, the metal hydrides used for our three-stage hydrogen thermal compression system are: stage 1 -$LaNi_{4.85}Al_{0.15}$, stage 2 -$LaNi_{4.9}Cu_{0.1}$, stage 3 -$MmNi_{4.05}Fe_{0.95}$. These alloys were prepared by arc melting technique and their pressure composition isotherms measured by a house-built volumetric Sievert installation, are shown in figure 2.

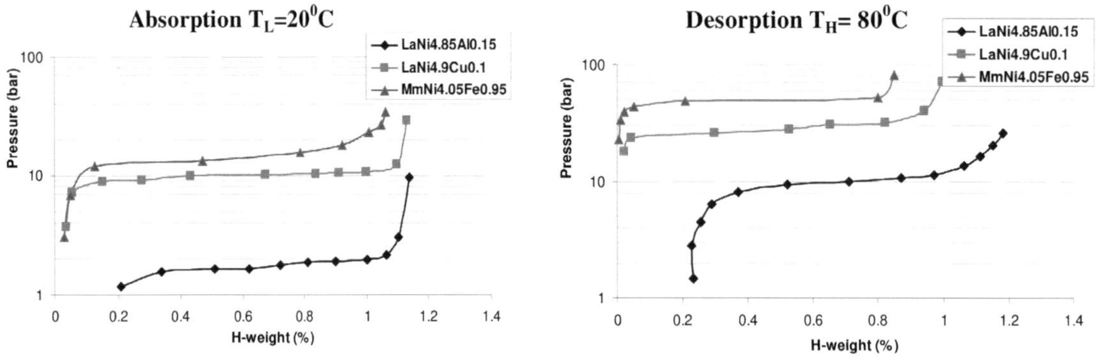


**Figure 2.** Pressure composition isotherms of selected alloys for a three stage hydrogen thermal compressor.

Table 1 presents the characteristics, pressure limit for absorption and desorption and the maximum reversible sorption capacity for selected alloy.

**Table 1.** Characteristics of $LaNi_{4.85}Al_{0.15}$, $LaNi_{4.9}Cu_{0.1}$ and $MmNi_{4.05}Fe_{0.95}$.

| Metal alloy | Absorption Pressure at 20°C (bar) | Desorption Pressure at 80°C (bar) | Storage capacity (H-wt%) |
|---|---|---|---|
| **$LaNi_{4.85}Al_{0.15}$** | 2 | 12 | 1.25 |
| **$LaNi_{4.9}Cu_{0.1}$** | 10 | 24 | 1.22 |
| **$MmNi_{4.05}Fe_{0.95}$** | 20 | 60 | 1.18 |

## 4. Prototype development

The prototype of the three-stage hydrogen thermal compressor (HTC) is designed to operate by cycling between two temperatures assured by thermostatic water baths. The schematic design of HTC and van't Hoff plots for a three-stage hydrogen compressor are shown in figure 3.

The HTC consists of three fast metal hydride reactors (FMHR) able to provide fast hydrogen absorption-desorption cycle and to process reasonable hydrogen flow rates. FMHR combine the effective increase of the thermal conductivity, optimized solutions to dissipate more effectively the heat of reaction and good permeability to hydrogen gas; each reactor is filled with 120 g of hydride alloy powder of starting grain size ~0.4 mm. Gas connections are built with Swagelok connectors with 6 mm pipes and hydrogen pressure inlet and outlet of each reactor are measured by four pressure transducers. Five needle valves and four one-way valves manage the hydrogen path through compressor and a mass flow meter measures the hydrogen flow rate. The hydrogen is delivered in a 3 l volume storage tank. Cold and hot water provided by two circulating baths are swapped between metal hydride reactors, depending on their absorption/desorption status. The experimental setup for the testing of the HTC is equipped with an Omni Logic Data logger for data acquisition regarding hydrogen pressure and flow rate, the cold and hot water temperature.

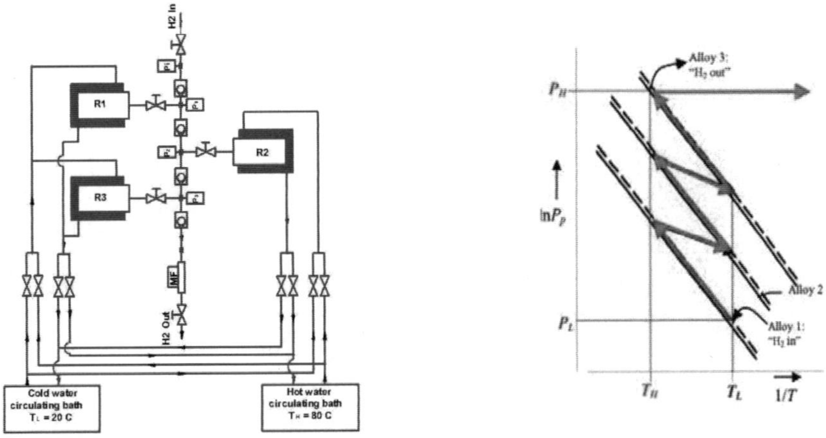


a b

**Figure 3.** Three-stage thermal hydrogen compressor: a) schematic design; b) van't Hoff plots.

## 5. Experimental

Thermal cycling experiments on the prototype shown in figure 4 were focused on the parameters which affect the performances of each stage of hydrogen thermal compressor: temperature and flow rate of cold and hot water, hydrogen inlet pressure and cycle duration.

**Figure 4.** Prototype of the three-stage hydrogen thermal compressor based on metal hydrides.

The temperatures for cold and hot water necessary for thermal cycling are 20°C and 80°C respectively, and the inlet hydrogen pressure is below 2 bars. While the hydrogen transfer rate between two FMHR is driven by pressure difference between the desorption plateau of an alloy and the absorption plateau of the one that occupies the next compressor stage, the heat of reaction tends to reduce the pressure gradient by changing the equilibrium conditions and the reaction slows down.

The absorption-desorption times were chosen to allow almost complete reaction, near 1 H-wt % in the hydrides alloys. Using FMHR the duration of a absorption-desorption cycle can be reduced down to 120 s, providing a good hydrogen absorption capacity of 0.85% H-weight, and a satisfactory flow rate. Thermal cycling behaviour of each stage of the compressor was studied measuring the absorption/desorption pressures and the flow rate of hydrogen inlet/outlet.

## 6. Results and discussion

The results obtained after the thermal cycling experiments on the prototype shows that for each stage of the compressor a fairly good compression ratio can be reached for a reduced duration of absorption-desorption cycle, t = 120 s. Figure 5 shows the evolution of pressure variation for each stage of the thermal compressor. It can seen that for the stage I, at an absorption pressure under 2 bars, after some cycles, the delivery pressure of first metal hydride reactor amounts to10 bar, which correspond to a compression ratio 5:1. For the second stage, the compression ratio is 2:1, and for the third stage is 2.8:1, corresponding to an outlet pressure about 56 bars.

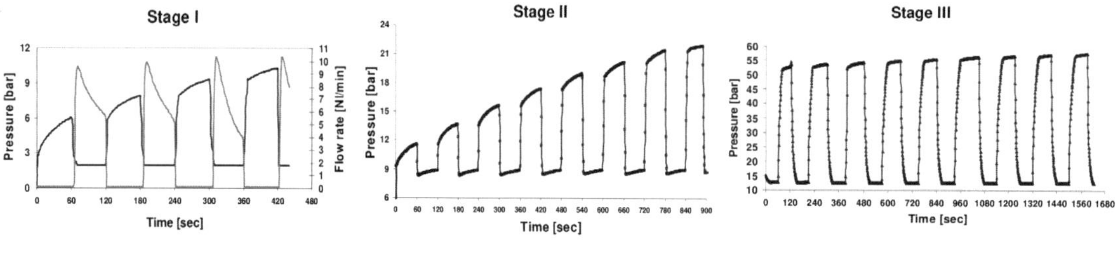


**Figure 5.** Pressure variation plots of the compressor's metal hydride reactors for a 120 sec recurrent absorption desorption cycle.

For an inlet pressure $P_L$=2 bars, the compressor can deliver hydrogen at the pressure $P_H$=56 bars, corresponding to an overall compression ratio 28:1, in the conditions of a reduced cycle duration t=120 s. This allows to obtain a satisfactory hydrogen flow rate q=10 l/cycle which ensures a hydrogen flow rate Q=300 l/hour using a small quantity of hydride alloys, about 360 g, with small energy input.

## 7. Conclusions

An analysis of a three-stage hydrogen thermal compressor performance tests shows that: (*i*) the alloys selected and developed bring out suitable properties for hydrogen storage and compression; (*ii*) technical solutions built in HTC prototype design offer promising conditions to develop efficiently novel compressors that operates in conjunction with advanced hydrogen production technologies and improves the efficiency and economics of the compression process.

## Acknowledgments

This work has received financial support from National Authority for Scientific Research Romania, through the National Plan for research, development and innovation 2007-2013, project no. 21-023.

## References

[1] National Academy of Engineering. Board on Energy and Environmental Systems 2004 *The Hydrogen Economy, Opportunities, Costs, Barriers, and R&D Needs* www.nap.edu.
[2] Zhang J, Fisher T S, Ramachradran P V, Gore JP and Mudawar I 2005 *Journal Heat Transfer* **127** 1391-9
[3] Dehouce Z, Grimard N, Laurencelle J and Bose T K 2005 *J. Alloys Compd.* **399** 224-36
[4] Golben M and DaCosta D H 2002 *Proceedings of the 2001 DOE Hydrogen Program Review*, www1.eere.energy.gov/hydrogenandfuelcells
[5] Jin-Kyeong K, Il-Seok P, Kwang J K and Keith G 2008 *Int. J. Hydrogen Energy* **33** 870-7
[5] Sakintuna B, Lamari F and Hirscher M 2007 *Int. J. Hydrogen Energy* **32** 1121-40.

Journal of Physics: Conference Series **182** (2009) 012054 doi:10.1088/1742-6596/182/1/012054

# Metal hydrides reactors with improved dynamic characteristics for a fast cycling hydrogen compressor

**G Popeneciu, I Coldea, D Lupu, I Misan and O Ardelean**

National Institute for Research and Development of Isotopic and Molecular Technologies, 65-103 Donath, 400293 Cluj-Napoca, Romania

E-mail: gabriel.popeneciu@itim-cj.ro

**Abstract**. This paper presents an investigation of coupled heat and mass transfer process in metal hydrides hydrogen storage reactors. Hydrogen storage and compression performance of our designed and developed reactors are studied by varying the operating parameters and analyzing the effects of metal hydride bed parameters. The metal alloy selected to characterize the cycling behaviour of reactors is $LaNi_5$, material synthesized and characterized by us in the range 20-80°C. Four types of metal hydride reactors were tested with the aim to provide a fast hydrogen absorption-desorption cycle, able to be thermally cycled at rapid rates. Some new technical solutions have been studied to make a step forward in reducing the duration of the reactors cycle, which combines the effective increase of the thermal conductivity and good permeability to hydrogen gas. Dynamic characteristic of developed fast metal hydride reactors is improved using our novel mixture metal hydride-CA conductive additive due to the increased effective thermal conductivity of the alloy bed. The advanced hydride bed design with high heat transfer capabilities can be thermally cycled at a rapid rate, under 120 seconds, in order to process high hydrogen flow rates.

## 1. Introduction

The major challenges for hydrogen to become a successful energy vector in the future consist in developing of new and efficient technologies in hydrogen production, storage and in the end uses. Technologies considered under this goal include: improving hydrogen production technologies, developing innovative ideas for hydrogen transport and storage, able to be employed in very practically purpose like engine supplies and fuel cells [1,2].

It is known that hydrogen storage represent a technological challenge in developing hydrogen economy. Among the different hydrogen storage methods, reversible metal hydrides are considered as an efficient volume medium for hydrogen storage, offering safe, low pressure and high energy density. Whatever particular hydrogen storage method is selected, the efficient hydrogen compression is an unavoidable step. Hydrogen compressors based on the reversible absorption/desorption ability of metal hydrides are now intensively studied [3,4], and can reach high hydrogen delivery pressure with many advantages: high reliability and low maintenance, low energy costs and conserve the purity of the hydrogen. Usage of the metal hydrides to hydrogen thermal compression, employment the thermodynamic characteristics of hydrogen absorbent alloys: the equilibrium pressure $P_{eq}$ variation with temperature T is descript by van't Hoff equation:

$$\ln P_{eq} = \frac{\Delta H}{RT} - \frac{\Delta S}{R} \quad (1)$$

where: $\Delta H$ and $\Delta S$ are the enthalpy and entropy variation and R is the ideal gas constant.

The selection of alloys employed in hydrogen storage and compression depend on their properties. If for storage devices, the alloys with high absorption capacity and a good hydrogen absorption-desorption kinetics are appropriate, in the case of compression devices based on metal hydrides the most important properties are: the good hydrogen absorption-desorption rate, a smaller process enthalpy, great structural stability during the cycles. In this case are needed alloys with big differences between the hydrogen absorption and desorption pressures at equilibrium at small temperature differences. The hydrogen thermal compression using a metal hydride is essentially a two-step process. Figure1 shows the principle of operation for a single stage compressor.

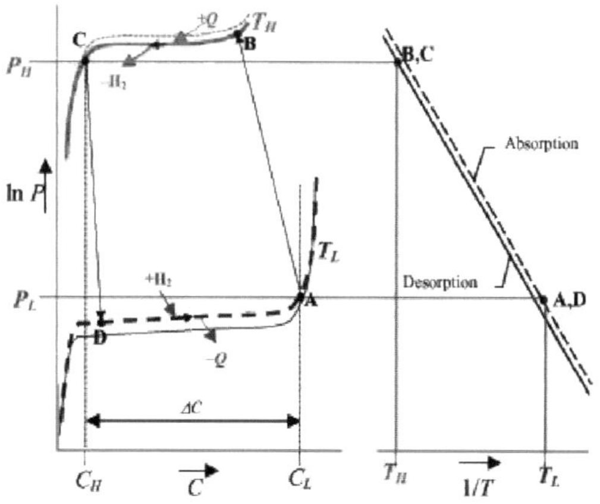


**Figure 1.** Principle of operation of a single stage hydrogen metal hydride compressor.

To realize an efficient hydrogen thermal compressor, the main goal of our researches, it is essential to develop fast cycling metal hydride reactors (FMHR). The most important characteristics of a hydrogen storage reactor suitable for a metal hydride based compressor are: a high thermal conductivity of the hydrogen storage bed and a good permeability for hydrogen gas. To aim a short duration of an absorption-desorption cycle, FMHR requires a rapid mass (hydrogen gas) and heat transfer. These can be previously hindered by the following constraints: (i) heat transfer through a metal hydride powdered bed is inherently poor; (*ii*) hydrogen pressure drops through the hydride beds may become so excessive that gas transfer is seriously impeded.

## 2. Experimental

The metal alloy selected to characterize the cycling behavior of FMHR is $LaNi_5$, synthesized by IMNR Bucharest and characterized by us, in the range 20-80°C. The pressure composition isotherm (p-c-T) for $LaNi_5$, measured by a house-built volumetric Sievert installation, is shown in figure 2. We designed and built four types of metal hydride reactors which were tested with the aim to provide a fast hydrogen absorption-desorption cycle, capable to be thermally cycled at rapid rates, in order to process reasonable hydrogen flow rates. Each type of reactor combines the effective increase of the thermal conductivity, optimized solutions to dissipate more effectively the heat of reaction and the good permeability to hydrogen gas. The FMHR, with the characteristics presented in figure 3, have the same active volume, capable to accommodate 120 g of hydride alloy powder.

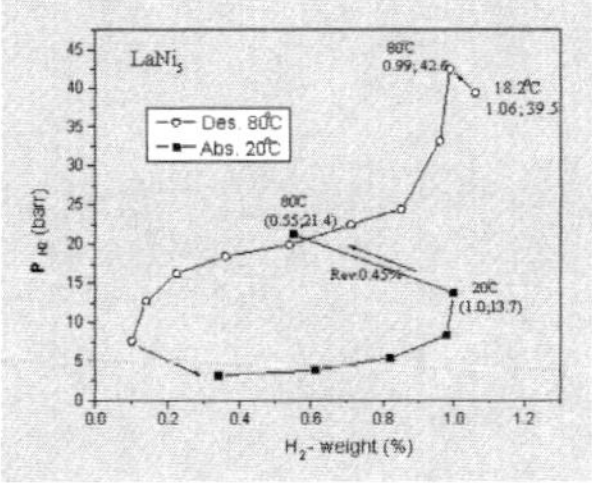


**Figure 2.** p-c-T curves of $LaNi_5$.

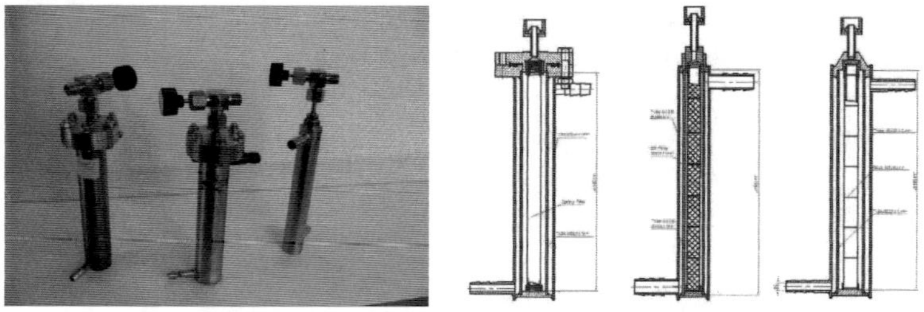

**Figure 3.** Design of the fast metal hydride reactors (FMHR).

The experimental set-up shown in figure 4 is designed to provide thermal absorption-desorption cycles for the FMHR and a Data Logger measure the main parameters of the process [5].

The thermal cycling experiments were focused on the study of the parameters which affect the FMHR performances: (*i*) operating parameters as cooling and heating water flow rate and temperature, (*ii*) overall heat transfer coefficient improved by conducting additives added to $MH_x$, (*iii*) metal hydride bed thickness and (*iv*) the increment of heat exchange surface reactor.

The absorption and desorption duration were chosen to allow almost complete reaction according to p-c-T curves of $LaNi_5$.

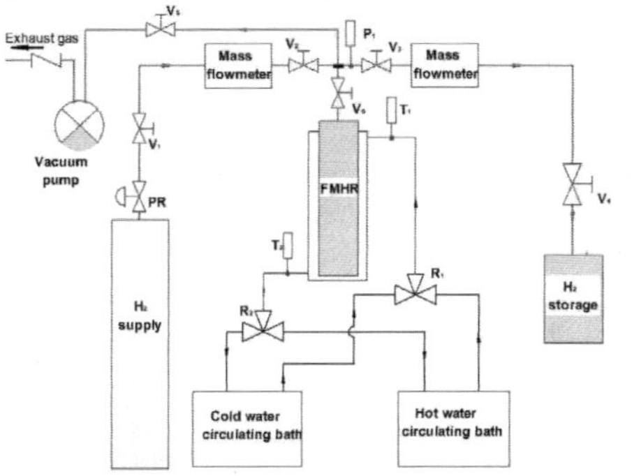


**Figure 4.** Set up for FMHR experiments: PR–2 stages pressure regulator; V1-V5- needle valves; R1, R2-3 ways ball valves; T1,T2-"K" thermocouples; P1-pressure transducer.

Applying this strategy, the experiment has been made following three main directions:

(*i*) In parallel, two FMHR type I were filled, one with 120 g $LaNi_5$ and the other with a mixture 120g $LaNi_5$ plus 5.25 g of our conducting additive (CA) were tested by varying operating parameters regarding temperature and flow rate of cooling and heating fluids;

(*ii*) Four type II FMHR were filled, one with 120 g $LaNi_5$ and for the others three adding to hydrogen storage powdered alloy an amount of some different conductive additives (CA based on carbon, adsorbent carbon and graphite) to increase the effective thermal conductivity of the bed, and cycled with different operating parameters to determinate their cycling behaviour;

(*iii*) FMHR type III and FMHR type IV with increased heat exchange surface, filled with the same quantity of mixture $LaNi_5$ + CA, having the bed alloy thickness $\Delta bed_{III}$ = 4mm and $\Delta bed_{IV}$ = 2.5mm were tested about 3000 absorption-desorption cycles varying the cycle duration.

## 3. Results and discussion

Figure 5 shows the time dependence of hydrogen absorption capacity of FMHR type I on the flow rate of cold fluid at a given absorption temperature $T_L$ = 20°C. It is observed that the duration of a full hydrogen absorption cycle decrease for high flow rates.

Figure 6 shows that that for a given supply pressure, hydrogen delivery and storage pressure increases with the hot fluid temperature, due to the increase in equilibrium pressure of the hydride bed with bed temperature according to van't Hoff equation. Thus, all later experiments were performed in the temperature range 20-80°C, using 12 l/min cold-hot water flow rate.

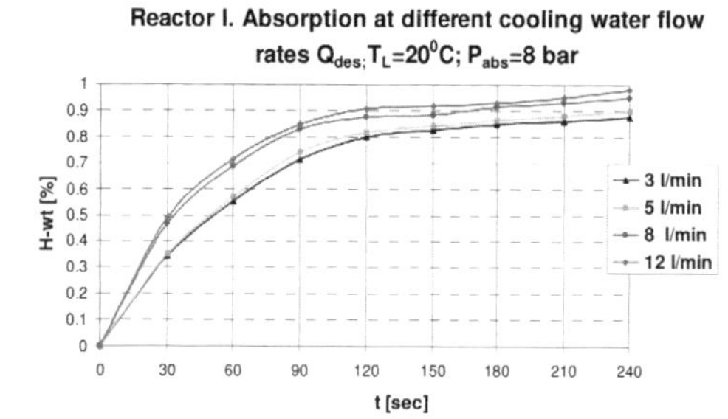


**Figure 5.** Variation of absorption cycle with cold fluid flow rate.

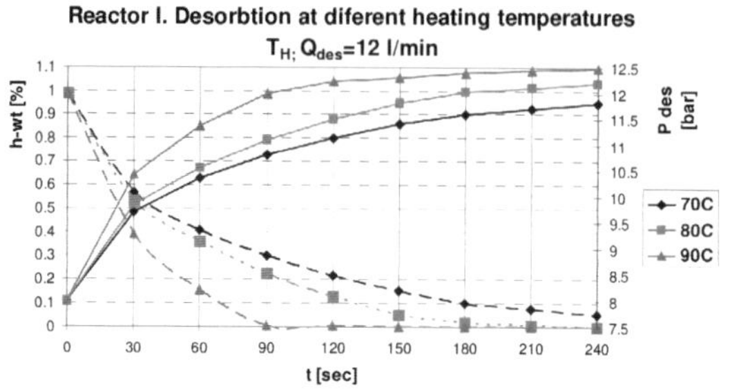


**Figure 6.** Variation of pressure and hydrogen desorption with fluid temperature.

Figure 7 shows the effect of CA conducting additive mixed with metal hydride on the duration of absorption-desorption cycle due to the increased effective thermal conductivity of the alloy bed. It was found that an amount of 4.3% conductive additive added to the $LaNi_5$ alloy does not affect the hydrogen absorption rates, but increases the reaction kinetics and could reach the target of 120 seconds cycle life.

Figure 8 shows the superior behavior of our CA conductive additive versus other carbon based additives like adsorbent carbon and graphite.

Figure 9 shows the influence of bed design and heat exchange unit optimization on the dynamic characteristics of FMHR. It is observed that a smaller bed thickness and an extended heat exchange surface of metal hydride reactor type IV increase the reaction kinetics.

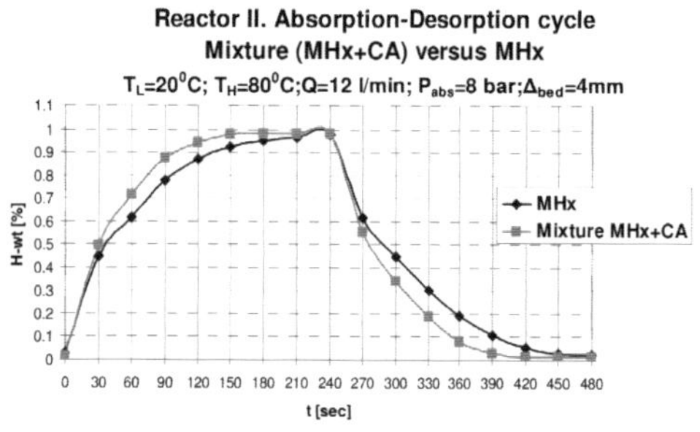


**Figure 7.** Effect of CA about the duration of a cycle.

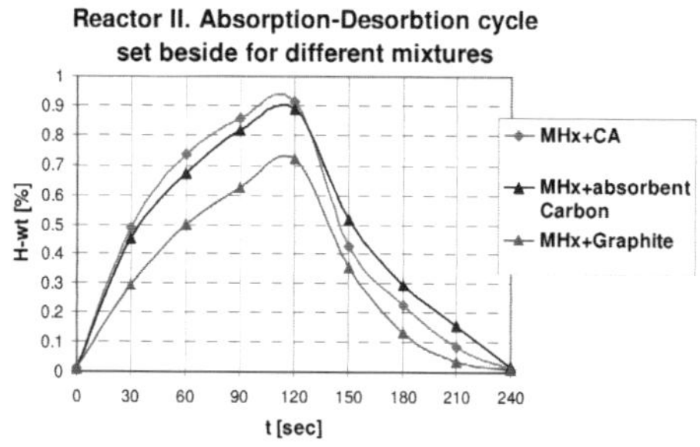


**Figure 8.** Behavior of some additives in $LaNi_5$.

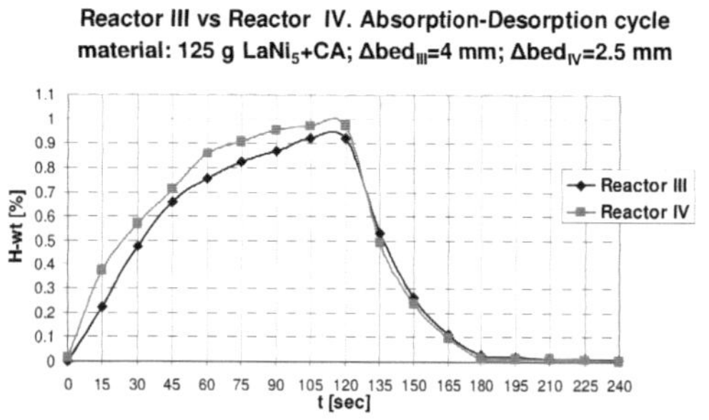


**Figure 9.** The influence of reactor bed design.

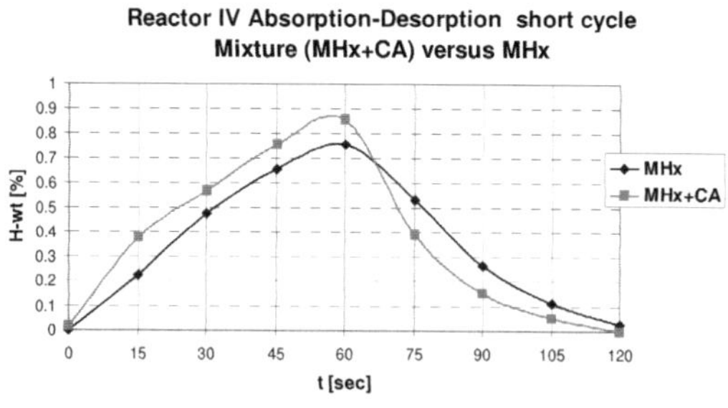


**Figure 10.** Similitude of two thermal cycle.

Thermal cycles of a FMHR type IV, shown in figure 10 using as hydriding materials $LaNi_5$ and ($LaNi_5$+CA) mixture, show that the reactor can be cycled at a rapid rate of 120 seconds providing a good hydrogen absorption capacity of 0.85% H-weight, and a satisfactory flow rate, 10 N liters/cycle.

## 4. Conclusions

The design and development of fast metal hydrides reactors has been studied, for an investigation of the coupled heat and mass transfer process that take place in hydrogen storage/compression devices. Dynamic characteristics of the developed FMHR define that the absorption/desorption cycle duration is improved using our novel mixture metal hydride-CA conductive additive due to the increased effective thermal conductivity of the alloy bed. The advanced hydride bed design with high heat transfer capabilities can be thermally cycled at a rapid rate, under 120 seconds. in order to process high hydrogen flow rates.

## Acknowledgments

This work has received financial support from National Authority for Scientific Research Romania, through the National Plan for research, development and innovation 2007-2013, project 21-023.

## References

[1] European Hydrogen and Fuel Cell Technology Platform 2005 *HFP strategic overview 2005* http://www.hfeurope.org/hfp/keydocs
[2] Lupu D 2004 oral presentation at *JRC-IE training workshop on Mapping European knowledge on Hydrogen Storage* http://ie.jrc.ec.europa.eu
[3] Golben M and DaCosta D H 2001 *Proceedings of the 2001 DOE Hydrogen Program Review* http://www1.eere.energy.gov/hydrogenandfuelcells
[4] Laurencelle F, Dehouche Z, Goyette J and Bose TK 2006 *Int. J. Hydrogen Energy* **31** 762-8
[5] Almasan V, Biris A, Coldea I, Lupu D, Misan I and Popeneciu G 2007 *Proc. 8th Int. Congr. Catalysis EuropaCat* p 11.

Journal of Physics: Conference Series **182** (2009) 012055 doi:10.1088/1742-6596/182/1/012055

# $TiO_2$ - based systems for photoelectrochemical generation of solar hydrogen

**Teofil D Silipas[1], Emil Indrea[1], Simina Dreve[1], Ramona-Crina Suciu[1], Marcela Corina Rosu[1], Virginia Danciu[2], Veronica Cosoveanu[2] and Violeta Popescu[3]**

[1] National Institute for Research and Development of Isotopic and Molecular Technologies, 65-103 Donath, 400293 Cluj-Napoca, Romania

[2] Babes-Bolyai University, Faculty of Chemistry and Chemical Engineering, 11 Arany Janos, 400028 Cluj-Napoca, Romania

[3] Technical University of Cluj-Napoca, Department of Materials Science and Engineering, 103-105 Muncii, 400641 Cluj-Napoca, Romania

E-mail: dan.silipas@itim-cj.ro

**Abstract.** In the present work our attention was focused on the obtaining photoelectrodes for photoelectrochemical cell that incorporate two electrodes, one of which has been titania ($TiO_2$) coated on a transparent conducting oxide (TCO), referred to as the primary electrode and the other, the counter electrode, a non-corrosive metal such as a thin layer of platinum. A thin layer of a nanoporous $TiO_2$ semiconductor was deposited onto a sheet of ITO conducting glass (sheet resistance ~ 30 $\Omega cm^{-2}$). Several complementary investigation techniques like BET, SEM and XRD were used to follow the influence of the reactants molar ratio and thermal treatment on the $TiO_2$ photoanode. The nanocrystalline $TiO_2$/ITO conducting glass electrode seems to be a promising photoanode in a photoelectrochemical cell for hydrogen generation by water splitting.

## 1. Introduction

Hydrogen generated from the splitting of water using solar energy, termed *solar-hydrogen*, represents a sustainable fuel that is environmentally safe [1]. There is a growing awareness that titania ($TiO_2$) and $TiO_2$-based oxide systems are the most promising candidates for the development of photoelectrodes for photoelectrochemical cell (PEC) for solar-hydrogen production [1]. The PEC cell is equipped with a single photoelectrode (photoanode) and cathode, both of which are immersed in an aqueous electrolyte. Exposure of the photoanode to sunlight results in charge transport within the PEC and evolution of gases at the photoanode and cathode. The oxygen energy level ($O_2/H_2O$) should be above the valence band of the photoanode in order to allow electron transfer and, for the same reason, the hydrogen energy level ($H^+/H_2$) should be below the Fermi level of the metal cathode (or the conduction band of a semiconducting photocathode).

The most critical issue in the development of solar-hydrogen is the development of a special class of new photosensitive materials for efficient and clean conversion of solar energy [2]. The development of these materials, which must exhibit sophisticated functional properties, will require the application of the most recent progress in the science of material interfaces and the solid state science. Because of their ultra low density and high surface area, $TiO_2$ aerogels are attractive for applications in solar energy conversion [3, 4].

In this work, we investigate the preparation of a novel nanocrystalline $TiO_2$/ITO conducting glass electrode that seems to be a promising photoanode in a PEC for generation of solar hydrogen. Nanocrystalline semiconductor films are constituted by a network of mesoscopic $TiO_2$ particles witch are interconnected to allow for electronic conduction to take place. The nanoporous $TiO_2$ film was investigated by several complementary techniques like SEM and XRD.

## 2. Experimental

### *2.1. Samples preparation*

$TiO_2$ gels were prepared with the acid ($HNO_3$)-catalyzed sol-gel method by using titanium isopropoxide (TTIP), $HNO_3$, EtOH and $H_2O$ with different molar ratios. The gels were allowed to age in the mother liquor for a week to several months. Before drying, the gels were successively washed with excess of fresh alcohol, at least four times. $TiO_2$ aerogels were obtained by supercritical evacuation of solvent from gels prepared by acid-catalyzed sol-gel method [5].

### *2.2. Electrodes preparation*

The nanocrystalline titania films photoelectrodes for PEC preparation involves the production of a titanium dioxide suspension and then the application of this suspension onto conducting ITO glass. The titania slurry was made from a mixture of: 2 g $TiO_2$ powder, 4 ml of distilled water, 5 drops of Triton X-100 (Sigma-Aldrich) and 0.2 ml of acetylacetonate (Sigma-Aldrich). By adjusting the quantity of Triton X-100 and water the desired viscosity was obtained. This slurry was mixed using a mortar and pestle until uniform. The solution was then applied to a piece of ITO conducting glass (sheet resistance ~ $30\Omega cm^{-2}$) and a semiconductor film was produced. The application technique was the doctor blade method. Greater thickness of the $TiO_2$ film is produced by repeating the doctor blade method with drying at 50°C in between coats. Repeating the doctor blade method with calcining between each coat can produce multiple coats. This is called repetitive combustion coating technique.

In our case the first layer was produced utilizing commercial nanostructured $TiO_2$ (Degusa P25) and the second one utilizing $TiO_2$ aerogel suspensions. The $TiO_2$/ITO conducting glass electrode is then dried at room temperature for 45 min and cut with a scalpel into a 0.5 x 0.5 cm square. The working electrode is then calcined at 450°C for 2 h.

### *2.3. Samples measurements*

The SEM images were collected with a JEOL 7601F electronic microscope with a field emission tip that operated at 10 kV.

The specific surface area of the as prepared samples was determined using Brunauer-Emmett-Teller (BET) method, in a partial pressure range of $0.05 < P/P0 < 0.3$. Krypton adsorption was carried out at 77 K. Before each measurement, the samples were heat-cleaned at 333 K for 2 h.

The X-ray diffraction patterns were obtained using a standard BRUKER D8 Advance X-ray diffractometer, working at 40 kV and 40 mA. The Cu Kα radiation was monochromatized with a germanium monochromator placed in the incident X-ray beam. The data of the X-ray diffraction patterns were collected in a step-scanning mode with $\Delta 2\theta = 0.01^0$ steps.

## 3. Results and discussion

### *3.1. $TiO_2$ aerogels*

Titanium dioxide is the most commonly used wide band-gap semi-conductor as photoelectrodes for solar-hydrogen [1]. The nanoporous $TiO_2$ aerogel layer has a high specific surface area. The increases of the contact area of the photoelectrodes with light raise the light energy conversion efficiency of the cell. Through scanning electron microscopies (SEM) the morphology of the $TiO_2$ aerogel was determinated to have two length scales, one for crystalline nanoparticles and other for mesoaggregates constituted by anatase closely packed (figure 1) [6].

**Figure 1.** SEM image of the as prepared TiO2 aerogel (sample T5). Note that the mesospheres dimension is of about 30 nm.

The $TiO_2$ aerogels were heat-treated at 400, 500 and 600°C for 2 h under argon atmosphere and these aerogels were referred to as T4, T5, and T6, respectively. The X-ray powder diffraction technique was used for structural characterization of the $TiO_2$ heat-treated aerogels.

The microstructural information obtained by the single X-ray profile Fourier analysis of the $TiO_2$ heat-treated T4, T5, and T6 $TiO_2$ aerogels are: the $D_{eff}$ (nm) effective crystallite mean size, the $\langle\varepsilon^2\rangle^{1/2}_{hkl}$ root mean square (rms) of the microstrains averaged along [hkl] direction, and dislocation density $\rho$ [7, 8]. Dislocation density $\rho$ (lines $m^{-2}$), as defined by Hirsch [9] is the length of dislocation lines per unit volume of the crystal.

Since both the crystallite size and residual strain are effects of the dislocation networks in the structure, the dislocation density can be expressed as [9]: $\rho = \sqrt{12}\left\langle \varepsilon^2 \right\rangle^{1/2} / (D_{eff} \cdot d)$ where $\langle\varepsilon^2\rangle^{1/2}_{hkl}$ is the rms of the microstrains, $D_{eff}$ (nm) is the effective crystallite mean size and d(nm) the interplanar spacing.

The Warren-Averbach X-ray profile Fourier analysis of the (200) and the (220) $TiO_2$ anatase peak profiles were processed by a XRLINE [8] computer program. Table 1 summarizes the microstructural parameters of the T4, T5, and T6 samples. It can be observed that higher temperature treatment favor the growth of larger crystallite size and lower crystalline imperfections.

**Table 1**. Microstructural parameters and BET surface area of the $TiO_2$ heat-treated T4, T5, and T6 $TiO_2$ aerogels.

| $TiO_2$ aerogel sample | Deff [nm] | $\langle\varepsilon^2\rangle^{1/2}_{hkl} \times 10^3$ | $\rho \times 10^{14}$ [lines $m^{-2}$] | BET surface area [$m^2$/g] |
|---|---|---|---|---|
| T4 | 4.9 | 3.432 | 13.63 | 685 |
| T5 | 9.1 | 0.819 | 1.75 | 593 |
| T6 | 14.8 | 0.609 | 0.81 | 466 |

### *3.2. $TiO_2$ nanocrystalline electrode*

The nanocrystalline $TiO_2$/ITO conducting glass electrode for photoelectrochemical generation of solar hydrogen was prepared by repetitive combustion coating. The first compact layer was produced utilizing commercial nanostructured $TiO_2$ (Degussa P25) and the second one utilizing $TiO_2$ aerogel suspensions. The $TiO_2$/ITO conducting glass electrode is then dried at room temperature for 45 min and annealed at 450°C for 2 h.

The X-ray powder diffraction technique was used for structural characterization of the nanocrystalline $TiO_2$/ITO conducting glass annealed photoelectrode [10].

X-ray diffraction pattern of the nanocrystalline $TiO_2$ photoelectrode suggests the formation of a major anatase-like crystalline network structure (83.77 vol %) with a = b = 0.378(4) nm and c = 0.951(4) nm lattice parameters and a minor rutile-like crystalline phase (16.23 vol. %) with a = b = 0.459(9) nm and c = 0.295(6) nm lattice parameters, quite close to the ASTM reported values. Unit cell parameters and $TiO_2$ anatase/rutile crystalline phase composition were calculated through Rietveld refinement using the PowderCell software [11].

The microstructural parameters of the nanocrystalline $TiO_2$ photoelectrode after heating two hours at 450ºC showed a average size of the $TiO_2$ anatase of 5 nm and a root mean square of the microstrains $<\varepsilon^2>^{1/2} = 8.2 \times 10^{-2}$.

## 4. Conclusions

The common strategy of research on photoelectrodes for photoelectrochemical generation of solar-hydrogen involves the application of $TiO_2$-based materials that are modified using several procedures, including $TiO_2$ aerogels that combine the aerogel properties with the photocatalytical property of the titania.

The nanocrystalline $TiO_2$/ITO conducting glass electrode seems to be a promising photoanode in a PEC for hydrogen generation by water splitting.

Structural and dimensional investigations of $TiO_2$ aerogels heat-treated for two hours at temperatures between 400 and 600ºC have been performed by means of XRD techniques. Correlations between the morphology and spectral characteristics of the annealed aerogels have been evidenced by assuming the occurrence of the finite size effect of nanocrystallites.

## Acknowledgements

This work was supported by the Ministry of Education and Research through MENER program, HYDROSOL project No. 710/24.072006. We thank to DEGUSSA A.G.- GERMANY for sponsoring the research team with Degussa P25 titania.

## References

[1] Nowotny J, Bak T, Nowotny M K and Sheppard L 2007 *Int. J. Hydrogen Energy* **32** 2609
[2] Nowotny J and Sheppard L R 2007 *Int. J. Hydrogen Energy* **32** 2607
[3] Yao N, Cao S and Yeung K L 2009 *Microporous and Mesoporous Materials* **117** 570
[4] Baia L, Peter A, Cosoveanu V, Indrea E, Baia M, Popp J and Danciu V 2006 *Thin Solid Films* **511** 512
[5] Peter A, Baia L, Cosoveanu V, Baia M, Constantin D, Indrea E, Popescu I C and Danciu V 2009 *J. Optoelectron. Adv. Mat. Symposia* **1** 6
[6] Danciu V, Cosoveanu V, Peter A, Marian I, Marginean P and Indrea E 2003 *Ann. West Univ. Timisoara: Chemistry* **12** 1029
[7] Indrea E and Barbu A 1996 *Appl. Surf. Sci.* **106** 498
[8] Aldea N and Indrea E 1990 *Comput. Phys. Commun.* **60** 155
[9] Hirsch P G 1956 *Prog. Math. Phys.* **6** 236
[10] Indrea E *et al* 2008 *J. Optoelectron. Adv. Mater.* **10** 2213
[11] Kraus W and Nolze G 1996 *J. Appl. Crystallogr.* **29** 301.

# Local structure information by EXAFS analysis using two algorithms for Fourier transform calculation

**N Aldea[1], S Pintea[1], V Rednic[1], F Matei[2], Hu Tiandou[3] and Xie Yaning[3]**

[1] National Institute for Research and Development of Isotopic and Molecular Technologies, 65-103 Donath, 400293 Cluj-Napoca, Romania

[2] University of Agricultural Sciences and Veterinary Medicine, 3-5 Calea Manastur, 400372 Cluj-Napoca, Romania

[3] Beijing Synchrotron Radiation Facilities of Beijing Electron Positron Collider National Laboratory, People's Republic of China

E-mail: nicolae.aldea@itim-cj.ro

**Abstract**. The present work is a comparison study between different algorithms of Fourier transform for obtaining very accurate local structure results using Extended X-ray Absorption Fine Structure technique. In this paper we focus on the local structural characteristics of supported nickel catalysts and $Fe_3O_4$ core-shell nanocomposites. The radial distribution function could be efficiently calculated by the fast Fourier transform when the coordination shells are well separated while the Filon quadrature gave remarkable results for close-shell coordination.

## 1. Introduction

The X-ray absorption spectroscopy (XAS) can yield several characteristic of the studied material: the electronic and structural information about the local environment around a specific atomic constituent in the amorphous materials [1-2], the location and chemical state of any catalytic atom on any support [3] and the nanoparticle size of the transition metal oxides [4-5]. Extended X-ray absorption fine structure (EXAFS) is a scattering technique in which a core electron ejected by an X-ray photon probes the local environment of the absorbing atom. The ejected photoelectron backscattered by the neighboring atoms located around the absorbing atom interferes constructively with the outgoing electron wave depending on its energy. The energy of the photoelectron is equal to the difference between the X-ray photon energy and a threshold energy associated with the ejection of the electron.

## 2. Theoretical background

The interference between the outgoing and backscattered electron waves has the effect of modulating the X-ray absorption coefficient. The EXAFS function $\chi(k)$ is defined in terms of the atomic absorption coefficient by:

$$\chi(k) = \frac{\mu(k) - \mu_0(k)}{\mu_0(k)} \tag{1}$$

where $k$ is the electron wave vector, $\mu(k)$ refers to the absorption by an atom from the material of interest and $\mu_0(k)$ refers to the absorption by a free atom. Theories of EXAFS based on the single scattering approximation of the ejected photoelectron by atoms in immediate vicinity of the absorbing atom give the following expression for $\chi(k)$ [6]:

$$\chi(k) = \sum_j A_j(k)\sin\left[2kr_j + \delta_j(k)\right] \tag{2}$$

where the sum extends over $j^{th}$ coordination shell, $r_j$ is the radial distance from the $j^{th}$ shell and $\delta_j(k)$ is the total phase shift function. The amplitude function $A_j(k)$ is given by:

$$A_j(k) = \left(\frac{N_j}{kr_j^2}\right)F(k, r_j, \pi)\exp\left[2r_j / \lambda_j(k) - k^2\sigma_j^2\right] \tag{3}$$

In this equation $N_j$ is the number of atoms in the $j^{th}$ shell, $\sigma_j$ is the root means squares deviation of distance about $r_j$, $F(k,r,\pi)$ is the backscattering amplitude and $\lambda_j(k)$ is the mean free path function for inelastic scattering. The backscattering factor and the phase shift depend on the kind of atom responsible for scattering and its coordination shell [7]. The analysis of EXAFS data for obtaining structural information [$N_j$, $r_j$, $\sigma_j$, $\lambda$ $(k)$] generally proceeds by employing the Fourier transform. The radial structure function (RSF) can be derived from $\chi(k)$. The single shell may be isolated by Fourier transform:

$$\Phi(r) = \int_{-\infty}^{\infty} k^n \chi(k) WF(k)\exp(-2ikr)dk. \tag{4}$$

The EXAFS signal is weighted by $k^n$ ($n = 1, 2, 3$) to get the distribution function of atom distances. Different apodization windows $WF(k)$ are available as Gauss or Hanning. For experimental reasons it is not possible to cover the full $(-\infty, +\infty)$ wave vector interval. If we take into consideration $\Delta r$ and $\Delta k$ as mesh values for variables $r$ and $k$ and we consider $N$ values of $k$ variable, then between meshes $\Delta r$ and $\Delta k$ there is following relation:

$$\Delta r = \frac{1}{N\Delta k}. \tag{5}$$

With these assumptions the general relation for discrete Fourier transform is given by

$$\Phi(n\Delta r) = \Delta k \sum_{m=-N/2}^{N/2-1} h(m\Delta k)\exp\left(-\frac{2\pi imn}{N}\right) \tag{6}$$

where $h = k^n \chi WF$ and $n = \overline{0, N-1}$. The fast Fourier transform algorithm [8] was developed based on relation (6) and was successfully used in many applications. Advantages and limits of this algorithm are treated in reference [9]. Beside this algorithm we also implemented in our software package a generalized Filon quadrature [10] for direct and inverse Fourier transform calculation. The following relations give the real and imaginary parts of the Fourier transform:

$$\int_{x_0}^{k_{2n}} h(k)\cos 2\pi krdk \approx \Delta k\left[\alpha(2\pi r\Delta k)\left(h_{2n}\sin 2\pi k_{2n}r - h_0 \sin 2\pi k_0 r\right) + \beta(2\pi r\Delta k)C_{2n} + \gamma(2\pi r\Delta k)C_{2n-1}\right] \tag{7}$$

$$\int_{x_0}^{k_{2n}} h(k)\sin 2\pi krdk \approx \Delta k\left[\alpha(2\pi r\Delta k)\left(h_0\cos 2\pi k_0 r - h_{2n}\cos 2\pi k_{2n} r\right) + \beta(2\pi r\Delta k)S_{2n} + \gamma(2\pi r\Delta k)S_{2n-1}\right] \tag{8}$$

where

$$C_{2n} = \sum_{i=0}^{n} h_{2i} \cos(2\pi r k_{2i}) - \frac{1}{2}\left[h_{2n} \cos 2\pi r k_{2n} + h_0 \cos 2\pi r k_0\right],\quad C_{2n-1} = \sum_{i=1}^{n} h_{2i-1} \sin 2\pi r k_{2i-1},$$

$$S_{2n} = \sum_{i=0}^{n} h_{2i} \sin(2\pi r k_{2i}) - \frac{1}{2}\left[h_{2n} \sin(2\pi r k_{2n}) - h_0 \sin(2\pi r k_0)\right],\ S_{2n-1} = \sum_{i=1}^{n} h_{2i-1} \sin(2\pi r k_{2i-1})$$

$$\alpha(\theta) = \frac{1}{\theta} + \frac{\sin 2\theta}{2\theta^2} - \frac{2\sin^2\theta}{\theta^3}, \beta(\theta) = 2\left(\frac{1+\cos^2\theta}{\theta^2} - \frac{\sin 2\theta}{\theta^3}\right), \gamma(\theta) = 4\left(\frac{\sin\theta}{\theta^3} - \frac{\cos\theta}{\theta^2}\right).$$

If $0 < \theta < 1$, functions $\alpha, \beta$ and $\gamma$ can be successfully approximated by:

$$\alpha \approx \frac{2\theta^3}{45} - \frac{2\theta^5}{315} + \frac{2\theta^7}{4725}, \beta \approx \frac{2}{3} + \frac{2\theta^2}{15} - \frac{4\theta^4}{105} + \frac{2\theta^6}{567}, \gamma \approx \frac{4}{3} - \frac{2\theta^2}{15} + \frac{\theta^4}{210} - \frac{\theta^6}{11340}.$$

## 3. Experimental

The transmission EXAFS measurements were carried out in the 4W1B beamlines at the Beijing Synchrotron Radiation Facilities (BSRF) operating at 50-80 mA and 2.2 GeV at room temperature. The features of 4W1B beamline are: energy range of 3.5-22 keV, energy resolution of ΔE = 0.5-2 eV at E = 10 keV, Bragg angle range of 5-70°, the crystals Si(111), Si(220) and Si(311) can be alternatively used. $Fe_3O_4$ powders with 99.98% purity and Ni nanoclusters supported on $Cr_2O_3$ treated at 350°C were used as test samples. Atomic concentration of Ni and Cr were 85%, respectively 15at%. Harmonics were rejected by detuning of monochromator. Absorption coefficients of Fe and Ni K edges were determined using a Si(111) double-crystal monochromator. Ionization chambers monitored the X-ray intensities of incident and transmitted beams. Special care was taken in sample preparation, especially for the thickness and homogeneity of the samples, in order to obtain high quality absorption spectra. All samples were ground to fine powder and homogeneously dusted on scotch tape. We used for absorption coefficients measurements, the energy scanning range from 6994 eV to 8108 eV and 8200 eV to 9300 eV for $Fe_3O_4$ and $Ni/Cr_2O_3$ nanosystems, respectively. The EXAFS analysis of the absorption coefficients was processed with computer codes EXAFS51 to EXAFS56 [11] from our library.

## 4. Results and discussion

The structure of $Fe_3O_4$ nanoclusters is an interesting problem from EXAFS perspective because it involves three types of absorbers: the tetrahedral $Fe^{3+}$, the octahedral $Fe^{3+}$ and octahedral $Fe^{2+}$.

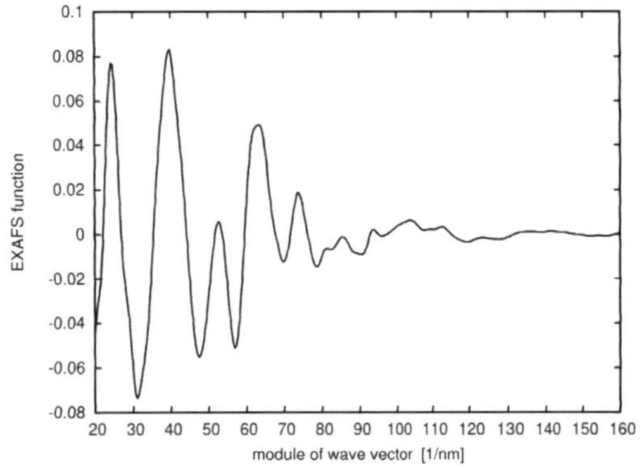


**Figure 1.** The EXAFS signal of Fe K edge for investigated sample.

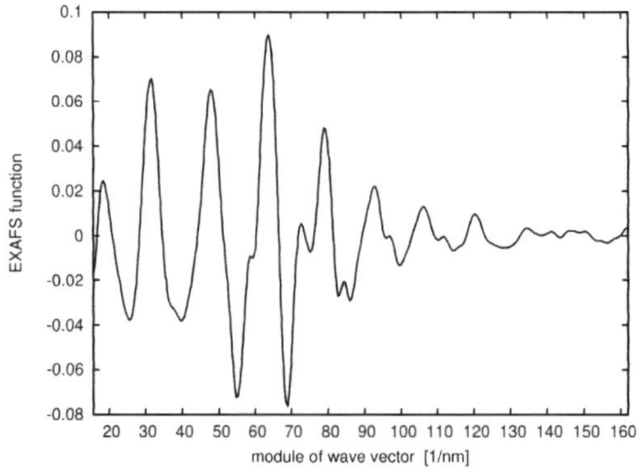


**Figure 2.** The EXAFS signal of Ni K edge for investigated sample.

The extraction of EXAFS signals requires several steps: calculation of the threshold energy of Fe and Ni K-edges, background removal through pre-edge and after-edge base line fitting with different possible modelling functions, and $\mu_0(k)$ and $\mu(k)$ evaluation. In accordance with equation (1) EXAFS signals were processed for intervals 37.5-160 $nm^{-1}$ and 15.5-161.9 $nm^{-1}$ for $Fe_3O_4$ and Ni as test samples.

In order to obtain the RSF we use equation (4). The mean Fe-O distances of the first and the second coordination shell for standard sample at room temperature are close to values of $R_1$ = 0.189 nm and $R_2$ = 0.209 nm. By taking into account this very small difference between $R_1$, $R_2$ and the discrimination steps, $\Delta k$ and $\Delta r$ of $\chi(k)$ and $\Phi(r)$ functions given by relation (5), we conclude that it is not possible to obtain a reliable resolution for the RSF. This means that the first and second shells are overlapped or only partial separated.

To avoid this disadvantage we developed the Filon algorithm for calculation of the Fourier transform described in relations (7-8). Based on this procedure the Fourier transforms of $k^3\chi(k)WF(k)$ performed in the interval 0.01-0.5 nm, are shown in figure 3 for standard $Fe_3O_4$ powder. Each peak from $|\Phi(r)|$ is shifted from the true distance due to the phase shift function that is included in the EXAFS signal. In the standard sample, the iron ions $[Fe^{3+}]$ and $[Fe^{2+}, Fe^{3+}]$ from tetrahedral and octahedral structure are surrounded by four and six oxygen anions $[O^{2-}]$, respectively. In the EXAFS measured spectra these three contributions are averaged.

By analyzing figure 3 we can observe the advantages of Filon quadrature in comparison with the fast Fourier transform algorithm. The peaks associated with the coordination shells have a much better resolution than FFT algorithm. Another important advantage consists in the large number of point pairs for each peak. This result is very important, for statistical reasons, in the numerical analysis of the coordination number and the radius values of the coordination shells based on the inverse Fourier transform technique.

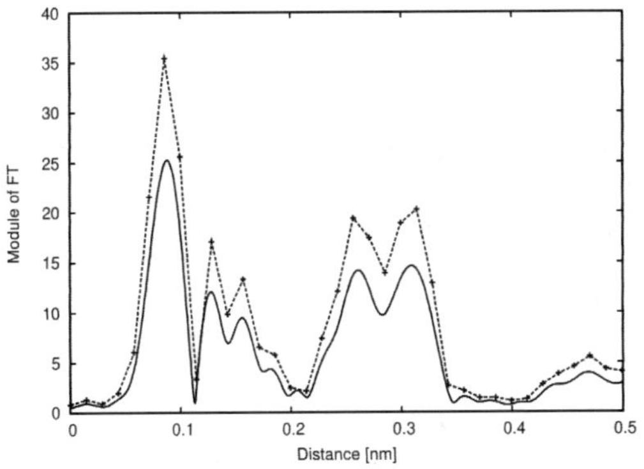


**Figure 3.** The Fourier transforms of the EXAFS spectra of $Fe_3O_4$ system (Filon quadrature -solid line, FFT algorithm - dashed line and +).

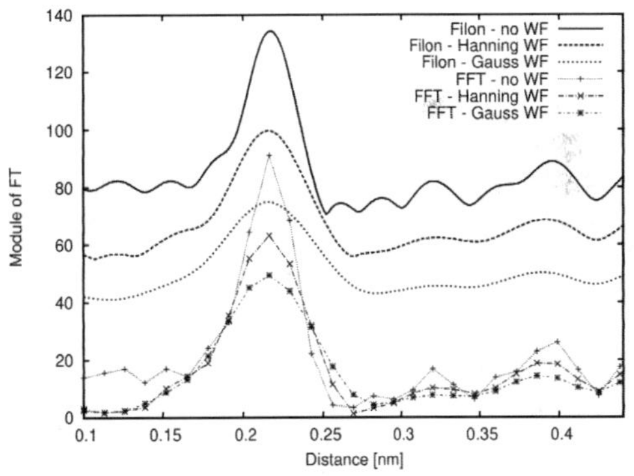


**Figure 4.** The Fourier transforms of the EXAFS spectra of supported Ni catalyst calculated by Filon quadrature and FFT algorithm.

The Fast Fourier transform method gives poor results when analyzing the close coordination shells because the peaks of the radial structure function from figure 3 have a poor resolution. Therefore they cannot be used in the calculation of the local structure parameters for systems that have close coordination shells. They contain one or more types of atoms distributed at two or more unique distances differing by 0.05 nm or less. Using Filon quadrature we obtained for the first and second coordination shells $N_1 = 3.11 \pm 0.009$ atoms, $r_1 = 0.189 \pm 0.001$ nm and $N_2 = 5.15 \pm 0.01$ atoms, $r_2 = 0.209 \pm 0.001$ nm, respectively. Generally speaking, such subshells are not resolved in the fast Fourier transform of the EXAFS spectra and can not be well analyzed by Fourier filtering technique. Other methods such as the regularization and the beat methods [12-14] in combination with the fast Fourier transform of RSF are reported in literature. The results based on the last methods are more or less reliable.

Figure 4 shows the radial distribution function for Ni supported on $Cr_2O_3$ nanocrystallites calculated by both Filon quadrature and fast Fourier transform algorithms. Because the first and the second coordination shells are not close, $R_1$ = 0.2491 nm and $R_2$ = 0.3524 nm, they can be well separated by both algorithms. By using the similar procedure for the local structural parameters we obtained $N_1 = 9.32 \pm 0.11$ atoms, $r_1 = 0.25 \pm 0.06$ nm, $N_2 = 4.46 \pm 0.02$ atoms, $r_2 = 0.352 \pm 0.01$ nm, respectively.The meaning of the window function for each algorithm is also presented. Various types of window functions such as Hanning or Gaussian reduce the spurious errors due to integral truncation but they involve a broadening for each contribution.

## 5. Conclusions

In the present contribution we showed how various Fourier transform algorithms with their specific advantages apply to EXAFS analysis and what extra information can they add to understanding the nanostructure of the magnetite and supported nickel catalysts. The following conclusions can be drawn from these studies: (*i*) Filon quadrature can be successfully used for determination the local structure of the close-shell systems as magnetite while the fast Fourier transform algorithm gives unreliable results for this case; (*ii*) the classical fast Fourier transform method gives good results when the investigated samples are not close-shell systems as supported metal catalysts.

## References

[1] Lytle W F, Sayers E D and Stern A E 1989 *Physica* B **158** 701
[2] Stern A E 1988 *X-Ray Absorption: Principles, Applications, Techniques of EXAFS, SEXAFS and XANES* (New York:Wiley)
[3] Sinfelt H J, Via H G and Lytle W F 1984 *Catal. Rev. Sci. Eng.* **26** 81
[4] Chen X L, Liu T, Thurnauer C M, Csencsits R and Rajh T 2002 *J. Phys. Chem.* B **106** 8539
[5] Turcu R, Darabont Al, Nan A, Aldea N, Macovei D, Bica D, Vekas L, Pana O, Soran L M, Koos A A and Biro P L 2006 *J. Optoelectron. Adv. Mater.* **8** 643
[6] Aldea N, Gluhoi A, Marginean P, Cosma C and Yaning X 2000 *Spectrochim. Acta* B **55** 997
[7] McKale G A, Veal W B, Paulikas P A, Chan K S and Knapp S G 1988 *J. Am. Chem. Soc.* **110** 3763
[8] Brigham O E 1974 *The Fast Fourier Transform* (New Jersey: Prentice-Hall Inc.)
[9] Walker S J 1997 *Fast Fourier Transform* (New York/London/Tokyo: CRC Boca Raton)
[10] Abramowich M and Stegun A 1968 *Handbook of Mathematical Functions* (New York: Dover)
[11] Aldea N and Indrea E 1990 *Comput. Phys. Commun.* **60** 145
[12] Babanov Yu *et al* 2005 *Phys. Scr.* **T115** 194
[13] Bayanov I B, Bunker G and Morrison I T 1966 *J. Synchrotron Rad.* **3** 120
[14] Gurman J S 1995 *J. Synchrotron Rad.* **2** 56.

# Synthesis of carbon nanotubes from acetylene on the FeCoMgO catalytic system obtained by ball milling

**A R Biris[1], A S Biris[2], E Dervishi[2], Z Li[2], F Watanabe[2], S Simon[1], D Lupu[1], I Misan[1] and M Lucaci[3]**

[1] National Institute for Research and Development of Isotopic and Molecular Technologies, 65-103 Donath, 400293 Cluj-Napoca, Romania

[2] UALR Nanotechnology Center, University of Arkansas, 2801 S University Ave, Little Rock, AR 72204, USA

[3] National Institute for Research and Development in Electrical Engineering ICPE-CA 313 Splaiul Unirii, 030138 Bucharest, Romania

E-mail: alexandru.biris@itim-cj.ro

**Abstract.** Highly crystalline multi wall carbon nanotubes have been synthesized by RF-CVD from acetylene at 850°C over a Fe:Co:MgO catalyst. The catalytic system was obtained by mixing for 100 h Fe, Co and MgO powders in a ball milling device under petroleum ether environment, followed by oxidation in air at 500°C for 24 h. Most of the nanotubes had external diameters in order of dozens of nm and lengths of microns, resulting in an aspect ration of over 1000. Their external to internal diameter ratio varied between 2.5 and 3.

## 1. Introduction

Given their remarkable electric, mechanic, and optical properties, the carbon nanotubes are continuously more attractive for various advanced applications. For the synthesis of carbon nanotubes, there are currently employed three major methods: electric arc discharge [1], laser ablation [2], and chemical vapour deposition [3]. Until now, the CVD method is the mostly commonly used because it allows the control of the morphological properties of the nanotubes by modifying the reaction parameters. This process is based on the pyrolysis of various gaseous hydrocarbons ($C_2H_2$, $CH_4$, $C_2H_4$), carbon monoxide CO or liquid (alcohols, benzene, hexane), typically between 700-1000°C over a catalytic system [4, 5]. The type of nanotubes (SW or MW), diameter, chirality, or the synthesis yield are very sensitive to the type, composition and the structure of the catalyst used in the reaction. Generally, the catalyst used for the growth of carbon nanotubes are mono or bi-metallic of transitional metals Fe, Co, Ni, or Mo on various inorganic supports MgO, $Al_2O_3$, $SiO_2$, usually obtained by impregnation [6] and co-precipitation [7]. It has also been previously reported the use of hydrogen uptaking alloys such as $AB_2$ (Zr-based), $AB_5$ (Mm-based) [8], or $LaNi_5$ [9].

In this contribution, we present the synthesis of multi wall carbon nanotubes from acetylene at 850°C over the Fe:Co:MgO (2.5:2.5:95 % wt) catalyst, which was obtained by mixing Fe, Co, and Mg powders in a ball milling device for 100 h. The mixing process was done under petroleum ether, followed by an oxidation in air at 500°C for 24 h. Previous studies [10] showed that by employing the same synthesis conditions on Fe:Co:MgO (2.5:2.5:95 % wt) catalyst, obtained by impregnation, favours the formation of single-wall carbon nanotubes.

## 2. Experimental Details

### *2.1. Catalyst Preparation*

The catalyst, Fe:Co:MgO (2.5:2.5:95 % wt), used in this report was obtained in two steps as follows: First the poweders Mg (purity 99%, particle size < 125 μm), Fe (purity 99%, particle size <63 μm) and Co (purity 98.5%, particle size < 45 μm) have been mixed together in a Fritsch type „Pulverized 6" ball milling device. This process was done in a stainless steel bowl with steel balls of 2, 4, and 8 mm in diameter, having the balls-powders ratio of 5.5-1. The mixing of powder was realized for 100 h, and at an angular rotation of 400 rot/min, under petroleum ether which ensures a good homogeneity and prevents the agglomeration of powder. The loading process and the sample collection was accomplished under an inert environment of argon in a glove box. The second step, the mixture was oxidized in air at 500°C for 24 h. The final product was a grey puffy powder and was characterized by measuring its surface area BET (Micromeritics ASAP 2020), scanning electron microscopy (SEM) (JEOL 7000F) and X-ray diffraction (Bruker D8 Discovery).

### *2.2. Carbon nanotubes synthesis*

The carbon nanotubes growth was done by RF-CVD, which was previously described [11] at 850°C using acetylene as the carbon source. 100 mg of catalyst powder have been placed in a graphite susceptor located in the center of a water-cooled quartz tubing with a diameter of 30 mm. The reactor was rinced with argon (100 ml/min ) for 10 min and was heated up to 500°C for 15 min, in order to desorb the gases present on the graphite surface. Ultimately, the temperature was raised to 850°C followed by the administration of acetylene (2 ml/min ) for 10 min. The cooling was done in argon flow until the reactor reached the room temperature.

To remove the catalyst, the final product was sonicated in HCl (1:1) for 1 h and kept in acid for 4 days. Finally after filtration, the product was washed with distilled water and dried at 120°C overnight. The nanotubes were characterized by SEM, Transmission Electron Microscopy (TEM) (JEOL-2100F TEM), thermogravimetric analysis (TGA) (Mettler Toledo TGA/SDTA 851e).

## 3. Results and Discussions

Figure 1a shows the SEM image of the catalytic system after oxidation. From the SEM analysis, it can be observed the presence of particles with dimensions of less than 3-4 μm, of irregular shapes with a sponge-like appearance. The X-Ray analysis indicating the presence of various phases and their relative concentrations (Rietveld refinement) are presented in figure 1b and table 1.

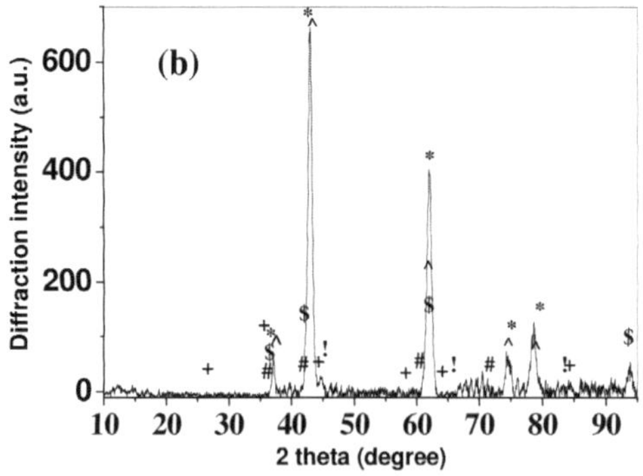


**Figure 1.** SEM image (a) and the X-Ray diffraction spectrum (b) of the catalyst powder after its oxidation at 500°C.

**Table 1.** The structural phases present in the catalyst system and their relative concentrations as analyzed by X-Ray diffraction studies.

| Identified Phases | Identification of the figure 1b | Relative Concentration (%) |
|---|---|---|
| MgO | * | 70.4% |
| $(MgO)_{0.91}(FeO)_{0.09}$ | ^ | 20.5% |
| CoO | $ | 6.8% |
| CoFeOx | ! | 2.9% |
| $(MgO)_{0.239}(FeO)_{0.761}$ | # | 3.8% |
| $(Mg_{0.19}Co_{0.81})(Mg_{0.68}Co_{1.32})O_4$ | + | 2.4% |

The BET surface area analysis performed by using nitrogen adsorption at 77K was of 68.84 $m^2/g$. The total volume of the pores was of 0.27 $cm^3/g$, with an average pore diameter of 162 Å.

Figure 2 presents the thermogravimetric analysis curves of the „as prepared" and the acid purified carbon nanotubes products. The measurements were done in air flow (150 ml/min)at a heating rate of 5°C/min. Based on the „as prepared" TGA curve it was determined a reaction efficiency of 16%. The purified nanotubes TGA curve shows the presence of two distinctive domains, first from 250 to 375°C corresponding to a mass loss of 3.5% attributed to the burning of the amorphous carbon mass [12], and the second one of 93% in the 400 and 680°C range representing the thermal mass loss of the crystalline graphitic structures of the nanotubes. Figure 2 inset shows the differential thermal analysis (DTA) curve for the purified nanotubes indicating a thermal desintegration temperature of 528.2°C.

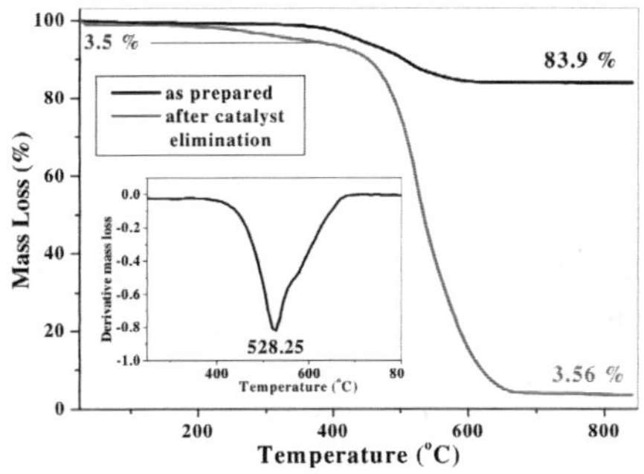


**Figure 2.** TGA curves for the „as prepared" (black curve) and „purified" (red curve) carbon nanotube samples. The inset presents the DTA analysis of the purified nanotubes.

The SEM image of the „as prepared" carbon nanotubes is presented in figure 3a. This image indicates the presence of a high density of entangled multi wall carbon nanotubes which covers uniformly the catalyst particles. The Nanotubes have lengths of douzens of μm and diameters in the order of douzens of nm, which results in an aspect ratio of ~ 1000. Low resolution TEM image of a carbon nanotube with a diameter of 12 nm and the visible length of about 1 μm is presented in figure 3a. Both the images presented in figures 3a and b indicate a high diameter uniformity for the studied nanotubes.

Figure 4 a and b represent the high magnification TEM images of two nanotubes with external diameters ($D_{ext}$) of 25.5 nm and 13.2 nm, respectively. The internal diameters (Dint) for the two nanotubes were of 10.5 nm and 4.2 nm. Also it was observed that the nanotubes have a relatively low number of highly crystalline graphitic walls with an inter-wall distance of 0.34 - 0.35 nm. Also the nanotubes present a small amount of amorphous carbon over their external surfaces, fact also confirmed by the thermogravimetric analysis studies. In figure 4a the arrow indicates a catalyst particle of a spherical shape with an diameter of 5 nm encapsulated within a 3 wall nanotube ($D_{ext}$ = 6.5 nm and $D_{int}$ = 5 nm). The $D_{ext}/D_{int}$ ratio of over 50 analyzed nanotubes ranges between 2.5 and 3.0.

Journal of Physics: Conference Series **182** (2009) 012057 doi:10.1088/1742-6596/182/1/012057

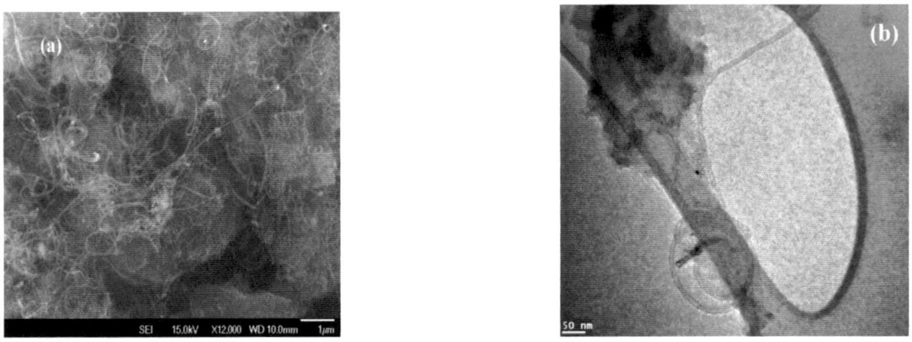


**Figure 3.** SEM images of the „as prepared" carbon nanotubes (a) and a low magnification TEM image of a single nanotube with a diameter of 12 nm and of an aproximative length of 1 μm.

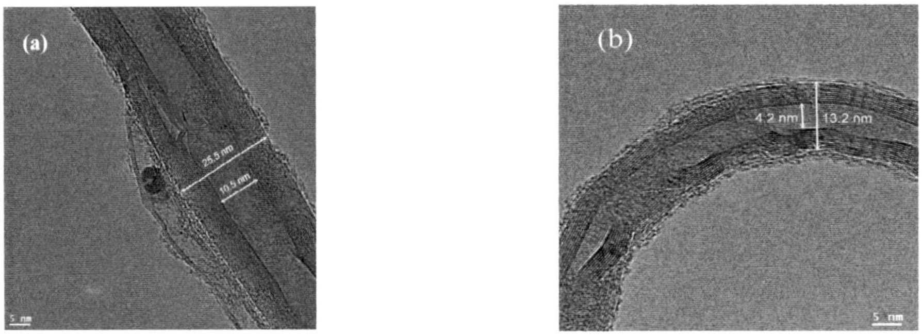


**Figure 4.** High magnification TEM images of two carbon nanotubes after the removal of the metallic catalyst.

## 4. Conclusions

Multi wall carbon nanotubes have been synthesized from acetylene over the Fe:Co:MgO (2.5:2.5:95 % wt) catalyst at 850°C by RF-CCVD. The catalyst was obtained by mixing Fe, Co and Mg powders in a ball milling device for 100 h and the final mixture was oxidized in air at 500°C for 24 h. The majority of the nanotubes had external diameters in the range of dozens of nm, lengths of dozens of μm with an „aspect ratio" of over 1000, a $D_{ext}/D_{int}$ ratio of 2.5-3 and a high degree of crystallinity.

## References

[1] Ebbesen T W and Ajayan P M 1992 *Nature* **358** 220
[2] Thess A *et al* 1996 *Science* **273** 483
[3] Terrones M *et al* 1997 *Nature* **388** 52
[4] Ren Z F, Huang Z P, Xu J W, Wang J H, Bush P, Siegal M P and Provencio P N 1998 *Science*, **282** 1105
[5] Kong J, Soh H T, Cassell A M, Quate C F and Dai H 1998 *Nature* **395** 878
[6] Liu B C, Lyu S C, Lee T J, Choi S K, Eum S J, Yang C W, Park C Y and Lee C J 2003 *Chem. Phys. Lett.* **373** 475
[7] Biris A R *et al* 2008 *Chem. Mater.* **20** 3466
[8] Shaijumon M M and Ramaprabhu S 2003 *Chem. Phys. Lett.* **374** 513
[9] Yi S, Zhang H, Pei L, Zhu Y, Chen X and Xue X 2006 *J. Alloys Comp.* **420** 312
[10] Dervishi E *et al* 2009 *Part. Sci. Technol.* **27** 222
[11] Biris A R, Biris A S, Lupu D, Trigwell S, Dervishi E, Rahman Z and Marginean P 2006 *Chem. Phy. Lett.* **429** 204
[12] Ramesh P, Okazaki T, Sugai T, Kimura J, Kishi N, Sato K, Ozeki Y and Shinohara H 2006 *Chem. Phys. Lett.* **418** 408.

# *Ab initio* study of DNA nucleotides sandwiched between Au(111) electrodes

**Diana Bogdan**[1], **Radu Brătfălean**[1], **Radu Isai**[1,2] **and Cristian Morari**[1]
[1] National Institute for Research and Development of Isotopic and Molecular Technologies, 65-103 Donath, 400293 Cluj-Napoca, Romania
[2] Babeş-Bolyai University, 1 Mihail Kogălniceanu, 400084 Cluj-Napoca, Romania
E-mail: cristian.morari@itim-cj.ro

**Abstract.** Using first-principles calculations, we study the electronic properties of the four DNA nucleotides sandwiched between two Au(111) electrodes. The geometrical structure of the systems is a realistic model of the recently proposed devices for DNA sequencing. For these metal-molecule-metal systems, we calculate the total and local density of states (DOS, LDOS), and the metal-molecule charge transfers. Our results suggest that the qualitative differences between the four systems are sufficient to ensure the recognition of the DNA bases by the proposed device. Nevertheless, the full investigation of the quantitative features of the current-voltage curves is needed to decide if the practical use is possible.

## 1. Introduction

The determination of a pacient's DNA sequence can, in principle, reveal an increased risk to fall ill with a particular disease [1, 2] and helps to design 'personalized medicine' [3]. Moreover, statistical studies and comparison of genomes [4] of a large number of individuals are crucial for the analysis of mutations [5] and hereditary diseases, paving the way to preventive medicine [6]. DNA sequencing is currently still a vastly time-consuming and very expensive task [4], consisting of pre-processing steps, the actual sequencing using the Sanger method, and post-processing in the form of data analysis [7]. Recently a new approach that relies on functionalized nanopore-embedded electrodes to achieve an unambiguous distinction of the four nucleic acid bases in the DNA sequencing process was proposed. It consists of measuring the current-voltage (I-V) characteristics of each nucleotide. This represents a significant improvement over previously studied designs [8, 9] which cannot reliably distinguish all four bases of DNA.

We calculated the electronic and structural properties of Au-X-Au systems (here X = deoxyadenosine monophosphate (dAMP), thymidine MP (TMP), deoxyguanosine MP (dGMP), and deoxycytidine (dCMP), respectively) by using the density functional theory (DFT) code SIESTA [10]. SIESTA uses pseudopotentials and expands the wave functions of valence electrons by linear combinations of numerical atomic orbitals. We used a double-zeta polarized (DZP) basis set with an energy cutoff of 25 meV. We performed Gamma-point calculations, by using the generalized gradient approximation (GGA) for the exchange and correlation functional, as parameterized by Perdew, Burke and Ernzerhof [11]. We used periodic boundary conditions. The investigated geometries for the Au-X-Au systems were built as follows (see figure 1): the structure of the isolated DNA nucleotides and of the bulk Au, respectively, were first optimized. For the latter, the equilibrium lattice constant was found to be 4.17 Å in good agreement with the experimental value. The calculated value is used to set the dimension of our hexagonal supercell for the Au(111) surface. To keep a minimum of about 10 Å between repeated images of the DNA nucleotides we adopted a 5×5 surface cell resulting in lattice constant of 14.77 Å. The length of the supercell in the Z direction was 27.62 Å. This value leads to a distance of 15.5 Å between the two gold surfaces. This value is a compromise resulting from which best accomodates all four nucleobases geometries (see n section two). The orientation and the distance of the molecule with respect to the surface were then optimized until the maximum force was less than 0.04 eV/Å.

We are aware that this is just a first approach to the realistic description of the experimental device. Additional studies are needed to understand the dependence of the electronic structure with respect to the distance between the two surfaces. These studies should provide also a qualitative description of the defects on the electrode's surfaces (i.e. defects of the atomic layers may lead to different distances between the electrodes in different regions of the nanojunction). This work is in progress in our group.

## 2. Results

The minimum distances between the organic atoms and the metallic surface resulting after the relaxation are given in table 1. One can see that our geometrical model describes fairly good all the four systems. On one hand, we kept the same distance between the gold plates; this corresponds to a given geometry of the sequencing device. On the other hand, the organic molecule-metal surface distances are within the limits of 2.10 Å and 2.65 Å which ensures a reliable description of the metal-molecule interaction.

**Table 1.** Distances (in Å) between Au plates and DNA nucleotides after the relaxation. Top values correspond to distances between the Au electrod and the nucleobase. Bottom: distances between phosphatic moiety of the nucleotide and the metallic surface.

| dAMP | TMP | dGMP | dCMP |
|---|---|---|---|
| 2.329 | 2.647 | 2.148 | 2.528 |
| 2.103 | 2.244 | 2.298 | 2.181 |

In order to give a qualitative description of the electronic transport accross the Au-DNA-Au devices, we concentrated our investigations on three relevant physical properties: (*i*) total density of electronic states (DOS) (*ii*) localized density of states (LDOS) and (*iii*) charge transfer.

The DOS corresponding to the four nucelobases . as well as to their associate metal-molecule-metal systems are given in figure 2. To compare the DOS for the metallic surface with that of the free molecules we set the Fermi level ($E_F$) to zero for both the free gold surface and the Au-X-Au systems. Next, we set the same value of the Hartree potential in vacuum for the free molecule and the gold surface. By inspecting figure 2 we note that the DOS of the gold(111) at $E_F$ is relatively low. This may indicate a relatively low tunneling probability for electrons at the metal-molecule interface. The analysis of the DOS for the four nucleotides (free and in contact with the metallic surfaces) shows that for dGMP the largest DOS peak occurs at $E_F$. In addition we note that this peak is almost unchanged when the gold surface is present (*i.e.* the red and the black curves are almost identical). As a consequence, we expect a strong signature of the guanine when the device is connected to a bias voltage. In the case of TMP the HOMO is located at about -0.6 below $E_F$, which decrease the electric signal. Finally, for dAMP and dCMP an intermediate situation occurs. In these cases, HOMO orbitals are located around -0.25 eV below $E_F$, in the presence of Au electrodes.

In order to explore the properties of the states close to the Fermi level for the four metal-molecule-metal systems, we compute their LDOS, i.e. $\rho(\mathbf{r}, E) = \sum_i |\psi_i(\mathbf{r})|^2 \delta(E_i - E)$, where $\psi_i$

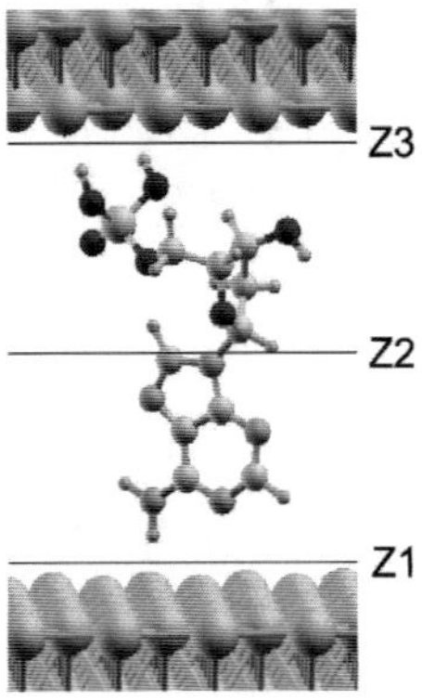


**Figure 1.** Geometrical structure used in calculation (for the system Au−dAMP−Au; similar models were used for the other three systems). The distance between the Au surfaces was set constant for the four systems (15.5 Å).

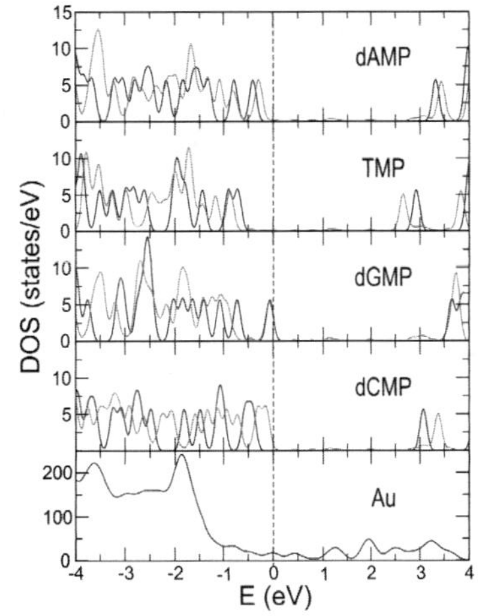


**Figure 2.** Top (four panels): The DOS for dAMP, TMP, dGMP, dCMP free (black) and attached to Au(111) electrodes (red). Bottom: the DOS for the clean Au(111) surface.

denote the eigenstates of the system and $E_i$ the corresponding eigenvalues. In general, although the LDOS does not fully determine the conductivity, the regions where the LDOS is depressed are expected to be barriers for electronic transmission, while large values of the LDOS indicate a high tunneling probability. Non-zero values of LDOS are equivalent with the presence of molecular orbitals with energies in the region under investigation. We compute the LDOS of the four systems, on an one volt range centered on $E_F$. This corresponds to a total bias of up to 1 eV applied to the device. The results are given in figure 3. First we note that the LDOS is non-zero over the dGMP for both energy ranges under investigation. Second, dAMP has a non-zero LDOS for the energy range $(E_F - 0.5V, E_F)$. For TMP and dCMP, LDOS is concentrated over different chemical residues in the molecule, which does not favour the recognition of the DNA bases by measuring the I-V characteristics of the devices. These results are consistent with the DOS analisys.

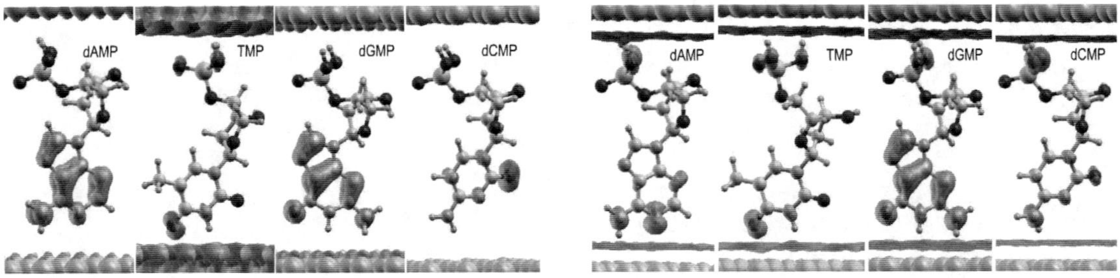


**Figure 3.** The LDOS integrated between Fermi level ($E_F$) and $E_F - 0.5$ V (left panel) and $E_F + 0.5$ V (right panel), respectively, for the four nucleotides.

Finally, we investigated the charge transfer bewteen each molecule and two gold surface. To this end, we computed the electronic density of the metal-molecule-metal device ($\rho_D$), that of the gold surface ($\rho_{Au}$) and that of the molecule ($\rho_M$), respectively. For the two late cases the geometry of the system was derived from that of the relaxed full system. Next we computed the electronic density associated with the charge transfer $\Delta\rho = \rho_D - \rho_{Au} - \rho_M$. By artificially dividing the cell with planes parallel with the gold surfaces (see figure 1) and integrating the $\Delta\rho$ quantity, we directly estimated the total amount of electric charge across the region where the molecule is located. While this method does not provide any information on the chemical behavior, it has the advantage to provide accurate physical data, since the electronic density is accurately computed within DFT. Precisely, we computed the total positive and negative charge transfered between the molecule and surface across the region where the DNA base is located ($Z_1 - Z_2$), and across the region where the phosphate moiety is located ($Z_2 - Z_3$) (see the location of the two planes defining the three regions of interest in figure 1). The results are given in table 2. They clearly show that for all systems the phosphate moiety has the tendency to lose electrons, while the DNA bases have a small tendency of accumulating electrons. The strongest manifestation of this effect is present for TMP and dGMP. Its consequence will be the assymetry of the I-V curve of the device, with respect to the bias polarization.

**Table 2.** Total charge transfer (in $e$) between Au plates and DNA nucleotides.

| | dAMP | TMP | dGMP | dCMP |
|---|---|---|---|---|
| $Z_1 - Z_2$ | -0.04/0.46 | -0.06/0.59 | -0.10/0.48 | -0.02/0.45 |
| $Z_2 - Z_3$ | -0.55/0.03 | -0.68/0.05 | -0.62/0.04 | -0.55/0.03 |

## 3. Conclusion

We studied the electronic properties of the DNA nucleotides (dAMP, TMP, dGMP, dCMP) sandwiched between Au(111) electrodes, using a DFT approach. By analizing the total DOS, the LDOS and the charge transfer between the DNA nucleotides and the gold plates we found that the proposed device is suited for the recognition of guanine and adenine. The asymmetry of the I-V characteristics with respect to the bias polarization may lead to a distinct signature for TMP unit also. To conclude, the qualitative analysis presented in this paper suggests that the experimental setup based on our model can be used for DNA sequencing. Nevertheless, this analysis does not give any indication on the magnitude of the total current for each nucleobase. Further work is needed to decide if current values are large enough for practical applications.

## Acknowledgments

We acknowledge financial support from the NASC (project PN 09-440101, contract no 44N/2009). The calculations were performed in the DataCentre of NIRDIMT.

## References

[1] Futreal P A, Kasprzyk A, Birney E, Mullikin J C, Wooster R and Stratton M R 2001 *Nature* **409** 850-2
[2] Jimenez-Sanchez G, Childs B and Valle D 2001 *Nature* **409** 853-5
[3] Chakravarti A 2001 *Nature* **409** 822-3
[4] Church G M 2006 *Scientific American* **294** 47-56
[5] Stoneking M 2001 *Nature* **409** 821-2
[6] Hood L, Heath J P, Phelps M E and Lin B 2004 *Science* **306** 640-3
[7] International Human Genome Sequencing Consortium, Initial sequencing and analysis of the human genome 2001 *Nature* **409** 860-921
[8] Zwolak M and Di Ventra M 2005 *Nano Lett.* **5** 421-4
[9] Lagerqvist J, Zwolak M and Di Ventra M 2006 *Nano Lett.* **6** 779-82
[10] Soler J M *et al* 2002 *J. Phys. Condens. Matter* **14** 2745-79
[11] Perdew J P, Burke K and Ernzerhof M 1996 *Phys. Rev. Lett.* **77** 3865-8

# Synthesis, characterization and drug delivery application of the temperature responsive pNIPA hydrogel

**I Craciunescu[1], A Nan[1], R Turcu[1], I Kacso[1], I Bratu[1], C Leostean[1] and L Vekas[2]**

[1] National Institute for Research and Development of Isotopic and Molecular Technologies, 65-103 Donath, 400293 Cluj-Napoca, Romania

[2] Romanian Academy, Timisoara Branch, Magnetic Fluids Laboratory, Timisoara, Romania

E-mail: izabella.craciunescu@itim-cj.ro

**Abstract:** Temperature – sensitive poly (N-isopropyl acrylamide) pNIPA gels were synthesized with nano size iron oxide ferrofluids ($Fe_3O_4$) using N, N'-methylene bis-acrylamide (BIS) as the crosslinking agent, obtaining a new type of material, named ferrogel. The swelling characteristics of these ferrogels at distinct compositions, as a function of temperature were investigated. It was found that the pNIPA ferrogels demonstrated the temperature – responsive nature, and it also showed good reversibility. The FTIR spectra of pNIPA ferrogel samples, showed the absorption region of the specific chemical groups associated with pNIPA and the $Fe_3O_4$ magnetic nanoparticles. The drug release experiments were performed using atenolol (AT) in simulated gastric and intestinal media. The amount of released drug was determined by UV spectroscopy at characteristic wavelength of the drug, using a calibration curve obtained from a series of drug solution with standard concentration.

## 1. Introduction

Hydrogels are one of the most promising types of polymers being used for new material development and are polymeric material that do not dissolve in water at a physiological temperature and pH, but swell considerably in an aqueous medium [1]. Hydrogels show reversible volume phase transition depending on external stimuli such as temperature, solvent, ionic strength, pH, electrical or magnetic field and this properties made them promising materials in biotechnology, pharmaceutical and industrial application [2].

Hydrogels have been used extensively in the development of the smart delivery system, because the hydrogels can protect the drug from hostile environments like the presence of enzyme and low pH in the stomach. Hydrogels can also control drug release by changing the gel structure in response to environmental stimuli [3]. Among theme, temperature and pH responsive hydrogels have been the most widely studied, because these two factors have a physiological significance.

Temperature responsive hydrogels exhibit a volume phase transition at a certain temperature; wich causes a sudden change in the solution state. These polymers become insoluble upon heating and have a so-called lower critical solution temperature (LCST) [4]. Poly (N-isopropyl acrylamide) (PNIPA) is probably the most extensively used because of its lower LCST in the range of 25 - 32° C, close to the body temperature [5].

By combination of this temperature sensitive polymers with nano solid or fluid magnetic materials displays novel and often enhanced properties compared to the traditional materials. Combination of magnetic and elastic properties leads to a number of striking phenomena in response to impressed magnetic fields. Magnetic field responsive materials are specific subsets of smart materials that can adaptively change their physical properties due to external magnetic fields [7]. Due to their sensitivities to both magnetic fields and temperature, this magnetic-field sensitive hydrogels, named ferrogels, offer a high potential application in the design of a targeting drug delivery system which is considered a safe and effective way for tissue specific release of drug.

In the present study we prepared PNIPA ferrogels to develop a dual sensitive system against both temperature and magnetic field. This study deals with the synthesis, characterization and the temperature dependence of the swelling behavior for PNIPA ferrogels.

Atenolol was encapsulated into magnetic hydrogel and the drug release behaviour of the magnetic sensitive hydrogel for drug eluting stents application was investigated in this paper.

## 2. Experimental

### *2.1. Preparation of the pNIPA ferrogels*

Different ratios of monomer N-isopropyl acrylamide (NIPA) and N, N'-methylene bis-acrylamide (BIS), like catalyst were dissolved in 3 ml of deionized water. After $N_2$ gas bubbling for 10 min to deoxygenate the solution, different concentrations of ferrofluid were added and sonicated for 30 minutes. 1 wt % ammonium persulphate (APS) as redox accelerator was added to the solution. The mixture was kept to react for 2 h at 60°C and after that, 24 h at room temperature. After the gelation was completed, the ferrogels was immersed in an excess amount of deionized water for 2 days to remove the residual unreacted monomers. The pNIPA ferrogels were dried for 3 days at 60°C to a constant weight.

### *2.2. Swelling experiments of the hydrogels*

The swelling ratio was measured gravimetrically in two different phosphate buffer solutions, a neutral solution with pH = 7 and an acid solution with pH = 2. The dry ferrogel samples were immersed in solution with pH 7 and 2 at different temperatures, until they swelled to equilibrium. After excessive surface water removed with filter paper, the swollen samples were weighed time to time. The swelling ratio can be calculated as a function of time [9].

$$Swelling\ ratio\,(\%) = \frac{W_s - W_d}{W_d} \cdot 100 \tag{1}$$

where $W_s$ represents the weight of the swollen state of the sample at a given time and $W_d$ is the weight of the sample in the dry state.

### *2.3. Characterization*

Fourier transform infrared spectroscopy (FTIR) was used to confirm the structure of PVA and PNIPA hydrogels and ferrogels. Infrared absorption spectra were recorded with a FTIR JASCO – 6100 spectrometer on pressed pellet prepared from dry hydrogels and ferrogels, embedded in KBr, in the 500-4000 $cm^{-1}$ spectral range. The amount of released AT was determined at 225 nm with a UV Jasco V-550 spectrometer.

### *2.4. Drug release experiment*

The release rate experiments were performed at 25°C under unstirred condition in pH 7.0 phosphate buffer solution. The ferrogels containing a known amount of drug (AT) were added to the release medium (50 ml) at a given time interval, the samples were withdrawn and assayed for the amount of released AT as a function of time. The amount of released AT was determined by UV-VIS spectroscopy at 225 nm using a calibration curve constructed from a series of AT solutions with standard concentrations. The results are expressed as cumulative release ratios.

## 3. Results and discussion

### *3.1. FTIR spectrum of the PNIPA ferrogel*

FTIR spectra of NIPA ferrogels are presented in figure 1, and the FTIR spectrum of NIPA hydrogel in the inset.

From the FTIR spectrum of hydrogel (inset figure 4) there is a carbonyl stretching vibration (amide I) at 1650 cm$^{-1}$, N–H bending vibration (amide at 1545 cm$^{-1}$) and two typical peaks of C–H vibration of $-CH(CH_3)_2$ at 1372 cm$^{-1}$ and 1385 cm$^{-1}$ are confirmed. A broad peak between 3440 cm$^{-1}$ and 3290 cm$^{-1}$ is also observed due to N–H stretching of NIPA repeating units. Also the presence of the characteristic bands of magnetite ($Fe_3O_4$) located around 600 cm$^{-1}$ is observed.

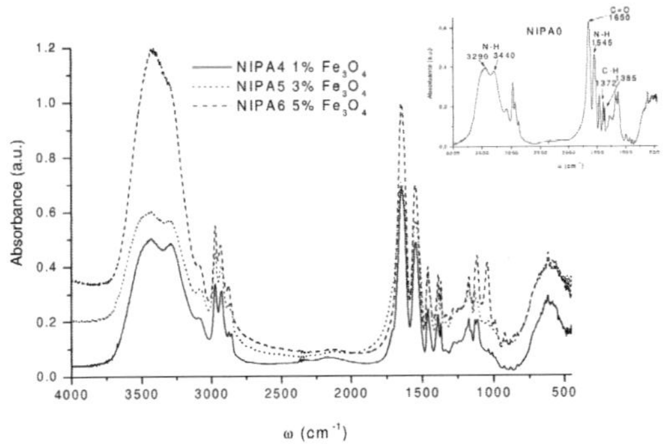


**Figure 1.** FTIR spectra of NIPA ferrogels. Inset: FTIR spectrum of NIPA hydrogel.

### *3.2. Swelling studies*

In the previous stage of the polymer dissolution the solvent molecules go into the polymeric mass, swell it until chain separation and is incorporated into the solution, in opposition to the intermolecular forces (e.g. van der Waals, electrostatic, etc.) that exist between them. However, if the polymer is not soluble, a solvent diffusion to the polymeric mass and from the polymeric mass to the medium will only exist. Then, the polymer swelling will occur. The equilibrium swelling will be reached when both solvent diffusion are equal [10].

The water content at different times can be calculated from equation (1). Temperature responsive hydrogels exhibit a volume phase transition at a lower critical solution temperature (LCST), which causes a change in the solvation state. The change in the hydratation state, which causes the volume phase transition, reflects competing hydrogen bonding properties, where intra- and inter- molecular hydrogen bonding of the polymer molecules are favorised compared to a solubilisation by water [4].

The time dependent swelling behavior of the pNIPA ferrogels at pH = 7 buffer solution and three different temperature are shown in figure 2. It can be observed that the swelling ratio of the ferrogels decrease with the increasing of temperature. This behavior is attributed to the collapse of the pNIPA chain around of 37°C, leading the hydrogel to assume a more hydrophobic state.

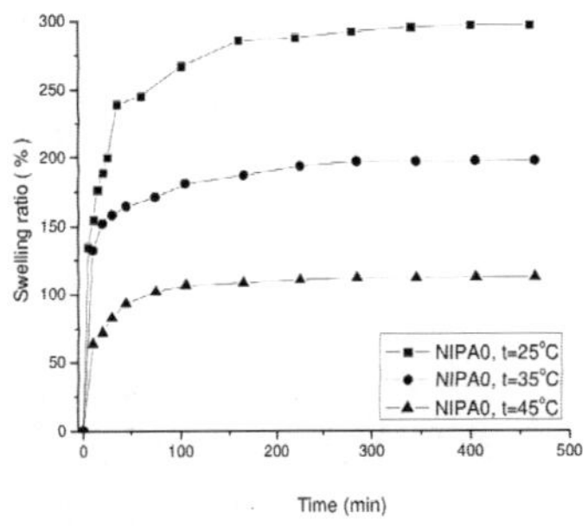


**Figure 2.** Swelling ratios for pNIPA ferrogels as a function of time for different temperatures.

### *3.3. Drug release studies*

All the hydrogels and ferrogels containing a known amount of atenolol were added to the release medium (50 mL of phosphate buffer solution pH = 7). The total concentration of atenolol in this medium is calculated at 0.02 mM.

The amount of drug at any selected time was calculated from the atenolol (AT) calibration curve. Figure 3 shows the calibration curve for AT at 225 nm determined by UV-VIS spectroscopy.

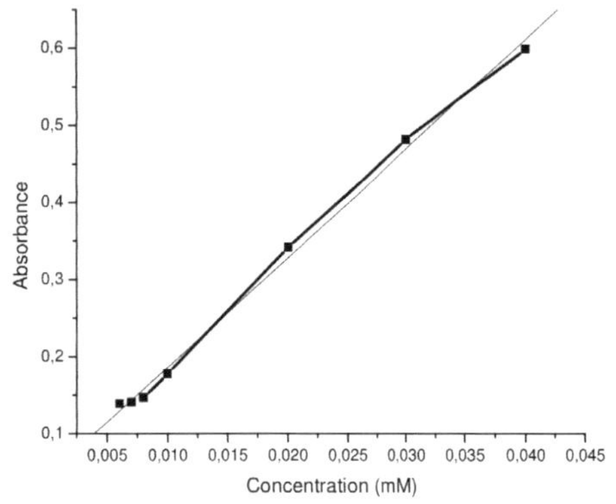


**Figure 3.** Calibration curve for atenolol (AT).

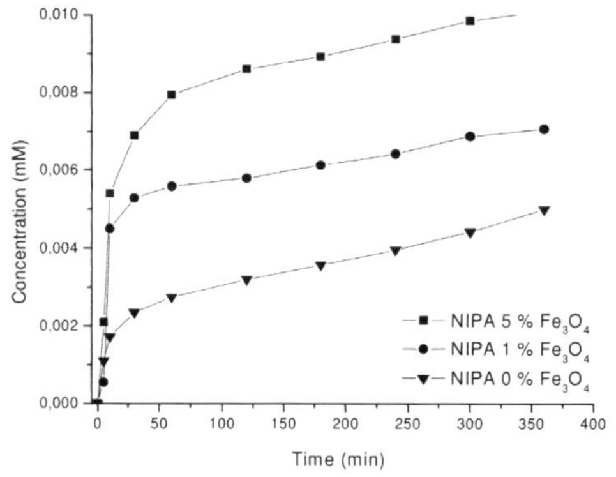


**Figure 4.** The release profile of atenolol of pNIPA.

Figure 4 shows the release profile (delivered atenolol concentration) of the prepared samples of pNIPA in physiological fluid (buffer solution of pH = 7) that is very similar to the tear fluid at 37°C.

By studying the release profile of atenolol (figure 4) one can observe that in pNIPA ferrogels a higher release rate (max. 50 %) is obtained at 25°C and pH = 7. The reason of this release behavior seems to be the nature of the ferrogels. The drug in the ferrogels could be released as a result of the ferrogels volume change and the interaction between the polymer network and atenolol. Because the experiments are made at pH = 7, there is electrostatic repulsion between -OH of atenolol and $-COO^-$ of pNIPA ferrogels that accelerates the atenolol release [12]. Also, it can be observed that the fractional released is directly proportional to the concentration of magnetic nanoparticles. This result indicates that the higher concentration of magnetic nanoparticles in the ferrogels creates larger surface areas to diffuse the drug.

## 4. Conclusions

pNIPA hydrogels and ferrogels (based on ferrofluid, a stable dispersion of magnetite nanoparticles in water) were synthesized. The polymerization of pNIPA containing the surface modified magnetic nanoparticles was confirmed by Fourier transform spectroscopy (FTIR). Its swelling behavior showed better temperature sensitivity in buffer solution at different temperature. The Atenolol release from the hydrogels and ferrogels was affected by the nature of the polymeric matrix and is directly proportional to the concentration of magnetic nanoparticles. All these results indicate that the pNIPA ferrogels can be used as a temperature responsive orally administered drug delivery system.

## References

[1] Ratner B D 1981 *Biomedical application of hydrogel. Review and critical appraisal* ed D F Williams (Florida: CRC Press Boca Raton) vol II p 146
[2] Tanaka Y, Gong J P and Osada Y 2005 *Prog. Polym. Sci.* **30** 1
[3] Qin J and Park K 2001 *Adv. Drug Delivery Rev.* **53** 321
[4] Schmaljohann D 2006 *Adv. Drug Delivery Rev.* **58** 1655
[5] Ebara M, Aoyagi T, Sakai K and Okama T 2000 *Macromolecules* **33** 8312
[6] Martien F L 1986 in *Encyclopedia of Polymer Science and Engineering* vol 17 (New York: Wiley)
[7] Filipcsei G, Csetreki I, Szilagyi A and Zrinyi M 2007 *Adv. Polym. Sci* **206** 137
[8] Bica D, Vekas L and Rasa M 2002 *J. Magn. Magn. Mater.* **252** 10
[9] Garcia D M, Escobar J L, Noa Y, Bade N, Hernaez E and Katime I 2004 *Eur. Polym. J.* **40** 1683
[10] Friends G and McGee J 1993 *J. Appl. Polym. Sci* **49** 1869
[11] Lee W F and Chain Y J 2001 *J. Appl. Polym. Sci.* **82** 2487
[12] Guo B L and Gao Q Y 2007 *Carbohydr. Res.* **342** 2416.

# FTIR spectroscopic study of some lead germanate glasses

**E Culea, L Pop, M Bosca, T Rusu, P Pascuta and S Rada**

Technical University of Cluj-Napoca, Physics Department, 15 Constantin Daicoviciu, 400020 Cluj-Napoca, Romania

Email: lidia.pop@phys.utcluj.ro

**Abstract.** The structures of two lead germante glasses – $yGeO_2·(100-y)PbO$ and $xNd_2O_3(100-x)[GeO_2·PbO]$ – are investigated by the Fourier transform infrared (FTIR) spectroscopy. The structural role of germanium, lead and neodymium ions is discussed. The presence of $GeO_4$, $GeO_6$ and $PbO_4$ structural units was evidenced in the studied glass networks. In the case of the $xNd_2O_3(100-x)[GeO_2·PbO]$ glasses it was shown that the ratio of the mentioned structural units depends on the $Nd_2O_3$ content of the samples. Therefore $Nd_2O_3$ play the network modifier role in studied glasses.

## 1. Introduction

In last years the need to find out new methods and materials for electronics, laser or nuclear technology in order to support the global economy became very important. It is known that, in general, the oxide glasses present interesting electrical, magnetic and optical properties as well as a good chemical stability and mecanical resistance. Among the oxide glasses, lead germanate are of interest because of several characteristcs that make them promising materials for applications as new lasing materials and upconverters. Recently, more and more researchers have been started to study such glasses due their good mechanical strength, high thermal stability, good chemical durability and high refractive index (~2) [1-5]. Also, germanate glasses doped with rare-earth ions have been investigated extensively because their physicochemical, optical and spectroscopic properties are advantageous for optoelectronic applications [6-8].

The purpose of this paper was to obtain information on the basic structural polyhedra which form the glass network in the $GeO_2$-PbO and the $GeO_2$-PbO-$Nd_2O_3$ glass systems with the aid of FTIR spectroscopy.

## 2. Experimental

The glass samples having the general chemical formula $yGeO_2·(100-y)PbO$, with y=10, 20, 30, 40, 50, 60, 70 mol% and $xNd_2O_3(100-x)[GeO_2·PbO]$ with x = 0, 1, 3, 5, 7, 10 mol% have been prepared by the melt quenching technique. Required quantities of analar grade $GeO_2$, PbO and $Nd_2O_3$ were mixed and miled together with the aim to obtain a fine powder. The mixture was melted in sintered corundum crucibles in an electric furnance a temperature of 1100°C for 15 min.

X-ray diffraction investigation was performed by a XRD-6000 SHIMADZU diffractometer using the Cu Kα radiation ($\lambda$ = 1.54056 Å), a working voltage of 40 kV, a working current of 30 mA and the scanning speed of 2°/min.

The FTIR spectra of the studied glasses were registered 400-2000 $cm^{-1}$ wave number range with a resolution of 4 $cm^{-1}$ at room temperature using a JASCO FTIR 6200 type spectrometer and the KBr pellet technique.

## 3. Results and discussion

X-ray diffraction patterns of all the studied glasses have shown broad halo patterns with no crystalline peaks. Thus, for example, figure 1 presents the X-ray diffractograms for the $xNd_2O_3(100-x)[GeO_2{\cdot}PbO]$ glass system and shows only broad humps typical of amorphous materials.

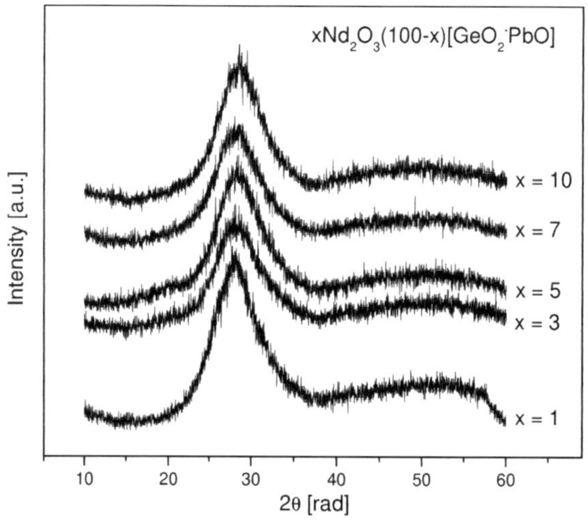


**Figure 1.** XRD patterns of the $xNd_2O_3(100-x)[GeO_2{\cdot}PbO]$ glass system.

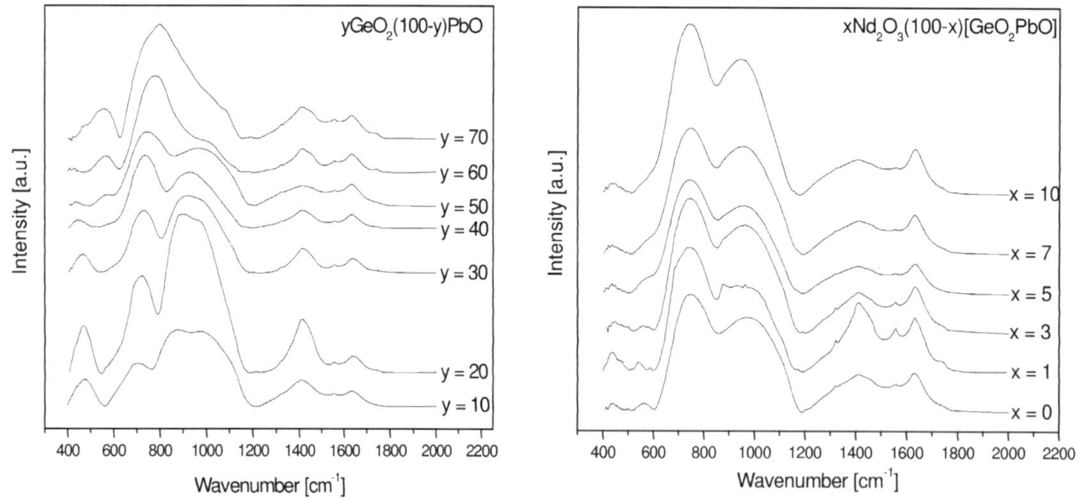


**Figure 2.** FT-IR spectra of $yGeO_2{\cdot}(100-y)PbO$ (left) and $xNd_2O_3(100-x)[GeO_2{\cdot}PbO]$ (right) glass systems.

Figure 2 shows the FTIR spectra of the $yGeO_2{\cdot}(100-y)PbO$ and the $xNd_2O_3(100-x)[GeO_2{\cdot}PbO]$ glasses. The positions of the absorption bands evidenced in the FTIR spectra and their assignments are summarized in table 1. The assignments were made based on data from the literature and comparing the experimental data of glasses with those of the related crystalline compounds [9-16]. Note that there were some absorption features located at 1300 and 1540 $cm^{-1}$, but based on the present data it was not possible to assign them.

**Table 1.** FTIR bands and there assigned in case of the $yGeO_2{\cdot}(100\text{-}y)PbO$ glass system.

| Wavenumber [$cm^{-1}$] | | Assignments |
|---|---|---|
| y = 10 | y = 70 | |
| 472 | 465 | symmetric bending vibration mode of Pb–O in $PbO_4$ units [15, 16] |
| 625 | 570 | symmetric stretching mode of Ge–O–Ge bonds [9, 12, 13]<br>asymmetric bending vibration modes of Pb–O–Pb bonds [15] |
| 705 | 695 | stretching mode of Ge–O–Ge bonds in $GeO_6$ units [11, 14] |
| 823 | 755 | asymmetric stretching mode of Ge–O–Ge bridges connecting $GeO_4$ units [12, 13] |
| 892 | 830 | asymmetric stretching mode of Ge–O–Ge bonds from $GeO_4$ units[9, 12] |
| 1015 | 954 | symmetric stretching vibration modes of Pb–O bonds [15, 16] |
| 1100 | 1075 | asymmetric stretching vibration modes of Pb–O bonds [15] |
| 1300 | 1365 | - |
| 1410 | 1430 | asymmetric stretching vibration modes of Ge–O–Ge bonds [9] |
| 1540 | 1545 | - |
| 1630 | 1630 | OH bending mode of vibration |

In the case of the $yGeO_2{\cdot}(100\text{-}y)PbO$ glass system, a simple inspection of the spectral features (figure 2) shows that the bands characteristic of the lead oxide decrease with increasing the germanium oxide content. It was shown that the germanium ions appear in the glass network in both the $GeO_4$ and $GeO_6$ polyhedra. On the other hand, the lead ions are incorporated in the glass network in the $PbO_4$ structural units for all the investigated glasses.

The $xNd_2O_3(100\text{-}x)[GeO_2{\cdot}PbO]$ glasses spectra (figure2) show that no dramatic changes occur with increasing the $Nd_2O_3$ content in the samples. In this glass system a new band was observed around 585 $cm^{-1}$. This band increases in intensity with x and was attributed to the Nd–O stretching vibrations [10]. In addition, the intensity of the peak assigned to germanium oxide increases with the increase of the $Nd_2O_3$ content. This suggests that the presence of $Nd_2O_3$ favourises the $GeO_4$ and $GeO_6$ units while the Pb–O vibration mode is affected.

The structural changes observed by varying the $Nd_2O_3$ content in the $xNd_2O_3(100\text{-}x)[GeO_2{\cdot}PbO]$ glass system by using the FTIR investigation suggest that the neodymium ions play a network modifier role in these glasses.

## 4. Conclusions

Glasses of the $yGeO_2{\cdot}(100\text{-}y)PbO$ and $xNd_2O_3(100\text{-}x)[GeO_2{\cdot}PbO]$ systems were obtained within y = 10÷70 mol% and x = 0÷10 mol%. The FTIR spectra of these glasses were registered. It was shown that for all the samples the glass matrix consists on $GeO_4$, $GeO_6$ and $PbO_4$ structural units. In the glasses containing $Nd_2O_3$, the proportion of the mentioned structural units depends on the neodymium ions content. Therefore, we can conclude that the neodymium ions play the network modifier role in the studied glasses.

## References

[1] Bomfim F A, Martinelli J R, Kassab L R P, Wetter N U and Neto J J 2008 *J. Non-Cryst. Solids* **354** 4755
[2] Cereyon A, Champagnon B, Martinez V, Maksimov L, Yanush O and Bogdanov V N 2006 *Optical Materials* **28** 1301
[3] Gackowska J, Gazda M, Trzebiatowski K and Kusz B 2006 *J. Non-Cryst. Solids* **352** 4242
[4] Kassab L R P, Hora W G, Lozano W B, de Oliveira M A S and Maciel G S 2007 *Optical*

*Comm.* **269** 356
[5] Klimesz B, Dominiak-Dzik G, Lisiecki R and Ryba-Romanowski W 2008 *J. Non-Cryst. Solids* **354** 515
[6] Wachtler M, Spegini A, Gatterer K, Fritzer H P, Ajo D and Bettinelli M 1998 *J. Am. Ceram. Soc.* **81** 2045.
[7] Kassab L R P, Hora W G, Martinelli J R, Sene F F, Jakutis J and Wetter N U 2006 *J. Non-Cryst. Solids* **352** 3530
[8] Dominiak-Dzik G and Ryba-Romanowski W 2008 *J. Alloys and Compd.* **451** 586
[9] Terakado N and Tanaka K 2008 *J. Non-Cryst. Solids* **354** 1992
[10] Sharaf El-Deen L M, Al Salhi M S and Elkholy M M 2008 *J. Alloys and Compd.* **465** 333
[11] Pascuta P and Culea E 2008 *Mater. Lett.* **62** 4127
[12] Kim Y, Saienga J and Martin S W 2005 *J. Non-Cryst. Solids* **351** 3716
[13] Santos L F, Wondraczek L, Deubener J and Ameida R M 2007 *J. Non-Cryst. Solids* **353** 1875
[14] Simon S, Ardelean I, Filip S, Bratu I and Cosma I 2000 *Solid State Commun.* **116** 83
[15] Trettenhahn G L J, Nauer G E and Neckel A 1993 *Vibrat. Spectroscopy* **5** 85
[16] Bosca M, Pop L, Borodi Ghe, Pascuta P and Culea E 2009 *J. Alloys and Compd.* **479** 579.

# Structural and physical characteristics of $xGd_2O_3(100\text{-}x)[Bi_2O_3\,B_2O_3]$ glasses

**Eugen Culea, Lidia Pop, Maria Bosca, Viorel Dan, Petru Pascuta and Simona Rada**

Technical University of Cluj-Napoca, Physics Department, 15 Constantin Daicoviciu, 400020 Cluj-Napoca, Romania

Email: lidia.pop@phys.utcluj.ro

**Abstract.** A study on the $xGd_2O_3(100\text{-}x)[Bi_2O_3\ B_2O_3]$ glass system where $0 < x \leq 30$ mol% was carried by using X-ray diffraction, FTIR spectroscopy and density measurements. The relationships between composition and properties were demonstrated. The results indicate that $Gd_2O_3$ play a network modifier role and stabilizes the structure of the vitreous matrix.

## 1. Introduction

Nowadays, glasses containing rare-earth ions has been intensively studied due their applications in important technological fields like optical communications, laser technology, optical storage and immobilization of radioactive materials. These applications are possible as a consequence of some of the physico-chemical properties of the glasses with rare earth such as the heat and mechanical resistance, the chemical stability and the interesting optical and magnetic properties [1-3].

It is well known that the $Bi_2O_3$ is not a classical glass network former. Nevertheless, glasses based on $Bi_2O_3$ show interesting physical properties such as a high density, a high linear and non-linear refractive index enabling their extensive various applications in optics and optoelecronics [4].

On the other hand, the boric oxide, $B_2O_3$, acts as one of the most important glass formers. The $B_2O_3$ is incorporated into various kinds of glass systems as a flux material in order to attain materials haveing specific physical and chemical properties. In their crystalline form, borates with various compositions, are of exceptional importance due to their interesting linear and nonlinear optical properties [5].

Taking into account the mentioned arguments, the properties of glasses of the $Bi_2O_3$-$B_2O_3$ system have attracted much interest [6-8].

The main objective of the present work was to study the structural and vibrational properties of some boro-bismuthate glasses containing gadolinium ions.

## 2. Experimental

Glasses with the $xGd_2O_3(100\text{-}x)[Bi_2O_3\ B_2O_3]$ composition were prepared where x increased from 1 to 30 mol% $Gd_2O_3$. High purity $Bi_2O_3$, $B_2O_3$ and $Gd_2O_3$ powders were used as starting materials. The mentioned oxides were mixed in suitable proportions to obtain the desired compositions. The mixtures were melted at 1200°C for 15 min. The glass samples were obtained by pouring the melts on a stainless steel block.

Density measurements were performed using the picnometric method with distilled water as the reference immersion liquid. The estimated error for the determined density values was less than $0.02 g/cm^3$.

X-ray diffraction investigation was performed by a XRD-6000 SHIMADZU diffractometer using the Cu Kα radiation (λ = 1.54056 Å), a working voltage of 40 kV, a working current of 30 mA and the scanning speed of 2°/min.

The FTIR absorption spectra of the glasses in the 400-2000 $cm^{-1}$ wave number range with a resolution of 4 $cm^{-1}$ were registered at room temperature using a JASCO FTIR-6200 type spectrometer and the KBr pellet technique. The obtained spectra are deconvoluted to enable us to remark the structural units that build up the glass matrix.

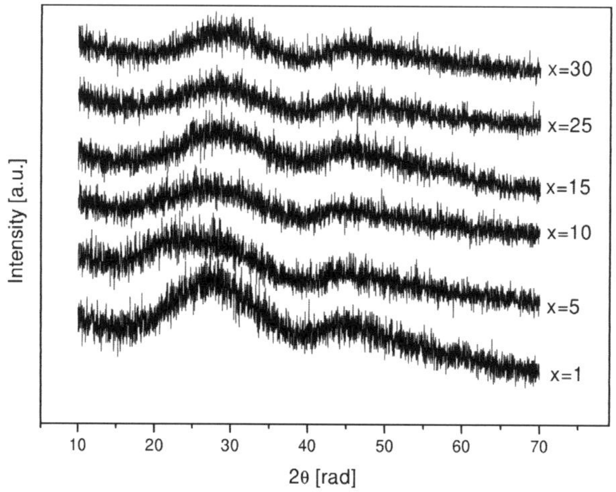


**Figure 1.** X-ray diffractograms of the $xGd_2O_3(100-x)[Bi_2O_3\ B_2O_3]$ glasses.

## 3. Results and discussion

The X-ray diffractograms of the samples show only a broad diffuse scattering at low angles. This type of scattering is characteristic of long-range structural disorder and confirms the amorphous nature of the samples in the studied composition range (figure 1).

In order to elucidate the structure of the $xGd_2O_3(100-x)[Bi_2O_3\ B_2O_3]$ glass system, we studied this system by density measurements and FTIR spectroscopy.

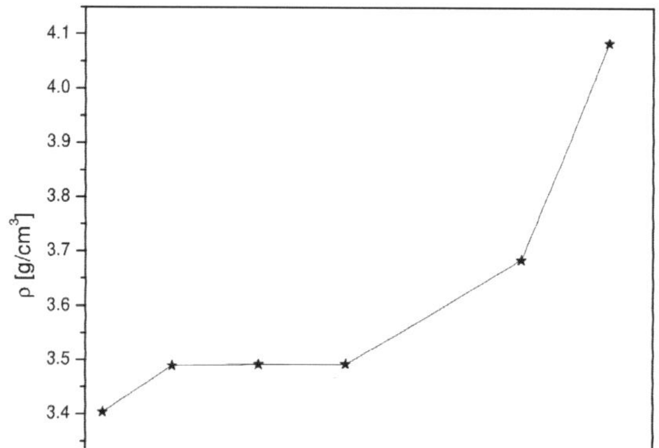


**Figure 2.** Density evolution of the $xGd_2O_3(100-x)[Bi_2O_3\ B_2O_3]$ glasses.

The density measurement is a simple but efficient tool capable of exploring the changes occurring in the structure of glasses. Thus, the density is affected by the structural softening/compactness, the changes in geometrical configuration, coordination number of the ions and the dimension of interstitial spaces of the glass. As shown in figure 2, the density of the studied glasses increases with increasing the $Gd_2O_3$ content of the samples.

From density data we calculated the molar volume and the Poisson's ratio (table 1) using the Makishima and Mackenzie model [9], that permitted to follow the compositional evolution of the packing degree of the atoms in the studied glasses. The Poisson's ratio, $\sigma$, is small if the atoms are loosely packed in the oxide glass whereas tightly packed glass has a higher Poisson's ratio.
The Poisson's ratio [9] was calculated from the expressions:

$$\sigma_{cal} = 0.5 - \frac{1}{7.2V_T} \quad (1)$$

$$V_T = \frac{\rho}{M}\sum_i V_i X_i \quad (2)$$

$$V_i = 6.023\times10^{23}\frac{4\pi}{3}\left(mR_A^3 + nR_O^3\right) \quad (3)$$

where $R_A$ and $R_O$ are respectively the Pauling's ionic radius of metal A and oxygen O, $\rho$ is the glass density, $M$ the glass molecular weight, $X_i$ the mole fraction of the oxide component $i$, $V_i$ is the packing factor of oxide $A_mO_n$, $V_t$ is the packing density of glass and $\sigma_{cal}$ is the theoretically calculated Poisson's ratio.

**Table 1.**

| x [mol%] | ρ [g cm$^{-3}$] | $V_m$ [cm$^3$] | $\sigma_{calc}$ |
|---|---|---|---|
| 1 | 3.4039 | 156.836 | 0.035 |
| 5 | 3.4894 | 151.009 | 0.045 |
| 10 | 3.4915 | 148.439 | 0.049 |
| 15 | 3.4924 | 145.923 | 0.058 |
| 25 | 3.6848 | 133.606 | 0.087 |
| 30 | 4.0843 | 118.419 | 0.140 |

The molar volume of the studied samples increases with increasing their $Gd_2O_3$ content. The compositional evolution of the molar volume and Poisson's ratio indicates that there might be some structural changes of the glass matrix, namely that the gadolinium ions play an important role in these glasses influencing the coordination number of the Bi and B ions. It is know that the addition of modifier oxides results in an increase of σ due to the creation of non-bridging oxygens (NBO) followed by a reduction in dimensionality [10]. Therefore, in the studied glasses, the changes in Poisson's ratio with the increase of the gadolinium oxide content can be attributed to the fact that $Gd_2O_3$ play the network modifier role. The behaviour of Poisson's ratio is related to the change in the type of the bonds from covalent (characteristic for the borate and bismuthate glasses) into ionic bonds.

The infrared absorption spectra of the $xGd_2O_3(100-x)[Bi_2O_3\ B_2O_3]$ glasses are shown in figure 3. Table 2 summarized the major absorption bands observed in the investigated glasses and their assignments in agreement with the references [1, 10, 12-14].

The $Bi_2O_3$ containing glasses have two fundamental vibrations in the IR spectral regions, namely at ~845 and ~ 630 cm$^{-1}$. The vibrational modes of the borate network are mainly active in three infrared spectral regions: a region around 700 cm$^{-1}$ due to the bending of the B–O–B linkage, another region at 800-1200 cm$^{-1}$ due to the B–O bond stretching of tetrahedral $BO_4$ units, and a third region at 1200-

1600 $cm^{-1}$ due to stretching of the trigonal $BO_3$ units. These bands are shifted under the influence of the surrounding cations, the extent and direction of the shift depending on the type of the cation.

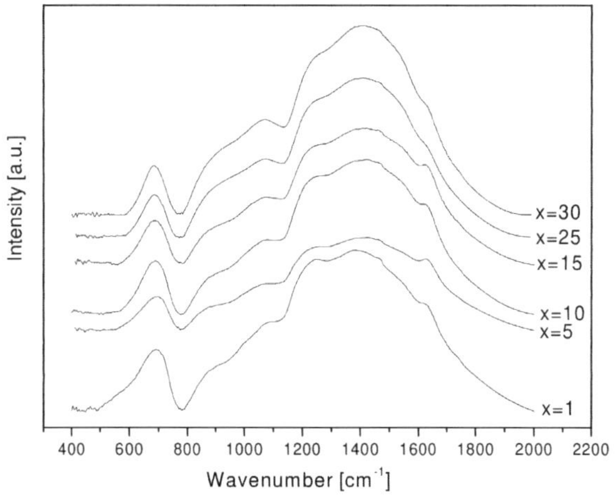


**Figure 3.** FTIR spectra of the $xGd_2O_3(100-x)[Bi_2O_3 B_2O_3]$ glasses.

**Table 2.** FTIR absorption bands observed in the $xGd_2O_3(100-x)[Bi_2O_3 B_2O_3]$ glasses and their assignments.

| Wavenumber [$cm^{-1}$] | Assignments |
|---|---|
| 630 | Bi–O stretching vibrations (of non-bridging oxygen) in $BiO_6$ units |
| 700 | The bending vibration of B–O–B linkages of $BO_4$ units |
| 845 | Symmetrical stretching vibration of the Bi–O bonds in the $BiO_3$ units |
| 900 | Stretching vibrations of $BO_4$ units that are connected by the bismuth cations |
| 1060 | B–O streching vibrations in $BO_4$ units from tri-, tetra- and pentaborate groups |
| 1205 | B–O stretching vibration in $BO_3$ units from pyro- and ortoborate groups |
| 1350 | B–O stretching vibrations of the trigonal $BO_3$ units in meta-, pyro- and orthoborate groups |
| 1465 | Asymmetrical stretching vibrations with three (NBOs) of B–O–B groups |
| 1635 | OH bending mode of vibration |

As the concentration of $Gd_2O_3$ is increased, the bands attributed to stretching vibration in the $BO_4$ seem to became narrower. The band attributed to the symmetrical stretching vibration in the $BiO_3$ units moves to higher wavenumbers. This suggests the conversion of $BO_4$ and $BiO_3$ units into $BO_3$ and $BiO_6$ units, respectively. Since the addition of gadolinium oxide causes some structural changes in the $Bi_2O_3$ $B_2O_3$ vitreous matrix, this suggests that the gadolinium ions act as network modifiers in the studied glass system.

## 4. Conclusions

Glasses of the $xGd_2O_3(100-x)[Bi_2O_3 B_2O_3]$ system with x from 1 to 30 mol% were prepared and studied by means of X-ray, FTIR and density measurements. The compositional variation in density, molar volume and Poisson's ratio indicate that the effect of $Gd_2O_3$ on the boro-bismuthate glass structure dependens on its concentration. By analyzing the infrared spectra of the $xGd_2O_3(100-x)[Bi_2O_3 B_2O_3]$ glass system, it was found that the main structural units that build up the glass network are the $BiO_3$, $BiO_6$, $BO_3$ and $BO_4$ units. The $Gd_2O_3$ play a network modifier role and stabilizes the glass structure of the studied vitreous system.

## References

[1] Cheng Y, Xiao H and Guo W 2008 *Ceram. Int.* **34** 1335
[2] Klimesz B, Dominiak-Dzik G, Solarz P, Zelechower M and Ryba-Romanowski W 2005 *J. Alloys Compd.* **403** 76
[3] Pascuta P, Borodi G and Culea E 2008 *J. Non-Cryst. Solids* **354** 5475
[4] Culea E, Pop L, Pascuta P and Bosca M 2009 *J. Mol. Struct.* **924-926** 192
[5] Yousef E S, El-Adawy A and El-KheshKhany N 2006 *Solid State Commun.* **139** 108
[6] Bale S, Rahman S, Awasthi A M and Sathe V 2008 *J. Alloys Compd.* **460** 699
[7] Rada S, Culea E, Bosca M, Muntean R and Pascuta P 2008 *Vibr. Spectros.* **48** 285
[8] Chen Y, Huang Y, Huang M, Chen R and Luo Z 2004 *Opt. Mat.* **25** 271
[9] Makishima A and Mackenzie J D 1975 *J. Non-Cryst. Solids* **17** 147
[10] Saddeek Y B and Gaafar M S 2009 *Mater. Chem. Phys.* **115** 280
[11] Saritha D, Markandeya Y, Salagram M, Vithal M, Singh A K and Bhikshamaiah G 2008 *J. Non-Cryst. Solids* **354** 5573
[12] Sharma G, Singh K, Manupriya, Mohan S, Singh H and Bindra S 2006 *Radiat. Phys. Chem.* **75** 959
[13] Ardelean I, Cora S, Ciceo Lucacel R and Hulpus O 2005 *Sol. State Sci.* **7** 1438
[14] Doweidar H and Saddeek Y B 2009 *J. Non-Cryst. Solids* **355** 348.

# Magnetic studies of $TeO_2$ - $Fe_2O_3$ glass systems obtained by the sol - gel method

**Adriana Dehelean[1] and Eugen Culea[2]**

[1] National Institute for Research and Development of Isotopic and Molecular Technologies, 65-103 Donath, 400293 Cluj-Napoca, Romania

[2] Technical University of Cluj-Napoca, Physics Department, 15 Constantin Daicoviciu, 400020 Cluj-Napoca, Romania

E-mail: abenta@itim-cj.ro

**Abstract.** Glass of the $xFe_2O_3(1\text{-}x)TeO_2$ systems with x = 0.03-0.40 were obtained by sol-gel method and studied by magnetic susceptibility measurements. The temperature dependence of the reciprocal magnetic susceptibility follows a Curie law till 10 mol% $Fe_2O_3$ and over this concentration follows a Curie- Weiss law with the negative paramagnetic Curie temperature. The main result coming out from this behaviour is: at concentrations lower than x = 10 mol% $Fe_2O_3$ iron ions are isolated or participates at dipolar interactions and at concentrations larger than 10 mol% $Fe_2O_3$ a higher part of iron ions participates at superexchange magnetic interactions.

## 1. Introduction

Glasses are materials that have an unlimited theoretical variety of compositions. This is the reason why their properties are very different, fact that leads to a large area of applications of glasses and these applications are opened at new concepts.

During recent years, there has been increasing interest in the synthesis, structure and properties of heavy metal oxide glasses containing $TeO_2$, due to their high refractive index (n), $n \approx 2$, high IR transparency, high non-linear third order optical susceptibility (50 times higher than the one of $SiO_2$ systems) [1-3].

The processing route mainly adopted for producing tellurite glasses is a melting and quenching technique, which is difficult to produce homogeneous multicomponent tellurite glasses due to their relatively high volatility, limited stability and ready contamination by crucibles. Sol-gel process is an attractive alternative to overcome these drawbacks [4].

The sol-gel method is a particularly attractive way of synthesizing these materials, since it offers a possibility of control over the microstructure and composition of the host matrix and the opportunity to prepare homogenous glasses with large heavy metals ions concentrations [5].

The study of the properties of tellurite glasses with iron ions represents an important thing due to the possibility of those existing in the valence state $Fe^{2+}$ and $Fe^{+3}$ [6], which determinates the magnetic properties of interest due to the antiferromagnetic behaviour.

In this present work we proposed to prepare and to study magnetic behaviour of iron ions in the tellurite glass systems obtained by the sol-gel method.

## 2. Experimental

A tellurite glass systems containing iron oxide was prepared by using the sol-gel method. The samples were obtained in the system: tellurium tetraethoxide /acetic acid /ethanol /water /iron nitrate. Tellurium tetraethoxide (85% Alfa Aesar) and iron nitrate (99.9% Alfa Aesar) were used as precursors. For the precursors reactivity control acetic acid were used.

Samples with the $xFe_2O_3(1-x)TeO_2$ composition (where x = 0.03, 0.05, 0.10, 0.20, 0.30, 0.40) were obtained.

In order to perform the magnetic susceptibility measurements, powder samples were pressed at a pressure of 10 atm. Pieces detached from these pressed samples were weighted and were used for magnetic measurements. Magnetic susceptibility measurements were performed in the 80-300 K temperature range using a Weiss type magnetic balance.

## 3. Results and discussion

The temperature dependence of the inverse magnetic susceptibility, $\chi^{-1}$, is presented in figure 1 for the studied tellurite glass systems containing iron ions. The collapse of data to straight lines indicates that the susceptibility per mole of magnetic ion follows a Curie-Weiss type behaviour [7]:

$$\chi^{-1} = \frac{T - \theta_p}{C} \qquad (1)$$

where $\chi$ represents the molar magnetic susceptibility, $T$, the temperature, $C$ is the molar Curie constant and $\theta_p$ is the paramagnetic Curie temperature. By fitting the experimental data acoording to equation (1), we determined the values of $C$ and $\theta_p$.

The paramagnetic Curie temperature, $\theta_p$, is a rough indicator of magnetic interaction between iron ions.

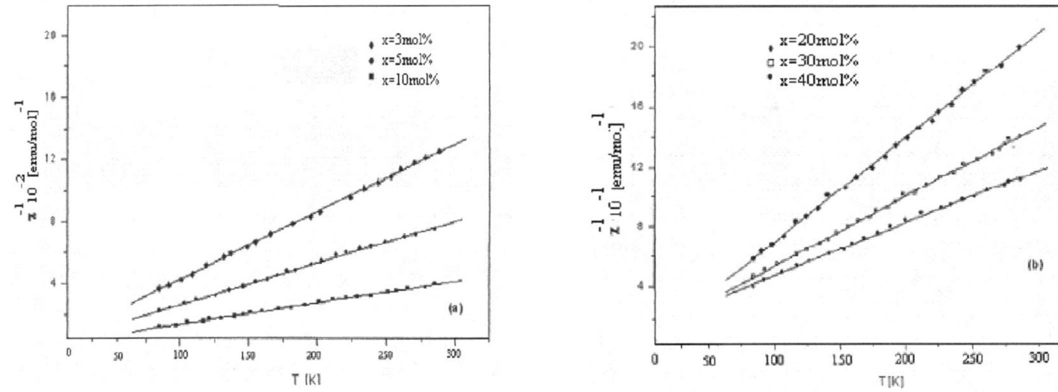


**Figure 1.** The temperature dependence of $\chi^{-1}$ for $xFe_2O_3(1-x)TeO_2$ glasses with (a) $3 \leq x \leq 10$ mol% and (b) $20 \leq x \leq 40$ mol%.

The temperature dependence of the reciprocal magnetic susceptibility follows a Curie law till $x \leq$ 10 mol% $Fe_2O_3$ and over this concentration follows a Curie-Weiss law with the negative paramagnetic Curie temperature. This fact suggest that at concentrations lower than x = 10 mol% $Fe_2O_3$ iron ions are isolated or participates at dipolar interactions and at concentrations x > 10 mol% $Fe_2O_3$ a higher part of iron ions participates at superexchange magnetic interactions. The nature of these superexchange magnetic interactions is antiferromagnetic because $\theta_p$ has a negative value.

Also, we have calculated Curie molar constants, the magnetic moment and molar fractions of $Fe^{3+}(x_1)$ and $Fe^{2+}(x_2)$ ions in studied samples (table 1).

**Table 1.** Curie molar constants, magnetic moment and molar fractions of $Fe^{3+}$($x_1$) and $Fe^{2+}$($x_2$) ions.

| x [mol% $Fe_2O_3$] | $C_M \cdot 10^2$ [emu/mol] | $\mu_{eff}$ [$\mu_B$] | $x_1$ [mol% $Fe_2^{3+}O_3$] | $x_2$ [mol% $Fe_2^{2+}O_3$] |
|---|---|---|---|---|
| 3 | 21.85 | 5.50 | 1.7 | 1.2 |
| 5 | 37.10 | 5.42 | 2.4 | 2.5 |
| 10 | 73.26 | 5.39 | 4.3 | 5.6 |
| 20 | 143.87 | 5.35 | 8.1 | 11.7 |
| 30 | 215.61 | 5.35 | 14.0 | 19.0 |
| 40 | 285.30 | 5.33 | 15.2 | 24.8 |

We observe that the magnetic moment decrease from 5.50 $\mu_B$ at 3 mol% $Fe_2O_3$ to 5.33 $\mu_B$ for 40 mol% $Fe_2O_3$. Knowing that the magnetic moments of $Fe^{3+}$ and $Fe^{2+}$ ions in free state are $\mu_{Fe^{3+}} = 5.92\mu_B$ and $\mu_{Fe^{2+}} = 4.90\ \mu_B$ and giving the fact that in many paramagnetic salt [8] and in oxide glasses [9] the magnetic moments of $Fe^{3+}$ and $Fe^{2+}$ ions are equal to the ones in free state, we assume that in all the studied samples these iron ions coexist [10-12].

The data presented in table 1 permitted the calculation of the amount of the iron ions in their 3+ ($x_1$) and 2+ ($x_2$) valence state. Given that in both valence states iron ion is magnetic, we calculated the amount of $Fe^{3+}$ and $Fe^{2+}$ ions by solving the equations:

$$\begin{cases} x\mu_{\exp}^2 = x_1\mu_{Fe^{3+}}^2 + x_2\mu_{Fe^{2+}}^2 \\ x = x_1 + x_2 \end{cases} \tag{2}$$

Thus, in the studied samples, iron ions are found in both valence states and at concentrations lower than x= 10 mol% iron ions are isolated or participates at dipolar interactions and at concentrations x> 10 mol% $Fe_2O_3$ a higher part of iron ions participates at superexchange magnetic interactions. The nature of these superexchange magnetic interactions is antiferromagnetic.

## 4. Conclusion

Tellurite glass systems doped with iron ions have been prepared by the sol-gel method and magnetic behaviour of these systems were studied. The temperature dependence of the reciprocal magnetic susceptibility follows a Curie law for low concentrations of magnetic ions (x≤10 mol % $Fe_2O_3$), iron ions are isolated or participates at dipolar interactions, respectively a Curie- Weiss law for high concentrations of magnetic iron ions (x>10 mol % $Fe_2O_3$), a part of iron ions participates at superexchange magnetic interactions for the $xFe_2O_3(1-x)TeO_2$ (where x= 0.03÷0.40) system.

## References

[1] Jeansannetas B *et al* 1999 *J. Solid State Chem.* **146** 329
[2] Prakash G V, Rao D N and Bhatnagar A K 2001 *Solid State Commun.* **119** 39
[3] El-Mallawany R A 1992 *J. Appl. Phys.* **72** 1774
[4] Weng L, Hodgson S, Bao X and Sagoe-Crentsil K 2004 *Mater. Sci. Eng.* B **107** 89
[5] Brinker C J and Scherer G W 1990 *Sol-Gel Science: the Physics and Chemistry of Sol-Gel Processing* (San Diego: Academic Press)
[6] Ardelean I 1978 *Solid State Commun.* **27** 679
[7] Culea E, Pop A and Cosma I 1996 *J. Magn. Magn. Mater.* **157/158** 163
[8] Chowdari B V R and Rong Z 1995 *Solid Sate Ionics* **78** 133
[9] Burzo E and Ardelean I 1979 *Phys. Chem. Glasses* **20** 15
[10] Ardelean I, Hong- Hua Qiu and Sakata H 1997 *Mater. Lett.* **32** 335
[11] Ardelean I, Petreanu M, Simion V, Filip S, Ciorcas F and Todor I 1999 *J. Magn. Magn. Mat.* **196/197** 257
[12] Ardelean I, Ilonca Gh, Cozar O, Simion V and Filip S 1994 *Mater. Lett.* **21** 321.

# Magnetic behaviour of europium ions in some tellurite glasses obtained by the sol - gel method

**Adriana Dehelean[1] and Eugen Culea[2]**

[1] National Institute for Research and Development of Isotopic and Molecular Technologies, 65-103 Donath, 400293 Cluj-Napoca, Romania

[2] Technical University of Cluj-Napoca, Physics Department, 15 Constantin Daicoviciu, 400020 Cluj-Napoca, Romania

E-mail: abenta@itim-cj.ro

**Abstract.** In the paper, tellurite glasses doped with europium ions have been prepared by the sol-gel method and were studied by magnetic susceptibility measurements. The authors found that $xEu_2O_3(1-x)TeO_2$ system shows a Curie-Weiss type magnetic behaviour. The magnetic europium ions are present in the host glass matrix in both their 2+ and 3+ valence states. The magnetic interactions between the europium ions are antiferomagnetic.

## 1. Introduction

Tellurite glass systems containing different rare earth oxides have attracted a great attention in the last years because of their interesting properties such as low glass transition temperature, high refractive index (in the range from 1.85 to 2.2), high transmittance from ultraviolet to near infrared, high non-linear third order optical susceptibility (50 times higher than the one of $SiO_2$ systems) [1-3]. Due to these properties tellurium glasses are ideal materials for applications in many important fields like telecommunication technology, laser and optical fibre technology, optical device technology, etc. [4, 5].

The tellurium oxide ($TeO_2$) is a conditional glass former. The main structural units present in the structure of the tellurite glasses are the $TeO_4$ trigonal bipyramids, or if we take into account the $5s^2$ lone pair of tellurium atoms (E), a distorted TeO4E bipyramid. In this bipyramid, the two equatorial oxygen atoms are separated from the tellurium atoms by distances shorter than the sum of the covalent radii of oxygen (0.73 Å) and tellurium atoms (1.35 Å), and the two axial oxygen atoms by distances longer than value (figure 1) [6, 7].

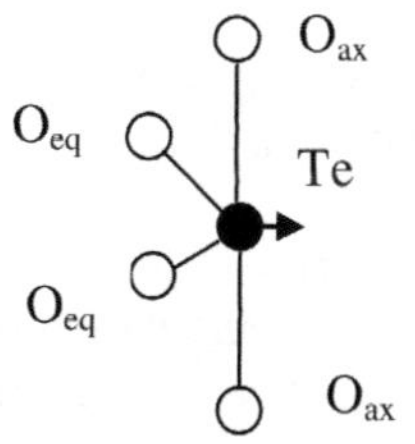


**Figure 1**. Model for the $TeO_4$ structural unit.

The type and the concentration of network modifying ions modifies the structure of the host glass matrix [8].

The traditional processing route adopted for producing tellurite glasses consists in mixing and melting appropriate amounts of the component oxides followed by the quenching of the melts. This technique presents some important disadvantages such as the difficulty to produce homogeneous multicomponent tellurite glasses due to their relatively high volatility, limited stability and contamination by crucibles. These disadvantages impair the optical properties of the tellurite glasses. The intensification of the investigation of different properties of the tellurite glass systems containing different metal ions (mainly rare earth or transitional metal ions) started in the last years, due to the development of sol-gel process. The sol-gel method is an attractive alternative to offer advantages of a better control of physical and chemical properties of the target materials [9].

The goal of this paper was to study the magnetic behaviour of europium ions in the tellurite glass systems obtained by the sol-gel method.

## 2. Experimental

A tellurite glass systems containing europium oxide was prepared by using the sol-gel method. The samples were obtained in the system: tellurium tetraethoxide ($Te(OEt)_4$) /acetic acid ($CH_3COOH$) /ethanol (EtOH) /water ($H_2O$) /europium nitrate ($Eu(NO_3)_3$). Tellurium tetraethoxide (85% Alfa Aesar) and $Eu(NO_3)_3$ (99,9% Alfa Aesar) were used as precursors. For the precursors reactivity control acetic acid were used.

Samples with the $xEu_2O_3(1\text{-}x)TeO_2$ composition (where x = 0.07, 0.11, 0.15) were obtained.

In order to perform the magnetic susceptibility measurements, powder samples were pressed at a pressure of 10 atm. Pieces detached from these pressed samples were weighted and were used for magnetic measurements. Magnetic susceptibility measurements were performed in the 80-300 K temperature range using a Weiss type magnetic balance.

## 3. Results and discussion

The temperature dependence of the inverse magnetic susceptibility, $\chi^{-1}$, for the studied tellurite glass system containing europium ions is presented in figure 2. The data collapse to straight lines indicating that the magnetic susceptibility follows a Curie-Weiss type behaviour:

$$\chi^{-1} = (T - \theta_p)/C \tag{1}$$

where $C$ is the Curie constant and $\theta_p$ is the paramagnetic Curie temperature.

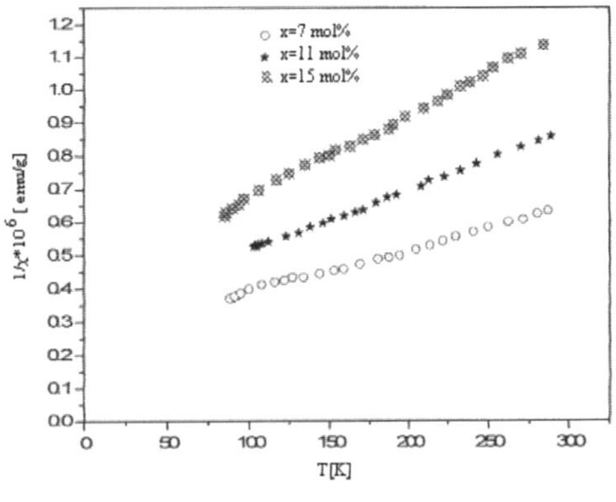


**Figure 2.** Temperature dependence of $\chi^{-1}$ for $xEu_2O_3(1\text{-}x)TeO_2$ glasses.

The temperature dependence of the reciprocal magnetic susceptibility follows a Curie-Weiss law with negative paramagnetic Curie temperatures. Because the obtained paramagnetic Curie temperatures, $\theta_p$, present negative values, the nature of these superexchange magnetic interactions is presumed to be antiferomagnetic. The paramagnetic Curie temperatures, the molar Curie constants, the magnetic moment per europium ion and the $Eu^{2+}/Eu^{3+}$ ions fractions are presented in table 1.

**Table 1.** Paramagnetic Curie temperature, molar Curie constants, magnetic moment per europium ion and the $Eu^{2+}/Eu^{3+}$ ratio for $xEu_2O_3(1-x)TeO_2$ glasses.

| x [mol% $Eu_2O_3$] | $-\theta_p$ [K] | $C_M$ [emu/mol] | $\mu_{eff}$ [$\mu_B$/atom] | $Eu^{2+}/Eu^{3+}$ |
|---|---|---|---|---|
| 7 | 40 | 0.17 | 3.2 | 0.11 |
| 11 | 160 | 0.37 | 3.6 | 0.14 |
| 15 | 165 | 0.58 | 3.9 | 0.19 |

In general, the europium ions appear in oxide glasses in two valence states, namely $Eu^{2+}$ and $Eu^{3+}$. The magnetic moments of the free $Eu^{3+}$ and $E^{2+}$ ions are $\mu_{Eu}^{3+} = 3.4\ \mu_B$ and respectively, $\mu_{Eu}^{2+} = 7\ \mu_B$ [10]. Consequently, the magnetic properties of the samples studied can be explained by the presence of both $Eu^{3+}$ and $Eu^{2+}$ ions, as well as by the occurrence of superexchange magnetic interactions between the europium ions. Due to these interactions the occurrence of $Eu^{2+}$–$Eu^{2+}$, $Eu^{3+}$–$Eu^{3+}$ and respectively, $Eu^{3+}$–$Eu^{2+}$ pairs is possible.

The data presented in table 1 permitted the calculation of the amount of the europium ions in their 3+ ($x_1$) and 2+ ($x_2$) valence state. Given that in both valence states europium ion is magnetic, we calculated the amount of $Eu^{3+}$ and $Eu^{2+}$ ions by solving the equations:

$$\begin{cases} x\mu_{\exp}^2 = x_1\mu_{Eu^{3+}}^2 + x_2\mu_{Eu^{2+}}^2 \\ x = x_1 + x_2 \end{cases} \quad (2)$$

We note that the $x_2/x_1$ ratio increases with increasing the europium ion content of the samples.

Figure 3 shows the variation of the paramagnetic Curie temperatures with respect to the europium oxide content of the samples.

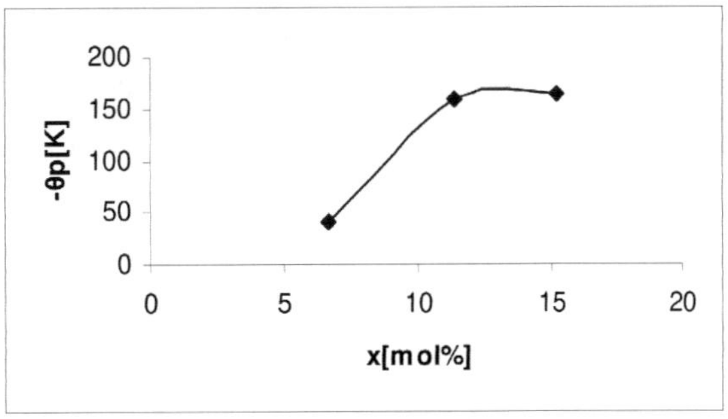


**Figure 3.** The variation of the paramagnetic Curie temperatures with respect to the europium oxide content of $xEu_2O_3(1-x)TeO_2$ glasses.

The increase of $\theta_p$ value shows the intensification of the superexchange magnetic interactions between europium ions with the increase of their amount. A certain reduction of the $\theta_p$ value at high concentrations of the content of europium ions can be due to the interactions between pairs of europium ions with different valences ($Eu^{3+}$–$Eu^{2+}$) which are weaker than the interactions between the same valent ions ($Eu^{2+}$–$Eu^{2+}$ or $Eu^{3+}$–$Eu^{3+}$).

Thus, in the studied samples, europium ions are found in both valence states, they interacting with each other through a superexchange mechanism. These interactions are of antiferomagnetic nature. Magnetic behaviour of the studied samples is probably mictomagnetic [11], the europium ions being present as both isolated and coupled species.

Journal of Physics: Conference Series **182** (2009) 012064 doi:10.1088/1742-6596/182/1/012064

## 4. Conclusion

Tellurite glasses doped with europium ions have been prepared by the sol-gel method and were studied by magnetic susceptibility measurements. The obtained data for the $xEu_2O_3(1-x)TeO_2$ (where $x = 0.07 \div 0.15$) system show a Curie-Weiss type magnetic behaviour. The magnetic europium ions are present in the host glass matrix in both their 2+ and 3+ valence states. The magnetic interactions between the europium ions are antiferomagnetic.

## References

[1] Kim S H, Yoko T and Sakka S 1993 *J. Am. Ceram. Soc.* **76** 2486
[2] Jeansannetas B *et al* 1999 *J. Sol. St. Chem* **146** 329
[3] Prakash G V, Rao D N and Bhatnagar A K 2001 *Solid State Commun.* **119** 39
[4] Clare A 1994 *Key Eng. Mater.* **94-95** 161
[5] Coste S, Lecomte A, Thomas P, Merle-Mejean T and Champarnaud-Mesjard J C 2007 *J. Sol-Gel Sci. Technol.* **41** 79
[6] Rolli R, Gatterer K, Wachtler M, Bettinelli M, Speghini A and Ajo D 2001 *Spectrochim. Acta* A **57** 2009
[7] Mirgorodsky A P, Merle-Méjean T, Champarnaud J C, Thomas P and Frit B 2000 *J. Phys. Chem. Solids* **61** 501
[8] Rada S, Culea E, Boşca M and Păşcuţă P 2007 *Proc. 21$^{st}$ Int. Congr. on Glass* (Strasbourg) p 27
[9] Brinker C J and Scherer G W 1990 *Sol-Gel Science. The Physics and Chemistry of Sol-Gel Processing* (Boston, MA: Academic Press)
[10] Burzo E and Ardelean I 1979 *Phys. Chem. Glasses* **20** 15
[11] Burzo E, Ardelean I and Ursu I 1996 *Mat. Lett.* **26** 103.

Journal of Physics: Conference Series **182** (2009) 012065 doi:10.1088/1742-6596/182/1/012065

# Chitosan-based delivery systems for diclofenac delivery: preparation and characterization

**Simina Dreve, Irina Kacso, Ioan Bratu and Emil Indrea**

National Institute for Research and Development of Isotopic and Molecular Technologies, 65-103 Donath, 400293 Cluj-Napoca, Romania

Email: simina.dreve@itim-cj.ro

**Abstract**. The preparation and characterization of novel materials for drug delivery has rapidly gained importance in development of innovative medicine. The paper concerns the uses of chitosan as an excipient in oral formulations and as a drug delivery vehicle for burnt painful injuries. The use of chitosan (CTS) as base in polyelectrolyte complex systems, to prepare liquid release systems as hydrogels and solid release systems as sponges is presented. In this paper the preparation of CTS hydrogels and sponges carrying diclofenac (DCF), as anti-inflammatory drug is reported. The immobilization of DCF in CTS is done by mixing the CTS hydrogel with the anti-inflammatory drug solutions. The concentration of anti-inflammatory drug in the CTS hydrogel generating the sponges was of 57 mg/l, 72 mg/l and 114 mg/l. The CTS sponges with anti-inflammatory drugs were prepared by freeze-drying at -610°C and 0,09 atm. The characterization of the hydrogels and sponges was done by infrared spectra (FTIR) and ultraviolet-visible spectroscopy (UV-VIS). The results indicated the formation of CTS-DCF intermediates. The DCF molecules are forming temporary chelates in CTS hydrogels and sponges and they are compatible with skin or some of biological fluids with satisfactory results.

## 1. Introduction

The concept of designing a specified delivery system to achieve selective drug targeting has been originated from the perception of Paul Ehrlich, who proposed drug delivery to be as a ‘magic bullet’ [1], where a drug-carrier complex/conjugate, delivers drug(s) exclusively to the preselected target cells in a specific manner. The objective of drug targeting is to achieve desired pharmacological response at a selected site without undesirable interactions at other sites. Among the novel biomaterials, chitin and chitosan are intensively studied due to their many potential applications as pharmaceutical drug carriers. CTS, a natural polysaccharide, is being widely used as a pharmaceutical excipient. CTS comprise a series of polymers varying in their degree of deacetylation, molecular weight, viscosity, pKa etc. The presence of a number of amino groups permits CTS to chemically react with anionic systems, thereby resulting in alteration of physicochemical characteristics of such combinations [2]. Modern biocompatible systems target not only infectious diseases, but also autoimmune disorders, allergies, chronic inflammatory diseases and cancer [3]. The study was aimed to develop and characterize a novel polyelectrolyte complex (PEC) CTS with Tween-80 and oleic acid as drug carrier for controlled drug delivery, with possible use in skin burnt painful injuries. The PEC CTS complexes were prepared by coacervation method using the same ratios of Tween-80, oleic acid and CTS. DCF, used as model drug, is one of the most useful non-steroidal anti-inflammatory drugs (NSAIDs) and is a practically insoluble compound in acidic solution (pKa 4.0), however, it dissolves in intestinal fluid [4]. Small molecular weight anti-inflammatory drugs that contain chemical groups able to establish temporary physical or chemical bonding with the amino group of the CTS linkage are suitable for the targeted and controlled drug CTS-based delivery systems. In support of this presumption the chemical formulae of CTS and anti-inflammatory DCF drug used in the experiment are illustrated in figures 1-2.

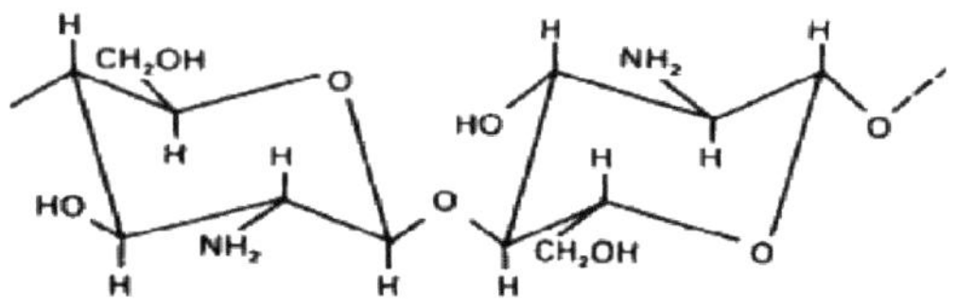

**Figure 1.** Structural formula of chitosan.

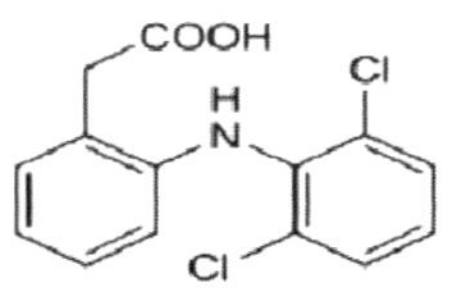

**Figure 2.** Structural formula of diclofenac.

## 2. Experimental

### *2.1. Materials*
Tween-80 was purchased from Sigma–Aldrich Chemie (Steinhaim, Germany), Diclofenac sodium was obtained by courtesy from Terapia-Ranbaxy SA, Cluj-Napoca. High molecular weight CTS grade (Sigma # 419419) was purchased from Sigma–Aldrich Chemie (Steinheim, Germany). At 86% deacetylation degree it had a 5.99% loss on drying, 0.21% of ashes, and 0.48% of insoluble matter (at 1% in 1% acetic acid) as specified by the manufacturer.

### *2.2. Preparation of CTS-based PEC solution*
CTS hydrogel 0.25% was prepared in double distilled water containing 1% (v/v) acetic acid with continuous stirring, at 300C, adding at final 2%v/v Tween-80 and 0.2% v/v oleic acid. After 6 h the solution is pale-yellow, with a homogenous consistence and aspect.

### *2.3. Preparation of CTS-based drug loaded nano/microparticles*
To the CTS PEC solution, split in 16 equal volumes, 10 ml each, different quantities of DCF were added to give PEC 1.1, PEC 1.2 and PEC 1.3. The concentration of anti-inflammatory drug in the CTS hydrogel generating the sponges was of 57 mg/l, 72 mg/l and 114 mg/l. The solutions were further stirred for 30 min and then ultrasonically treated in an Elmasonic E60H ultrasonic bath for 10 min. In order to prepare the sponges the hydrogel was freeze-dried in a CHRIST ALPHA 1-4 LD Plus Freeze Dryer, at -600°C and 0.09 atm. Particles from native CTS (NC) were also developed similarly.

### *2.4. Characterization of PEC*
CTS-based polyelectrolyte complex with and without drug were characterized by FTIR (FTIR JASCO 6100 spectrometer) and ultraviolet-visible spectroscopy, UV-VIS (ABLE&JASCO 550V).

## 3. Spectroscopic analysis of CTS-based DCF loaded materials

### *3.1. FTIR spectral studies*
Figure 3 shows the FTIR spectra of plain CTS, DCF, CTS+DCF (physical mixture) combination and 1.1, 1.2, 1.3 combined microspheres. As it can be seen in the FTIR spectra the maximum at 1100 $cm^{-1}$ corresponding to the C–O stretching vibration mode in CTS molecule is shifted to the 1108 $cm^{-1}$ peak due to the steric modifications imposed by the CTS-DCF PEC microspheres. The sharp peak at 1452 $cm^{-1}$, corresponding to the vibration mode of $–CH_2–$ of DCF molecule, is shifted to 1462 $cm^{-1}$ which correspond to –NH and C=O stretching vibration, respectively, thereby confirming intermolecular CTS-DCF grafting reaction.

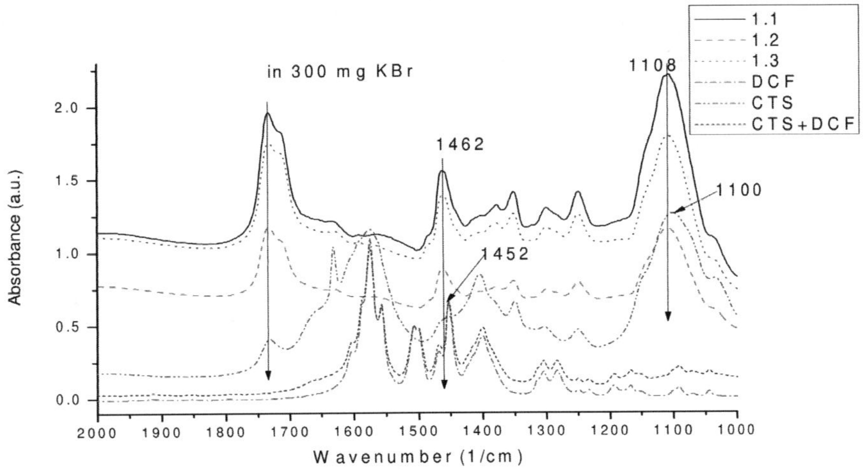


**Figure 3.** FTIR spectra of plain CTS, DCF, and 1.1, 1.2, 1.3 PEC combined microspheres, 2000-1000 cm$^{-1}$ spectral region.

The new peak at 1560 cm$^{-1}$ corresponding to C–N bending vibration further supports the grafting reaction. The other identified vibrations for each sample are presented in table 1.

**Table 1.** The assignments of the most intense FTIR bands for CTS, DCF and CTS+DCF PEC microspheres.

| Wavenumber (cm$^{-1}$) | CTS | DCF | CTS+DCF PEC |
|---|---|---|---|
| 767 vs; 748vs | | δ(CH) | |
| 1020 | intramolecular C–O stretching | | |
| 1050 | intramolecular C–O stretching | | |
| 1073 | C–O stretching | | |
| 1100 | C-O stretching | | |
| 1108 | | | C-O stretching |
| 1145 | C–O–C stretching | | |
| 1355 | O–H bending | | O–H bending |
| 1379 | amide-III | | |
| 1594 | amide-II | | |
| 1650 | amide-I | | |
| 1380 | $CH_3$ symmetrical deformation | | |
| 1390 s | | $\nu_s(COO^-)$ | |
| 1410 | $CH_3$ symmetrical deformation | | |
| 1453 | | $\delta CH_2$- | |
| 1453 vs | | $\delta(CH_2)$ | |
| 1464 | | | –NH and C=O stretching |
| 1506 sh; 1557 s | | ring str. | |
| 1560 | | | C–N bending |
| 1576 vs | | $\nu_{as}(COO^-)$ | |
| 1590 s | | δ(NH)+ring | |
| 2845 | C–H stretching | | |
| 2919 | C–H stretching | | |
| 3440 | O–H stretching | | O–H stretching |
| 2920 w; 2980 m | | $\nu(CH_2)$ | |
| 2929 | Aliphatic C–H stretching | | Aliphatic C–H stretching |
| 3085 m; 3040 m | | ν(CH) | |
| 3260 br; 3388 s | | ν(NH) | |
| 3450 | amine N–H symmetric stretching | | |

### *3.2. UV-VIS spectroscopy*
Figure 4 shows the UV–VIS absorption spectra of CTS and CTS-DCF in 1% dilute acetic acid.

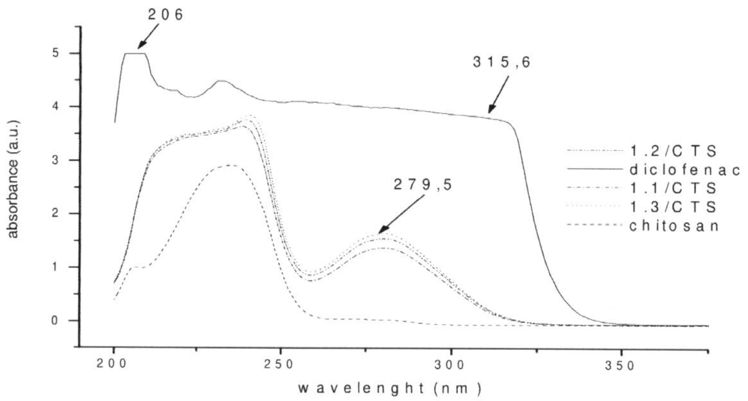


**Figure 4.** UV–VIS absorption spectra of CTS and CTS-DCF PEC microspheres in dilute acetic acid solution.

CTS only yields absorption peak at 210 nm. DCF curve has a broad absorption band between 200 nm and 315.6 nm. The spectral absorption at 279.5 nm is specific for CS-DCF PEC microspheres.

## 4. Conclusions
CTS-DCF polyelectrolyte complex microspheres were prepared and investigated by FTIR and UV-VIS spectroscopy. The results are in agreement to similar data already reported [7, 8]. The increased absorbance at 279 nm and the FTIR shifted vibrational bands can be assigned to the formation of CTS-DCF PEC supramolecular assemblies. Further experiments on the complexation mechanism are in progress in order to elucidate a detailed structure of the complex.

## Acknowledgments
This work received financial support from the Romanian Research and Education Ministry under the Core Project PN-09-44 02 01.

## References
[1] Ehrlich P 1902 *Collected papers of Paul Ehrlich: Immunology and cancer research* (London: Pergamon Press) p 442
[2] Jollès P and Muzzarelli R A A ed 1999 *Chitin and Chitinases Birkhäuser* (Basel)
[3] Svirshchevskaya E V, Alekseeva L G, Reshetov P D, Phomicheva N N, Parphenyuk S A, Ilyina A V, Zueva V S, Lopatin S A, Levov A N, Varlamov V P 2009 *Eur. J. Med. Chem.* **44** 2030-37
[4] *Martindale: The complete drug reference* 2002 ed C S Sweetman (London: Pharmaceutical Press) pp 30-32
[5] Peng T, Yao K D, Chen Y and Goosen M F 1994 *J. Polym. Sci.* A: *Polym. Chem.* 32 591-96
[6] Sannan T, Kurita K, Ogura K and Iwakura Y, 1978 *Polymer* **19** 458-59
[7] AL-Kahtani A, Bhojya Naik H S and Sherigara B S 2009 *Carbohydr. Res.* **344** 699-706
[8] Bratu I, Astilean S, Ionescu C, Indrea E, Huvenne J P and Legrand P 1998 *Spectrochim. Acta* A **54** 191-96.

# Structural characterisation of binary $SiO_2/TiO_2$ nanoparticle aerogels by X-ray scattering

**Emil Indrea[1], Anca Peter[2], Danut T Silipas[1], Simina Dreve[1], Ramona-Crina Suciu[1], Marcela Corina Rosu[1], Virginia Danciu[2] and Veronica Cosoveanu[2]**

[1] National Institute for Research and Development of Isotopic and Molecular Technologies, 65-103 Donath, 400293 Cluj-Napoca, Romania

[2] Babes-Bolyai University, Faculty of Chemistry and Chemical Engineering, 11 Arany Janos, 400028 Cluj-Napoca, Romania

E-mail: emil.indrea@itim-cj.ro

**Abstract.** The effect of $SiO_2$ to $TiO_2$ aerogels content on their local structure is evidenced from X-ray scattering measurements. $TiO_2$-$SiO_2$ aerogels were prepared by sol-gel method followed by supercritical drying with liquid $CO_2$ and heat treatment. The local structure of $TiO_2$-$SiO_2$ system was investigated by analysing the atomic pair distribution function PDF , G(r) , obtained from X ray scattering data The modified coordination radius values obtained for the first coordination spheres by addition of $SiO_2$ denotes a diminution of the atom packing in this system. The PDF of the $300^0$C heat treated $TiO_2$ aerogel sample shows well-defined features allowing an unambiguous structure search and refinement. In particular, the first peak in the PDF of the precursor phase is quite narrow and centered at about 1.9 Å, which is the Ti-O distance in Ti-$O_6$ octahedra. This observation shows that well-defined Ti-$O_6$ octahedra are already formed during the homogeneous precipitation stage of the chemical process. Inspecting the data from the analysis of the atomic pair distribution function G(r) one remarks the occurrence of Si–O pair as a first coordination formation at a distance of 1.49 Å and the interatomic distances at 2.18 Å for the first Si---Si and 3.58 Å for the second Si---Si bond.

## 1. Introduction

Silica-titania materials $TiO_2$-$SiO_2$ aerogels have received significant attention as high performance photocatalysts because of their larger specific surface area and higher porosity. These porous solids combine the mechanical properties of silica with the chemical properties of titania. Anderson and Bart [1] have demonstrated that a mixed oxide of $TiO_2$ and $SiO_2$ produced by a sol-gel method was a more efficient photocatalysts for the photodecomposition of rhodamine-6G than $TiO_2$ alone. The increase in efficiency was attributed to the presence of $SiO_2$ as adsorbent, which enhances the concentration of the organic material near to the $TiO_2$ sites. The heat treatment temperature influences the structure of the $TiO_2$-$SiO_2$ aerogels and also their photoactivity. In this work, we prepared $TiO_2$-$SiO_2$ aerogels with different contents of Ti and Si, which are subsequently heat treated at different temperatures. We studied the local structure particularities of the prepared $TiO_2$-$SiO_2$ aerogels by means of an X-ray scattering experiment. Along with the neutron scattering the X-ray scattering from disordered materials is largely used to investigate their local structure [2, 3]. The non-crystalline state of a compound can occur in a large structural variety depending on the preparation conditions. This is due to the fact that during the sample preparation the "freezing" process of the atoms arrangement has an important effect on the atom bonds and atom relaxation energy barriers. The XRD pattern for the $TiO_2$-$SiO_2$ aerogels, however, is very diffuse in nature and could not be analyzed in the traditional way. That is why the local structure investigation was considered in terms of the atomic pair distribution function (PDF) analysis, an approach that has recently proven to be very successful in structure studies of heavily disordered oxide materials [4].

## 2. Experimental

In order to study the influence of Si content and thermal treatment temperature on the properties, $SiO_2$-$TiO_2$ aerogels with different $TiO_2$ content were synthesized by sol-gel method followed by low temperature supercritical drying and different temperatures thermal treatment [5].

The sols were prepared from Ti(IV) isopropoxide (Merck), tetramethoxysilane (Merck), anhydrous ethanol (Fluka), water and nitric acid reagent (Aldrich). The molar ratio of reactants was: [Ti/Si]:[$H_2O$] = 1:3.5, [Ti/Si]:[$C_2H_5OH$] = 1:23.33, [Ti/Si]:[$HNO_3$]= 1:0.08.

The as prepared gels were kept for 4 weeks in closed vessels and then were dried using supercritical $CO_2$ in a SAMDRI – 790 A (Tousimis) critical point dryer. Annealing, at 300°C, 600°C and 700°C for 2 h was performed to obtain crystalline phases of $TiO_2$ and to remove residues of organic substances [5].

The X-ray diffraction patterns were obtained using a standard BRUKER D8 Advance X-ray diffractometer, working at 40 kV and 40 mA. The Cu Kα radiation was monochromatized with a germanium monochromator placed in the incident X-ray beam. The data of the X-ray diffraction patterns were collected in a step-scanning mode with $\Delta 2\theta = 0.02°$ steps.

The specific surface area of the as prepared samples was determined using Brunauer-Emmett-Teller (BET) method, in a partial pressure range of $0.05 < P/P0 < 0.3$. Kripton adsorption was carried out at 77 K. Before each measurement, the samples were heat-cleaned at 333 K for 2 h.

In table 1 are presented the Ti composition and the BET specific surface area of the $TiO_2$-$SiO_2$ aerogels.

**Table 1.** The Ti molar percentage content of the $TiO_2$-$SiO_2$ aerogels and the specific surface area of the $TiO_2$-$SiO_2$ aerogels.

| Sample | Sample type | Ti [%mol] | $S_{BET}$ [$m^2/g$] |
|---|---|---|---|
| A300 | $TiO_2$ aerogel<br>Heat treated for 2 h at 300°C | 100% | 56 |
| A600 | $TiO_2$ aerogel<br>Heat treated for 2 h at 600°C | 75% | 52 |
| A700 | $TiO_2$ aerogel<br>Heat treated for 2 h at 700°C | 75% | 44 |

## 3. Results and discussion

Titania, $TiO_2$, is used in a number of important technological applications such as functional ceramics, reinforced polymers, pigments, photocatalysts, gas sensors, solar energy conversion [6, 7]. It occurs in three crystalline polymorphs: rutile, anatase, and brookite. All are built of Ti-O6 octahedra; the way the octahedra couple to each other into a 3D network is slightly different in the different polymorphs. Bulk rutile is the most thermodynamically stable polymorph of titania, and an irreversible phase transition from brookite to anatase and then to rutile usually occurs with increasing temperature. The thermodynamic stability of titania polymorphs, however, also depends on the crystallite size, and because most applications require fine titania powders, it is actually the anatase polymorph that is most frequently used. Anatase, however, exhibits a relatively large band gap (~ 3.2 eV), which is unfavorable to some applications such as photocatalysis. This has inspired an intensive search for fine crystallite titania powders with a narrower band gap. The titania powders were found to exhibit photocatalytic activity in the sequence of pure brookite > mixture of rutile and brookite/anatase > pure rutile [8].

The local structure of TiO2-SiO2 system was investigated by analyzing the atomic pair distribution function PDF, G(r), obtained from X ray scattering data using a PEDX program [9].

The frequently used pair distribution function PDF, is defined as:

$$G(r) = 4\pi r\,[\rho(r) - \rho_0] \tag{1}$$

where $\rho(r)$ and $\rho_0$ are the local and average atomic number densities, respectively, and r is the radial distance [9]. As defined, the PDF G(r) is a one-dimensional function that oscillates around zero and shows positive peaks at distances separating pairs of atoms, *i.e.* where the local atomic density exceeds the average one. The atomic pair distribution function G(r) shows maxima and their position, width and area are determined by the distribution of atom pairs in the structural disordered sample (figure 3).

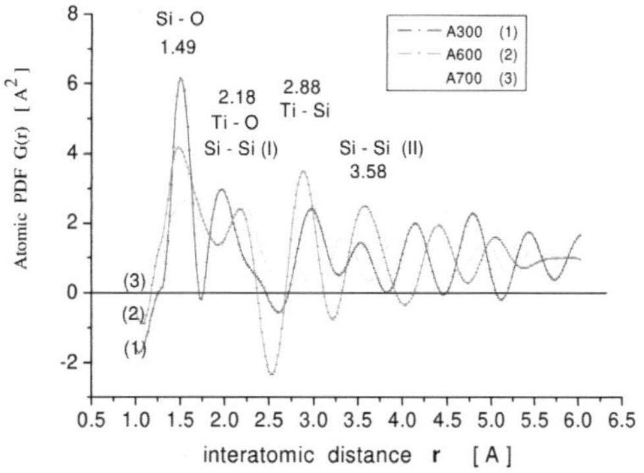


**Figure 1**. The atomic pair distribution function G(r) for $TiO_2$-$SiO_2$ aerogels; A300, A600 and A700 samples.

The real space distance corresponding to the maxima determined from the data obtained in this study may be compared with results reported for other similar systems [8, 10] and allow to identify the atom pairs orderly disposed in the investigated $TiO_2$-$SiO_2$ system. They are summarised in table 1.

**Table 1**. The atom pairs orderly disposed in the vitreous system, identified from the analysis of atomic pair distribution function G(r).

| Coordination sphere | Atom pairs | Interatomic distance [Å] |
|---|---|---|
| I | Si – O | 1.49 |
| | Ti – O | 1.96 |
| II | Ti – O | 2.18 |
| | Si – Si | 2.18 |
| III | Ti – Si | 2.88 |
| IV | Si – Si | 3.58 |
| | Si – Si | 3.58 |

Inspecting the data obtained for the A300 sample from the analysis of the atomic pair distribution function G(r) one remarks the occurrence of Si – O pair as a first coordination formation at a distance of 1.49 Å. The interatomic distances at 2.18 Å for the first Si---Si and 3.58 Å for the second Si---Si bond agree well with data on bond lengths in the current literature [11].

The PDF of the A300 sample shows well-defined features at about 1.96 Å, which is the Ti---O bond distance in Ti-O6 octahedra. A distortion of the $Ti$-$O_6$ octahedra was observed in the A600 and

A700 samples indicated by the rise of the Ti---O bond distance to 2.18 Å. This observation shows that well-defined Ti-$O_6$ octahedra are already formed during the homogeneous precipitation stage of the chemical process employed by Yin *et al* [12] It prompted us to narrow the search for the precursor phase structure to models built of Ti-$O_6$ octahedra [8]. A well-defined feature at 2.88 Å that appears in the PDF function of the $TiO_2$-$SiO_2$ aerogel samples indicates the formation of a Ti---Si bond in the system. This bond distance gets shorter in the A700 sample. The new $TiO_2$-$SiO_2$ system formed after heat treatment at 700°C has the lowest photocatalytic activity [5].

## 4. Conclusion

In this work, we prepared $TiO_2$-$SiO_2$ aerogels with different contents of Si, which are subsequently heat treated at different temperatures. The morpho-structure of the as obtained silica-titania aerogels was studied by X-ray diffraction and specific surface area measurements.

X-ray scattering results indicate changes induced in the local structure of $TiO_2$ aerogels containing $SiO_2$. Total X-ray scattering and atomic pair distribution function PDF data analysis can be successfully employed to determine the atomic scale local structure of heavily disordered oxides. The disordered phase is found to be built of nanosize layers/fragments of Ti-$O_6$ octahedra arranged according the coupling scheme found in lepidocrocite. The well-defined Ti-$O_6$ octahedra are already formed during the homogeneous precipitation stage.

## Acknowledgements

This work was supported by the Ministry of Education and Research through MENER program, HYDROSOL project 710/24.072006.

## References

[1] Anderson C and Bart A J 1997 *J. Phys. Chem.* **B 101** 2611
[2] Petkov V, Gateshki M, Choi J, Gillan E G and Ren Y 2005 *J. Mater.Chem.* **15** 4654
[3] Gateshki M, Hwang S-J, Park D H, Ren Y and Petkov V 2004 *J. Phys.Chem.* B **108** 14956
[4] Petkov V, Gateshki M, Niederberger M and Ren Y 2006 *Chem. Mater.* **18** 814
[5] Peter A, Baia L, Cosoveanu V, Baia M, Constantin D, Indrea E, Popescu I C and Danciu V 2009 *J. Optoelectron. Adv. Mat. - Symposia* **1** 6
[6] Nowotny J, Bak T, Nowotny M K and Sheppard L 2007 *Int. J. Hydrogen Energy* **32** 2609
[7] Baia L, Peter A, Cosoveanu V, Indrea E, Baia M, Popp J and Danciu V 2006 *Thin Solid Films* **511** 512
[8] Gateshki M, Yin S, Ren Y and Petkov V 2007 *Chem. Mater.* **19** 2512
[9] Proffen Th and Billinge S J L 1999 *J. Appl. Crystallogr.* **32** 572
[10] Swenson J, Börjeson L, McGreevy R L and Howells W S 1997 *Phys. Rev.* B **55** 11236
[11] Keay A, Wells A, Owens A and McCarthy K J 1997 *Nucl. Instrum. Methods Phys. Res.* B **124** 615
[12] Yin S, Aita Y, Komatsu M, Wang J, Tang Q and Sato T J 2005 *J. Mater. Chem.* **15** 674.

# Synthesis of New Pyrrole-containing Biomolecules as Building Blocks for Functionalized Polypyrroles in Nanobiotechnology

**S Karsten[1], A Nan[2], R Turcu[2] and J Liebscher[1]**

[1] Department of Chemistry, Humboldt-Universität zu Berlin, Brook-Taylor-Str. 2, 12489 Berlin, Germany
[2] National Instiute of Research and Development for Isotopic and Molecular Technologies, Donath 65-103, RO-400293 Cluj-Napoca, Romania

E-mail: sebastian.karsten@chemie.hu-berlin.de, liebscher@chemie.hu-berlin.de

**Abstract**. New pyrroles were synthesized wherein biomolecules such as α-amino acids, peptides or carbohydrates are attached to position 1 or 3 via suitable linkers. These conjugates were copolymerized with pyrrole by chemical polymerisation alone giving corresponding copolymers or in the presence of magnetic iron oxid nanoparticles resulting in functionalized core-shell-nanoparticles. Such products are interesting for biosensing, imaging or separation of biomaterials. They were characterized by several methods such as TEM, HRTEM, FTIR and measuring of magnetization.

## 1. Introduction

Synthesis of conducting polypyrroles is especially promising for many commercial applications because of its unique optical, electric, thermal and mechanical properties.[1] Intrinsically conducting polypyrrole can be used for corrosion protection,[2,3] accumulators,[4] OLED's,[5] as infrared polarizer,[2] as packing material in liquid chromatography,[6,7] as electrode material for amperometric detections,[8-10] for artificial muscles[11,12] and for tissue engineering.[13-15] As a result of the conductivity and the biocompatibility *in vitro* and *in vivo*[16] polypyrrole is the most used conducting polymer in biosensing.[17] Biosensors are based on direct assembling of an immobilised biologically active system (e.g. DNA, enzymes, antibodies or antigens) with an electrode and a signal amplifier. The analyte in solution interacts with immobilized biomolecules and causing a measurable electronic signal. Several strategies are available for the immobilization of biomolecules. Effective immobilizations were achieved by adsorption,[18] by affinity interaction,[19] by insertion in[20] or covalent attachment to a polymer matrix.[21]

Attempts to construct nanostructures consisting of polypyrrole were reported in the literature. So far nanowires have been established for biorecognition,[22] nanotubes for the detection of glucose[23] and nanoparticles for application in diagnostics.[24,25] The use of polypyrrole in nanotechnology seems to be a promising area of application. Hence the synthesis and research of new polypyrrole-containing nanomaterials is of great interest.

## 2. Results and discussion

We became interested in developing syntheses of new pyrrole-biomolecule derivatives in which the biomolecule is covalently attached to pyrrole and to use them as starting materials for novel functionalized polypyrroles.[26] For the covalent attachment of biomolecules linkers with terminal functional groups (e.g. amino, carboxyl, thiol or hydroxyl group) had to be introduced. By reactions of these functional groups with biomolecules new functionalized pyrrole monomers were synthesized. Later on these monomers were to be polymerized with pyrrole to get copolymers. Therefore the α-positions of the pyrrole ring had to be unsubstituted. Thus two positions remained as attachment points of the functionalized linkers (figure 1). These are 1-position (nitrogen atom) and β-positions (3-, 4-position). Fixing substituents to the β-positions often faces problems of regioselectivity because the α-positions are also attacked.

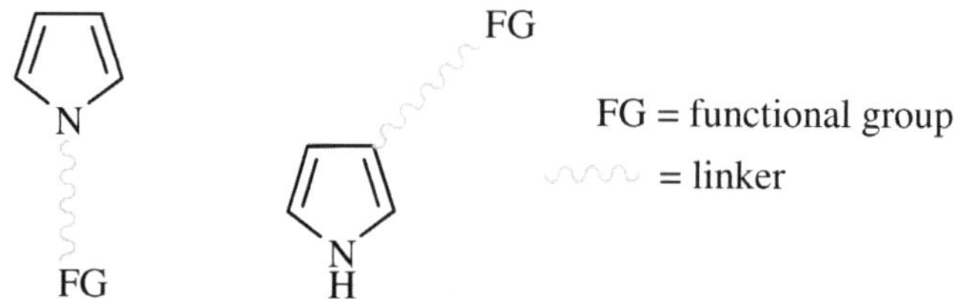


**Figure 1.** Possibilities of introduction of linkers

Syntheses of pyrroles wherein biosensing proteins are attached to the nitrogen atom in position 1 are scarcely reported in literature.[27,28] We focussed on the attachment of biomolecules in 1-position and 3-position and to investigate the explicit properties of copolymers derived from them. So far polypyrroles with attached biomolecules have not been obtained by copolymersation of corresponding pyrroles but by introduction of the biomolecule into an existing properly functionalized polypyrrole.[29-31] The latter were mainly synthesized by electropolymerization. We attached biomolecules (e. g. proteins, enzymes, biotin, carbohydrates, polysaccharides or DNA) covalently to the monomeric pyrrole to have a better chance for regulation of the ratio in the copolymer. Liquid phase synthesis as well as synthesis on solid supports were used to generate the functionalized monomer **1**. These monomers were transformed into copolymer layers **2** by chemical polymerization. The resulting polypyrrole layers can interact with other biomolecules in solution to detect structure-response relationships **3** (figure 2).

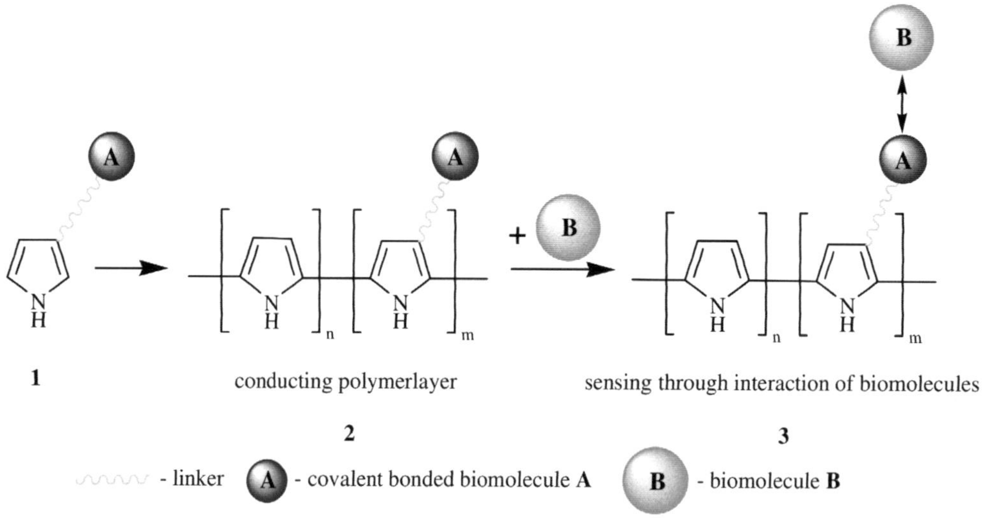


**Figure 2.** Application of functionalized polypyrroles

Alternatively, the monomers **1** were investigated as building blocks for coating of nanoparticles. Because of their functionalized surface these nanoparticles can be used for analytical, medical or biomedical applications. Core-shell-nanoparticles containing a polypyrrole-shell were described already in earlier works. While the shell consisted of polypyrrole with carboxylic groups in 1-position, the core was either made of silica-particles[24] or of a polystyrene-polyethylene mixture.[25] These cores are completely unreactive and only serve as a template for the growth of the functionalized shell.

In our case, functionalized polypyrroles form the shell of superparamagnetic magnetite ($Fe_3O_4$) core-shell-nanoparticles. Since the core exhibits magnetic properties it does not only serve as a template, but provides additional possibilities of separation and manipulation, too (figure 3). Such nanoparticles are biocompatible and non-toxic. Magnetic nanoparticle achieve a growing interest for application in imaging, separation techniques and drug delivery.[32,33] The magnetic polypyrrole core-shell-nanoparticles were characterized by several methods, such as TEM, HRTEM, FTIR, magnetic measurements.

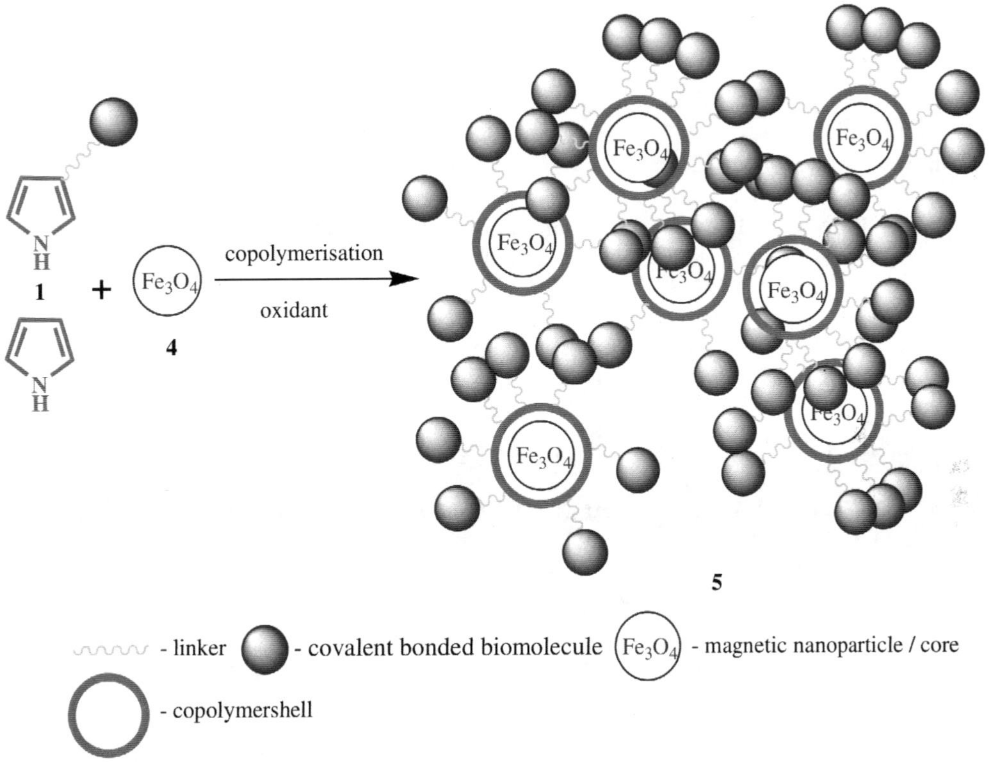


**Figure 3.** Preparation of functionalized magnetic nanoparticles

## References

[1] Wang L X, Li X G and Yang Y L 2001 *React. Funct. Polym.* **47** 125-139.
[2] Stenger-Smith J D 1998 *Progress in Polymer Science* **23** 57-79.
[3] Kousik G, Pitchumani S and Renganathan N G 2001 *Prog. Org. Coat.* **43** 286-291.
[4] Rehahn M 2003 *Chem. uns. Zeit* **37** 18-30.
[5] Roth H K and Schrodner M 2003 *Materialwiss. Werkstofftech.* **34** 254-261.
[6] Ge H, Gilmore G and Wallace G G 1994 *J. Liq. Chromatogr.* **17** 1301-1316.
[7] Lee H S and Hong J 2000 *J. Chromatogr.* **868** 189-196.
[8] Wallace G G, Maxwell K E, Lewis T W, Hodgson A J and Spencer M J 1990 *J. Liq.*

*Chromatogr.* **13** 3091-3110.
[9] Lu W and Wallace G G 1997 *Electroanalysis* **9** 454-460.
[10] Deore B A, Shiigi H and Nagaoka T 2002 *Talanta* **58** 1203-1211.
[11] Otero T F and Sansinena J M 1995 *Bioelectrochem. Bioenerg.* **38** 411-414.
[12] Baughman R H 1996 *Synth. Met.* **78** 339-353.
[13] Khor E, Li H C and Wee A 1996 *Biomaterials* **17** 1877-1879.
[14] Ateh D D, Vadgama P and Navsaria H A 2006 *Tissue Eng.* **12** 645-655.
[15] Ateh D D, Navsaria H A and Vadgama P 2006 *J. Roy. Soc.* **3** 741-752.
[16] Wang X D, Gu X S, Yuan C W, Chen S J, Zhang P Y, Zhang T Y, Yao J, Chen F and Chen G 2004 *J. Biomed. Mater. Res.* **68A** 411-422.
[17] Geetha S, Rao C R K, Vijayan M and Trivedi D C 2006 *Anal. Chim. Acta* **568** 119-125.
[18] Azioune A, Siroti F, Tanguy J, Jouini M, Chehimi M M, Miksa B and Slomkowski S 2005 *Electrochim. Acta* **50** 1661-1667.
[19] Dubois M-P, Gondran C, Renaudet O, Dumy P, Driguez H, Fort S and Cosnier S 2005 *Chem. Commun.* 4318-4320.
[20] Li C M, Chen W, Yang X, Sun C Q, Gao C, Zheng Z X and Sawyer J 2005 *Front. Biosci.* **10** 2518-2526.
[21] Dong H, Li C M, Chen W, Zhou Q, Zeng Z X and Luong J H T 2006 *Anal. Chem.* **78** 7424 7431.
[22] Yang H-H, Zhang S-Q, Tan F, Zhuang Z-X and Wang X-R 2005 *J. Am. Chem. Soc.* **127** 1378 1379.
[23] Tsai Y-C, Li S-C and Liao S-W 2006 *Biosens. Bioelectron.* **22** 495-500.
[24] Azioune A, Ben Slimane A, Hamou L A, Pleuvy A, Chehimi M M, Perruchot C and Armes S P 2004 *Langmuir* **20** 3350-3356.
[25] Pich A, Lu Y and Adler H-J P 2006 *Polymer* **47** 6536-6543.
[26] Yon-Hin B F Y, Smolander M, Crompton T and Lowe C R 1993 *Anal. Chem.* **65** 2067-2071.
[27] Brahim S, Wilson A M, Narinesingh D, Iwuoha E and Guiseppi-Elie A 2003 *Microchim. Acta* **143** 123-137.
[28] Shimomura M, Miyata R, Kuwahara T, Oshima K and Miyauchi S 2007 *Eur. Polym. J.* **43** 388 394.
[29] Ryder K S, Morris D G and Cooper J M 1996 *Langmuir* **12** 5681-5688.
[30] Zelikin A N, Lynn D M, Farhadi J, Martin I, Shastri V and Langer R 2002 *Angew. Chem. Int. Ed.* **41** 141-144.
[31] Freitas J M, Duarte M L, Darbre T and Abrantes L M 2005 *Synth. Met.* **155** 549-554.
[32] Safarik I and Safarikova M 2004 *Biomagn. Res. Technol.* **2** 1-17.
[33] Petri-Fink A, Chastellain M, Juillerat-Jeanneret L, Ferrari A and Hofmann H 2005 *Biomaterials* **26** 2685-2694.

# Surface initiated ring-opening polymerization of lactones on iron oxide nanoparticles

**Alexandrina Nan[1], Rodica Turcu[1], Izabell Craciunescu[1], Cristian Leostean[1], Ioan Bratu[1] and Jürgen Liebscher[2]**

[1] National Institute for Research and Development of Isotopic and Molecular Technologies, 65-103 Donath, 400293 Cluj-Napoca, Romania

[2] Institute of Chemistry, Humboldt-University Berlin, Brook-Taylor 2, 12489 Berlin, Germany

E-mail: alexandrina.nan@itim-cj.ro

**Abstract**. In this work we report the synthesis of hybrid magnetic core-shell nanostructure with inorganic core consisting of magnetite surface-functionalized by glycolic acid or serine. For the attachment of the polymer chain on the preformed functionalized magnetite we applied the "grafting-from" strategy, where the polymerization is initiated directly from the particles surface to give a high number of end attached polymer chains. The surface initiated ROP of ε-caprolactone and lactic acid was initiated by the common catalysts, tin (II) 2-ethylhexanoate ($Sn(oct)_2$) produced poly(ε-caprolactone), polylactic acid, respectively, containing one hydroxyl chain. Moreover we performed the surface initiated ROP of ε-caprolactone microwave assisted in the presence of $Sn(oct)_2$. The morphology of magnetic core-shell nanostructure was determined by TEM. The chemical structure of the obtained polyesters coated magnetic nanoparticles was characterized by FT-IR.

## 1. Introduction

The functionalization of inorganic nanoscopic particles with a polymeric shell is a promising approach for the design of well-defined composite materials and dispersions [1-4]. The rapidly growing interest in biodegradable materials, polyesters as polycaprolactone (PLC) and polylactic acid (PLA) are one of the biodegradable industrial polyesters with excellent mechanical strength, biocompatibility and non-toxicity. PCL is suitable for controlled drug release because of its high permeability to many drugs and biocompatibility in physiological environments. Generally, polyesters are prepared by catalyzed ring-opening polymerization (ROP) of ε-caprolactone (CL) and lactic acid (LA) [5-8]. Tin compounds (including salts and complexes) have proven to be an extremely versatile and highly efficient group of catalysts or initiators. Among the numerous catalytically active tin compounds, tin(II) 2-ethylhexanoate ($SnOct_2$) plays a predominant role, because it is the most widely used initiator for research purposes and for the technical production of polyesters and related copolyesters [9-22].

In this work, we report the classical ring opening polymerization of CL and lactic acid in the presence of the magnetic nanoparticle and the formation of PCL and PLA block copolymer on the iron oxide nanoparticles surface. We also describe about the solvent-free controlled microwave irradiation of CL in the presence of the magnetic nanoparticles. The cores are selected to possess superparamagnetic behaviour, while the polymeric shell serves as a protective layer, which can be functionalized later on.

## 2. Experimental details

### *2.1. Synthesis and Modification of Magnetite Surface*

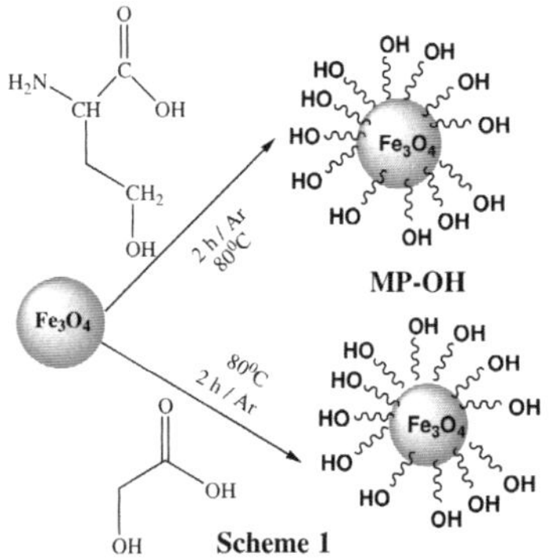
Scheme 1

The synthesis of magnetite and the adsorption of serine and glycolic acid on the magnetite surface were performed under inert ($N_2$) atmosphere (scheme 1). Magnetite nanoparticles were prepared by coprecipitation in basic conditions of a ferrous and ferric chloride (molar ratio 1:2) aqueous solution. The resulting magnetite particles were surface-stabilized *in situ* by the addition of *L*-serine or glycolic acid strongly stirred for 2 h at 80°C. The particles were then repeatedly washed with water and dried proving a brown dark powder (**MP-OH**).

### *2.2. Classical polymerization of CL and LA in the precense of magnetic nanoparticles*

The surface-initiated ROP of ε-CL and LA onto magnetite was carried out was carried out according to the procedure shown in Scheme 2. The stabilized magnetic nanoparticles (**MP-OH**) were mixed with freshly distilled CL, lactic acid, respectively and strongly shacked at room temperature for 30 minutes at 1200 rotation/min. After this $SnOct_2$ was added as a catalyst and the reaction mixture was stirred in a shaker at $100^0$C for 5 h at 1200 rot./min.. After reaction the particles (**MP-PCL** and **MP-PLA**) was repeatedly washed with chloroform and methanol for removing the catalyst and then dried.

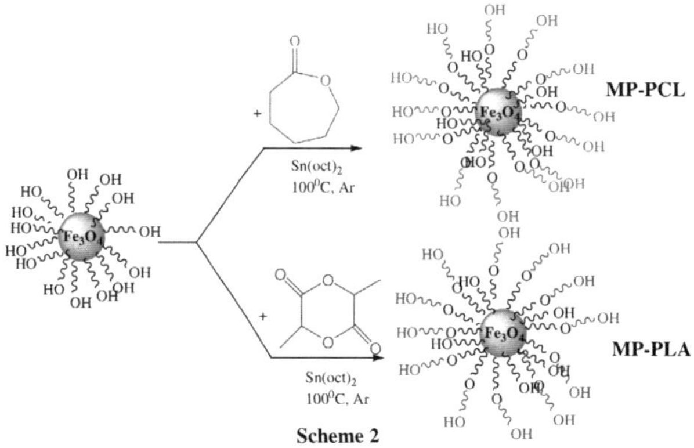
Scheme 2

### *2.3. Formation of block copolymer on the iron oxide nanoparticles*

The surface-initiated ring-opening copolymerization of CL and LA (Scheme 3) in the presence of magnetic nanoparticles is a two step reaction and will direct us to obtain hybrid material containing a block copolymer and iron oxide (**MP-PCL-b-PLA**):

- First step is the ring opening polymerization of CL, the reaction condition was mentioned before
- Second step is polymerization of lactic acid in the same reaction condition as was mentioned before

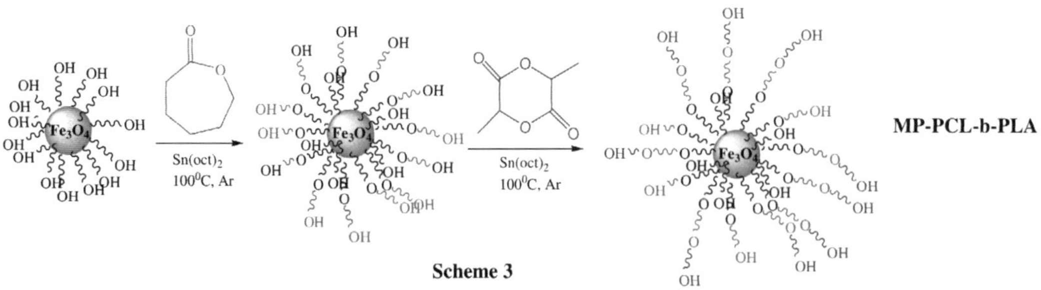
Scheme 3

*2.4. Polymerization of CL onto magnetite nanoparticles by microwave irradiation*

A mixture of magnetic nanoparticles, CL and $Sn(oct)_2$ was place in a dried glass ampoule with a magnetic stirrer and was vacuum sealed. The reaction mixture was then subject to microwave irradiation and the temperature of reaction was recorded every second. After being irradiated for a period of time, the ampoule was cooled in an ice bath and the product (**MP-MW-PCL**) was dispersed in chloroform. The reaction was performed at the (200°C) at a microwave power of 200 W for 7 min while the pressure reached 8 bar. The energy uptake finished after 100 s.

## 3. Characterization of the hybrid material

The morphology of magnetic core-shell nanostructures based on PCL and PLA was determined by TEM using 1010 JEOL microscope and FT-IR spectra were carried out on a JASCO FTIR 610 spectrophotometer.

*3.1. Morphologic characterization*

The TEM images (figure 1) of magnetic nanoparticles modified with serine **MP-OH** (figure 1a) and the hybrid material MP-PCL presented in figure 1b, prove the core-shell structure. The shapes of the magnetic nanoparticles with the modified surface are nearly spherical with diameters between 11 nm - 16 nm. For the hybrid material the shape is approximately the same with diameters between 22-27 nm. These diameters of the magnetite cores are well below the critical domain size of magnetite and therefore a superparamagnetic behaviour was expected for the functionalized magnetic nanoparticles.

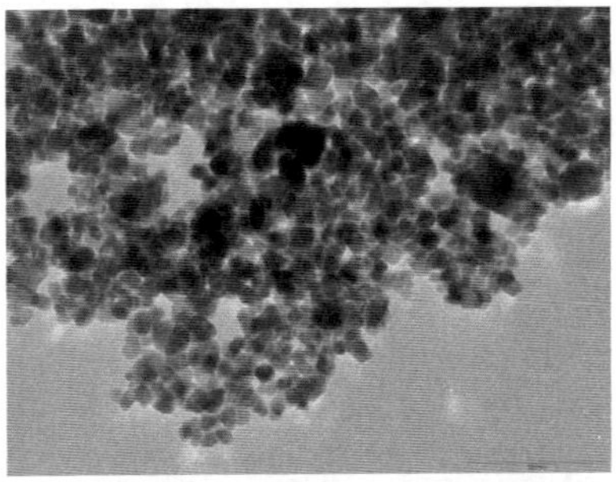

(a)

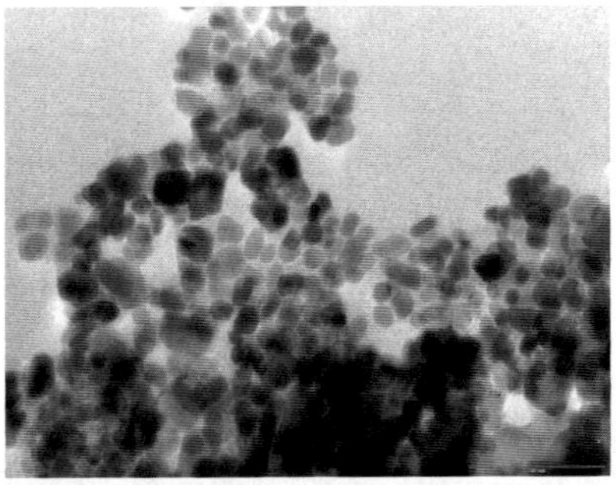

(b)

**Figure 1.** (a) the TEM image of MP-OH, (b) the TEM image of MP-PCL (bar size 100 nm).

*3.2. FTIR-investigation*

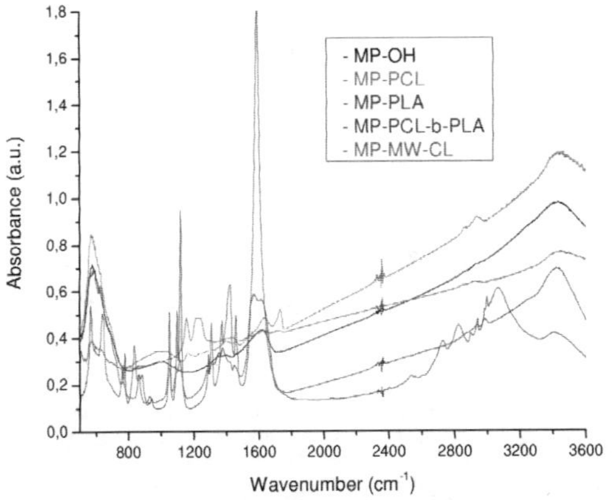


**Figure 2**: FTIR spectra of MP-OH, MP-PCL, MP-PLA, MP-PCL-b-PLA, MP-MW-CL

FTIR spectra, shown in figure 2, exhibited characteristic adsorption bands of both component materials, namely the MP-OH and hybrid materials. The intense band located at 572 cm$^{-1}$ is characteristic of magnetite. The characteristic bands of PCL appear between 1157 cm$^{-1}$ and 2949 cm$^{-1}$. This characteristics vibration bands are significantly shifted to lower frequencies in the spectrum of all hybrid materials based on PCL. In the case of **MP-PLA** spectra, the characteristic –C=O band present in PLA spectra at 1746 cm$^{-1}$, is found at 1645 cm$^{-1}$. This shift of the –C=O band demonstrates the grafting of PLA on the iron oxide surface.

## 4. Conclusions

We obtained for the first time block copolymers based on PCL and PLA on the iron oxide nanoparticles surface. The $Sn(oct)_2$ catalyzed surface-initiated ROP of ε-CL in the presence of surface modified magnetite was successfully conducted by microwave irradiation for the first time. The effect of microwave irradiation on the polymerization involves the enhancement of chain propagation of PCL and the restraint of growing centre formation and allowed to finish the polymerisation in 10 min. In addition, solvents are not necessary in this process.

The FTIR spectra contain the characteristic absorption bands of both constituents, namely, hybrid materials and magnetite. TEM revealed the the size of the magnetite nanoparticles increased by the ROP from 13-16 nm to 22-27 nm.

## References

[1] Carrot G, Rutot-Houze D, Pottier A, Degée P, Hilborn J and Dubois P 2002 *Macromolecules* **35** 8400
[2] El Harrak A, Carrot G, Oberdisse J, Eychenne-Baron C and Boué F 2004 *Macromolecules* **37** 6376
[3] Pyun J, Shijun J, Kowalewski T, Patterson G D and Matyjaszewski K 2003 *Macromelcules* **36** 5094
[4] Fendler J H 1996 *Chem. Mater.* **8** 1616
[5] Kowalski A, Duda A and Penczek S 2000 *Macromolecules* **33** 689
[6] Storey R F and Sherman J W 2002 *Macromolecules* **35** 1504
[7] Storey R F and Taylor A E 1998 *J. Macromol. Sci.-Pure Appl. Chem.* A **35** 723
[8] Lőfgren A, Albertsson A C, Dubois P and Jérôme R 1995 *Rev. Macromol. Chem. Phys.* C **35** 379
[9] Vasantharamari R and Pennings A J 1983 *Polymer* **24** 175
[10] Kohn F E, van Ommen J G and Feijen J E 1983 *Polym. J.* **19** 1081
[11] Nijenhuis A J, Grijpman D W and Pennings A J 1991 *Polym. Bull.* **26** 71
[12] Nijenhuis A J, Grijpma D W and Pennings A J 1992 *Macromolecules* **25** 6419
[13] Dahlmann J and Rafler G 1993 *Acta Polym.* **44** 103
[14] Kricheldorf H R and Meier-Haack J 1993 *Makromol. Chem.* **194** 715
[15] Kricheldorf H R, Kreiser-Saunders I and Boettcher C 1995 *Polymer* **35** 219
[16] Du Y J, Lemstra P J, Nijenhuis A, van Aert H A M and Bastiaansen C 1995 *Macromolecules* **28** 2124
[17] Zhang X, MacDonald M, Goosen F A and Auley K B 1994 *J. Polym. Sci., Part A: Polym. Chem.* **32** 2965
[18] In't Veld P J A, Velner E M, von de Witte P, Harnhuis J, Dijkstra P J and Feijen J 1997 *J. Polym. Sci., Part A: Polym.Chem.* **35** 219
[19] Schwach G, Courdano J, Engel R and Vert M 1997 *J. Polym. Sci., Part A.: Polym. Chem.* **35** 3431
[20] Witzke D R, Narayan R and Kolstad J J 1997 *Macromolecules* **30** 7075
[21] Storey R F and Taylor A E 1998 *J. Macromol. Sci. Pure Appl. Chem.* **35** 723
[22] Kowalski A, Duda A and Penezek S 1998 *Macromol. Rapid Commun.* **19** 567.

# Synthesis and characterization of LSMO nanoparticles covered with Au having a core-shell structure

**O Pana, R Turcu, M L Soran, S Macavei and C Leostean**

National Institute for Research and Development of Isotopic and Molecular Technologies, 65-103 Donath, 400293 Cluj-Napoca, Romania

E-mail: ovidiu.pana@itim-cj.ro

**Abstract.** In this work we report the synthesis and characterization of the nanomagnetic perovskite Sr manganites covered with Au shells having different thickness. $La_{2/3}Sr_{1/3}MnO_3$ (LSMO) manganite nanoparticles were first prepared by a sol-gel procedure. The LSMO manganite nanoparticles were chemically covered with gold to produce the core-shell samples. TEM, HRTEM and atomic emission spectroscopy techniques were used to determine the morphology and structure of the LSMO@Au nanoparticles. The magnetization behaviors were also determined by VSM technique.

## 1. Introduction

During the last years the core-shell nanostructured particles has drawn the attention of physicist, chemists and material scientists from both fundamental and applicative point of view. By taking advantage of the combined properties of both core and shell's surface one provides an increased number of applications like catalysis, coatings, hyperthermia and magneto-resistant materials. The gold coating of the magnetic nanoparticles enhances the chemical stability by protecting the core from oxidation and corrosion, and to exhibit good biocompatibility and affinity *via* amine / thiol terminal groups [1-6]. Among magnetic materials, perovskites manganites have attracted much attention due to the colossal magnetoresistance [7, 8]. In this paper we report the synthesis and characterization of perovskite manganite nanoparticles covered with gold namely $La_{2/3}Sr_{1/3}MnO_3$(core)@Au(shell) (LSMO@Au).

## 2. Experimental

The $La_{2/3}Sr_{1/3}MnO_3$ (LSMO) manganite nanoparticles were prepared by a sol-gel procedure using diethylenetriaminepentaacetic acid as the gelificant agent [7-10]. The corresponding metallic oxides and $Mn(CH_3COO)_2 \cdot H_2O$ in the necessary stoichiometric ratio were used as precursors. By using $HNO_3$ aqueous solution the precursors were brought to the corresponding nitrates. The final annealing was made for 2 h at 700°C. The coating of the pre-synthesized manganites with gold shells resulting in the LSMO@Au sample (LSMA) was performed by the seeding method [11-13]. After reaction, the solution was cooled to room temperature and ethanol was added in the reaction solution. The composition of the sample was determined by ICP-AES with an FMD 07 ICP-AES SpectroFlame (Spectro Analytical) (table 1).

The morphology of the LSMO@Au nanoparticles was investigated by HRTEM using a Hitachi H9000NAR transmission electron microscope. The structural characterization of the sample was

Journal of Physics: Conference Series **182** (2009) 012071 doi:10.1088/1742-6596/182/1/012071

performed by X-ray diffraction (XRD) by using a Bruker D8 Advance X-Ray diffractometer set up with $CuK_\alpha$ radiation. The magnetic behavior of LSMO and core-shell LSMO@Au nanoparticles was recorded at room temperature by using a Vibrating Sample Magnetometer produced by Cryogenic.

**Table 1.** The synthesis conditions for the LSA samples together with the samples composition (*%wt*).

| Sample | Mole ratio $Au(OOCCH_3)_3$:LSMO | Concentration in the synthesis solution (wt%) | | | Concentration determined by ICP-AES (wt%) | | | |
|---|---|---|---|---|---|---|---|---|
| | | Au | Sr | Mn | Au | Sr | Mn | LSMO |
| LSA | 5 | 81.47 | 2.40 | 4.58 | 83.23 | 1.79 | 3.09 | 12.64 |

## 3. Results and discussion

The HRTEM image of the LSMA sample is presented in figure 1. One can see the formation of the Moiré fringes. This is an indication that two different crystallites, possible the inner core of LSMO and the outer shell of Au, having almost the same constant are superimposed. Also the images show that inside of some globular gold particles could be several different core nanocrystallites of LSMO.

The XRD diffraction patterns of LSMO bare cores are presented in figure 2 together with the diffraction patterns of the core-shell LSMA sample. The indexing of the XRD patterns from figure 2 was done in the orthorhombic system. The resulting constants for LSMO were found as $a = 3.862$ Å, $b = 3.857$ Å and $c = 3.854$ Å respectively. This observation is accordance with the fact that, as it was reported in the literature, the usual rhombohedral structure of $La_{1-x}Sr_xMnO_3$ is valid only for x doping levels up to 0.3 and by further increase of the x content, $0.3 < x < 0.5$, the structure becomes almost cubic [14]. In figure 2 the gold lines of the LSMA sample are indicated by indexes while the LSMO manganite lines are marked by stars.

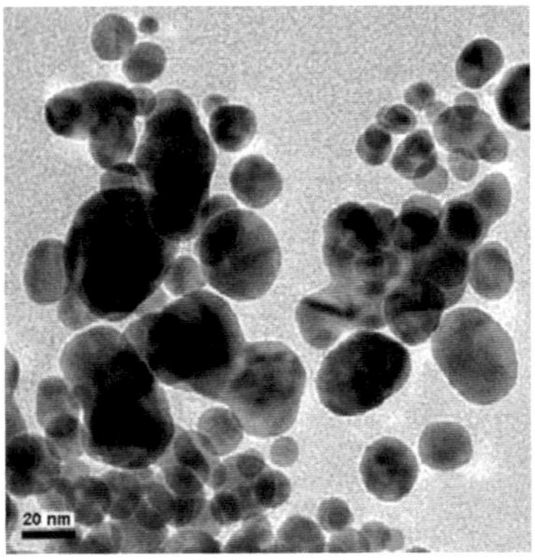


**Figure 1.** HRTEM image of nanoparticles of LSMA sample from table 1.

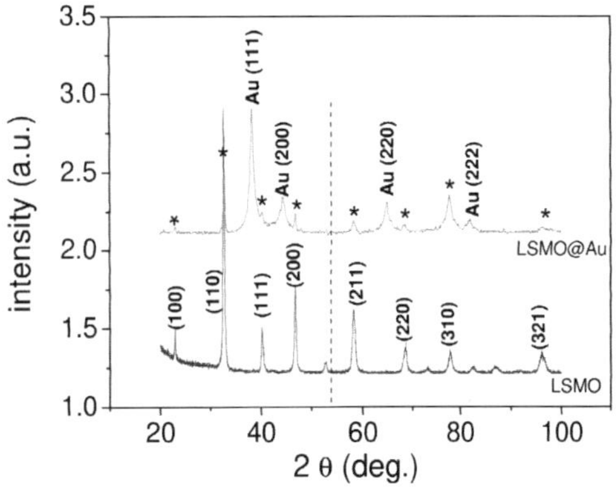


**Figure 2.** The diffraction patterns of the core-shell three LSMO@Au sample together with diffraction patterns of bare LSMO manganite nanoparticles. Among the LSMO@Au diffraction patterns the LSMO manganite peaks are marked by stars.

A comparison of the magnetization curves *vs.* the applied magnetic field, $M = f(H)$ of the bare LSMO nanoparticles as well as of LSMO@Au sample is shown in figure 3. The experimental values of the magnetizations, are divided to the content of corresponding magnetic material as shown in table 1. The $M = f(H)$ curves have no hysteresis loops, indicating a superparamagnetic behavior for the investigated nanoparticles

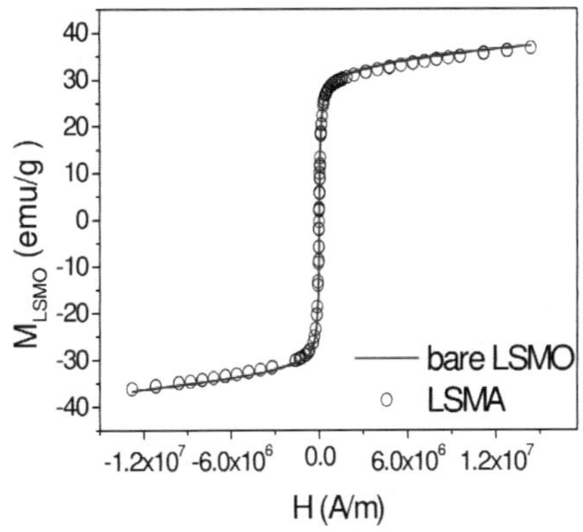


**Figure 3.** Magnetization *vs.* applied magnetic field at room temperature for the LSMA sample and bare manganite sample.

## 4. Conclusions

Manganite nanoparticles $La_{0.67}Sr_{0.33}MnO_3$ having mean diameter of 8.4 nm were obtained by using the sol-gel procedure. The surfactant layer on the manganite nanoparticle surfaces cannot avoid their aggregation due to the dipolar interactions. By using the so called seeding method, the LSMO nanoparticles were covered with gold forming core-shell like nanostructures. Therefore, by covering of the LSMO cores with a properly adjusted gold shell one can significantly diminish the tendency of the nanoparticles to form clusters. The missing hysteresis loop in the magnetization vs. applied magnetic field represents a clear evidence for the superparamagnetic behavior for the core-shell LSMO@Au nanoparticles.

## References

[1] Zhou W, Kumbhar A, Wiemann J, Fang, J, Carpenter E E and O'Connor C J 2001 *J. Solid State Chem.* **159** 26
[2] Carpenter E E, Sangregorio C and O'Connor C J 1999 *IEEE Transaction on Magnetics* **35** 3496
[3] Bao Y and Krishnan K M 2005 *J. Magn. Magn. Mater.* **293** 15
[4] Zhou W L, Carpenter E E, Lin J, Kumbhar A, Sims J and O'Connor C J 2001 *Eur. Phys. J.* D **16** 289
[5] Pham T T H, Cao C and Sim S J 2008 *J. Magn .Magn. Mater.* **320** 2049
[6] Sounderya N and Zhang Y 2008 *Recent Patents on Biomedical Engineering* **1** 34
[7] Huang Y H, Yan C H, Wang Z M, Liao C S and Xu G X 2001 *Solid State Commun.* **118** 541
[8] Huang Y H, Xu Z G, Yan C H, Wang Z M, Zhu T, Liao C S, Gao S and Xu G X 2000 *Solid State Commun.* **114** 43
[9] Mahía J, Vázquez-Vázquez C, Mira J, López-Quintela M A, Rivas J, Jones T E and Oseroff S B 1994 *J. Appl. Phys.* **75** 6757
[10] Sánchez R D, Rivas J, Vázquez- Vázquez C, López-Quintela M A, Causa M T, Tovar M and Oseroff S 1996 *Appl. Phys. Lett.* **68** 134
[11] Wang L, Luo J, Fan Q, Suzuki M, Suzuki I S, Engelhard M H, Lin Y, Kim N, Wang J Q, Zhong C J 2005 *J. Phys. Chem.* **109**, 21593
[12] Wang L, Luo J, Maye M M, Fan Q, Rendeng Q, Engelhard M H, Wang C, Lin Y, Zhong C J 2005 *J. Mater. Chem.* **15** 1821
[13] Lyon J L, Fleming D A, Stone M B, Schiffer P and Williams M E 2004 *Nano Lett.* **4** 719
[14] Gaudon M, Laberty-Robert C, Ansart F, Stevens P and Rousset A 2002 *Solid State Sci.* **4** 125.

Journal of Physics: Conference Series **182** (2009) 012072 doi:10.1088/1742-6596/182/1/012072

# Preparation and structural characterization of some $Fe_2O_3$-$B_2O_3$-ZnO glasses and glass ceramics

**Petru Pascuta[1], Gheorghe Borodi[2], Maria Bosca[1], Lidia Pop[1], Simona Rada[1] and Eugen Culea[1]**

[1] Technical University of Cluj-Napoca, Physics Department, 15 C Daicoviciu, 400020 Cluj-Napoca, Romania

[2] National Institute for Research and Development of Isotopic and Molecular Technologies, 65-103 Donath, 400293 Cluj-Napoca, Romania

E-mail: petru.pascuta@phys.utcluj.ro

**Abstract.** X-ray diffraction (XRD) and Fourier transform infrared (FTIR) spectroscopy have been employed to investigate the $(Fe_2O_3)_x\cdot(B_2O_3)_{(60-x)}\cdot(ZnO)_{40}$ samples, with $0 \leq x \leq 20$ mol%. The samples have been prepared by melting at 1100 °C for 10 minutes followed by rapid cooling at room temperature. The structure of samples was analyzed by means of XRD. XRD pattern show that the sample without iron ions and the samples contained 3 and 5 mol% were glasses. For samples contain 10, 15 and 20 mol% $Fe_2O_3$ it obtains glass ceramics. These glass ceramics contains a unique crystalline phase, zinc ferrite ($ZnFe_2O_4$), embedded in an amorphous matrix. FTIR spectroscopy data suggest that for the glass samples the iron ions play the network modifier role. These data show that the glass network consists of $BO_3$ and $BO_4$ structural units. For glass ceramics samples FTIR spectra show characteristic bands of $ZnFe_2O_4$.

## 1. Introduction

Zinc borate glasses exhibit low melting temperatures, so they are of technological interest owing to their applications in different fields of electronic products [1-4]. This is due to the fact that the higher the polarizability of an ion, the lower the melting temperature of the substance. Low melting glasses have been widely used for lowering the sintering temperature and optimizing coefficient thermal expansion in the field of electric devices such as multi layer ceramic capacitor, low temperature co-fired ceramics, plasma display panels, cathode ray tube, electric modules, etc. [1-4]. On the other hand iron ions have strong bearing on electrical, optical and magnetic properties of glasses. In general the present of iron ions in glasses are considered to assemble together and form clusters which exhibit super paramagnetic behaviour and below the freezing temperature, individual spins are frozen in random directions because of antiferomagnetic interaction between nearby ions [5-8]. A general condition for semiconducting behaviour of borate glasses containing iron ions is the coexistence of iron ions in more than one valence state, $Fe^{2+}$ and $Fe^{3+}$ [6]. Having in view that the addition of iron ions in glasses usually induces significant changes in their magnetic, electric and optical behaviour, opening thus opportunities in the finding of new applications, the careful structural investigations of glasses containing $Fe_2O_3$ becomes necessary. The purpose of the present work was to investigate by

XRD and FTIR spectroscopy measurements the zinc borate glasses and glass ceramics containing $Fe_2O_3$ in order to establish the crystallization behaviour and the structural changes induced by iron oxide addition in these samples.

## 2. Experimental

Samples of the $(Fe_2O_3)_x{\cdot}(B_2O_3)_{(60-x)}{\cdot}(ZnO)_{40}$ system were prepared using $Fe_2O_3$, $H_3BO_3$ and ZnO of high purity (99.9%) in suitable proportion. The mechanically homogenized mixtures were melted in sintered corundum crucibles at 1100 °C, in an electric furnace. The samples were put into the electric furnace at the mentioned temperature. After 10 minutes, the molten material was quenched at room temperature by pouring onto a stainless-steel plate.

The X-Ray diffraction measurements of the samples were made with a XRD – 6000 Shimadzu diffractometer, with a monochromator of graphite for Cu Kα radiation (λ = 1.54 Å) at room temperature.

The FTIR absorption spectra of the samples in the 360-1600 $cm^{-1}$ spectral range were obtained with a JASCO FTIR 6200 spectrometer. The IR absorption measurements were done using the KBr pellet technique. In order to obtain good quality spectra the samples were crushed in an agate mortar to obtain particles of micrometer size. This procedure was applied every time to fragments of bulk glass to avoid structural modifications due to ambient moisture.

## 3. Result and discussion

### *3.1. XRD data*

The XRD patterns of $(Fe_2O_3)_x{\cdot}(B_2O_3)_{(60-x)}{\cdot}(ZnO)_{40}$ samples with various content of iron oxide ($0 \le x \le$ 20 mol%) are presented in figure 1.

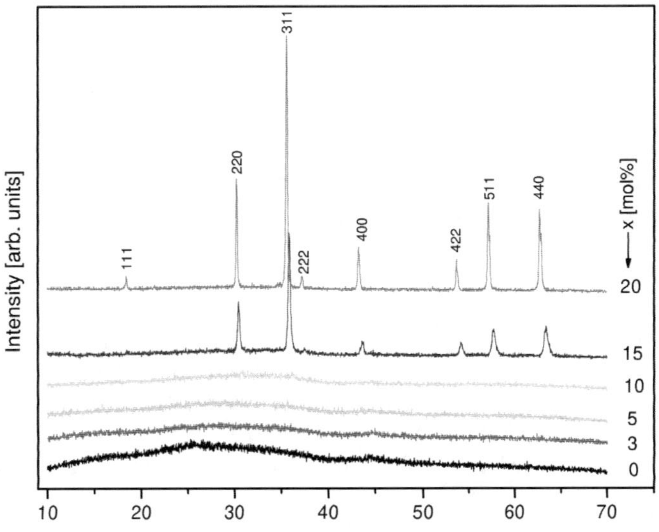


**Figure 1.** The XRD patterns of the $(Fe_2O_3)_x{\cdot}(B_2O_3)_{(60-x)}{\cdot}(ZnO)_{40}$ samples.

These patterns show that for the samples containing 0, 3 and 5 mol% $Fe_2O_3$ amorphous phase is present. Crystalline phase is not detected but is possible to be in small quantity. For sample containing 10 mol% $Fe_2O_3$ a small amount of crystalline phase appears which grow up as the content of iron ions increases. For sample containing 20 mol% $Fe_2O_3$ the crystalline phase becomes a majority one. The crystalline phase was identified as being $ZnFe_2O_4$ which crystallizes in a cubic system. Lattice parameter increases with increasing of iron ions content from a = 8.293 Å for sample with 10 mol% $Fe_2O_3$ at a = 8.319 Å for sample containing 15 mol% $Fe_2O_3$ and respectively at a = 8.381 Å for sample containing 20 mol% $Fe_2O_3$.

### *3.2. FTIR data*

The experimental FTIR spectra of $(Fe_2O_3)_x \cdot (B_2O_3)_{(60-x)} \cdot (ZnO)_{40}$ samples with various content of iron oxide ($0 \le x \le 20$ mol%) were presented in figure 2. Obtained bands and their assignments are summarized in table 1.

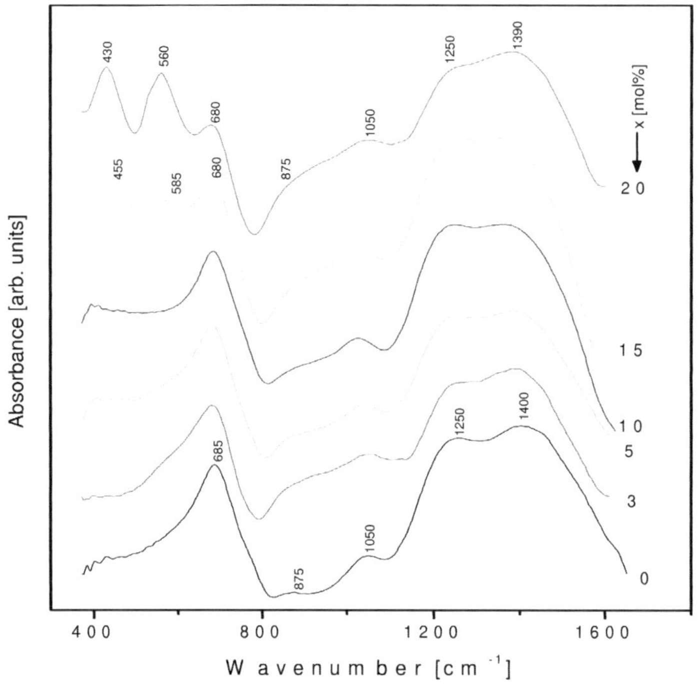


**Figure 2.** The FTIR spectra of the $(Fe_2O_3)_x \cdot (B_2O_3)_{(60-x)} \cdot (ZnO)_{40}$ samples.

FTIR spectrum of studied samples contains al the important bands of vitreous $B_2O_3$, but shifted from their original position. So, in all FTIR spectrum there is an intense band at 680-685 $cm^{-1}$ corresponding to the band at around 720 $cm^{-1}$ from vitreous IR spectrum of $B_2O_3$ and is attributed to B–O–B bending vibrations [9-12]. The intensity of this band decreases with the increasing of $Fe_2O_3$ concentration. The absorptions from around 875 $cm^{-1}$ can be attributed to the stretching vibrations of B–O bonds in $BO_4$ units from diborate groups [9-12] and the band from around 1050 $cm^{-1}$ is attributed to the stretching vibrations of B–O bonds in $BO_4$ units from tri-, tetra-, and penta-borate groups [9-12].

**Table 1.** Wavenumbers and their assignments for FTIR spectra of $(Fe_2O_3)_x \cdot (B_2O_3)_{(60-x)} \cdot (ZnO)_{40}$ samples.

| Wavenumber ($cm^{-1}$) | Assignment |
|---|---|
| 430-455 | The vibration of the chemical bond ($Fe^{3+}$–$O^{2-}$) in B location of the octahedron |
| 560-585 | The vibration of the chemical bond ($Zn^{2+}$–$O^{2-}$) in A location of the tetrahedron |
| 680-685 | B-O-B bending vibrations |
| 875 | B-O stretching vibrations of tetragonal ($BO_4$) units in di-borate groups |
| 1050 | B-O stretching vibrations of tetragonal ($BO_4$) units in tri-, tetra- and penta-borate groups |
| 1250 | B–O stretching vibrations of trigonal ($BO_3$) units in boroxol rings |
| 1390-1400 | B-$O^-$ stretching vibrations of trigonal ($BO_3$) units in different borate groups |

The intensity of these bands is increasing with the increasing of iron ions content. The absorption band from around 1250 $cm^{-1}$ is attributed to the asymmetric stretching vibrations of B–O bonds in $BO_3$ units from in boroxol rings [9-12] and the band from 1390-1400 $cm^{-1}$ was attributed to the stretching vibrations of $B–O^-$ in $BO_3$ units from different borate groups [9-12]. The intensity of this band increases with the increasing of $Fe_2O_3$ concentration till 15 mol%, and for higher concentrations the intensity of this band decreases. The presence of $ZnFe_2O_4$ in studied samples for x > 10 mol% is confirmed by the appearances of two absorption pecks. That of sample containing 15 mol% $Fe_2O_3$ appears at around 585, 445 $cm^{-1}$, and that of sample containing 20 mol% $Fe_2O_3$ at around 560, 430 $cm^{-1}$, which are basically consistent with the literature [12]. The two strong absorption peaks are attributed to the vibration of the chemical bond ($Zn^{2+}–O^{2-}$) in A location of the tetrahedron and the vibration of the chemical bond ($Fe^{3+}–O^{2-}$) in B location of the octahedron respectively [12]. The intensity of these bands increases with the increasing of iron ions content.

## 4. Summary

Homogeneous glasses of the $(Fe_2O_3)_x \cdot (B_2O_3)_{(60-x)} \cdot (ZnO)_{40}$ system were obtained within $0 \le x \le 5$ mol%. For x = 10, 15 and 20 mol% the XRD patterns shows that were obtained glass ceramics with contain a unique crystalline phase, identified as been $ZnFe_2O_4$, which crystallizes in a cubic system. Lattice parameter increases with increasing of iron ions content from a = 8.293 Å for x = 10 mol% at a = 8.381 Å for x = 20 mol%.

The FTIR study shows that $B_2O_3$ play the network former role and the iron ions play a network modifier role in the studied samples. The glasses network consists of $BO_3$ and $BO_4$ structural units. From FTIR spectra, the bands from 430-455 $cm^{-1}$ and from 560-585 $cm^{-1}$ confirm the presence of $ZnFe_2O_4$.

**Acknowledgements**

This work was supported by a Grant of Romanian Ministry of Research and Education, CNCSIS PNII-IDEI 226/ 2008 grant.

**References**

[1] Kim D W, Lee D G and Hong K S 2001 *Mater. Res. Bull.* **36** 585
[2] Lu S G, Kwok K W, Chan H L W and Choy C L 2003 *Mater. Sci., Eng.* B **99** 491
[3] Raju G N, Reddy M S, Sudhakar K S V and Veeraiah N 2008 *Opt. Mater.* **29** 1467
[4] Zhou L, Lin H, Chen W and Lua L 2008 *J. Phys. Chem. Solids* **69** 2499
[5] Del Nery S M, Pontuschka W M, Isotani S and Rouse C G 1994 *Phys. Rev.* B **49** 3760
[6] Ardelean I, Pascuta P and Ioncu V 2001 *Mod. Phys. Lett.* B **15** 1445
[7] Srinivasarao G and Veeraiah N 2002 *J. Phys. Chem. Solids* **63** 705
[8] Ardelean I and Pascuta P 2004 *Int. J. Mod. Phys.* B **18** 1525
[9] Kamitsos E I, Karakassides M A and Chyssikos GD 1987 *J. Phys. Chem.* **91** 1073
[10] Abo-Naf M, El Batal F H and Azooz M A 2002 *Mater. Chem. Phys.* **77** 846
[11] Kumar A, Rai S B and Rai D K 2003 *Mater. Res. Bull.* **38** 333
[12] Yang G Q, Han B, Sun Z T, Yan L M and Wang S Y 2002 *Dyes and Pigments* **55** 9.

# Ferromagnetic resonance measurements on Co nanowires

**A Popa[1], M Jivanescu[2], D Toloman[1], O Raita[1] and L M Giurgiu[1]**

[1] National Institute for Research and Development of Isotopic and Molecular Technologies, 65-103 Donath, 400293 Cluj-Napoca, Romania

[2] Semiconductor Physics Laboratory, Katholieke Universiteit Leuven, 3001 Leuven, Belgium

E-mail: adriana.popa@itim-cj.ro

**Abstract**. Co nanowires grown by electrodeposition technique in polycarbonate membrane have been analysed. The nanowires have 25 μm length and 60nm diameter and are oriented perpendicular to the sample plane. The study focuses on the ferromagnetic resonance (FMR) measurements in Q band (34 GHz), study which allow the identification of the anisotropy effects in Co nanowires.

## 1. Introduction

Magnetic nanowires are an interesting physical system, with many potential applications, such as high density magnetic recording and sensors [1]. This systems offer the opportunity for the studies of static and dynamic magnetic properties due to controllable and reproducible characteristics and properties.

For such nanostructured materials it is possible to study both, a large ensemble of wires as well as isolated single wires. Although studies on single wires are important to understand the mechanism of magnetization reversing in a single wire, it is also of importance for possible device applications to quantify the distribution of the internal fields for wire arrays. The internal fields include the domain-wall nucleation field and the depinning field (the magnetization reversal by domain wall propagation) as well as the anisotropy field.

The anisotropies fields of nanowires arrays can be investigated by various techniques [2]. One of these techniques is the ferromagnetic resonance spectroscopy (FMR) due to the fact that the resonance field depends directly on the anisotropy. The measurements had been performed on Co nanowire arrays with diameter of 60 nm which are embedded in a polycarbonate membrane. The room temperature FMR experiments were done in Q band, frequency f = 34.4 GHz. The FMR technique has still the advantage of yielding well- defined spectra from which a quantitative analysis of the angular dependence of the resonance field could be performed.

## 2. Experiment

The Co nanowires arrays were fabricated by an electrodeposition method using polycarbonate membrane with a diameter of 5 mm and the pore sizes of 60 nm. This method consists in the deposition of a metal inside the membrane pores under the current action [3]. Prior to deposition, a metallic layer of 1 μm Au serving as the cathode is deposited by evaporation on one face of the membrane. Then, the membrane is introduce inside an electrolytic bath (pH = 5) witch contains the metals ions and a Pt anode. The solution pH must be kept constant since his variation may induces changes in the magnetic behaviour of the nanowires. The pH of the electrolytic bath can be adjusted by adding $H_2SO_4$ or NaOH [4].

A current is applied between the cathode and the anode. During the deposition the current variation between the cathode and the anode is measured. This variation must be constant and when it increases the membrane pores are filled.

The room temperature FMR experiments were carried out in Q band, frequency f = 34.4 GHz. In the experiments presented here, the microwave pumping field h had a frequency f = 34.4 GHz (Q band) and was always oriented perpendicular to the applied bias field $H_0$ and to the wire axes as shown in figure 1.

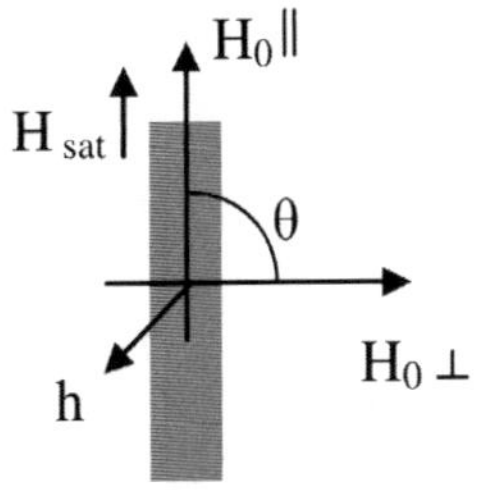


**Figure 1.** A schematic picture showing the orientation of the applied bias field, $H_0$, and the microwave pumping field, h.

The angle $\theta$ of the applied bias field was then changed in zero field such that the orientation of $H_0$ varied between parallel to the wire axes ($\theta = 0°$) and perpendicular to that axes ($\theta = 90°$).

## 3. Results and discussions

The experimental FMR spectra measured at different angles $\theta$ are shown in figure 2. It can be clearly observed a change in the line width of FMR spectra with the angle variation. By increasing the value of angle $\theta$, the resonance spectra change from a relatively narrow line, $\Delta H$ = 1760 kOe for $\theta = 0$ to a broader one, $\Delta H$ = 2300 kOe for $\theta$= 90.

For the resonance field, $H_R$, a sinusoidal dependence with the angle $\theta$ was found. This dependence is presented in figure 3. The minimum field of 5 kOe corresponds to the angles 0° and 180° while the maximum field (~ 14 kOe) to the angle 90°.

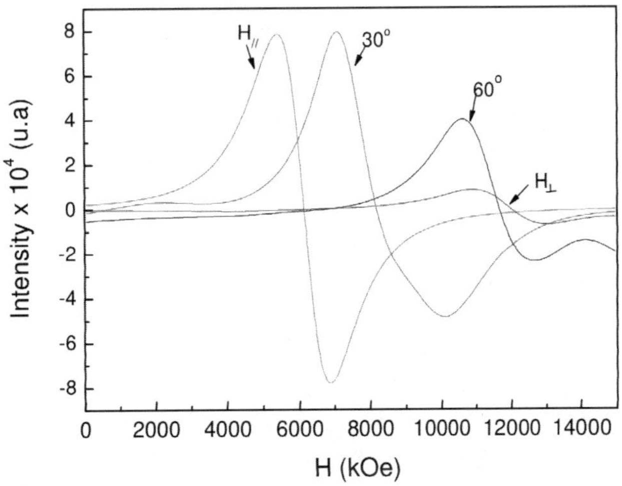


**Figure 2.** FMR spectra for a 60 nm Co-wire array at 34.4 GHz.

In order to explain the angular dependence of resonance field, a theoretical model has been used [5]. This model starts from the total energy of an infinite cylinder and considers that the resonance field could be obtained from the dispersion relation:

$$\left(\frac{\omega}{\gamma}\right)^2 = \left[\left\{H_{eff}\cos 2\theta_0 + H_0\cos(\theta_0-\theta)\right\}\times\left\{H_{eff}\cos^2\theta_0 + H_0\cos(\theta_0-\theta)\right\}\right] \quad (1)$$

where $\gamma$ is the giromagnetic ratio $\gamma = g\mu_B/\hbar$, $\theta_0$ is the equilibrium angle of the magnetization $M$ and $\theta$ is the applied field $H_0$ direction angle. $H_{eff}$ is the effective uniaxial field along the wire axis.

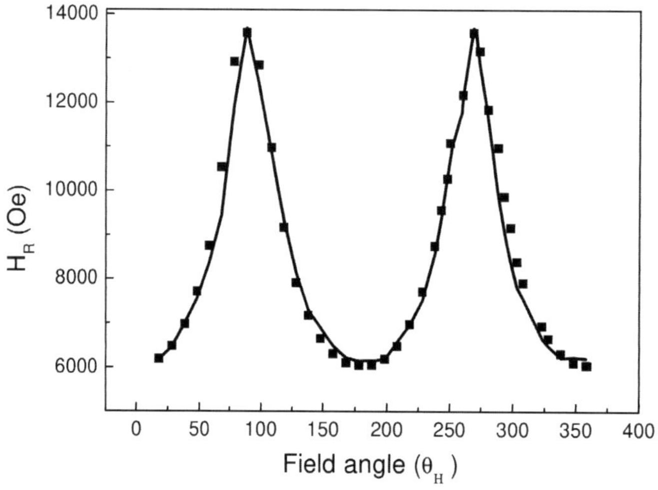


**Figure 3**. The experimental angular variation of the resonance field $H_R$ for Co nanowires at 34.4 GHz (full squares) and the calculated angular variation of the resonance field (full line).

The $\theta_0$ is obtained from equilibrium conditions for the static magnetization by using the first derivative of the total energy of the sample [5].

The best fit of the experimental data with the equation (1) is shown in figure 3, and was obtained for $H_{eff} = 4.7kOe$. The effective field contains three components: the shape anisotropy field, $H_{shape}$, the dipolar interaction field, $H_{dip}$, and an additional second order uniaxial anisotropy contribution, $H_u$. His expression is the following:

$$H_{eff} = H_{shape} - H_{dip} \pm H_u \quad (2)$$

In this equation the sign of the uniaxial anisotropy contribution is given by the orientation of the anisotropy field towards the wire axis, a positive sign for a parallel orientation and a negative one for a perpendicular orientation.

For the Co nanowires the shape anisotropy field is given by the equation: $H_{shape} = 2\pi M_S = 8.8kOe$, where $M_S$ is the saturation magnetization.

The second term in the equation (2) is the dipolar field described by the expression: $H_{dip} = 6\pi M_S f$, where $f$ is the filing factor. This expression was obtained considering two components: $4\pi M_S$ due to the magnetic charges situated on the wires end, and $2\pi M_S$ from the charges situated on the wires length. The calculated value for dipolar field corresponding to the Co nanowires is 800 Oe.

Considering all this arguments we obtained for the second order anisotropy contribution $H_u$ an average value of 3.3 kOe. Therefore, the magnetic anisotropy appears to be uniaxial with the easy axis almost perpendicular to the wire axis.

## 4. Conclusions

For an array of Co nanowires, the angular dependence of the resonance field was investigated by FMR. The fit of the resonance field position $H_R$ vs the bias field angle $\theta$ yields an effective anisotropy filed of $H_{eff} = 4.7 kOe$. This value is smaller than the sum between the anisotropy shape field and the dipolar field. Therefore, a small additional second order uniaxial anisotropy could exist in the nanowires, the easy axis being oriented almost perpendicular to the wire axis.

## References

[1] Rivasa J, Kazadi Mukenga Bantua A, Zaragozab G, Blancob M C and Lopez-Quintelab M A 2002 *J. Magn. Magn. Mater.* **249** 220
[2] Paulus P M, Luis F, Kroll M, Schmid G and de Jongh L J *J. Magn. Magn. Mater.* 2001 **224** 180
[3] Caffarena V R, Guimaraes A P, Folly W S D, Silva E M and Capitaneo J L 2008 *Mater Chem. Phys.* **107** 297
[4] Croll I M 1987 *IEEE Trans. Magn.* **MAG-23** 59
[5] Ramos C A, Vasallo Brigneti E and Vasquez M 2004 *Physica* B **354** 195.

# The local structure of gadolinium-borate-tellurate vitroceramics investigated by FTIR and EPR spectroscopy

**M Rada[1], E Culea[1], S Rada[1], P Pascuta[1], M Culea[2], V Dan[1], T Rusu[1], V Maties[1] and I Bratu[3]**

[1] Technical University of Cluj-Napoca, Physics Department, 103-105 Muncii, 400641 Cluj-Napoca, Romania

[2] Babes-Bolyai University, Faculty of Physics, 1 Mihail Kogalniceanu, 400084 Cluj-Napoca, Romania

[3] National Institute for Research and Development of Isotopic and Molecular Technologies, 65-103 Donath, 400293 Cluj-Napoca, Romania

E-mail: eugen.culea@phys.utcluj.ro, simona.rada@phys.utcluj.ro

**Abstract.** The present study provides interesting information concerning devitrification behavior of $xGd_2O_3\cdot(100-x)[6TeO_2\cdot4B_2O_3]$ vitreous system with $0 \leq x \leq 30$ mol%. The structural changes have been analyzed with increasing rare earth concentration. Two halos characteristic of the amorphous compounds can be observed in XRD diffraction pattern for sample with $x \geq 20$ mol% $Gd_2O_3$. These changes can be explained only if we admit that the adding of gadolinium oxide now participate in the network as $[GdO_n]$ structural units yielding a change from the continuous borate-tellurate network to the continuous gadolinium-borate-tellurate network with interconnected through Gd–O–B and Gd–O–Te bridges. The EPR spectra of the studied glass ceramics reveal an increase of the content of $Gd^{+3}$ ions in network former and modifier positions. Therefore, $Gd^{+3}$ ions will coordinate more with non-bridging oxygens leading the decreased of the number of individual $Gd^{+3}$ ions.

## 1. Introduction

Glasses containing rare earth ions have attracted a great deal of interest due to their important physical and chemical properties [1-5]. Glass ceramics based on partial crystallization of precursor glasses from the silicate or phosphate systems were developed for the possible solidification of fuel recycle waste or the immobilization of high level waste [6-9].

Design of the controlled crystallization process of gadolinium-borate-tellurate glasses in order to obtain glassy-crystalline materials has been the main goal of the work. Glass ceramics structural changes have been studied using the X-ray diffraction, FTIR and EPR spectroscopy.

## 2. Experimental

Glasses with composition $xGd_2O_3\cdot(100-x)$ $[6TeO_2\cdot4B_2O_3]$ ($0 \leq x \leq 30$ mol%) were prepared by conventional melt-quenching method, melting mixtures of $TeO_2$, $H_3BO_3$ and $Gd_2O_3$ reagents of purity grade in a corundum crucibles at 800°C for 60 min. Then, the glasses samples were subject to heat treatment applied at 475°C for 24 h, respectively.

The samples were analyzed by means of X-ray diffraction using a XRD-6000 Shimadzu diffractometer. The IR spectra were recorded in the range 400-1600 $cm^{-1}$ using a JASCO FTIR 6200 spectrophotometer. EPR measurements were performed at room temperature using ADANI Portable EPR PS 8400-type spectrometer, in X frequency band (9.1 Hz) and a field modulation of 100 kHz.

## 3. Results and discussion

The X-ray diffraction patterns reveal that the composition of the glass ceramics with the $Gd_2O_3$ concentration up to 10 mol% was found to consist mainly of the $TeO_2$ and $B_2O_3$ crystalline phases (figure 1). Surprising for samples with $x \geq 20$ mol% $Gd_2O_3$ can be observed two halos characteristic of the amorphous compounds. The role of such oxide in the geometrical arrangement of the building units of the glass ceramics network can be considered as follow (figure 2):

*i.* For low concentrations, $0 \leq x \leq 20$ mol%, the intensity of the band centered at ~780 $cm^{-1}$ increase with increasing the gadolinium ions concentration. This observation is consistent with the destruction of the [$TeO_4$] structural units. The band located at about ~680 $cm^{-1}$ corresponds to the stretching vibrations in [$TeO_4$] structural units [10-12]. The content of three-coordinated tellurium attains its maximum values of 1 mol% $Gd_2O_3$. This increase yields the apparition of the $TeO_2$ crystalline phase. It seems that the content of [$BO_4$] structural units become higher, because results to transformation of some [$BO_3$] units into [$BO_4$] structural units.

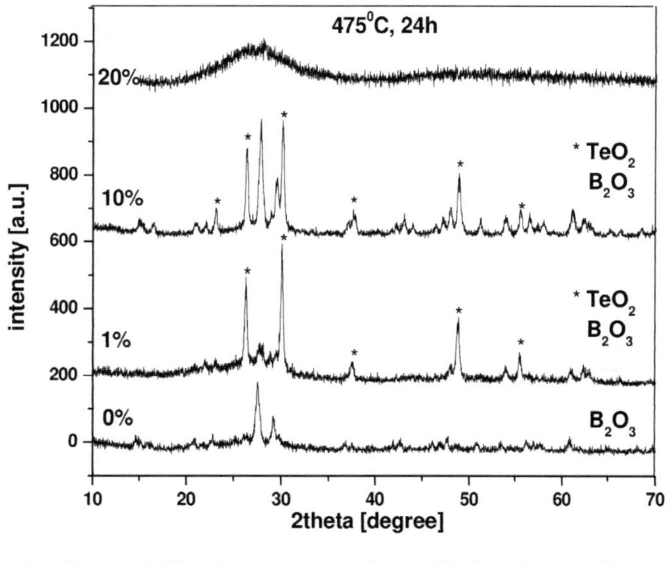


**Figure 1.** X-ray diffraction patterns for $xGd_2O_3 \cdot (100-x)[6TeO_2 \cdot 4B_2O_3]$ glass ceramics treated at 475°C with x = 20, 25 and 30 mol%.

*ii.* The band centred at around 450 $cm^{-1}$ is most often assigned to bending vibrations of Te–O–Te linkages [13]. The intensity of this band varies with the increasing amounts of gadolinium oxide. The effects suggest that doping with gadolinium ions deforms the Te–O–Te linkages and changes the distribution of ordered microregions within the glass ceramics [14, 15].

*iii.* The broad bands in the region 800-1200 $cm^{-1}$ are due to the stretching of B–O bonds of [$BO_4$] units. The features of the bands presented in the region 1200-1600 $cm^{-1}$ can be attributed to the [$BO_3$] structural units stretching vibrations, which are associated with the vibrational modes inside the various borate rings and the non-bridging B– bonds, respectively.

The gadolinium ions have a strong affinity towards these [$BO_3$] structural units containing non-bridging oxygens, which are negative-charged, because they are readily available for charge compensation. On the other hand, the gadolinium ions have also an affinity pronounced towards

[$TeO_3$] structural units yielding the deformation of the Te–O–Te linkages, the changed of the distribution of ordered microregions of the $TeO_2$ crystalline phase and pursuant to the intercalation of [$GdO_4$] entities in the [$TeO_4$] chain network [16]. These observations can be related by modifies of characteristic features of the bands located in the region ranges between 500-800 and 1200-1600 $cm^{-1}$.

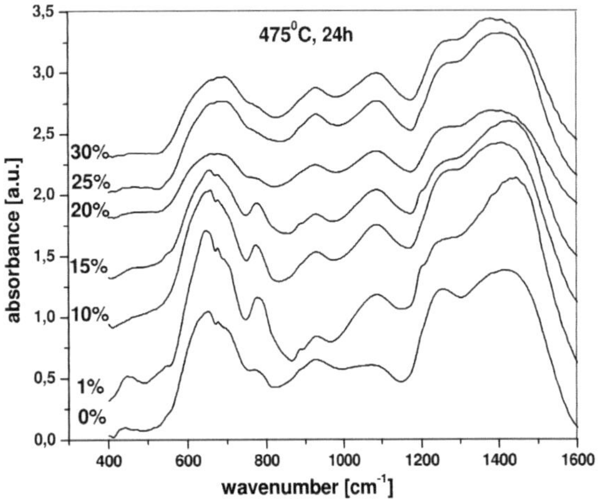


**Figure 2.** FTIR spectra of $xGd_2O_3{\cdot}(100-x)[6TeO_2{\cdot}4B_2O_3]$ glass ceramics treated at 475°C.

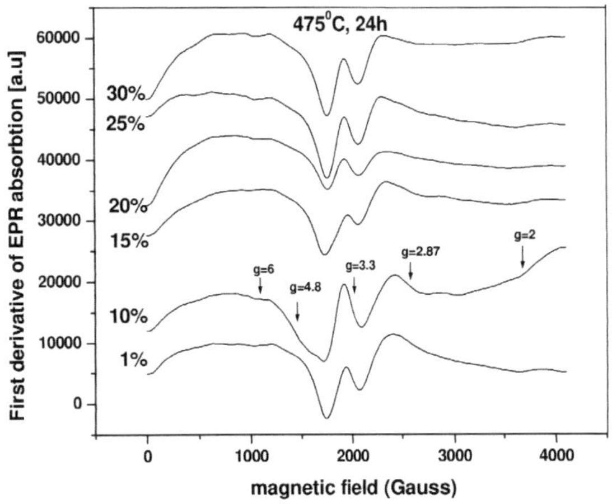


**Figure 3.** EPR spectra of $xGd_2O_3{\cdot}(100-x)[6TeO_2{\cdot}4B_2O_3]$ glass ceramics.

With the increase of $Gd_2O_3$ content up to 20 mol%, a sharp decreasing trend was observed in strength on the bands centered at about 780 and 1200 $cm^{-1}$, which correspond to the $TeO_2$ and $B_2O_3$ crystalline phases. This may be attributed to the electrostatic field of the strongly polarizing $Gd^{+3}$ ions which will yield the apparition of amorphous compounds by increasing of gadolinium ions.

The EPR spectra of $xGd_2O_3{\cdot}(100-x)[6TeO_2{\cdot}4B_2O_3]$ glass ceramics sample treated at 475°C show a significant increase of the resonance line attributed to the type of $Gd^{+3}$ ions which can be presented as network former in the glass ceramics ($g \approx 4.8$) (figure 3). Moreover, we observed a weak resonance line attributed the $g \approx 2$ due to formation of $Gd^{+3}$ clusters for all glass ceramics samples. The weak resonance line $g \approx 6$ was considered to be a characteristic feature of intermediate crystal field sites of axial symmetry and have attributed the broadened general appearance of the U-spectrum to isolated rare earth ions in wide varieties of sites. The $Gd^{+3}$ ions in these glass ceramics can be considered as isolated in the sense of the absence of clustering. The weakening intensity of the clusters EPR line could be due to the $Gd^{+3}$ migrations inside the glass network. It should lead to the appearance of more $Gd^{+3}$ content in network former positions. It can be pointed that the $Gd^{+3}$ ions are generally suspected to improve their environment [17, 18].

In brief, the gadolinium ions have a strong affinity towards [$BO_3$] groups containing non-bridging oxygens yielding the disappearance of the $B_2O_3$ crystalline phase. Then, the $Gd_2O_3$ behaves as a glass-former by means of the intercalation of [$GdO_4$] entities in the [$TeO_4$] chain network.

## 4. Conclusions

The structure of $xGd_2O_3 \cdot (100-x)[6TeO_2 \cdot 4B_2O_3]$ glass ceramics prepared by melt-quenching method (for x = 0-30 mol% Gd2O3) was studied by FTIR, EPR spectroscopy and X-ray diffraction. The increase of $Gd_2O_3$ content leads the disappearance of $B_2O_3$ and $TeO_2$ crystalline phases and causes a change from the glass ceramics network to the continuous gadolinium-borate-tellurate glass network with interconnected through Gd–O–B and Gd–O–Te bridges. The EPR spectra of the studied glass ceramics reveal the presence of line at $g \approx 4.8$ characterized for $Gd^{+3}$ ions in sites of strong crystal fields having a low coordination number. This type of $Gd^{+3}$ ions can be presented as network former in the glass.

## References

[1] Simon S, Ardelean I, Filip S, Bratu I and Cosma I 2000 *Solid State Commun.* **116** 83
[2] Ristoiu T and Culea E 2001 *J. Non-Cryst. Solids* **279** 93
[3] Ristoiu T, Culea E and Bratu I 1999 *Mater. Lett.* **41** 135
[4] Coroiu I, Borodi G, Vida-Simiti I, Darabont A, Culea E and Jumate N 2006 *J. Optoelectron. Adv. Mater.* **8** 529
[5] Coroiu I and Cristea I 2005 *J. Magn. Magn. Mater.* **293** 520
[6] Kim H. G and Komatsu T 1988 *J. Mater. Sci. Lett.* **17** 1198
[7] Sitarz M and Szumera M 2008 *J. Ther. Anal. Calorim.* **91** 255
[8] Lucaci P, Burzo E and Lupsa I 1996 *J. Alloys. Compd.* **L4** 238
[9] Donald W, Metcalfe B L and Taylor R N J 1997 *J. Mater. Sci.* **32** 5851
[10] Wright A C, Vedischeheva N M and Shakmatahin B A 1995 *J. Non-Cryst. Solids* **192-193** 92
[11] Rada S, Culea M and Culea E 2008 *J. Phys. Chem.* A **112** 11251
[12] Pan A and Ghosh A 2002 *Phys. Rev.* B **66** 012301
[13] Rada S, Culea M, Neumann M and Culea E 2008 *Chem. Phys. Lett.* **460** 196
[14] Jha A, Shen S and Naftaly M 2000 *Phys. Rev.* B **62** 6215
[15] Iton I E, Brodbeck C M, Suib S L, Stucky G D 1983 *J. Chem. Phys.* **79** 1185
[16] S Rada and E Culea 2009 *J. Mol. Struct.* **929** 141
[17] Culea E, Pop L and Simon S 2004 *Mater. Sci. Eng.* B **112** 59
[18] Brodbeck C M and Iton I E 1985 *J. Chem. Phys.* **83** 4285.

# Immobilization of gadolinium in borate-tellurate glasses

**S Rada, E Culea, M Rada, V Maties, M Bosca, L Pop, R Fechete, R Chelcea and D Moldovan**

Technical University of Cluj-Napoca, Physics Department, 103-105 Muncii, 400641 Cluj-Napoca, Romania

E-mail: simona.rada@phys.utcluj.ro, eugen.culea@phys.utcluj.ro

**Abstract.** Glasses in the $xGd_2O_3{\cdot}(100-x)[6TeO_2{\cdot}4B_2O_3]$ system, where x = 0, 5, 10, 15, 25 and 30 mol%, have been prepared from melt quenching method. The structural changes were studied by FTIR spectroscopy and DFT calculations. The FTIR spectroscopy data for these systems show that the gadolinium ions have a strong affinity towards the structural units containing non-bridging oxygens, which are negative-charged, because they are readily available for charge compensation. So that, it seems that the content of $[BO_4]$ structural units cannot become higher, because the modified $[BO_3]$ units containing one or more B–O–Gd bonds are unable to accept a fourth oxygen atom. On the other hand, the gadolinium ions have also an affinity pronounced towards $[TeO_3]$ structural units yielding the deformation of the Te-O–Te linkages. The compositional evolution of the network was accommodated with excess of oxygen by the formation of orthoborate structural units, disintegration of some boroxol rings and radical rearrangement of the network formed by the $[TeO_6]$ octahedral units.

## 1. Introduction

Immobilization of high level nuclear wastes by vitrification is a well established process that has been studied extensively over last 40 years. A suitable glass host is used to dissolve the high level nuclear waste to form a glassy (vitreous) homogeneous product that can be cast into suitable forms, including large glass blocks. Under suitable conditions, it is possible to incorporate up to 25-30 wt% high level waste into a glass [1-4].

The future is directed towards the identification of similar glasses and glass ceramics with improved durability and processing characteristics, which will mainly be achieved through compositional modifications of existing waste forms. Then, other types of glasses and glass ceramics waste forms with lower manufacturing temperature and easier processing route also offer potential for further study.

The present contribution investigates the borate-tellurate glasses as possible candidate for the immobilization of nuclear wastes. We propose than needful for alternative vitrification processes the borate glasses formulations because they appear to offer advantages of reduced process temperature and improved chemical durability compared to other glasses. Although there are many reports in the literature about glass ceramics, these were undesirable due to high processing temperature (>1400$^0$C) and poor chemical durability [4, 5]. This new formulation differs in the type and quantity of network modifiers added, resulting in a glass matrix with improved chemical durability and reduced melting temperature.

## 2. Experimental

Glasses with composition $xGd_2O_3 \cdot (100-x)[6TeO_2 \cdot 4B_2O_3]$ ($0 \leq x \leq 30$ mol%) were prepared by conventional melt-quenching method, melting mixtures of $TeO_2$, $H_3BO_3$ and $Gd_2O_3$ reagents of purity grade in a corundum crucibles at 800°C for 60 minute.

The samples were analyzed by means of X-ray diffraction using a XRD-6000 Shimadzu diffractometer. The FTIR absorption spectra of the studied glasses were measured for each glass sample over the range 400-1600 $cm^{-1}$ of wavenumbers.

The proposed model for borate-tellurate glass network was performed by direct optimization of the geometry with help of the B3PW91/CEP-4G/ECP method using Gaussian 03 program. Dangling bonds of outer atoms of the models were saturated with hydrogen atoms.

## 3. Results and discussion

### *3.1. FTIR spectroscopy*

We performed X-ray diffraction investigation of the samples and these data indicated unambiguous the amorphous character of the samples.

FTIR spectrum of the $6TeO_2 \cdot 4B_2O_3$ glass for the 400 and 1400 $cm^{-1}$ wavenumber range is shown in figure 1. Obtained bands and their assignments can be summarized as follow:

*i.* The large band centered at ~625 $cm^{-1}$ is assigned to the stretching mode of $[TeO_4]$ trigonal bypiramidal with bridging oxygens [7, 8]. The shoulder located at about ~780 $cm^{-1}$ indicates the presence of $[TeO_3]$ units [9, 10].

*ii.* The broad bands in the region 800-1200 $cm^{-1}$ are due to the stretching of B-O bonds of $[BO_4]$ units [11, 12]. The feature of band centered in the region ~1030 $cm^{-1}$ is due to the pentaborate, tetraborate, triborate and diborate vibrations. The IR absorption band at about 1263 $cm^{-1}$ was attributed to the B-O stretching vibrations of the $[BO_3]$ units from the boroxol rings [13, 14].

*iii.* The bands centered in the region between ~1200 and 1600$cm^{-1}$ can be attributed to the $[BO_3]$ units stretching vibrations, which are associated with the vibrational mode inside the various borate rings and the non-bridging $B–O^-$ bonds, respectively [12-15]. The absorption band located at about ~1200$cm^{-1}$ was assigned to $B–O^-$ asymmetric stretching vibrations in $BO_3$ units from pyro- and ortho-borate groups. This band responsible for non-bridging oxygen atoms increases in the borate groups with increase the gadolinium ions indicating that more non-bridging oxygen ions exist in borate glass.

The addition of $Gd_2O_3$ in the $6TeO_2 \cdot 4B_2O_3$ glass matrix (figure 1), gives some drastically changes in the FTIR spectra of these glasses as follows:

*i.* The intensity of the band centered at ~780 $cm^{-1}$ increase with increasing the gadolinium ions concentration. This observation is consistent with the destruction of the $[TeO_4]$ structural units.

*ii.* By increasing the $Gd_2O_3$ content between 5% and 25%, the characteristic features of the four absorption bands were included. The decreasing trends in wavenumber of the bands located at 1256 and 1384$cm^{-1}$ imply an inter-conversion process of the $[BO_3] \leftrightarrow [BO_4]$ structural units and the transformation of the large number of boroxol rings [13-15]. The increasing trend in wavenumber of the band centered in the regions ~680 $cm^{-1}$ can be explained considering a conversion process of the $[TeO_4]$ structural units into $[TeO_3]$ structural units.

*iii.* The decreasing trend of the bands located in the region between 900 and 1150 $cm^{-1}$ (1020 $cm^{-1}$) can be due to the formation of bridging bond of Te–O–B. Since the stretching force constant of Te-O bonding is substantially lower than that of the B–O, the stretching frequency of Te–O–B might trend to be lower.

*iv.* The responsible for this attack of the gadolinium ions seems to be the partial disappearance of the small band located in the range between 450$cm^{-1}$. This band is assigned the bending mode of Te–O–Te or O–Te–O linkages. The effect suggests that the doped with gadolinium ions deforms the Te–O–Te linkages and generate their destruction.

*v.* With the increase of $Gd_2O_3$ content up to 30 mol%, a new peak was observed at about ~1200 $cm^{-1}$, which was assigned to B-O$^-$ asymmetric stretching vibrations in $BO_3$ units from orthoborate groups This may be attributed to the electrostatic field of the strongly polarizing $Gd^{+3}$ ions. It seems that upon addition of the rare earth ions to the glass, some free B–O bonds are also generated. These simpler units containing a large number of non-bridging oxygen and boroxol rings (the band located at about 1254$cm^{-1}$ becomes sharper and pronounced), are expected to be created if $Gd_2O_3$ behaves as a glass modifier.

It seems that the content of [$BO_4$] structural units cannot become higher, because the modified [$BO_3$] units containing one or more B–O–Gd bonds are unable to accept a fourth oxygen atom [13]. Accordingly, the borate-tellurate glasses may be good hosts for the incorporation of high gadolinium ions contents at low temperature.

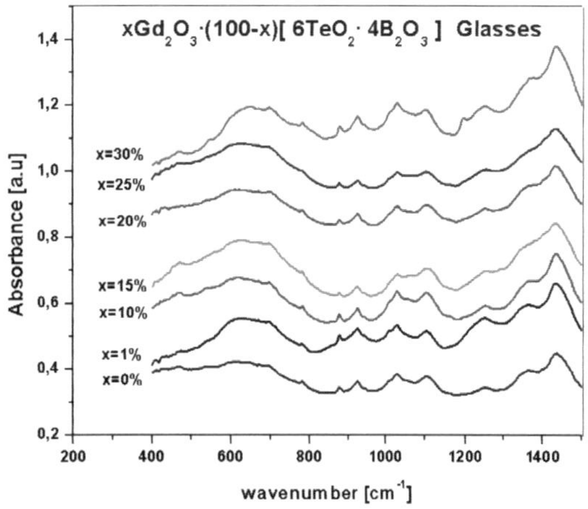


**Figure 1.** FTIR spectra of $xGd_2O_3{\cdot}(100-x)[6TeO_2{\cdot}4B_2O_3]$ glasses where $0 \le x \le 30$ mol%.

For this reason, it is assumed that the framework of glasses is built up of [$BO_4$] tetrahedra, [$BO_3$] triangles, trigonal bypiramids [$TeO_4$], pyramidal [$TeO_3$] and accidentaly at a high $B_2O_3$ content of [$TeO_6$] octahedral units. Accordingly, the increase in the number of non-bridging oxygen atoms would decrease the connectivity of the glass network and would necessity quite a radical rearrangement of the network formed by the [$TeO_6$] octahedral. Our results show the role of the gadolinium ions as a modifier on the glass network [14].

### *3.2. DFT calculations*

The IR data were used in the present research in order to compute a possible structural model of the $6TeO_2{\cdot}4B_2O_3$ glass network. Similar methodology has previously been reported to study other glasses [10, 13, 15, 16]. The tellurium atom in our model has six oxygen atoms in vicinity and the Te–O distances are 1.99, 2.01, 2.02, 2.05, 2.09 and 2.15 Å. Acceding to the data from this study, the Te–O distances vary from 1.99 to 2.15 Å and this octahedron is distorted [17]. Tellurate networks are flexible to form the appropiate coordination environments for groups of opposite charge such as [$BO_4$]$^-$ tetrahedra where change of the $TeO_n$ network units provides the needed number of oxygen neighbours. This result, namely the existence of the distorsed [$TeO_6$] structural units was previously reported based on neutron and X-ray diffraction experiments for the $P_2O_5$-$TeO_2$ glasses [18].

Our DFT calculations show that the increase in the number of non-bridging oxygen atoms would decrease the connectivity of the glass network, would depolymerize of borate chains and would necessity quite a radical rearrangement of the network formed by the [$TeO_6$] octahedral units. This is possible considering that tellurium dioxide brings stoechiometrically with two oxygen atoms in [$TeO_4$] and needs an additional oxygen atom for the formation of [$TeO_6$] octahedrons. This additional oxygen atom can be taken off also from the boron coordination and thus boron atoms transfer their [$BO_4$] coordination into [$BO_3$] coordination.

In these conditions, the compositional evolution of the network was accommodated with excess of oxygen by the formation of orthoborate structural units, disintegration of some boroxol rings and radical rearrangement of the network formed by the $[TeO_6]$ octahedral units. The reduced processing temperature makes borate-tellurate glasses suitable for the use of immobilization of high level hazardous waste, as well as for many alternative commercial glass and glass ceramic products.

## 4. Conclusions

Glasses with composition $xGd_2O_3{\cdot}(100\text{-}x)[6TeO_2{\cdot}4B_2O_3]$ ($0 \leq x \leq 30$ mol%) were prepared by conventional melt-quenching method. The investigations of FTIR spectroscopy show that the content of $[BO_4]$ structural units cannot become higher, because the modified $[BO_3]$ units containing one or more B–O–Gd bonds are unable to accept a fourth oxygen atom. The FTIR spectroscopy and DFT investigations show that the increase in the number of non-bridging oxygen atoms would decrease the connectivity of the glass network, would depolymerize of borate chains by the formation of orthoborate structural units and disintegration of some boroxol rings, respectively and would necessity quite a radical rearrangement of the network formed by the $[TeO_6]$ octahedral units.

## References

[1] Kehagias Th, Komninou Ph, Kavouras P, Chrissafis K, Nouet G and Karakostas Th 2006 *J. Eur. Ceram. Soc.* **26** 1141
[2] Jantzen C M 1986 *J. Non-Cryst. Solids* **84** 215
[3] Sales C B and Boatner A L 1984 *Science* **226** 45
[4] Sales C B and Boatner A L 1986 *J. Non-Cryst. Solids* **79** 83
[5] Donald W, Metcalfe B L and Taylor R N J 1997 *J. Mater. Sci.* **32** 5851
[6] Frisch M J *et al* 2003Gaussian 03, Revision A.1 (Pittsburgh PA: Gaussian, Inc)
[7] Kozhukharov V, Nikolav S, Marinov M and Troev T 1979 *Mater. Res. Bull* **14** 735
[8] Rada S, Culea E, Rus V, Pica M and Culea M 2008 *J. Mater. Sci.* **43** 3713
[9] Rada S, Culea M, Rada M and Culea E 2008 *J. Mater. Sci.* **43** 6122
[10] Rada S, Culea M and Culea E 2008 *J. Phys. Chem.* A **112** 11251
[11] Konijnendijk W L, Stevels J M 1975 *J. Non-Cryst. Solids* **18** 30
[12] Bhargava A, Snyder R L and Condrate R A 1987 *Mater. Res. Bull.* **22** 1603
[13] Rada S, Culea M, Neumann M and Culea E 2008 *Chem. Phys. Lett.* **460** 196
[14] Coroiu I, Borodi G, Vida-Simiti I, Darabont A, Culea E and Jumate N 2006 *J. Optoelectron. Adv. Mater.* **8** 529
[15] Rada S, Culea M and Culea E 2008 *J. Non-Cryst. Solids* **354** 5491
[16] Rada S, Pascuta P, Bosca M, Culea M, Pop L and Culea E 2008 *Vib. Spectrosc.* **48** 255
[17] Ciceo Lucacel R, Marcus C, Timar V and Ardelean I 2007 *Solid State Sci.* **9** 850
[18] Hope U, Gugov I, Burger H, Jovari P and Hannon A C 2005 *J. Phys: Condens. Matter* **17** 2365
[19] Hoppe U, Yousef E and Russel C 2002 *Solid State Commun.* **123** 407.

Journal of Physics: Conference Series **182** (2009) 012076 doi:10.1088/1742-6596/182/1/012076

# EPR investigations of $Sn_{1-x}Fe_xO_2$ nanopowders

**O Raita, D Toloman, A Popa, M Stan, Al Darabont and L MGiurgiu**

National Institute for Research and Development of Isotopic and Molecular Technologies, 65-103 Donath, 400293 Cluj-Napoca, Romania

E-mail: oana.raita@itim-cj.ro

**Abstract**. An X-band electron paramagnetic resonance (EPR) investigation of $Fe^{3+}$ ions in $Sn_{1-x}Fe_xO_2$ nanopowders with x = 0.005-0.07 is reported. These samples are interesting to investigate as Fe doping produce ferromagnetism in $SnO_2$, making it a promising ferromagnetic semiconductor at room temperature.

## 1. Introduction

$SnO_2$ is an attractive system for quite a wide variety of practical applications, being a chemically stable oxide semiconductor. It has been shown that Fe doping produces ferromagnetic semiconductor at room temperature [1]. This material, therefore, has the potential for use in spintronic devices such as spin transistors, spin light emitting diodes, very high density nonvolatile semiconductor memory and optical emitters. It is believed that oxygen vacancies and substitutional incorporation are important to produce ferromagnetism in semiconductor oxide doped with transition metal ions.

The present paper reports our electron paramagnetic resonance (EPR) studies on $Sn_{1-x}Fe_xO_2$ (x = 0.005-0.07) nanopowders in order to determine the oxidation state and site occupancy of the dopant ions in the host $SnO_2$ lattice, and also to detect a possible ferromagnetic ordering of these ions.

## 2. Experimental

A series of nanopowders of $Fe^{3+}$ ions-doped $SnO_2$ having the general formula $Sn_{1-x}Fe_xO_2$ (x = 0.005, 0.01, 0.03, 0.05 and 0.07) were obtained and investigated by EPR.

The samples were prepared using stannous chloride dihydrate ($SnCl_2 \cdot 2H_2O$ 95%) and oxalic acid dihydrate ($H_2C_2O_4 \cdot 2H_2O$ 99.4%) in stoichiometric amounts doped with iron by using $FeCl_3$ monocrystalline needles [2]. Appropriate quantities were dissolved in distilled water using a magnetic stirrer, and finally a homogenous solution was obtained. The homogeneous solution was then decomposed at 523 K in air until dry. The temperature is raised gradually over 2-3 days. After heat treatment this mass precursor was ground to give fine powder. Fractions of the obtained powders have been calcinated at 873 K for 2 h in air.

The EPR measurements on the samples were performed with a Bruker ELEXSYS 500 spectrometer at X-band.

## 3. Results and discussion

Figure 1 shows X-band EPR spectra recorded at room temperature for $Sn_{1-x}Fe_xO_2$ nanopowders with different Fe concentration x.

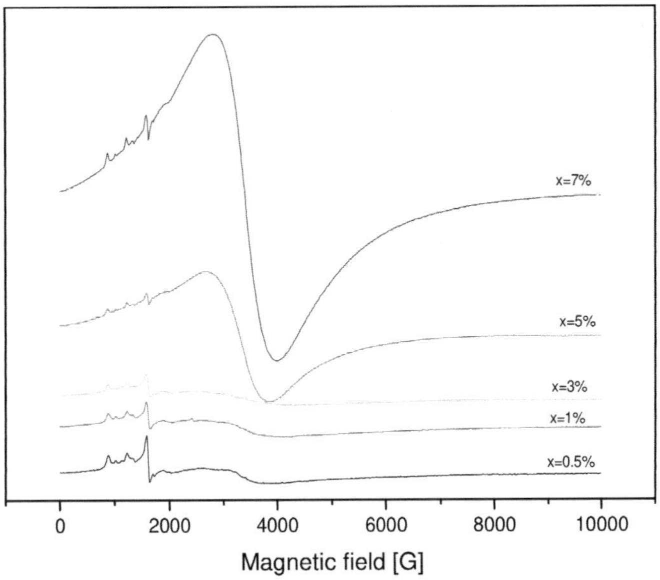


**Figure 1.** EPR spectra at room temperature for $Sn_{1-x}Fe_xO_2$ nanopowders as function of Fe concentration x.

From the experimental spectra (figure 1), one can distinguish that there exists an overlap of three types of Fe-related EPR spectra: (*i*) a sharp line (LS) arising from interstitially incorporated low-spin $Fe^{3+}$ ions, S = 1/2, (*ii*) a spectra (HS) due to the substitutionally incorporated high-spins $Fe^{3+}$ ions, S=5/2, which are probably situated in the interior of the nanodomains [1] and (*iii*) a broad signal (FM) which could be attributed to the ferromagnetically coupled $Fe^{3+}$ ions.

The assignment of the above contributions is based on the corresponding simulations which are shown in figure 2.

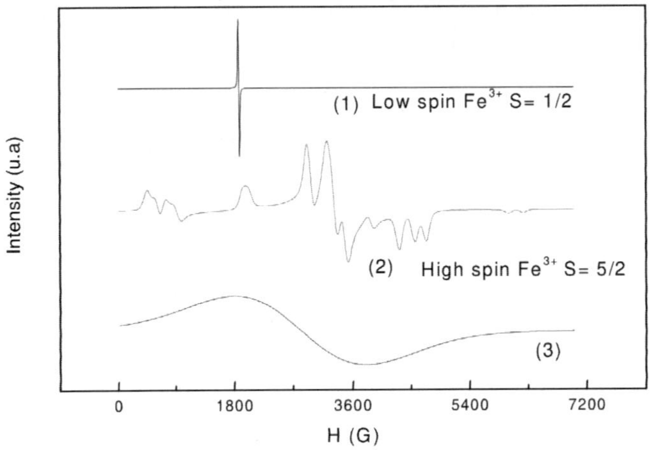


**Figure 2.** Simulated spectra for: (1) interstitial contribution (2) substitutional contribution and (3) ferromagnetic contribution.

In the spectra simulations, we have used the X-sophe software where the following spin Hamiltonians were considered:

$$H_S = \mu_B\, B\, g\, S, \text{ for the low spin (S=1/2)} \tag{1}$$

$$H = g\,\mu_B\, B\, S + D\,[S_z^2 - (1/3)S(S+1)] + E[S_x^2 - S_y^2], \text{ for high spin(S = 5/2)} \tag{2}$$

Here, D and E are axial and nonaxial parameters, respectively.

Table 1 summarize the spin Hamiltonian parameters, the line widths and the spin assignments used to simulate the observed three EPR spectra.

**Table 1.** Spin Hamiltonian parameters, line widths and spin assignments employed in the simulation of the experimental EPR spectra in $Sn_{1-x}Fe_xO_2$ nanopowders.

| | $g_x$ | $g_y$ | $g_z$ | D [G] | E [G] | ΔH [G] | S |
|---|---|---|---|---|---|---|---|
| LS | 3.612 | 3.612 | 3.612 | 0 | 0 | 20 | 1/2 |
| HS | 2.004 | 2.004 | 2.004 | 700 | -190 | 100 | 5/2 |
| FM | 2.412 | 2.412 | 2.412 | 0 | 0 | 2000 | 3/2 |

The values of the g-tensors for HS and LS spectra are very closed to the other values reported in the literature [3]. The effective g-factor for the broad signal (FM) is greater than 2 which represents a clear signature of ferromagnetism in $Sn_{1-x}Fe_xO_2$ nanopowders [4]. Therefore, the FM signal could be attributed to the ferromagnetic resonance arising from the exchange interactions between $Fe^{3+}$ ions.

One can observe in figure 3 that in the low concentrations domain ($0.005 \leq x \leq 0.03$) the line width ($\Delta H_{1/2}$) corresponding to FM signal increases with increasing the concentration. This tendency could be attributed to the dominance of the dipolar interactions. In the high concentration limit ($x > 0.03$) the line width decrease with increasing the concentration. It could means that in this range of concentrations the ferromagnetic behavior is present and the magnetic exchange interaction is dominant.

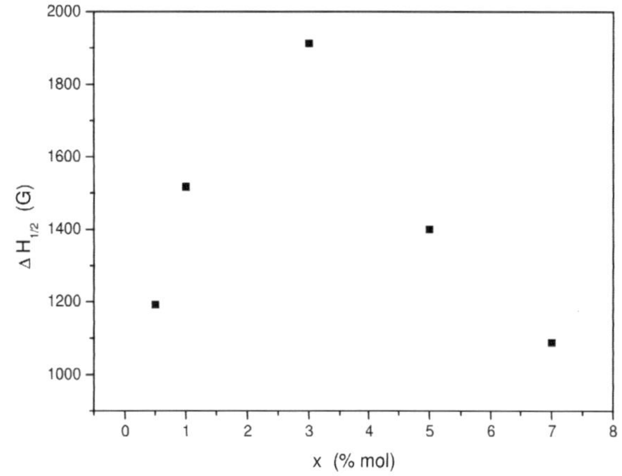


**Figure 3.** Concentration dependence of $\Delta H_{1/2}$ linewidth for FM resonance in $Sn_{1-x}Fe_xO_2$ nanopowders.

## 4. Conclusions

We have investigated by EPR the incorporation of Fe ions into $SnO_2$ nanopowders. EPR measurements reveal that Fe doping results in interstitial and substitutional incorporation of $Fe^{3+}$ ions into the $SnO_2$ lattice as well as in a ferromagnetic coupling of these ions. For higher x concentration of $Fe^{3+}$ ions the FM linewidth decreases with increasing the concentration due to the dominance of magnetic exchange interaction in $Sn_{1-x}Fe_xO_2$ nanopowders.

## References

[1] Misra S K and Andronenko SI 2007 *J. Appl. Phys.* **101** 09H120
[2] Parra R, Ramajo L A, Goes M S, Varela J A and Castro M S 2008 *Mater. Res. Bull.* **43** 3202
[3] Dusausoy Y, Ruck R and Gaite J M 1988 *Phys. Chem. Miner.* **15** 300
[4] Karmakar D, Mandal S K, Kadam R M, Paulose P L, Rajarajan A K, Nath T K, Das A K, Dasgupta I and Das G P 2007 *Phys. Rev.* B **75** 144404.

# X-ray photoelectron spectroscopy and magnetism of AlDyNi, $AlDyNi_4$ and $AlDy_3Ni_8$ compounds

**V Rednic[1,3], M Coldea[1], L Rednic[1], L G Pascut[1,2], N Aldea[3], S Pintea[1,3] and M Neumann[4]**

[1] Babes-Bolyai University, Faculty of Physics, 1 Mihail Kogalniceanu, 400084 Cluj-Napoca, Romania

[2] University of Bristol, H H Wills Physics Laboratory, Tyndall Avenue, Bristol BS8 1TL, UK

[3] National Institute for Research and Development of Isotopic and Molecular Technologies, 65-103 Donath, 400293 Cluj-Napoca, Romania

[4] University of Osnabrück, Fachbereich Physik, 49069 Osnabrück, Germany

E-mail: vasile.rednic@itim-cj.ro

**Abstract**. The electronic and magnetic properties of AlDyNi, $AlDyNi_4$ and $AlDy_3Ni_8$ are studied using X-ray photoelectron spectroscopy (XPS), X-ray diffraction (XRD), static and dynamic magnetic measurements. The three compounds AlDyNi, $AlDyNi_4$ and $AlDy_3Ni_8$ are single phases and crystallize in the $Fe_2P$, $CaCu_5$ and $CeNi_3$ structure types, respectively. All the investigated compounds order ferromagnetically below the corresponding Curie temperatures. The Curie temperature and the magnetic moments are 25 K and 6 $\mu_B$/f.u. for AlDyNi, 14 K and 6.9 $\mu_B$/f.u. for $AlDyNi_4$ and 22K and 19.2 $\mu_B$/f.u. for $AlDy_3Ni_8$. At high temperature the magnetic susceptibility obey the Curie-Weiss law. The paramagnetic Curie temperature and the effective magnetic moments are 30K and 10.88 $\mu_B$/f.u. for AlDyNi, 28 K and 10.94 $\mu_B$/f.u. for $AlDyNi_4$ and 18 K and 18.33 $\mu_B$/f.u. for $AlDy_3Ni_8$. XPS valence band and Ni 2p spectra indicated the presence of small magnetic moment on Ni sites in $AlDy_3Ni_8$ and $AlDyNi_4$, and a complete filling of Ni 3d band in AlDyNi.

## 1. Introduction

The In the ternary metallic system Dy-Ni-Al ten intermetallic compounds with different crystallographic structure were reported [1].

The nearest neighbourhood of Dy and Ni ions, as well as, the distances Ni-Ni, Dy-Dy, Dy-Ni are different in each compound. This leads to different states of Ni ions and also influences the Dy-Ni interaction in these compounds.

The electronic 3d band width of the transition metals (T) is determined by the overlap between the d orbitals of adjacent atoms and depends on the number of nearest neighbours and the hopping integral $J_h$, which is very sensitive to the T-T distances [2]. On the other hand, the hybridizations between the 3d Ni and 5d6s Dy states and 3d Ni and 3sp Al states may lead to a partial or complete filling of the Ni

3d band. In many rare-earth-nickel compounds, nickel atoms do not carry a magnetic moment because of charge transfer of rare-earth conduction electrons to the 3d band [3].

The aim of this paper is to study the magnetic properties of AlDyNi, $AlDyNi_4$ and $AlDy_3Ni_8$ and to correlate the results with the electronic structure obtained from XPS measurements. The three compounds AlDyNi, $AlDyNi_4$ and $AlDy_3Ni_8$ crystallize in the $Fe_2P$, $CaCu_5$ and $CeNi_3$ structure types, respectively [1].

## 2. Experimental

The investigated compounds of AlDyNi, $AlDyNi_4$ and $AlDy_3Ni_8$ were prepared by the standard arc-melting technique using a water-cold crucible in an argon atmosphere. The samples were melted repeatedly in the same atmosphere to ensure homogeneity. The weight loss of the final materials was found to be less than 1%. The purity of the starting materials was 99.9% for Dy, 99.99% for Ni and 99.999% for Al. X-ray powder diffraction measurement, performed using a Bruker D8 Advance diffractometer, showed that the compounds are single phases with the expected crystallographic structure type.

The magnetization measurements were performed with a vibrating sample magnetometer at temperature range 4 K and magnetic fields up to 10 T. The magnetic susceptibility measurements were performed using a Weiss balance starting from room temperature until 600 K. The AC susceptibility measurements were performed using a superconducting quantum interference device (SQUID magnetometer) from 4.2 to 300 K.

The XPS spectra were recorded using a PHI 5600ci ESCA spectrometer with monochromatized Al Kα radiation at room temperature. The pressure in the ultra-high vacuum chamber was in the $10^{-10}$ mbar range during the measurements. The samples were cleaved in situ. The surface cleanness was checked by monitoring the O 1s and C 1s core levels in the survey spectra.

## 3. Results and discussions

### *3.1. Magnetic measurements*

The magnetization as a function of magnetic field at T = 4 K for all the investigated compounds are shown in figure 1.

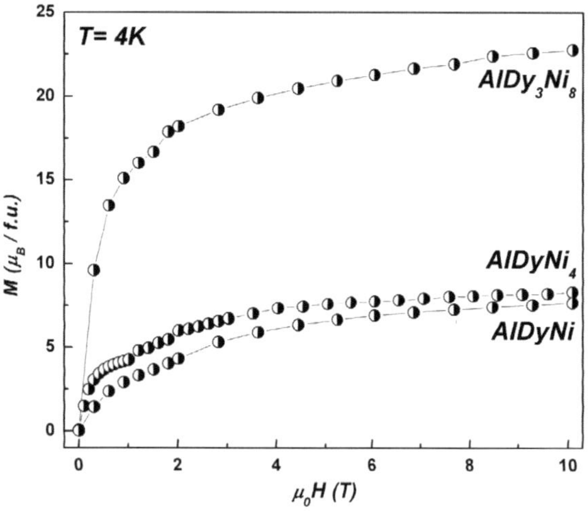


**Figure 1.** The magnetization dependence of the magnetic field at T = 4 K for AlDyNi, $AlDyNi_4$ and $AlDy_3Ni_8$ compounds.

The dependence of the magnetization with the magnetic field is characteristic for ferromagnetic materials. The spontaneous magnetization (table 1) was obtained by extrapolation to H→0 of the linear part (in the high fields range) of the M(H) curve. The corresponding values of the magnetic moments per Dy atom in the ordered state are smaller than for free $Dy^{3+}$ ion (10 $\mu_B$). This is characteristic for many Dy compounds [4] and may be explained taking into account the of crystal field effect; the ground state of $Dy^{3+}$ ion is $^6H_{15/2}$ [5].

The real part of the ac susceptibility (χ') is constant in the temperature range 75-300 K, therefore in figure 2 was plotted only the temperature interval 4-75 K, in order to order to obtain a better resolution for Curie temperature ($T_C$) determination. The values for the $T_C$ are given in table 1. The value of the $T_C$ for $AlDyNi_4$ is in good agreement with the earlier measurements [6].

In the high temperature range the magnetic susceptibility obey the Curie-Weiss law $\chi = C/(T-\Theta)$ (see figure 3).

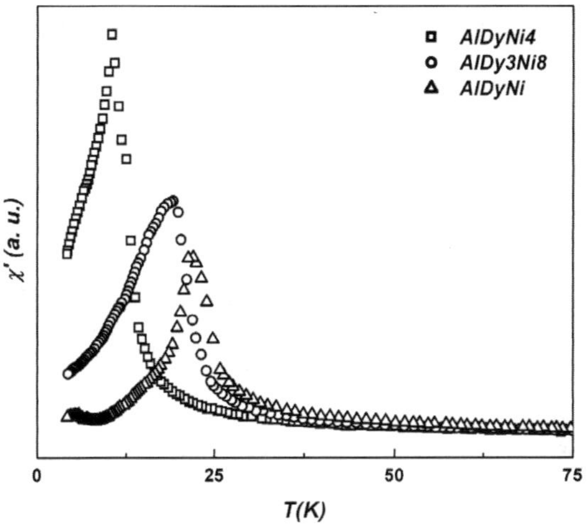


**Figure 2.** The χ' dependence as a function of temperature of AlDyNi, $AlDyNi_4$ and $AlDy_3Ni_8$ compounds.

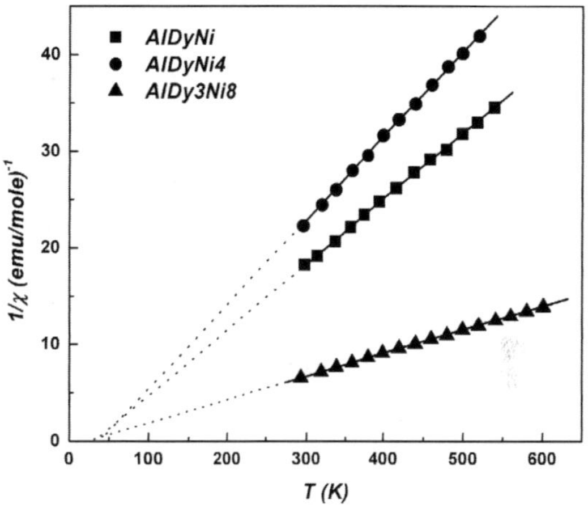


**Figure 3.** The reciprocal susceptibility versus temperature of AlDyNi, $AlDyNi_4$ and $AlDy_3Ni_8$ compounds.

The effective magnetic moment per Dy atom in AlDyNi compound, determined from the Curie constant, is bigger than the value for $Dy^{3+}$ free ion (10.6 $\mu_B$). This can be explained if we consider that the Dy 5d electrons are polarized by the exchange field from the localized 4f spins. Because this polarization is in the same direction as that of the localized 4f moments, an excess magnetic moment pe Dy atom would be expected. This is in agreement with many experimental findings that the 5d electrons contribute to the total moment per rare-earth atom [7]. In addition to this contribution in $AlDyNi_4$ and $AlDy_3Ni_8$ compounds the Ni atoms carry also a small magnetic moment, confirmed by the XPS spectra.

**Table 1.** Ferromagnetic ($T_C$) and paramagnetic ($\Theta$) Curie temperatures, magnetic moments per formula unit in the ordered ($\mu$) and paramagnetic ($\mu_{eff}$) states for all investigated compounds.

| | AlDyNi | $AlDyNi_4$ | $AlDy_3Ni_8$ |
|---|---|---|---|
| $T_C$ (K) | 25 | 14 | 22 |
| $\mu$ ($\mu_B$/f.u.) | 6 | 6.9 | 19.2 |
| Θ (K) | 30 | 28 | 18 |
| $\mu_{eff}$ ($\mu_B$/f.u.) | 10.88 | 10.94 | 18.33 |

*3.2. XPS measurements*

The main contribution to the valence band spectra is given by the Ni 3d and Dy 4f electrons. The Ni 3d band is situated close to the Fermi level while Dy 4f band is more localized and split in several components (figure 4).The reference compound AlNi is a Pauli paramagnet with the Ni 3d band completely filled [8].

Ni 3d band centroids are shifted towards higher binding energy relative to metallic Ni, suggesting a partial filling of the Ni 3d band due to a strong hybridization between Ni 3d, Dy 5sp and Al 3sp states. The decrease in the satellite structure intensity of Ni 2p line compared to pure Ni confirms the partial filling of the Ni 3d band (figure 5).

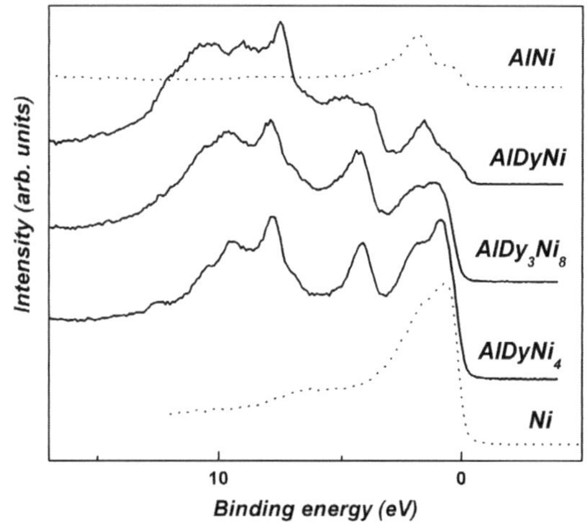


**Figure 4.** Valence band spectra of AlDyNi, $AlDyNi_4$, $AlDy_3Ni_8$ and of reference Ni metal and AlNi.

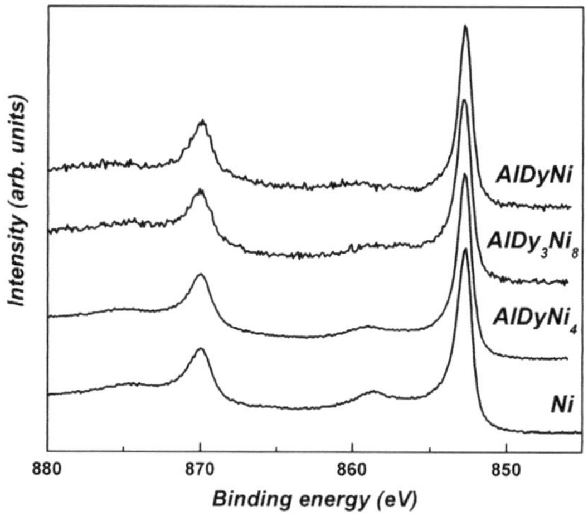


**Figure 5.** Ni 2p core-level lines of investigated compounds together with Ni metal.

These spectra indicate also that Ni atoms carry a magnetic moment in $AlDyNi_4$ and $AlDy_3Ni_8$ compounds, which in the ordered state is orientated antiparallel to the Dy magnetic moment due to the 4f-5d-3d coupling. In the earlier investigation concerning the magnetic properties of $AlDyNi_4$, the magnetic contribution of Ni atoms was neglected [6]. Ni 2p satellite structure for AlDyNi disappears, suggesting that Ni 3d band is completely filled and Ni atoms have no contribution to the total magnetic moment. This is also confirmed by the position of the valence band centroid at ~1.5eV, close to that observed for the Pauli paramagnet AlNi.

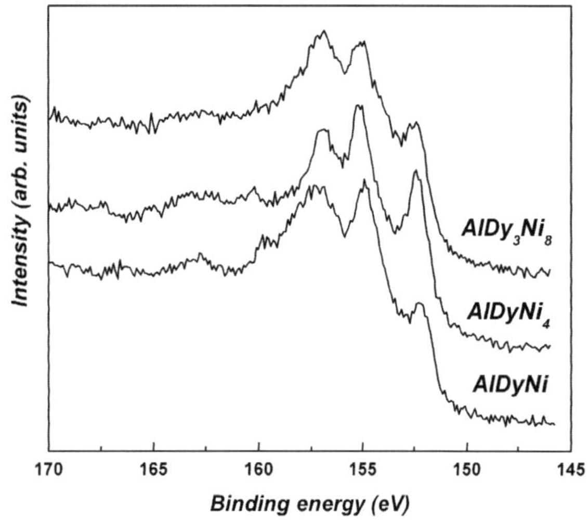


**Figure 6.** Dy 4d core-level lines of AlDyNi, $AlDyNi_4$ and $AlDy_3Ni_8$.

In figure 6 are shown the Dy 4d lines for all the investigated compounds. A chemical shift smaller than 0.4 eV comparing to pure Dy [9] was evidenced. Furthermore the exchange splitting of Dy 4d core level in all three compounds is similar with that found in pure Dy, suggesting that Dy 4f level is not affected by alloying.

## 4. Conclusions

The AlDyNi, $AlDyNi_4$ and $AlDy_3Ni_8$ compounds order ferromagnetically below the corresponding Curie temperatures. At high temperature the magnetic susceptibility obey the Curie-Weiss law. The strong hybridization between Ni 3d, Dy 5sp and Al 3sp states leads to a gradual filling of Ni 3d band. XPS core-level and valence band spectra indicated that Ni atoms carry a small magnetic moment in $AlDyNi_4$ and $AlDy_3Ni_8$ compounds. In AlDyNi compound the Ni 3d band is completely filled and Ni atoms do not carry a magnetic moment. The complete filling of Ni 3d band in AlDyNi compound is due to the high number of Al and Dy atoms in the Ni atoms first vicinity. The polarization of 5d electrons by the local exchange interaction 4f-5d gives rise to an excess magnetic moment of Dy ions in the paramagnetic state of AlDyNi compound.

## References

[1] Villar P and Calvert D L 1994 *Pearson's Handbook of Crystallographic Data for Intermetallic Phases* vol. 1 (Cleveland, OH: American Society of Metals)
[2] Mathon J 1991 *Contemporary Physics* **32** 143
[3] Burzo E, Chelkowski A and Kirchmayr H R 1990 *Landolt-Börstein New Series* **III/19d2** 248
[4] Buschow K 1979 *Rep. Prog. Phys.* **42** 1373
[5] Kittel C 1975 *Introduction to Solid State Physics* (New York: Wiley)
[6] Toliński T, Schäfer W, Kowalczyk A, Andrzejewski B, Hoser A and Szlaferek A 2004 *J. Alloys Compd.* **385** 28-32
[7] Coldea M, Pop V, Neumann M, Isnard O and Pascut G L 2005 *J. Alloys Compd.* **390** 16
[8] Müller Ch, Wonn H, Blau W, Ziesche P and Krivitshii P V 1979 *Phys. Status Solidi* B **95** 215
[9] Moulder F J, Stickle F W, Sobol E P and Bomben D K 1992 *Handbook of X–ray Photoelectron Spectroscopy* (Minnesota: Physical Electronics Inc)
[10] J S 1995 *J. Synchrotron Rad.* **2** 56.

# A spectroscopic study of dyes decomposition by irradiated nanocrystalline $TiO_2$

**M C Rosu, R C Suciu, I Kasco, S V Dreve, E Indrea and T D Silipas**

National Institute for Research and Development of Isotopic and Molecular Technologies, 65-103 Donath, 400293 Cluj-Napoca, Romania

E-mail: marcela.rosu@itim-cj.ro

**Abstract**. The photocatalytic activity of the $TiO_2$ films prepared by wet-chemical techniques and deposited on the ITO glass was tested by examining the decomposition of violet gentian (VG), methylen blue (MB) and methyl green (MG) aqueous solutions under UV light illumination. The photodegradation processes of dyes have been studied using UV–VIS spectrophotometry and spectrofluorimetry. The comparative results of the present study demonstrate the capability of UV irradiation of the $TiO_2$/ITO photocatalytic system to degrade studied organic dyes from aqueous solutions.

## 1. Introduction

Paper, textile, cosmetics, ink, ceramic, leather, food processing produces a large volume of dyes in wastewater. These dyes create human health and environmental pollution problems by releasing toxic substances into the aqueous phase, with carcinogenic potential [1].

Heterogeneous photocatalysis processes are one of the most efficient methods for destroying organic pollutants in aqueous media [2,3]. Recently, titanium dioxide is considered an excellent inorganic photocatalyst used in the advanced oxidation process for degradation of a number of different dyes by $TiO_2$ under UV irradiation [4].

In the present study, it was evaluated the photocatalytic degradation of triphenylmethane dye (violet gentian), tetramethylthionine chloride (methylen blue) and pyronin (methyl green) by the $TiO_2$/ITO photocatalytic system under the UV light by measuring the degradation rate of dye aqueous solutions.

## 2. Experimental

### *2.1. Materials and procedures*

Titania P-25 (ca. 80% anatase, 20% rutile) was kindly supplied by Degussa AG, Germany. The $TiO_2$ colloidal suspensions were prepared via wet-chemical techniques, using bi-distilled water, acetyl acetone (Merck, Germany), poly-propylene-glycol (Machery-Nagel, Germany) and Triton X-100 (Fluka, Switzerland) as additives [5]. Each sample (name samples NTC1, NTC2 and NTC3) was spread on a conductive indium tin oxide glass (ITO, 20 Ω/square) with Triton X-100 as the binder and then was dried in normal conditions and fired at 600°C in air for 90 min, with an increase rate of temperature of 5°C/min.

Violet gentian (Loba Chemie Fischamend, Austria), methylene blue (Fluka, Switzerland) and methyl green (May & Baker LTD, England) were used without any further purification. Solutions containing $3x10^{-5}$ M of every studied dyes in bi-distillated water were prepared by dissolution.

### *2.2. Instruments*

Optical characteristics of $TiO_2$ and the dyes solutions were examined with UV–VIS spectrophotometer and spectrofluorimeter. UV–VIS absorption spectra of the as-prepared photocatalysts were taken on a JASCO V-550 spectrometer. The fluorescence spectra were obtained using an ABL&Jasco V 6500 spectrofluorimeter with xenon lamp.

### *2.3. Irradiation experiments*

The photocatalytic degradation of violet gentian (VG), methylene blue (MB) and methyl green (MG) aqueous solutions 3 x $10^{-5}$ M have been studied in the presence of $TiO_2$/ITO photocatalytic system illuminated with a UV-Hg lamp (100 $mW/cm^2$) under continuous magnetic stirring. The reaction was performed in normal conditions. A 100 ml of dye solution was added in the photocatalytic system. Dye solutions samples were taken at different time intervals and were investigated by UV absorption spectroscopy. The irradiation time t was: 0, 20, 40, 60, 80, 100, 120, 140, 160, 180, 200 min.

The dyes degradation ratio was defined by the following formula (1):

$$\text{degradation\%} = \frac{ABS_0 - ABS_t}{ABS_0} \cdot 100 \quad (1)$$

where: $[ABS_0]$ is the initial absorbance of dyes; $[ABS_t]$ is the absorbance of dyes on time t.

## 3. Results and discussion

### *3.1. UV–VIS spectra*

Figure 1 shows UV–VIS absorption spectra of the $TiO_2$/ITO photocatalytic system and UV-VIS absorption spectra for dye solutions, before UV irradiation and after irradiation time t = 200 min.

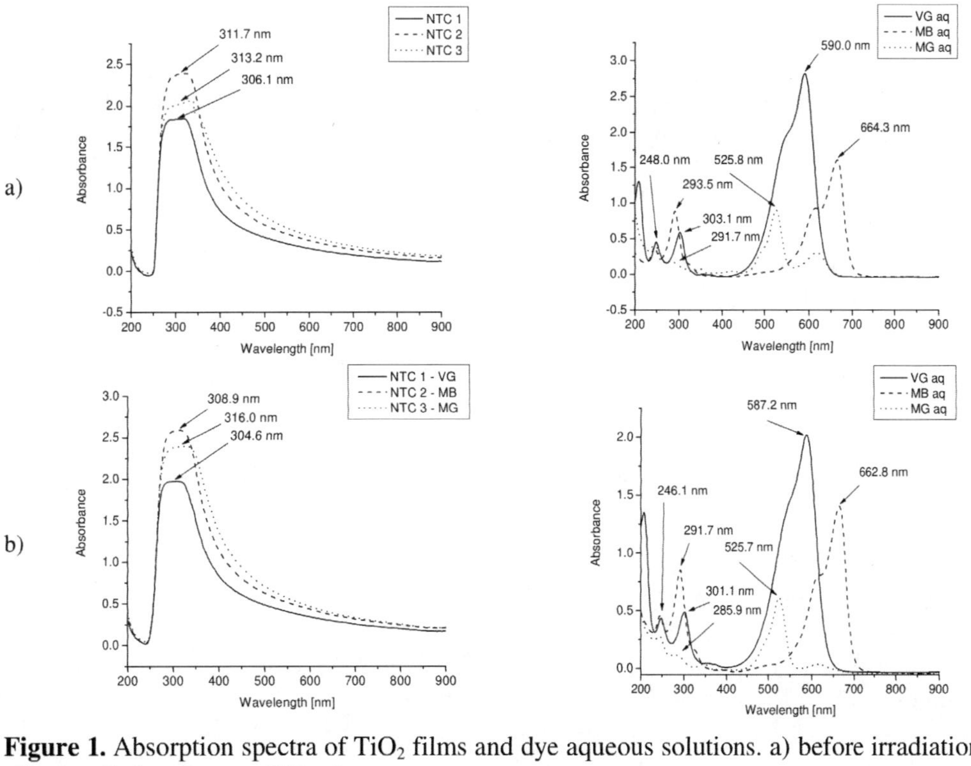


**Figure 1.** Absorption spectra of $TiO_2$ films and dye aqueous solutions. a) before irradiation. b) after irradiation time t = 200 min.

From figure 1 it can be seen that one absorbance peak (about 310 nm) appears for all of three $TiO_2$ films. After irradiation it was observed any peak displacements in UV-VIS spectra.

VG dye shows absorption peaks at 248.0 nm and 303.1 nm in UV region and one absorption peak at 590 nm in visible region. MB dye shows absorption peaks at 248.0 nm and 293.5 nm in UV region and one absorption peak at 664.3 nm in visible region. Also MG dye has an absorption peaks at 248 nm and 291.7 nm in UV region and one absorption peak at 525.8 nm in visible region. After irradiation no significant changes has been observed in the UV-VIS spectra of the dye aqueous solutions.

Figure 2 shows absorption changes when the dye aqueous solutions were degraded by the $TiO_2$ photocatalytic films. However, it is clear that a total degradation needs a longer reaction time.

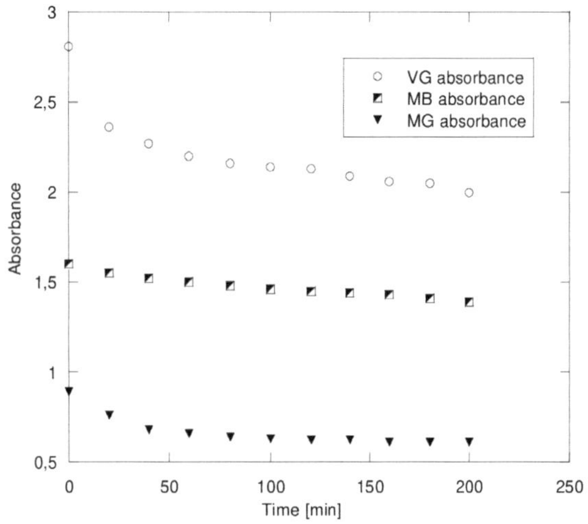


**Figure 2.** Absorbance time dependence under UV light irradiation of dye aqueous solutions (3 x $10^{-5}$M).

*3.2. Spectrofluorimetry spectra*

The photodegradation processes of dyes have been studied using spectrofluorimetry as well.

Figure 3a shows the spectrofluorimetric spectra of $TiO_2$ films before UV irradiation. Fluorescence spectra of the $TiO_2$ films are similar (fluorescence maxima 349-354 nm, excitation 318-333 nm). Figure 3b shows the spectrofluorimetric spectra of $TiO_2$ films after the contact of the dyes aqueous solutions under UV light.

The fluorescence excitation wavelengths and the fluorescence emission maxima of the $TiO_2$ films after irradiation are presented in table 1.

**Table 1.** The excitation wavelengths and the emission maxima of the $TiO_2$ films after UV irradiation time t = 200 min.

| Probe | Excitation wavelength [nm] | Emission maxima [nm] |
|---|---|---|
| NTC 1 – VG | 309 | 328.6; 345.9; 424.2; 469.1 |
| NTC 2 – MB | 316 | 343.5; 385.2; 414.9; 485.7 |
| NTC 3 – MG | 305 | 335.5; 392.9; 416.5; 463.5 |

*3.3. Study degradation ratio of the dyes*

Degradation efficiency of $TiO_2$/ITO photocatalytic system for the investigated dyes, evaluated with formula (1) is presented in figure 4. The catalytic activity after 200 min irradiation time followed the decreasing order 31.5%>28.8%>13.1%.

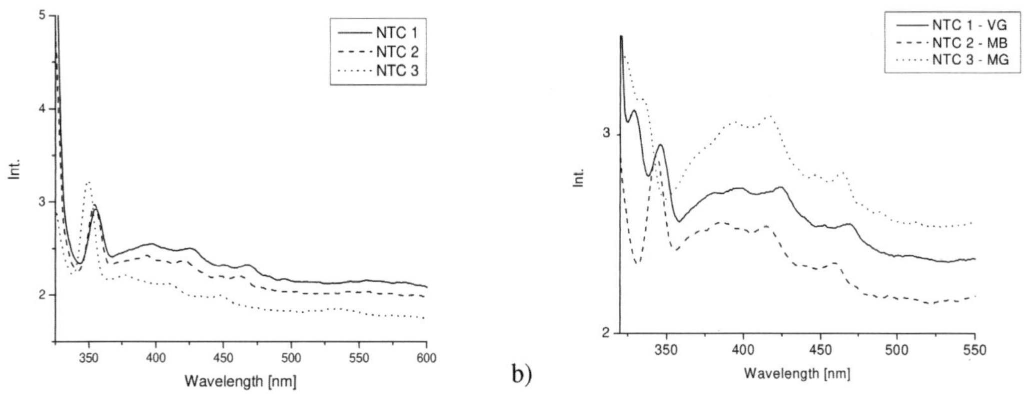


a) b)

**Figure 3.** Spectrofluorimetric spectra of $TiO_2$ films. a) before irradiation; b) after irradiation time t = 200 min.

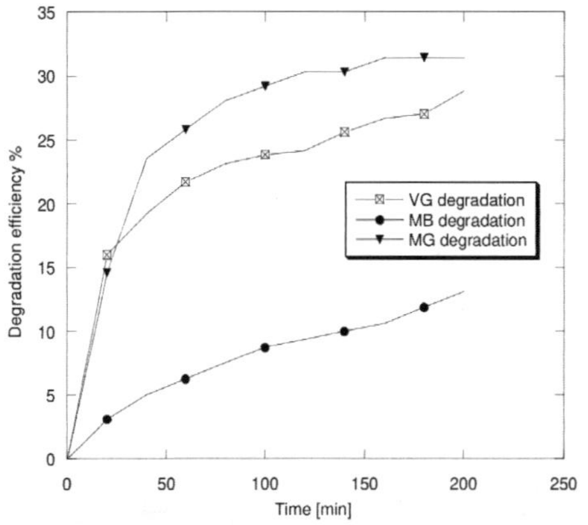


**Figure 4.** Degradation efficiency VG, MB and MG aqueous solutions with $TiO_2$ catalyst.

## 4. Conclusions

Experimental results indicated that the degradation of VG, MB and MG dyes increased with increasing the irradiation time. Fluorescence of the active fluorescent dyes at excitation at the UV-VIS maxima absorption was put also into evidence.

The results of the present study demonstrate the capability of $TiO_2$/ITO photocatalytic system under UV irradiation to degrade the same organic dyes.

## References

[1] Tezcanli-Guyer G and Ince N H 2003 *Ultrason. Sonochem.* **10** 235
[2] Andreozzi R, Caprio V, Insola A and Marotta R 1999 *Catal. Today* **53** 51
[3] Hermann J M 2005 *Top. Catal.* **34** 49
[4] Yamazaki S, Matsunaga S and Hori K 2001 *Water Res.* **35** 1022
[5] Myung Soon Lee, Ik Chan Cheon and Yeong Il Kim 2003 *Bull. Korean Chem. Soc.* **24** 18.

# Dispersion of carbon nanotubes by single-stranded DNA wrapping for advanced biomedical applications

**S Simon[1], A R Biris[1], D M Lupu[1], I Misan[1], S Clichici[2], T Mocan[2] and A S Biris[3]**

[1] National Institute for Research and Development of Isotopic and Molecular Technologies, 65-103 Donath, 400293 Cluj-Napoca, Romania

[2] University of Medicine and Pharmacy "Iuliu Hatieganu", Department of Physiology, 1-3 Clinicilor, 400006 Cluj-Napoca, Romania

[3] UARL Nanotechnologi Center, University of Arkansas, 2801 S University Ave, Little Rock, AR 72204, USA

E-mail: stefania.simon@itim-cj.ro

**Abstract.** Carbon nanotubes (CNT) are novel materials with unique electrical, thermal, and structural proprieties, which make them attractive for applications in composites, medicine, biology, electronics, and energy management. There are more and more attempts to use CNT, especially single-walled carbon nanotubes (SWNT), in various biomedical applications, which require highly purified samples, with no contents of catalyst residual metal particles, good dispersion in an aqueous solution at relatively high concentrations. This report presents all these three main requirements as proved by Raman spectroscopy, UV-Vis-NIR spectroscopy and thermogravimetrical analysis. The maximum achieved concentration of SWNTs dispersed in water when wrapped with single-stranded DNA (ss-DNA) was of 400 mg/l.

## 1. Introduction

Carbon nanotubes (CNTs) are novel structures formed from a hexagonal network of carbon atoms, rolled into a shape of a perfect cylinder, with diameters from 0.7 to a few nanometers. CNTs may have one, two or more walls, of one atom thick graphitic layers, depending on the synthesis conditions used. Due to their unique electronic and mechanical proprieties [1] and size, they have attracted considerable attention over the past decade, for their potential applications in various areas such as: biosensors [2], hydrogen storage media [3], composites [4] and actuators [5]. Recently [6], the possibility of photo-thermal destruction of cancer cells was demonstrated by using SWNTs. This potential biomedical application branches from the property of SWNTs to heat rapidly by absorption of infra-red radiation without harming the human tissue because it is transparent to infra-red radiation.

In biomedical applications it is imperative to examine their potential toxicity before further use. *In vitro* and *in vivo* tests for SWNTs cytotoxicity require aqueous media of well dispersed nanotubes which is difficult because of their hydrophobic nature and poor solubility in organic solvents [7]. Moreover, they are organized in bundles held together by strong van der Waals interactions so their individual separation and dispersion still remains a challenge. Several covalent [8] and non-covalent [9] surface functionalization methods have been used to make SWNTs water soluble and suitable for biological and medical research, especially in what concerns surfactants toxicity. The non-covalent approaches are more proper because it does not suppose the CNTs lattices interruption. Until now, stable individual dispersion of SWNT have been obtained with surfactants like: Triton X-100, NaDDBS, SDS [10] or biomolecules like BSA or chitosan [11], but the best results have been obtained when wrapped with DNA fragments [12]. Zheng *et al* [13], revealed that single-stranded DNA (ss-DNA) binding to CNT is extremely effective and facile at low concentrations.

The purpose of this paper is to obtain stable homogeneous and highly concentrated aqueous dispersions of individual SWNT by wrapping them with ss-DNA.

## 2. Materials and methods

SWNTs were synthesized using an original version of catalytic chemical vapor deposition method based on induction heating (RF-CCVD), previously developed by our group [14], from the pyrolysis of acetylene, at 850°C, over a bi-metallic catalyst system Fe–Mo supported on MgO. The Fe–Mo/MgO catalyst with the chemical weight ratio composition of Fe:Mo:MgO = 1.4:0.14:98.46 (%wt), was produced by using the co-precipitation method [15]. The carbonaceous products obtained were purified in a Soxhlet apparatus with HCl (1:1) for 24 h, washed with distilled water, and dried overnight at 150°C. After the purification, the SWNTs morphology was investigated by Raman spectroscopy, which was performed at room temperature using the Jasco NRS-3300 Raman Spectrometer with a 633 nm laser. Thermogravimetrical analyses (SDT Q600 TA Instruments) were performed in air at a flow rate of 100 ml/min and a heat rate of 5°C/min.

The ss-DNA from salmon testes, in water, was purchased from Sigma-Aldrich. Dispersion of SWNTs by ss-DNA was carried out as described previously [16]. Briefly, 5 mg purified SWNT was mixed with 5 ml ss-DNA 0.1% solution and sonicated with a cup-horn sonicator (Sonics, Vibra-Cell VC505, 500 W, 20 kHz) on an ice-water bath for 15 min at 30% amplitude. The mixture was then centrifugated (Sigma) 1 h at 4000 g, in order to remove undispersed SWNTs, and the supernatant was collected and kept for some time. Optical absorption measurements were recorded using a Jasco V-570 UV-Vis-NIR Spectrophotometer, in 1 cm quarts cuvette, at a dilution of 1/40, for both, ss-DNA 0.1% solution and the collected supernatant.

## 3. Results and discussion

Raman spectra for SWCNTs (figure 1) present bands in two characteristic spectral regions: at low frequency (100-400 $cm^{-1}$), or the radial breathing mode (RBM) region, and at high frequency D, G and G' bands. While D, G, and G' modes are also found in graphite, the RBM is a unique CNT mode. It is the real signature of the presence of carbon nanotubes in a sample and strongly depends on the diameter and chirality of the nanotubes. The RBM frequency (ω) is inversely proportional to the CNT diameter that can be calculated with equation (1) [32].

$$\omega\left(_{cm^{-1}}\right) = \frac{223.5}{d(nm)} + 12.5 \qquad (1)$$

The differences, in Raman spectra, between SWNT after the purification step and SWNT dispersed with ss-DNA consist only in the RBM region. In RBM band picks intensity is proportional with existing quantity in the sample. The ss-DNA selectively extracts certain SWNT diameters. Spectral position and corresponding diameter values for our samples (SWNTs before and after dispersion) are represented in Table 1. D band (between1305 and 1330 cm-1) is associated to the vacancies, defects, and carbonic impurities (amorphous carbon, glassy carbon) that destroy the symmetry, and in this case its intensity is low meaning that our SWNT contain almost no defects and no carbonic impurities [17].

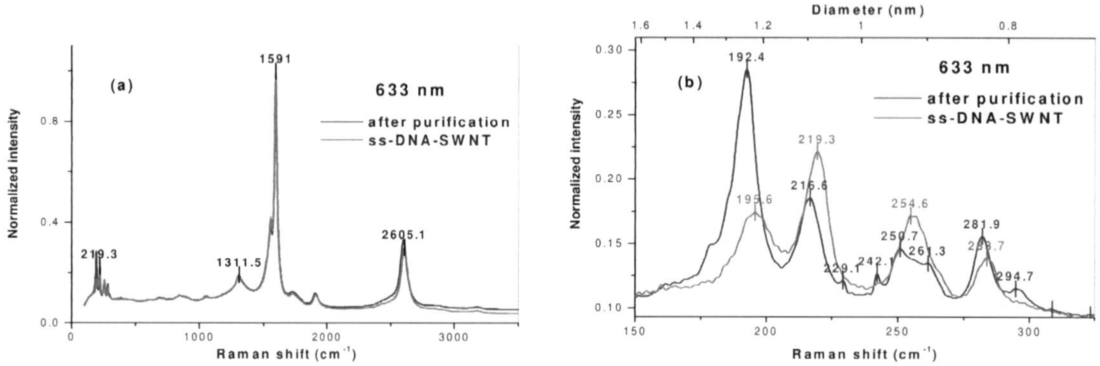


**Figure 1.** Raman analysis spectra at 633 nm (a) and magnified RBM band (b) for SWNTs right after the purification (blue) and after the dispersion with ss-DNA (red).

In figure 2, the thermogravimetrical analysis (TGA) was performed to confirm SWNT's purification step. The final purity of the SWNTs was of 96.97%. The TGA curves indicate the existence of a single mass loss profile for all reaction temperatures, which indicates that all the analyzed nanotubes samples are composed of a single carbonaceous type.

**Table 1.** Spectral position and corresponding diameter values from RBM band at $\lambda_{exc}$ = 633 nm.

| SWNT after purification | | SWNT dispersed with ss-DNA | |
|---|---|---|---|
| $\omega_{RBM}$ (cm$^{-1}$) | d (nm) | $\omega_{RBM}$ (cm$^{-1}$) | d (nm) |
| 192.4 | 1.24 | 195.6 | 1.22 |
| 216.6 | 1.09 | 219.3 | 1.08 |
| 229.1 | 1.03 | - | - |
| 242.1 | 0.97 | - | - |
| 250.7 | 0.94 | 254.6 | 0.92 |
| 261.3 | 0.90 | - | - |
| 281.9 | 0.83 | 283.7 | 0.82 |
| 294.7 | 0.79 | - | - |

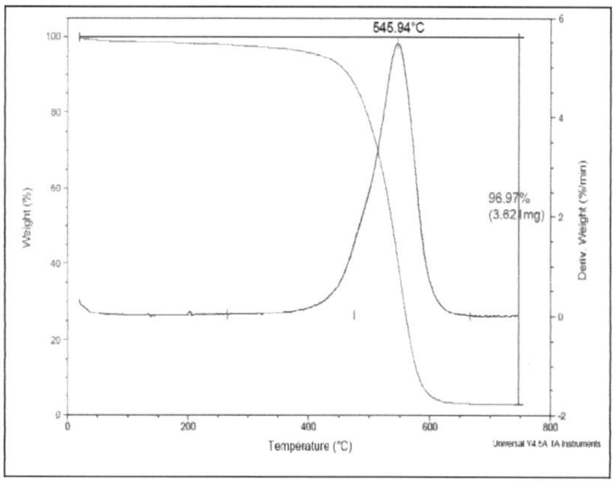


**Figure 2.** The TGA and the derivative TGA (dW/dT) profiles of the purified SWNTs.

The SWNTs concentration in the collected supernatants aqua solution after the dispersion process can be evaluated from the Beer-Lambert equation by deeming the absorbance [18, 19].

Concentration evaluation from the absorbance of the ~320 nm pick is not possible because it is due not only to the CNT absorption but also to the absorption of the $\pi$ plasmons and the ss-DNA. As it can bee seen in the absorption spectra of ss-DNA (figure 3), the ss-DNA absorbs at the same wavelength range of 200-320 nm [19]. The concentration obtained, from the absorption spectra of SWNT dispersed with ss-DAN (figure 3), in agreement with Ryabenko *et al* [19], is 400 mg/l for SWNT dispersed with ss-DNA.

The SWNT identified from the absorption spectra of ss-DNA dispersed SWNTs (figure 4), following the method described in literature [19, 20], have the diameters: 0.8 nm, 0.95 nm, 1.0 nm, 1.20 nm. Larger diameters can be identified from the absorption spectra but those tubs diameters do not appear in the Raman spectra because are not resonant at 633 nm laser frequency.

Journal of Physics: Conference Series **182** (2009) 012079 doi:10.1088/1742-6596/182/1/012079

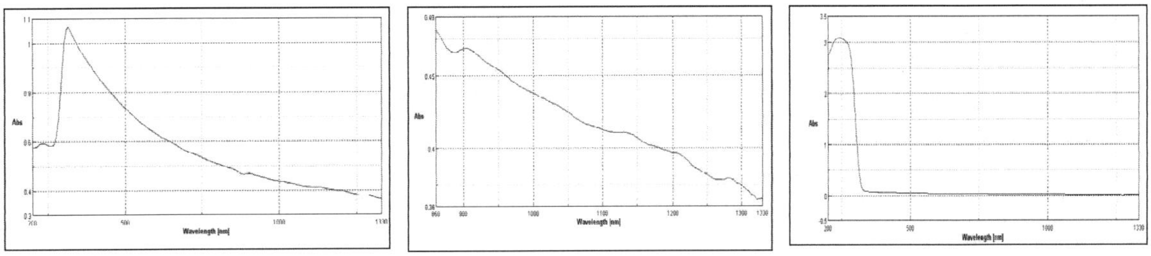

**Figure 3.** Absorption spectra for ss-DNA dispersed SWNTs (left). Fragment from the absorption spectra of ss-DNA dispersed SWNTs (center). Absorption spectra for ss-DNA (right).

## 4. Conclusions

The purpose of this research was to obtain stable homogeneous and concentrated aqueous dispersions of individual SWNT. The reported concentration of SWNTs dispersed in water with ss-DNA was found to be 400 mg/l and the suspension was stable for several months when stored at 4°C. It is currently being used to determine the SWNT potential toxicity and oxidative stress on culture cells and animal models. The SWNTs purity was demonstrated by thermo-gravimetrical analyses and confirmed once again by the low intensity of the D band in the Raman spectra.

## References

[1] Shuhui Q, Dongqi F, Warren T, Herrera J E and Weisman R B 2004 *Macromolecules* **37** 3965
[2] Hu C, Zhang Y, Bao G, Zhang Y, Liu M and Wang Z L 2005 *J. Phys. Chem.* B **109** 2007
[3] Lawrence J and Xu G 2004 *Appl. Phys. Lett.* **84** 918
[4] Bower C, Rosen R, Lin J, Han J and Zhou O 1999 *Appl. Phys. Lett.* **74** 3317
[5] Vohrer U, Kolaric I, Haque M H, Roth S and Detlaff-Weglikowska U 2004 *Carbon* **42** 1159
[6] Kam N V S, O'Connell M, Wisdom J A and Dai H 2005 *PNAS* **102** 11600
[7] Mickelson E T,Smalley R E, Hauge R H and Margrave J L 1999 *J. Phys. Chem. B* **103** 4318
[8] Yang W, Thordarson P, Gooding J J, Ringer S P and Braet F 2007 *Nanotechnology* **18** 1
[9] Zhao W, Song C and Pehrsson P E 2002 *J. Am. Chem. Soc.* **124** 12418
[10] Karajanagi S S, Yang H, Asuri P, Sellitto E, Dordick J S and Kane R S 2006 *Langmuir* **22** 1392
[11] Moulton S E, Minett A I, Murphy R, Ryan K P and Wallace G G 2005 *Carbon* **43** 1879
[12] Zheng M, Jagota A, Semke E D, Diner B A, Mclean R S, Lustig S R, Richardson R E and Tassi N G 2003 *Nat. Mat.* **2** 338
[13] Lupu D, Biris A R, Jianu A, Bunescu C, Burkel E and Mihailescu G 2004 *Carbon* **42** 503
[14] Lupu D 2003 *Studia Univ. Babes-Bolyai, Physica* **1** 142
[15] Biris A R, Biris A S, Lupu D, Trigwell S, Dervishi E and Rahman Z 2006 *Chem.Phys. Lett.* **429** 204
[16] Jorio A, Pimenta M A, Souza Filho A G, Saito R, Dresselhaus G and Dresselhaus M S 2003 *New J. Phys* **5** 139
[17] Jeong S, Kim K K, Jeong S J, An K H, Lee S H and Lee Y H 2007 *Synthetic Metals* **157** 570
[18] Landi B J, Ruf H J, Evans C M, Cress C D and Raffaelle R P 2005 *J. Phys. Chem.* B **109** 9952
[19] Ryabenko A G, Dorofeeva T V and Zvereva G I 2004 *Carbon* **42** 1523
[20] Lebetkin S, Hennrich F, Skipa T and Kappes M 2003 *J. Phys. Chem.* **107** 1949.

Journal of Physics: Conference Series **182** (2009) 012080 doi:10.1088/1742-6596/182/1/012080

# $TiO_2$ thin films prepared by sol - gel method

**R C Suciu[1], E Indrea[1], T D Silipas[1], S Dreve[1], M C Rosu[1], V Popescu[2], G Popescu[2] and H I Nascu[2]**

[1] National Institute for Research and Development of Isotopic and Molecular Technologies, 65-103 Donath, 400293 Cluj-Napoca, Romania

[2] Technical University of Cluj-Napoca, Physics Department, 15 C Daicoviciu, 400020 Cluj-Napoca, Romania

E-mail: ramona.suciu@itim-cj.ro

**Abstract**. There is a growing awareness that titania ($TiO_2$) and $TiO_2$-based oxide systems are the most promising candidates for the development of photoelectrodes for photoelectrochemical cell (PEC) for solar-hydrogen production [1]. The PEC is equipped with a single photoelectrode (photoanode) and cathode, both of which are immersed in an aqueous electrolyte. In this work we present a sol-gel method to prepare $TiO_2$ thin films on ITO using tetraisopropoxides of titanium, acetylacetone, 1-butanol and Tween 80 as surfactant. The films were deposited on ITO coated glass slides by spray pyrolysis. UV-VIS spectra and fluorescence measurements were made for the solutions and films. X-ray diffraction was used for structural investigations and the morphology of the film was studied by Scanning Electron Microscopy.

## 1. Introduction

$TiO_2$ has found wide and important applications in many fields of chemical engineering and materials engineering including either traditional catalysis, or photocatalysis, dye-sensitized solar cells, lithium-insertion-based devices, integrated circuits, gas sensors and in the paint industry [2]. Most of these applications are a consequence of its n-type semiconducting property and realized with micro or nano structured $TiO_2$ powders or thin films.

A number of methods have been employed to prepare $TiO_2$ films, including e-beam evaporation, sputtering, chemical vapor deposition and sol-gel process. The sol-gel conventional method uses the hydrolytic route, which involves the initial hydrolysis of the alkoxide precursor followed by continual condensations between the hydrolysed particles forming the gel.

This process is carried out at room temperature, and the desired morphological properties of the particles are obtained by controlling the conditions under which the synthesis is carried out. Moreover, sol-gel processing route is particularly attractive for the scaling-up of oxide thin films fabrication, since the liquid precursor can easily be applied on a substrate by dipping, spinning or spraying [3] and heat treated at lower temperatures.

The spray-coating technique potentially offers the advantage of conformal film deposition on non-planar structures (*e.g.* steps, stacks or trenches) on semiconductor chips [4].

In this paper, we report the preparation of $TiO_2$ films using the hydrolytic sol-gel process to obtain the oxide and the spray-coating technique to deposit the films on ITO glass substrates.

## 2. Experimental

### *2.1. Preparation of coating solution*
Firstly, 1-butanol sol was made, which include 0.496M commercial ultrapure titanium isopropoxide (TTIP, Fluka), 0.28M acetylacetone (99%, Alfa Aesar), 0.92M $H_2O$ and 1-butanol (absolute). Subsequently, Tween 80 and 1-butanol solvent were added to the 1-butanol sol. During the sol preparation, the alkoxide solution was vigorously stirred at room temperature, so as to keep a homogeneous mixture of the chemical compositions.

### *2.2. Deposition and thermal treatment of precursors films*
The film consists on five layers of $TiO_2$ was obtained by spray pyrolisis. The as obtained films were heat-treated at 600°C, 1 h.

### *2.3. Analysis*
UV–VIS absorption spectra of the $TiO_2$ coating solutions and thin films deposited on ITO glass were taken on a JASCO V-550 spectrometer.

X-ray diffraction (XRD) measurements were performed using a BRUKER D8 Advance X-ray diffractometer, working at 45 kV and 45 mA. The Cu $K_\alpha$ radiation, Ni filtered, was collimated with Soller slits. A germanium monochromator was used. The data of the X-ray diffraction patterns were collected in a step-scanning mode with steps of $\Delta 2\theta = 0.01°$. Pure silicon powder (standard sample) was used to correct the data for instrumental broadening. The microstructural informations obtained by single X-ray profile Fourier analysis of the $TiO_2$ anatase nanoparticles were the effective crystallite mean size ($D_{eff}$) and the root mean square (rms) of the microstrains, averaged along the [*hkl*] direction, $<\varepsilon^2>^{1/2}hkl$ [5]. The Warren-Averbach X-ray profile Fourier analysis of the (101) and (200) anatase peak profiles were processed by the XRLINE [6] computer program. The unit cell parameters were calculated by Rietveld refinement using the PowderCell software [7]. PowderCell program enables a quantitative phase (volume fractions) analysis method by comparison of the different scattering powers of the component materials.

The morphology of the films was investigated by Scanning Electron Microscopy using a JSM 5600 LV field emission – high resolution scanning electron microscope equipment with an Oxford INCA Crystal electron backscattering diffraction (EBSD) systems.

## 3. Results and Discussion

### *3.1. Precursor characterization*
The X-ray diffraction pattern of the dried (100°C) hydrolyzed precursors, shown in figure 1, evidences its amorphous nature with a slight crystallization tendency. The powder prepared by sol-gel method was quasi-amorphous to X-ray as long as it was calcined below 500°C. On heating to 500°C, for 1h the reflections corresponding to titania anatase are detected (figure 1).

Figure 2 shows the effective crystallite size distribution for the powder obtain by the heat treatment of the precursor solution at 500°C, respectively 600°C.

### *3.2. Films characterization*
The UV absorption property of $TiO_2$ films is a important factor for the photocatalyst. The UV spectra of $TiO_2$ films before and after calcinations were shown in figure 3. It has been reported that the band-gap electronic transition of anatase $TiO_2$ is indirect [8]. After heat treatment the intensities of absorption peaks of $TiO_2$ film increased and the peak position slightly shifted to a higher wavenumber, because formation of $TiO_2$.

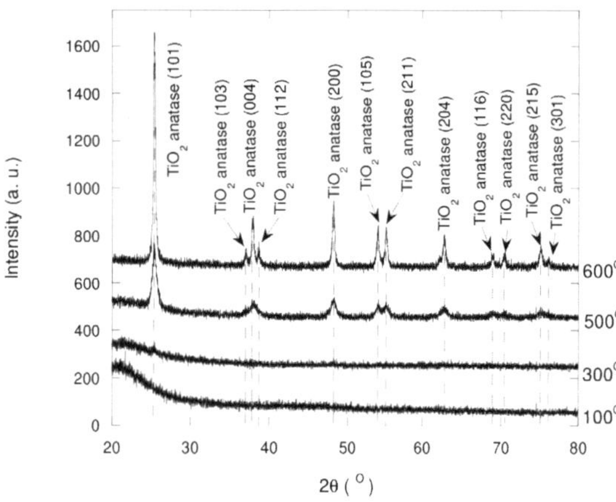


**Figure 1**. The X-ray diffraction pattern of the titania precursor dried at 100°C, and calcined at 300°C, 500°C and 600°C.

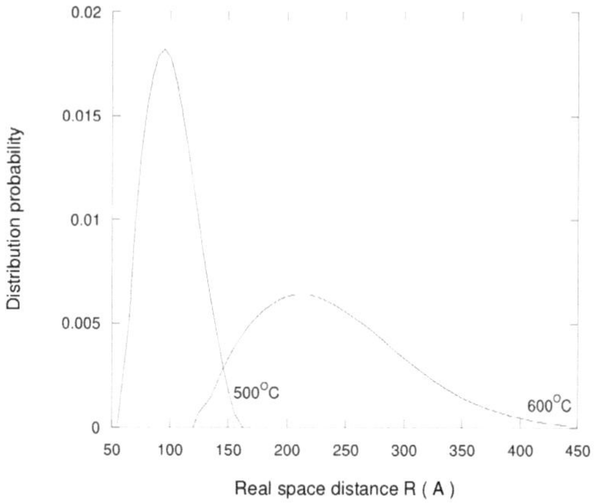


**Figure 2.** Effective crystallite size distribution along the [101] crystallographic direction for the $TiO_2$ anatase structure.

With regard to the relationship between the absorption coefficient α and the incident photon energy *hv* near the band edge, one can write out a good approximation [8] as $(\alpha \times hv)^{1/2} = A\ (hv - Eg\ )^{1/2}$, where the photon energy is *hv*, *h* being the Planck constant, and Eg is the indirect optical band-gap. From the function curve $(\alpha \times hv)^{1/2}$ vs *hv*, shown in figure 4, the band gap energy of indirect transition is calculated to be about 3.45 eV, which is larger than of 3.2 eV reported for the bulk $TiO_2$ anatase, indicating a quantum size effect [8].

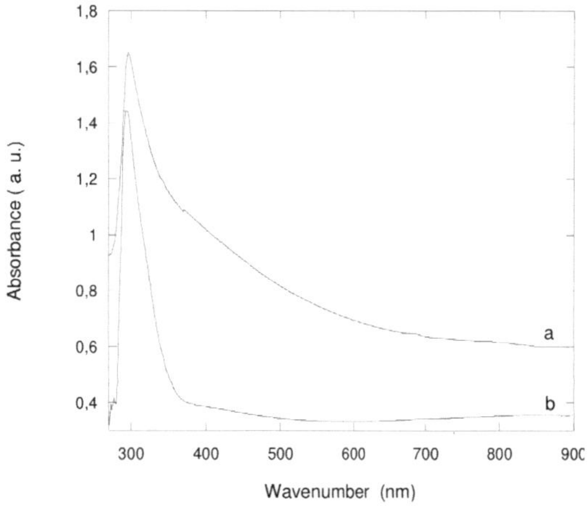


**Figure 3.** UV-VIS measurement of $TiO_2$ thin films before (a) and after (b) the heat treatment.

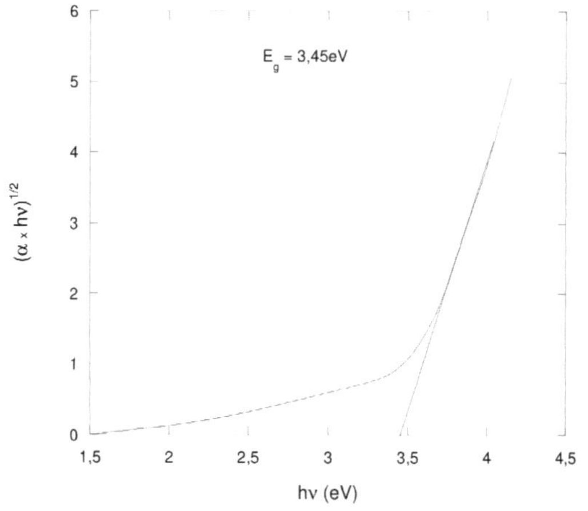


**Figure 4**. $(\alpha hv)^{1/2}$ as a function of *hv* for the $TiO_2$ film after the heat treatment.

The diffraction pattern of the $TiO_2$/ITO thin film (figure 5) exhibit the diffraction peaks owning to anatase phase however, thiny amount of the rutile phase (7.4% volume fraction) at 2θ of 27.43° was also detected. The value of the anatase particles size are $D_{eff}$ = 5.4 nm, indicating low crystallinity of the anatase phase.

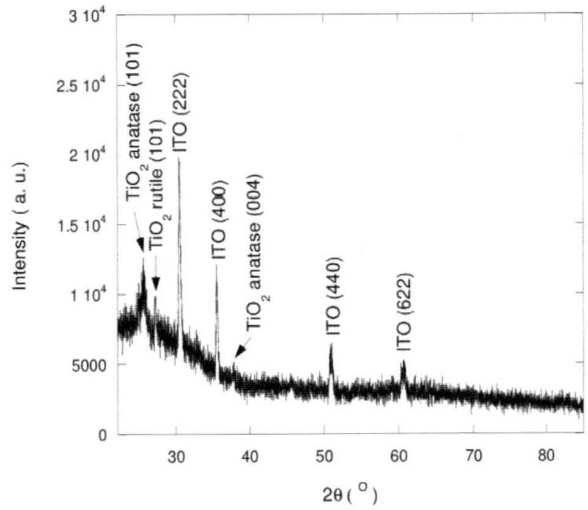


**Figure 5.** The XRD patterns of $TiO_2$/ITO thin film.

The films were free from the pinholes and cracks as generally observed in sprayed films due to its high deposition temperature (figure 6). Formation of was study SEM using quantitative analyses such as EDX (figure 7). This analysis confirms the $TiO_2$ thin films formation also.

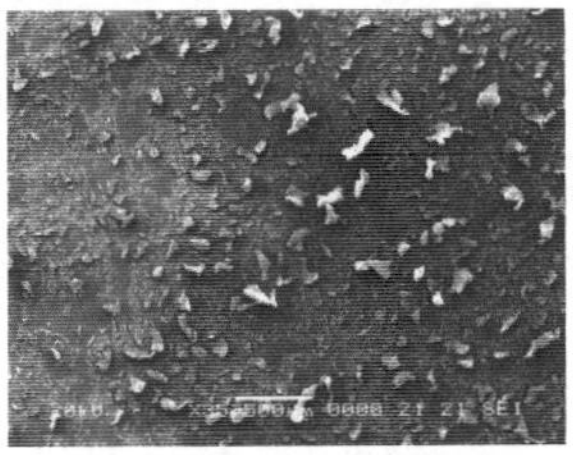

**Figure 6.** SEM images of anatase $TiO_2$.

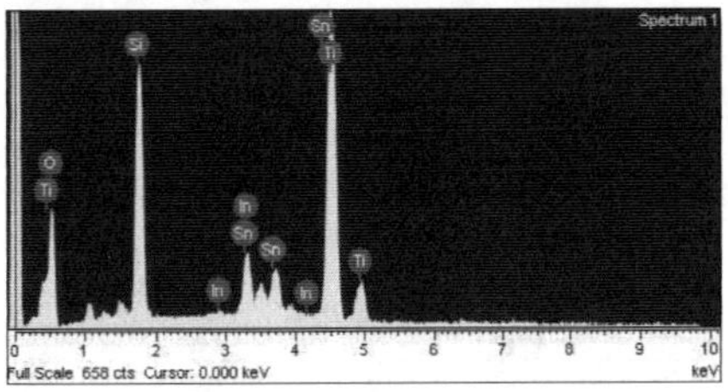


**Figure 7.** EDX measurements of $TiO_2$ thin film.

## 4. Conclusions

$TiO_2$ thin films were prepared by a sol-gel spray coating process using titanium alkoxide. XRD analysis of our titanium precursor powder shows that starting from 500°C annealing temperature the $TiO_2$ anatase is the main crystalline phase. The band gap energy of indirect transition of the $TiO_2$ thin films is calculated to be about 3.45 eV, indicating a quantum size effect. Our nano structured $TiO_2$ thin films which has a little red-shifted compared with the band-gap energy of the $TiO_2$ indirect electronic transition may be a more efficient candidates in the development of photoelectrodes for photoelectrochemical cell (PEC) used in solar-hydrogen production.

## References

[1] Nowotny J, Sorrell C C, Sheppard L R and Bak T 2005 *Int. J. Hydrogen Energy* **30** 521-44
[2] Abou-Helal M O and Seeber W T 2002 *Appl. Surf. Sci.* **195** 53-62
[3] Knoth K, Schlobach B, Hunne R, Schultz L and Holzapfel B 2005 *Physica* C 426-431
[4] Schwartz R W, Schneller T and Waser R 2004 *C. R. Chimie* **7** 433-461
[5] van Bercum J G M, Vermeulen A C, Delhez R, T H de Keijser and Mittemeijer E M 1994 *J. Appl. Phys.* **27** 345-353
[6] Aldea N and Indrea E 1990 *Comput. Phys. Commun.* **601** 55-159
[7] Kraus W and Nolze G 1996 *J. Appl. Crystallogr.* **29** 301-303
[8] Linsebigler A L, Lu G Q and Yates J T 1995 *Chem. Rev.* **95** 735-758.

# Smart composites based on magnetic nanoparticles and responsive polymers

**R Turcu, A Nan, I Craciunescu, O Pana, C Leostean and S Macavei**

National Institute for Research and Development of Isotopic and Molecular Technologies, 65-103 Donath, 400293 Cluj-Napoca, Romania

E-mail: rodica.turcu@itim-cj.ro

**Abstract.** A comparative study of the properties of smart composites obtained by the combination of $Fe_3O_4$ nanoparticles with biocompatible responsive polymers is reported. Two types of magnetic composites with different morphology were obtained: (*i*) as core-shell nanostructures consisting of $Fe_3O_4$ functionalized by serine and covered with biocompatible polymer shell poly(caprolactone); (*ii*) as smart ferrogels, by the encapsulation of $Fe_3O_4$ nanoparticles into chemically cross linked polymer like poly (N-isopropylacril amide). The composites formation and their characteristics as structure, chemical composition, magnetic behaviour were investigated by XRD, FTIR, XPS and magnetic measurements. The magnetic properties of the smart composites tailor as a function of magnetite content. Both types of magnetic composites show superparamagnetic behaviour. Higher magnetization values were obtained for the composites with the functionalized magnetite core covered by poly(caprolactone) as compared with the ferrogels.

## 1. Introduction

Smart materials obtained by mixing the magnetic nanoparticles with an amphiphilic biocompatible polymer attracted a special interest because of the large applicability: cellular separation, the increase of the contrast in NMR imaging, hyperthermia treatment of the cancerous tumors, protein purification, targeting the drug transport, fixing the retina damages, artificial muscles, temperature and pH sensors, micro-fluidic devices [1].

In this work we present a comparative study of the properties of smart composites obtained by the combination of $Fe_3O_4$ nanoparticles with biocompatible responsive polymers. Magnetic composites with different morphology are investigated: (*i*) as core-shell nanostructures consisting of $Fe_3O_4$ functionalized by serine and covered with biocompatible polymer shell poly(caprolactone), PCL; (*ii*) as smart ferrogels, by the encapsulation of $Fe_3O_4$ nanoparticles into chemically cross linked polymers like poly (N-isopropylacril amide), pNIPAM.

## 2. Experimental

### *2.1. Samples preparation*

The synthesis of core-shell nanostructures with magnetite core covered with polymer or copolymer shell consist first in the preparation of magnetite nanoparticles by alkaline hydrolysis of a ferrous and ferric chloride aqueous solution using a common precipitation process under inert ($N_2$) atmosphere. Magnetite particles were surface-functionalized *in situ* by the addition of *L*-serine. The magnetic nanoparticles were mixed with freshly distilled ε-caprolactone and strongly shacked for 30 min at 1200 rotation/min. The surface initiated ring opening polymerization (ROP) of ε-caprolactone was initiated by the common catalysts, tin (II) 2-ethylhexanoate ($Sn(oct)_2$), resulting in magnetite coated with poly(caprolactone), $Fe_3O_4$/PCL.

The smart ferrogels were prepared using ultrastable dispersions of $Fe_3O_4$ nanoparticles coated by double layer of lauric acid (LA) as ferrofluids and chemically cross linked polymer pNIPA. The monomer N-isopropylacrylamide (NIPA) and the N,N'-methylenbisacrylamide (BIS), like catalyst were dissolved in 3 ml of deionized water. After $N_2$ gas bubbling for 10 min to deoxygenate the solution, different concentrations of ferrofluid were added and sonicated for 30 min. Ammonium peroxydisulfate (APS), 1 wt% as redox accelerator was added to the solution. The mixture was kept to react for 2 h at 60°C and after that, 24 h at room temperature. After the gelation was completed, the ferrogels was immersed in an excess amount of deionized water for 2 days to remove the residual unreacted monomers. The pNIPA ferrogels were dried for 3 days at 60°C to a constant weight.

### *2.2. Characterization methods*

The X-ray diffraction (XRD) pattern of magnetic nanoparticles was obtained using a Diffractometer DRON 3M X-Ray, working at 35 kV and 35 mA. The Co $K_\alpha$ radiation, Fe filtered was collimated with Soller slits. FTIR spectra were carried out on a JASCO FTIR 610 spectrophotometer. X-ray Photoelectron spectra were recorded using a XPS spectrometer SPECS equipped with a dual-anode X-ray source Al/Mg, a PHOIBOS 150 2D CCD hemispherical energy analyzer and a multi-channeltron detector. The Mg X-ray source operated at 200 W was used for XPS investigations. The samples powder was pressed on an indium foil to allow the XPS measurements. The magnetic measurements were performed at room temperature using a Vibrating Sample Magnetometer Cryogenics.

## 3. Results and discussion

Figure 1 shows the XRD pattern of magnetic nanoparticles functionalized with serine. The characteristic peaks of magnetite are observed in the XRD spectrum from the figure 1. The microstructural parameters, namely effective crystallite mean size, $D_{eff}$(nm) and the root mean square (rms) of the microstrains averaged along the [hkl] direction, $<\varepsilon^2>^{1/2}_{hkl}$ were obtained by single X-ray profile Fourier analysis of the $Fe_3O_4$ sample [2]. The analysis of single (220) and (440) X-ray diffraction profiles gives the following values of the microstructural parameters for $Fe_3O_4$ functionalized with serine: $D_{eff}$ = 9.4 nm; $<\varepsilon^2>^{1/2}_{hkl} = 0.546 \times 10^2$.

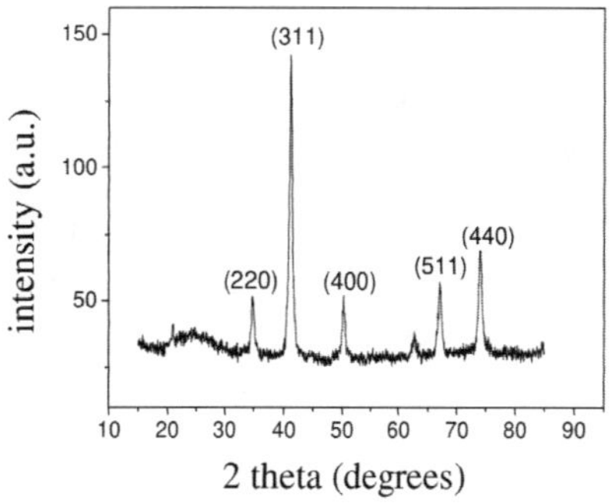


**Figure 1.** XRD pattern of magnetic nanoparticles functionalized with serine.

In the figure 2 are given the FTIR spectra of magnetite functionalized with serine ($Fe_3O_4$/serine), the composite $Fe_3O_4$/PCL and that of PCL respectively. The comparison of these spectra shows that the spectrum of the composite $Fe_3O_4$/PCL contains the characteristic absorption bands of magnetite and of PCL respectively, which evidence the composite formation.

From the FTIR spectrum of ferrogels, figure 3 one can observe the absorption bands characteristic of magnetite ($Fe_3O_4$) located around 580 $cm^{-1}$ and of NIPA respectively: a carbonyl stretching vibration (amide I) at 1650 $cm^{-1}$, N–H bending vibration (amide at 1545 $cm^{-1}$) and two typical peaks of C–H vibration of $-CH(CH_3)_2$ at 1372 $cm^{-1}$ and 1385 $cm^{-1}$, a broad peak between 3440 $cm^{-1}$ and 3290 $cm^{-1}$ due to N–H stretching of NIPA repeating units.

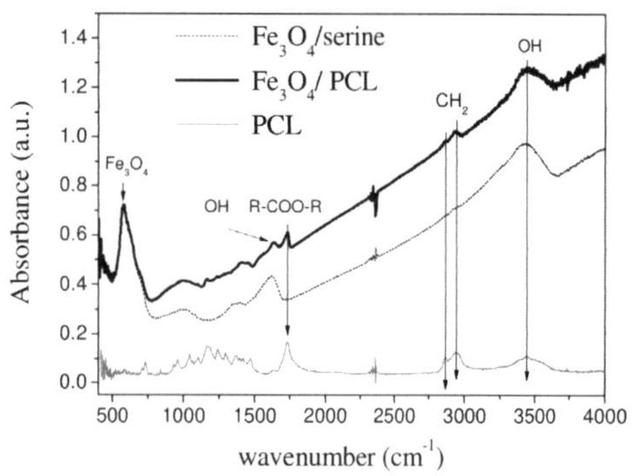


**Figure 2.** FTIR spectra of magnetite functionalized with serine ($Fe_3O_4$ /serine), the composite $Fe_3O_4$ /PCL and PCL.

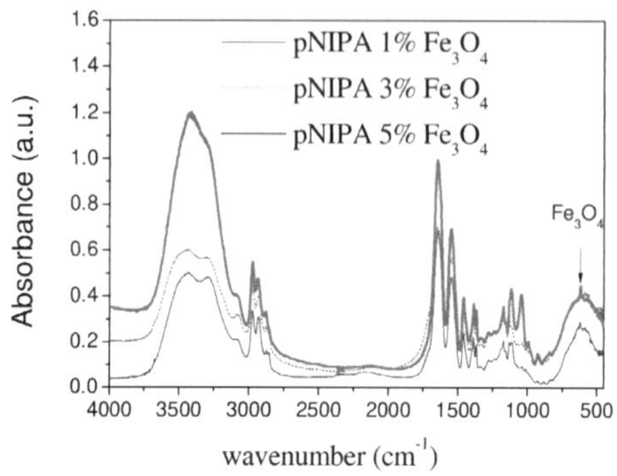


**Figure 3.** FTIR spectra of pNIPA ferrogels prepared with different concentrations of $Fe_3O_4$ from ferrofluid.

XPS measurements give the surface composition of functionalized magnetite nanoparticles. The high resolution spectra of C 1s, O 1s and Fe 2p for magnetite functionalized with serine are given in the figure 4. No contribution from N1s core level was observed in the XPS spectrum of magnetite functionalized with serine, due to the deamination of serine in the presence of iron salt. The high resolution spectrum for C 1s, figure 4, can be deconvoluted into three Gaussian peaks corresponding to carbon atoms from different groups. The higher binding energy component at 288.2 eV corresponds to carbon atoms in carboxyl. The most intense component at 286.2 eV corresponds to CH/C-C, and the lower binding energy component could be ascribed to carbon atoms from $CH_2$. The oxygen spectrum in the figure 4 also exhibit three components assigned to the oxygen atoms from OH, $COO^-$ and Fe-O.

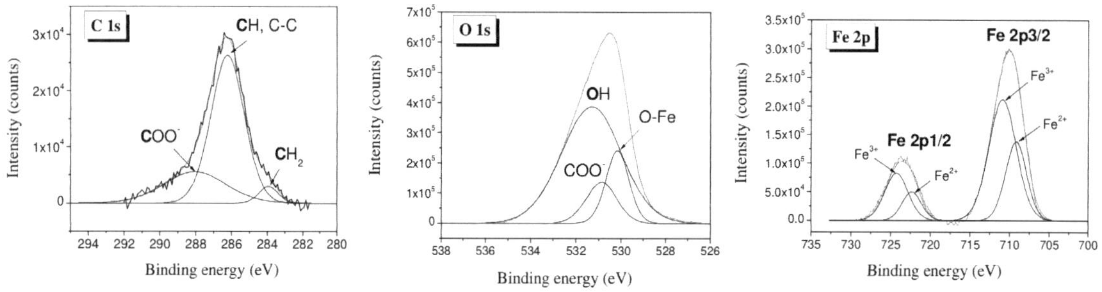


**Figure 4.** XPS spectra from C 1s, O 1s and Fe 2p of magnetite functionalized with serine.

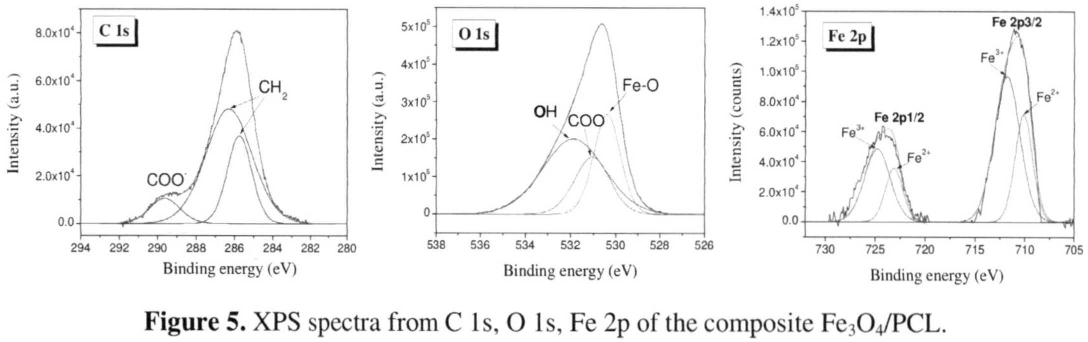


**Figure 5.** XPS spectra from C 1s, O 1s, Fe 2p of the composite $Fe_3O_4$/PCL.

The Fe 2p spectrum contains the doublet Fe 2p3/2 and Fe 2p1/2 with binding energy values of 710 and 723.6 eV, typical for magnetite [3]. The peaks from Fe 2p spectrum can be deconvoluted into two components corresponding to $Fe^{3+}$ and $Fe^{2+}$ ions from magnetite. These results indicate that functionalization with serine doesn't change the composition of the core magnetite. Figure 5 shows the high resolution spectra from C 1s, O 1s, Fe 2p of functionalized magnetite coated with poly(ε-caprolactone). The C 1s spectrum from the figure 5 contains the contribution of three components peaks located at 289.5, 286.5 and 285.7 eV, corresponding to carbon atoms from PCL [4].

The magnetic properties of the smart composites tailor as a function of magnetite content, figures 6 and 7. The value of saturation magnetization is $M_S$ = 69 emu/g for $Fe_3O_4$/serine that contains 85 % wt. $Fe_3O_4$ and it decreases slightly to the value $M_S$ = 54.9 emu/g for the composite $Fe_3O_4$/PCL containing 78 %wt $Fe_3O_4$. Lower values for the saturation magnetization were obtained for the pNIPA ferrogels as compared with functionalized magnetite coated with PCL due to the lower amount of magnetite that can be encapsulated into the ferrogels. The values of $M_S$ for the ferrogels are in the range 2.5-13 emu/g increasing with the increase of magnetite content. The superparamagnetic behavior of low dimensionality magnetite particles in the investigated composites is evidenced by the missing of hysteresis loop in the magnetization curves for all the investigated samples.

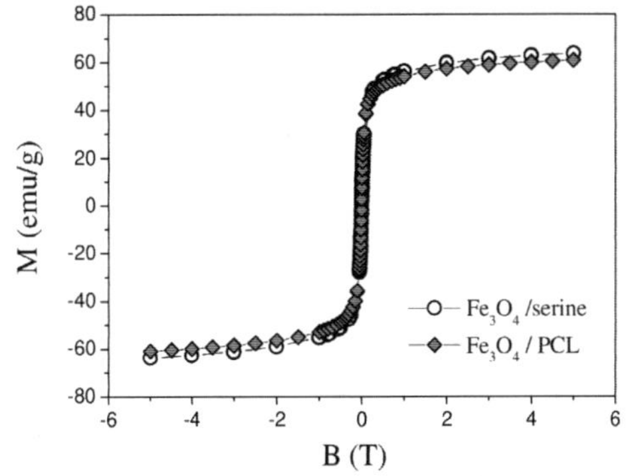


**Figure 6.** Magnetization vs. applied magnetic field at room temperature for magnetite nanoparticles functionalized with serine ($Fe_3O_4$/serine) and the composite $Fe_3O_4$/PCL.

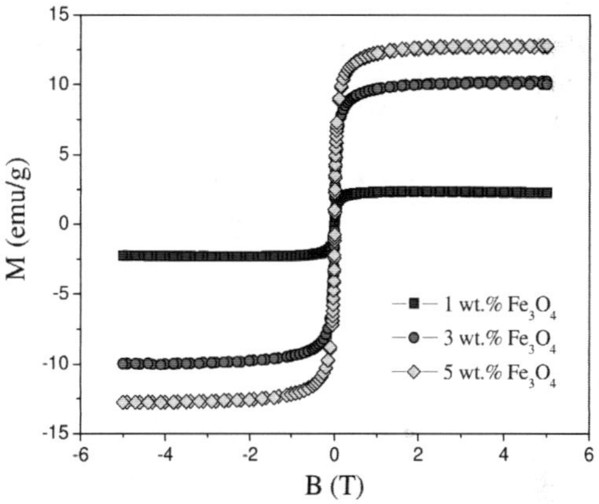


**Figure 7.** Magnetization vs. applied magnetic field at room temperature for pNIPA ferrogels containing different concentrations of $Fe_3O_4$ nanoparticles.

## 4. Conclusions

Smart composites based on magnetite and biocompatible polymers like poly(caprolactone) and pNIPA were obtained. The synthesis method influences the magnetic properties of the resulting composites. Higher magnetization values are obtained for the composites with the functionalized magnetite core covered by polymer as compared with the ferrogels.

The protective coating of $Fe_3O_4$ nanoparticles with a shell of the biocompatible polymer poly (ε-caprolactone) represents a promising route for the development of functional hybrid particles and ferrofluids for both technical and biomedical applications.

## References

[1] Pyun J 2007 *Polym. Rev.* **47** 231
[2] Aldea N and Indrea E 1990 *Comput. Phys. Commun.* **60** 155
[3] Bhargava G, Gouzman I, Chun C M, Ramanarayanan T A and Bernasek S L 2007 *Appl. Surf. Sci.* **253** 4322
[4] Beamson G and Briggs D 1992 *High Resolution XPS of Organic Polymers* (Chichester: Wiley).

# Ceramics with decorative aspect

**Cezara Voica**

National Institute for Research and Development of Isotopic and Molecular Technologies, 65-103 Donath, 400293 Cluj-Napoca, Romania

E-mail: cezara.voica@itim-cj.ro; cezara_voica@yahoo.com

**Abstract.** The last decades brought the development of bone china techniques used for producing the decorative articles. These products can be glazed with a transparent and thin glaze layer, even with more special (decorative) ones which gives new aesthetic aspect. The present article presents the results obtained after the studies performed for matte glazes for decorative bone china. As microcrystalization agent were used zinc oxide; the content of this oxide bring some changes of the basic glaze thus the chemical composition must be adjusted as the fluxes would present the desired properties after the heating process.

## 1. Introduction

The ceramic glazes are thin coatings covering the ceramic products. From structural point of view they are glass, non-crystalline masses, which confer gloss, smoothness, water resistance, as well as mechanical, aesthetic and hygienic properties to the product [1-3]. The diversity of glazes for fine ceramics is an actual goal that contributes to promotion of products on internal and external market. Special effects glazes are highly solicited in present days, for their special aesthetic aspect and for their superior thermal characteristics as well [4, 5].

The purpose of this research is to get certain microcrystalline glazes with a content of zinc excess, obtained through the current technology used in obtaining ceramic glazes to be deposited on the surface of bone china. Thus, the results of the research can be applied fast, and the firing process of the "effect" glazes does not need any changes, either in the diagram in the functioning of the kiln.

This study was reasearch informations regarding the fusibility range, the aspect of glazes after firing, the nature of formed components for the recipes of glazes with a variable content of zinc oxide as crystalizing agents.

Zinc oxide is an effective fluxing agent for glazes with a medium to high melting temperature, but only with other fluxing agents added in small amounts, it improves the gloss; used in larger amounts, however, it leads to opacification and matting of the glazes [6, 7]. The saturation of a glaze with zinc oxide is a method frequently used for obtaining a crystalline texture [8]. Zinc orthosilicate ($Zn_2SiO_4$) plays an important role in the fabrication of mat and crystalline glazes and sodium silicofluoride ($Na_2SiF_6$) promotes opacification of them [9].

## 2. Experimental

In order to study the glaze from the structure point of view and also the factors which influence the process of micro-crystallizing, it was started from a glaze' preparation destined to the bone china as a vitreous matrix, of a borosilicate type, partial fritted, made according to the clasic technology of glazes (batching, mixing, grinding, sitting, firing). Base glaze was made of raw materials currently used in industry: quartz, clay, feldspar, dolomite. The experiments have been done by using the basic glazes into which it has put compounds of zinc in different quantities, thus enabling the basic glaze to be devitrified [10].

Three sets of experiments were performed. First, was introduced zinc oxide in high proportions, up to 30%. The effect has been amplified by making some compositions with sodium hexafluorosilicate ($Na_2SiF_6$) and willhemite ($Zn_2SiO_4$) – as admixtures, in different proportions, going up to 25% (the second and the third sets) [11]. The operating conditions – technological flux – are those of the materials (the density, the degree of dispersion, the thickness of the layer of the glaze) have been carefully controlled in order to assure the reproducible data. Table 1 presents the oxidic compositions of experimental masses.

**Table 1.** Oxidic compositions of experimental masses.

| Experimental samples | Oxides / Oxidic compositions [%] | | | | | | | | | | |
|---|---|---|---|---|---|---|---|---|---|---|---|
| | $SiO_2$ | $TiO_2$ | $Al_2O_3$ | ZnO | $Fe_2O_3$ | CaO | MgO | $Na_2O$ | $K_2O$ | $B_2O_3$ | PC |
| Basic glaze | 56.31 | 0.16 | 9.18 | 9.07 | 0.24 | 7.28 | 1.50 | 3.00 | 1.80 | 2.96 | 8.50 |
| 10% ZnO | 50.14 | 0.15 | 8.17 | 19.07 | 0.21 | 6.48 | 1.33 | 2.68 | 1.60 | 2.67 | 7.57 |
| 20% ZnO | 49.50 | 0.14 | 8.07 | 20.07 | 0.21 | 6.40 | 1.32 | 2.63 | 1.59 | 2.60 | 7.47 |
| 25% ZnO | 40.83 | 0.12 | 6.66 | 34.07 | 0.17 | 5.28 | 1.08 | 2.18 | 1.30 | 2.14 | 6.17 |
| 30% ZnO | 37.73 | 0.11 | 6.15 | 39.07 | 0.16 | 4.88 | 1.00 | 2.01 | 1.20 | 1.99 | 5.70 |
| 25% ZnO + 2%, 4% $Na_2SiF_6$ | 39.73 | 0.12 | 6.31 | 34.44 | 0.21 | 5.66 | 1.01 | 1.88 | 1.23 | 2.90 | 6.50 |
| 30% ZnO + 2%, 4% $Na_2SiF_6$ | 36.71 | 0.11 | 5.83 | 39.44 | 0.19 | 5.22 | 0.94 | 1.74 | 1.14 | 2.67 | 6.01 |
| 33% ZnO + 2%, 4% $Na_2SiF_6$ | 34.89 | 0.1 | 5.54 | 42.44 | 0.18 | 4.96 | 0.89 | 1.65 | 1.08 | 2.54 | 5.73 |
| 15% ZnO + 15% $Zn_2SiO_4$ | 58.94 | 0.025 | 1.39 | 35.39 | 0.046 | 1.24 | 0.22 | 0.41 | 0.27 | 0.63 | 1.43 |
| 20% ZnO + 20% $Zn_2SiO_4$ | 60.29 | 0.018 | 1.38 | 44.03 | 0.023 | 1.12 | 0.14 | 0.20 | 0.18 | 0.40 | 0.93 |
| 25% ZnO + 25% $Zn_2SiO_4$ | 61.64 | 0.012 | 0.91 | 52.68 | 0.018 | 0.91 | 0.09 | 0.19 | 0.12 | 0.20 | 0.45 |
| 25% $Zn_2SiO_4$ | 61.45 | 0.047 | 2.61 | 27.68 | 0.08 | 2.33 | 0.42 | 0.77 | 0.50 | 1.19 | 2.69 |

## 3. Results and discussion

The research has intended to determine the characteristic and corrolation which is established with the ceramic support. Thus, the following has been observed: the thermic behaviour by observing the change of the aspect at the thermic microscope; establishing the optimum melting temperature; the thermal expansion and the compatibility with the ceramic mass; the aspect of the fired glazes and their phase composition.

To establish the technological firing conditions, the thermal analysis of glazes as performed, with a thermal analyzer type MOM Q 1500. The most samples have an almost similar thermal behaviour and results (table 2).

**Table 2.** Thermal processes of experimental glazes.

| Process | Type of process | Temperature (°C) |
|---|---|---|
| Loss of water | endothermal | 210 |
| Polymorphous modifications of silica | endothermal | 530 |
| Decomposition of $CaCO_3$ | endothermal | 855 |

The melting process of the glazes has been observed at the thermal microscope for high temperatures Leitz, emphasizing the optimum melting temperature, the melting interval, the softening interval and the viscosity in the melting phase of the melted mixture (table 3). In experimental glazes containing sodium hexafluorosilicate has been observed that: the softening range increases with the melting range, keep constant the content of sodium hexafluorosilicate and modify the content of zinc oxide; the softening range increases with decreasing melting range, keep constant the content of zinc oxide and modify the content of sodium hexafluorosilicate; the softening decreases with decreasing melting range, with the increasing content of ZnO and wilhemit.

The X-ray diffraction (XRD), performed by a DRON diffractometer with Cu $K_\alpha$ radiation has put in evidence the phase composition of the glazes treated thermically in the industrial kiln: zinc orthosilicate as majority, together with quartz and sometimes zinc oxide excess.

**Table 3.** The thermic behaviour of experimental glazes.

| **Experimental masses** | Beginning of softening [°C] | Beginning of melting [°C] | Beginning of flowing [°C] | Softening range [°C] | Melting range [°C] |
|---|---|---|---|---|---|
| Basic glaze | 1010 | 1060 | 1160 | 50 | 100 |
| 10% ZnO | 1010 | 1060 | 1160 | 50 | 150 |
| 25% ZnO | 1010 | 1040 | 1160 | 30 | 150 |
| 25% ZnO + 2% $Na_2SiF_6$ | 1035 | 1060 | 1100 | 25 | 65 |
| 25% ZnO + 4% $Na_2SiF_6$ | 1030 | 1055 | 1090 | 25 | 60 |
| 30% ZnO + 2% $Na_2SiF_6$ | 1050 | 1080 | 1145 | 30 | 95 |
| 30% ZnO + 4% $Na_2SiF_6$ | 1040 | 1090 | 1110 | 50 | 70 |
| 33% ZnO + 4% $Na_2SiF_6$ | 1060 | 1110 | 1140 | 50 | 80 |
| 15% ZnO + 15% $Zn_2SiO_4$ | 1040 | 1100 | 1140 | 60 | 100 |
| 20% ZnO + 20% $Zn_2SiO_4$ | 1050 | 1100 | 1150 | 50 | 100 |
| 25% ZnO + 25% $Zn_2SiO_4$ | 1230 | 1245 | 1260 | 15 | 30 |
| 25% $2ZnO{\cdot}SiO_2$ | 1020 | 1050 | 1100 | 30 | 80 |

The most important thermal property of a glaze is its coefficient of linear thermal expansion, the value of which is usually determined experimentally. The measurements of thermal expansion in this study show a correct compatibility between glazes and the mass of bone china; the coefficients of thermal expansion of studied glazes are lower than the bone china ($\alpha{\times}10^{-6}$ [$^0C^{-1}$]=7.62) (table 4).

**Table 4.** Thermal expansion of studied glazes.

| Experimental glazes | Thermal expansion $\alpha \times 10^{-6}$ [°$C^{-1}$] |
|---|---|
| 10% ZnO | 6.51 |
| 20% ZnO | 6.58 |
| 25% ZnO | 6.43 |
| 30% ZnO | 6.27 |
| 15% ZnO + 15% $Zn_2SiO_4$ | 5.62 |
| 20% ZnO + 20 % $Zn_2SiO_4$ | 5.56 |
| 25% $Zn_2SiO_4$ | 5.65 |
| 25% ZnO + 4% $Na_2SiF_6$ | 5.96 |
| 30% ZnO + 2% $Na_2SiF_6$ | 5.87 |

Table 5 shows the aspect of experimental fired masses. Glaze base has a high capacity to dissolve the zinc oxide. Among to the approximately 10% of zinc oxide from frit base, the addition of 20-25% in raw glaze does not cause crystallization diagram on current firing bone china. The glaze with the best mattness was the one with the addition of 25% willhemite. Glaze base has a high capacity to dissolve the zinc oxide. Among to the approximately 10% of zinc oxide from frit base, the addition of 20-25% in raw glaze does not cause crystallization diagram on current firing bone china.

**Table 5.** Aspect of experimental masses.

| Samples | Aspect |
|---|---|
| 10% ZnO | transparent, glossy |
| 20% ZnO | transparent, glossy |
| 25% ZnO | transparent, glossy |
| 30% ZnO | transparent, glossy |
| 20% ZnO + 2% $Na_2SiF_6$ | satined |
| 25% ZnO + 2% $Na_2SiF_6$ | satined |
| 25% ZnO + 4% $Na_2SiF_6$ | satined, lightly crystallized |
| 30% ZnO + 2% $Na_2SiF_6$ | crystals in relief, lightly satined |
| 30% ZnO + 4% $Na_2SiF_6$ | crystals in relief, lightly satinized |
| 33% ZnO + 2% $Na_2SiF_6$ | crystals in relief, satined background |
| 33% ZnO + 4% $Na_2SiF_6$ | rude crystallization, pronounced relief |
| 15% ZnO + 15% $Zn_2SiO_4$ | sugar-like |
| 20% ZnO + 20% $Zn_2SiO_4$ | sugar-like |
| 25% ZnO + 25 % $Zn_2SiO_4$ | sugar-like |
| 25% $Zn_2SiO_4$ | mat |

## 4. Conclusion

This study presents experimental works for obtaining ceramic glazes with a quality higher than those applied until now and a complex study of their properties for producing bone china articles. The study followed obtaining a rich scale of effect glazes, matte, satined, through introduction of raw materials that cause special aesthetic effects in vitreous mass: micro crystallizations.

The influence of ZnO on the aspect, phase composition and thermal behaviour in bone china glaze have been investigated, with aim at obtaining glazes with high aesthetic effect. The sodium hexafluorosilicate addition strengthens the micro-crystallization decorative of ZnO in the studied glazes. The measurements of thermal expansion show a correct compatibility between glazes and the mass of bone china. The glaze with the best mattness was the one with 25% of wilhemite. The glazes are capable of being used under the current flux technology.

These glazes are definitely expensive, but they can be used for certain products which require a more sophisticated aspect which justify their use.

## References

[1] Teoreanu I 1995 *Introduction in physical chemistry of solid phase, oxidic compunds* vol I (Bucureşti: Editura Didactică şi Pedagogică)
[2] Aliprandi G 1974 *Ceramurgia & Tecnologia Ceramica* (Genova-Corso Europa: Editioni Culturali Internazionali) pp 881-908
[3] Shaw K 1971 *Ceramic glazes* (Amsterdam, London: Elsevier)
[4] Escardino A, Amoros J L, Orts M J, Mestre S, Belda J, Marco J and Salas J J 2000 *Proc. World Congress on Ceramic Tile Quality* vol 1 93-109
[5] Dakkai S, Orlova L A and Miklailenko N 1999 *Glass and Ceramics* **5-6** 177-80
[6] Moreno A, Bon E, Bordes M C, Navarro M C, Deraedt J and Lerusse A 2000 Qualicer **VI** *Proc. World Congress on Ceramic Tile, Quality* vol 1 27-40
[7] Karosu B, Caki M and Turan S 2000 *J. Eur. Ceram. Soc.* **20** 2225-31
[8] Stefanov S and Batschwarov S 1988 *Ceramic glazes* (Wiesbaden: Bauverlag)
[9] Parmelee W C 1951 *Ceramic glazes* (Chicago Illinois: Industrial Publications Inc.)
[10] Rahaman M and Lakkwani S 1999 *Adv. Sci. Technol.* **14** 33-40
[11] Voica C, Gagea L and Literat L 2006 *Proc. Nat. Simp. Chem. Eng.* (Cluj-Napoca: Accent) p 78.

# AUTHOR INDEX


Aldea, N ... 236, 307
Aldea, Nicolae ... 217
Almasan, V ... 205, 213, 222
Al-Said, Said A Farha ... 58
Ardelean, C ... 194
Ardelean, I ... 49, 137
Ardelean, O ... 190, 198, 209, 222, 227
Ardelean, P ... 194
Avram, V ... 170
Badea, Codruta ... 49
Banakh, O ... 124
Bende, Attila ... 1
Bibire, Cristina ... 186
Biris, A R ... 137, 198, 209, 241, 316
Biris, A S ... 241, 316
Bîrlea, N M ... 85
Bîrlea, S I ... 85
Bitterli, Roland ... 89
Blanita, G ... 190, 198, 209
Blumenthal, H ... 42
Bocos-Bintintan, V ... 94
Bogdan, Diana ... 245
Bogdan, M ... 5, 33, 53
Borghese, Antonio ... 103, 159
Borodi, Gheorghe ... 10, 29, 38, 288
Bosca, M ... 253, 257, 288, 300
Bratfalean, Radu ... 245
Bratu, I ... 10, 15, 23, 38, 80, 249, 269, 281, 296
Brown, Steven P ... 107
Buimaga-Iarinca, Luiza ... 19
Charmet, J ... 89, 124
Chelcea, R ... 300
Ciontea, L ... 63
Clichici, S ... 316
Cobzac, Simona Codruta ... 67
Coldea, I ... 198, 209, 222, 227
Coldea, M ... 307
Cosma, V ... 94
Cosoveanu, Veronica ... 232, 273
Costinel, Diana ... 163
Cozar, Onuc ... 115
Craciunescu, I ... 249, 281, 324
Cruickshank, Dyanne ... 45
Culea, E ... 253, 257, 262, 265, 288, 296, 300
Culea, M ... 115, 296
Cuna, C ... 94
Cuna, Stela ... 94, 163
Dadarlat, Dorin ... 99
Dan, Monica ... 205
Dan, V ... 257, 296
Danciu, Virginia ... 232, 273
Darabont, Al ... 304
Dehelean, Adriana ... 154, 262, 265
Dervishi, E ... 241
Di Palma, Tonia M ... 103
Dragan, Felicia ... 23
Dreve, S ... 194, 232, 269, 273, 312, 320
Durante, Guido Spinola ... 89
Farcas, Sorin I ... 10
Farcas, Sorin I ... 15, 38
Fechete, R ... 300
Filip, Claudiu ... 29, 107, 141
Filip, Xenia ... 29, 107, 141
Filoti, G ... 190
Floare, C G ... 5, 33, 53
Gabor, M S ... 63
Garabagiu, Sorina ... 201
Geba, Maria ... 186
Giurgiu, L M ... 137, 292, 304
Glajar, Ioana C ... 167
Gorodyska, Ganna ... 89
Graf, B ... 124
Grecu, Voicu V ... 163
Hanelt, S ... 42
Haquette, H ... 89, 124
Hassanien, Reda ... 58
Hîncu, Lucian ... 23
Horovitz, O ... 33
Horrocks, Benjamin R ... 58
Horvát, Szabolcs ... 111
Horváth, E Ágnes ... 111
Houlton, Andrew ... 58
Indrea, E ... 194, 232, 269, 273, 312, 320
Iordache, Andreea Maria ... 115
Isai, Radu ... 245
Jeandupeux, L ... 124
Jivanescu, M ... 292
Kacso, I ... 10, 15, 38, 58, 249, 269
Káptalan, Erna ... 111
Karsten, S ... 277
Kasco, I ... 312
Keppner, H ... 89, 124
Kirillin, Mikhail ... 149
Knapp, Helmut ... 89
Kovács, Katalin ... 120, 178
Laux, E ... 89, 124
Lazar, M ... 198, 205, 213
Leostean, C ... 249, 281, 285, 324
Leuca, M ... 94
Li, Z ... 241
Liebscher, J ... 42, 277, 281
Lucaci, M ... 241
Lung, Ildiko ... 67, 76
Lupsa, N ... 94
Lupu, D ... 198, 209, 222, 227, 241, 316
Macavei, S ... 285, 324
Marginean, Petru ... 205, 217
Marincas, O ... 170
Máté, Gabriell ... 111
Matei, F ... 236

# AUTHOR INDEX


Maties, V ........ 296, 300
Mic, Mihaela ........ 45
Mihailescu, G ........ 194, 201
Mihet, Maria ........ 205, 213
Mirel, V ........ 94
Misan, I ........ 198, 209, 222, 227, 241, 316
Mocan, T ........ 316
Mocanu, A ........ 33
Moldovan, D ........ 300
Moldovan, Z ........ 167, 170, 174, 182, 194
Morari, Cristian ........ 245
Morariu, Vasile V ........ 19, 141
Mos, B ........ 63
Muresan-Pop, Marieta ........ 15
Nan, A ........ 249, 277, 281, 324
Nascu, H I ........ 320
Nasui, M ........ 63
Neamtu, Camelia ........ 72
Nechifor, Ruben ........ 49
Néda, Zoltán ........ 111
Neumann, M ........ 307
Niculescu, Violeta ........ 128
Noell, Wilfried ........ 89
Olenic, L ........ 58, 80
Oprean, Luminita ........ 29
Palade, P ........ 190
Pamula, A ........ 154
Pana, O ........ 285, 324
Pascalau, V ........ 198
Pascut, L G ........ 307
Pascuta, P ........ 253, 257, 288, 296
Paun, Nadia ........ 128
Peter, Anca ........ 273
Petrisor Jr, T ........ 63
Petrisor, T ........ 63
Petrov, P ........ 170
Pintea, S ........ 217, 236, 307
Pîrnau, Adrian ........ 5, 53, 128
Pop, L ........ 253, 257, 288, 300
Pop, Mircea N ........ 99
Popa, A ........ 137, 292, 304
Popeneciu, G ........ 198, 209, 222, 227
Popescu, G ........ 320
Popescu, V ........ 232, 320
Portuondo-Campa, Erwin ........ 89
Pruneanu, Stela ........ 58
Rada, M ........ 296, 300
Rada, S ........ 253, 257, 288, 296, 300
Raita, O ........ 137, 292, 304
Rednic, L ........ 307
Rednic, V ........ 217, 236, 307
Ristoiu, T ........ 63
Rosu, M C ........ 194, 232, 273, 312, 320
Rusu, T ........ 253, 296
Sadiq, S ........ 42
Shah, J ........ 42
Silipas, Danut T ........ 273
Silipas, T D ........ 194, 232, 312, 320
Simon, S ........ 241, 316
Soran, M L ........ 67, 285
Speranza, Luciano ........ 159
Stan, M ........ 304
Streza, Mihaela ........ 99
Suciu, R C ........ 194, 232, 273, 312, 320
Surducan, Emanoil ........ 67, 72, 76
Surducan, Vasile ........ 67, 72, 76
Szabo, Laszlo ........ 128
Tamaian, Radu ........ 128
Ternes, T ........ 170
Textor, Marcus ........ 89
Tiandou, Hu ........ 236
Toadere, Florin ........ 133
Toloman, D ........ 137, 292, 304
Tomoaia-Cotisel, M ........ 33
Tosa, N ........ 80
Tosa, V ........ 99
Tosa, V ........ 85, 120, 178
Tripon, Carmen ........ 29, 107, 141
Tudoran, Cristian D ........ 145
Tudoran, Lucian Barbu ........ 58
Turcu, I ........ 1, 45, 80, 149
Turcu, R ........ 249, 277, 281, 285, 324
Turdeanu, R ........ 80
Tusa, Florina ........ 94, 182
Varodi, Codruta ........ 67
Vekas, L ........ 249
Vlassa, M ........ 182, 198
Voica, Cezara ........ 154, 328
Vornicu, Nicoleta ........ 186
Vremera, Raluca ........ 163
Watanabe, F ........ 241
Yacob, Z ........ 42
Yaning, Xie ........ 217, 236